Statistics for Social Science and Public Policy

Advisors:
S.E. Fienberg   D. Lievesley   J. Rolph

Springer

New York
Berlin
Heidelberg
Barcelona
Hong Kong
London
Milan
Paris
Singapore
Tokyo

Statistics for Social Science and Public Policy

Robert L. Brennan

Generalizability Theory

 Springer

Robert L. Brennan
Iowa Testing Programs
University of Iowa
Iowa City, IA 52242-1529
USA

Advisors:

Stephen E. Fienberg
Department of Statistics
Carnegie Mellon University
Pittsburgh, PA 15213
USA

Denise Lievesley
Institute for Statistics
Room H.113
UNESCO
7 Place de Fontenoy
75352 Paris 07 SP
France

John Rolph
Department of Information and
  Operations Management
Graduate School of Business
University of Southern California
Los Angeles, CA 90089
USA

Library of Congress Cataloging-in-Publication Data
Brennan, Robert L.
    Generalizability theory / Robert L. Brennan
      p. cm. — (Statistics for social science and public policy)
    Includes bibliographical references (p.) and indexes.
    ISBN 0-387-95282-9 (alk. paper)
      1. Psychometrics. 2. Psychology—Statistical methods. 3. Analysis of
variance. I. Title. II. Series.
    BF39.B755 2001
    150'.1'5195—dc21   2001032009

Printed on acid-free paper.

Production managed by Allan Abrams; manufacturing supervised by Jeffrey Taub.
Photocomposed copy prepared by the author using LaTeX.
Printed and bound by Sheridan Books, Inc., Ann Arbor, MI.
Printed in the United States of America.

9 8 7 6 5 4 3 2 1

ISBN 0-387-95282-9   SPIN 10834003

Springer-Verlag   New York Berlin Heidelberg
A member of BertelsmannSpringer Science+Business Media GmbH

To Cicely

Preface

In 1972 a monograph by Cronbach, Gleser, Nanda, and Rajaratnam was published entitled *The Dependability of Behavioral Measurements.* That book incorporated, systematized, and extended their previous research into what came to be called *generalizability theory*, which liberalizes classical test theory, in part through the application of analysis of variance procedures that focus on variance components. Generalizability theory is perhaps the most broadly defined measurement model currently in existence, and the Cronbach et al. (1972) treatment of the theory represents a major contribution to psychometrics. However, as Cronbach et al. (1972, p. 3) state, their book is "complexly organized and by no means simple to follow" and, of course, it is nearly 30 years old.

In 1983, ACT, Inc. published my monograph entitled *Elements of Generalizability Theory*, with a slightly revised version appearing in 1992. That treatment is considerably less comprehensive than Cronbach et al. (1972) but still detailed enough to convey much of the richness of the theory and to facilitate its application. However, the 1983/1992 monograph is essentially two decades old, it does not cover multivariate generalizability theory in depth, and it does not incorporate recent developments in statistics that bear upon the estimation of variance components. Also, of course, there have been numerous developments in generalizability theory in the last 20 years.

This book provides a much more comprehensive and up-to-date treatment of generalizability theory. It covers all of the major topics that have been discussed in generalizability theory, as well as some new ones. In addition, it provides a synthesis of those parts of the statistical literature that are directly applicable to generalizability theory.

The principal intended audience is measurement practitioners and upper-level graduate students in the behavioral and social sciences, particularly education and psychology. Generalizability theory has broader applicability, however. Indeed, it might be used in virtually any field that attends to measurements and their errors. Readers will benefit from some familiarity with classical test theory and analysis of variance, but the treatment of most topics does not presume specific background. In particular, variance components are a central focus of generalizability theory, but it is not assumed that readers are familiar with them or with procedures for estimating them.

Although the statistical aspects of generalizability theory are undeniably important, perhaps the most distinguishing feature of the theory is its conceptual framework, which permits a multifaceted perspective on measurement error and its components. What makes generalizability theory both challenging and useful is that it marries this rich conceptual framework with powerful, but sometimes complicated, statistical procedures. This book gives substantial attention to both aspects of generalizability theory—the conceptual framework and the statistical machinery. However, the book per se is neither a treatise on the philosophy of measurement, nor a textbook on statistical procedures. Rather, it integrates those parts of both topics that bear upon generalizability theory.

Precursors to generalizability theory are evident in papers written as long ago as the 1930s. However, generalizability theory per se is relatively new, it is evolving, and there are a few somewhat different perspectives on the theory. Most of these perspectives are complementary, or might be viewed as special cases or extensions. Even so, I judged it necessary to adopt one principal perspective and maintain it throughout this book. That perspective is closely aligned with Cronbach et al. (1972), but there are some occasional differences. For example, except for the last chapter, this book does not emphasize regressed score estimates of universe scores nearly as much as Cronbach et al. (1972). Also, there are some notational differences, especially in those chapters that treat multivariate generalizability theory.

There are three sets of chapters in this book. They are ordered in terms of increasing complexity. The fundamentals of univariate generalizability theory are contained in Chapters 1 to 4. They might be used as part of a graduate-level course in advanced measurement. Additional, more challenging topics in univariate theory are covered in Chapters 5 to 8, and Chapters 9 to 12 provide my own perspective on multivariate generalizability theory.

The treatment of multivariate generalizability theory is inspired by the work of Cronbach et al. (1972), but there are noticeable differences in emphasis, coverage, and notational conventions. I have tried to provide the reader with different ways of thinking about multivariate generalizability theory, and I have tried to illustrate its similarities to and differences from univariate theory. An important goal of this book is to make multivariate generalizability theory more accessible to practitioners.

More consideration is given to reliability-like coefficients than is necessitated by the theory. However, in my experience, many students and measurement practitioners have great difficulty, at least initially, in appreciating the applicability and usefulness of generalizability theory unless they can relate some of its results to classical reliability coefficients. For this reason, such coefficients are actively considered, although the magnitudes of variance components, and particularly error variances, are clearly more important.

Many of the topics covered here could be treated using matrix operators. With the exception of one appendix, however, matrix operators are not employed, because doing so would render the content inaccessible to many students and practitioners who might benefit from the theory.

I am grateful to ACT, Inc., for permitting me to use parts of Brennan (1992a). That monograph clearly influenced my treatment of Chapters 2 to 5 and several appendices. Also, Chapter 1 is largely a revised version of Brennan (1992b) used with the permission of the publisher, the National Council on Measurement in Education, and parts of Section 5.4 are from Brennan (1998) used with permission of the publisher, Sage. I am also grateful to ACT, Inc. for permitting me access to ACT Assessment data used for various multivariate examples in the later chapters of this book, to Suzanne Lane for permitting me to use the QUASAR data referenced in Section 5.4, to Clare Kreiter for the opportunity to analyze data discussed in Section 8.3, and to Judy Hu at Iowa Testing Programs (ITP) for her assistance with ITP data.

I especially want to acknowledge the considerable benefit I have received over the last 30 years from numerous communications with Lee Cronbach. Also, I am particularly grateful to Michael Kane, whose research, insights, criticisms, and support have contributed greatly to my own thinking, research, and writings about generalizability theory. I have benefited as well from joint research with Xiaohong Gao, especially in the area of performance assessments.

Others who have influenced my work include David Jarjoura, Joe Crick, Richard Shavelson, Noreen Webb, Gerald Gillmore, and Dean Colton. Finally, I want to thank my students, especially Won-Chan Lee, Scott Bishop, Guemin Lee, Dong-In Kim, Janet Mee, Ping Yin, and Steven Rattenborg. They have assisted me in numerous ways. In particular, their questions and comments have often influenced how I think about and present the theory. My thanks to all of them. Finally, I am grateful to my secretary, Sue Wollrab, for her help in preparing the manuscript.

Iowa City, IA   Robert L. Brennan
January, 2001

Contents

Principal Notational Conventions

Operators

×	Crossed with
:	Nested within
E	Expectation

Miscellaneous

ˆ or Est	Estimate
γ	Confidence coefficient

Univariate G Studies and Universes of Admissible Observations

n	G study sample size for a facet
N	Size of a facet in the universe of admissible observations
α	An effect, or the indices in an effect
$\dot{\alpha}$	The indices not in α
ω	All indices in a G study design
X_ω	An observable score for a G study design
μ_α	Population and/or universe mean score for α
ν_α	Score effect for α
$\pi(\dot{\alpha})$	Product of sample sizes for indices not in α
$df(\alpha)$	Degrees of freedom for α
$SS(\alpha)$	Sum of squares for α
$MS(\alpha)$	Mean square for α

$EMS(\alpha)$ Expected mean square for α

$\sigma^2(\alpha)$ G study random effects variance component for α

$\sigma^2(\alpha|M)$ G study variance component for α given a model M

Univariate D Studies and Universes of Generalization

n' D study sample size for a facet

N' Size of a facet in universe of generalization

τ Object of measurement (often $\tau = p$)

$\bar{\alpha}$ An effect, or the indices in an effect; the bar emphasizes that interest focuses on mean scores in a D study

$\sigma^2(\bar{\alpha})$ D study random effects variance component for α

$\sigma^2(\bar{\alpha}|M')$ D study variance component for α given a model M'

$\sigma^2(\tau)$ Universe score variance

$\sigma^2(\delta)$ Relative error variance

$\sigma^2(\Delta)$ Absolute error variance

$\sigma^2(\overline{X})$ Error variance in using $\overline{X}$ as an estimate of μ

$ES^2(\tau)$ Expected observed score variance

$E\rho^2$ Generalizability coefficient

Φ Index of dependability

$\sigma^2(\delta_p)$ Conditional relative error variance for person p; that is, relative error variance for a specific person

$\sigma^2(\Delta_p)$ Conditional absolute error variance for person p

Multivariate G and D Studies

v, v' Fixed variables

ν, ξ Effects associated with v and v', respectively

$\bullet$ Linked facet

$\circ$ Independent facet

$S_v^2(\alpha)$ Observed score variance for v

$S_{vv'}(\alpha)$ Observed score covariance for v and v'

$SP_{vv'}(\alpha)$ Sum of products for v and v'

$MP_{vv'}(\alpha)$ Mean product for v and v'

$EMP_{vv'}(\alpha)$ Expected mean product for v and v'

$\sigma_v^2(\alpha)$ Random effects variance component for v

$\sigma_{vv'}(\alpha)$ Covariance component for v and v'

$\rho_{vv'}(\alpha)$ Disattenuated correlation for v and v'

Σ_α Variance-covariance matrix

w_v A priori (nominal) weight for v

a_v Estimation weight for v

b_v Raw score regression weight for v

β_v Standard score regression weight for v

$\sigma_C^2(\alpha)$ Variance component for composite

1
Introduction

The pursuit of scientific endeavors necessitates careful attention to measurement procedures, the purpose of which is to acquire information about certain attributes or characteristics of objects. However, the information obtained from any measurement procedure is fallible to some degree. This is evident even for a seemingly uncontroversial measurement procedure such as one used to associate a numerical value (measurement) with the length of an object. Clearly, the measurements obtained may vary depending on numerous conditions of measurement, such as the ruler used, the person who records the measurement, lighting conditions, and the like.

Although all measurements are fallible to some extent, scientists seek ways to increase the precision of measurement. To do so, they frequently average measurements over some subset of predefined conditions of measurement. This average measurement serves as an estimate of the "ideal" measurement that would be obtained (hypothetically) by averaging over all predefined conditions of measurement. A substantive question then becomes, "How many instances of which conditions of measurement are needed for acceptably precise measurement?" For example, if prior research has demonstrated that the choice of ruler has little influence on measurements of the length of certain objects, but considerable variability is associated with the persons who record measurements, then it is sensible to average measurements over many persons but few rulers.

Another way that scientists sometimes increase the precision of measurement is to fix one or more conditions of measurement. For example, a specific ruler might be used to obtain all measurements of the length of an object. However, the choice of a specific ruler for all measurements involves a restriction on the set of measurement conditions to which generalization is intended. In other words, fixing a condition of measurement reduces error and increases the precision of measurements, but it does so at the expense of narrowing interpretations of measurements.

It is evident from this perspective on measurement that "error" does not mean mistake, and what constitutes error in a measurement procedure is, in part, a matter of definition. It is one thing to say that error is an inherent aspect of a measurement procedure; it is quite another thing to quantify error and specify which conditions of measurement contribute to it. Doing so necessitates specifying what would constitute an "ideal" measurement (i.e., over what conditions of measurement is generalization intended) and the conditions under which observed scores are obtained.

These and other measurement issues are of concern in virtually all areas of science. Different fields may emphasize different issues, different objects, different characteristics of objects, and even different ways of addressing measurement issues, but the issues themselves pervade scientific endeavors. In education and psychology, historically these types of issues have been subsumed under the heading of "reliability." Generalizability theory liberalizes and extends traditional notions of reliability.

Broadly conceived, reliability involves quantifying the consistencies and inconsistencies in observed scores. It has been stated that "A person with one watch knows what time it is; a person with two watches is never quite sure!" This simple aphorism highlights how easily investigators can be deceived by having information from only one element in a larger set of interest. Generalizability theory enables an investigator to identify and quantify the sources of inconsistencies in observed scores that arise, or could arise, over replications of a measurement procedure.[1]

Generalizability theory offers an extensive conceptual framework and a powerful set of statistical procedures for addressing numerous measurement issues. To an extent, the theory can be viewed as an extension of classical test theory (see, e.g., Feldt & Brennan, 1989) through an application of certain analysis of variance (ANOVA) procedures to measurement issues. Classical theory postulates that an observed measurement can be decomposed into a "true" score T and a single undifferentiated random error term E. As such, any single application of the classical test theory model cannot clearly differentiate among multiple sources of error. By contrast, when Fisher (1925) introduced ANOVA, he

> revolutionized statistical thinking with the concept of the factorial experiment in which the conditions of observation are classified in several respects. Investigators who adopt Fisher's line of thought must abandon the concept of undifferentiated

---

[1]Brennan (in press) provides an extensive consideration of reliability from the perspective of replications.

error. The error formerly seen as amorphous is now attributed to multiple sources, and a suitable experiment can estimate how much variation arises from each controllable source (Cronbach et al., 1972, p. 1).

Generalizability theory liberalizes classical theory by employing ANOVA methods that allow an investigator to untangle multiple sources of error that contribute to the undifferentiated E in classical theory

The defining treatment of generalizability theory is a book by Cronbach et al. (1972) entitled *The Dependability of Behavioral Measurements*. A history of the theory is provided by Brennan (1997). In discussing the genesis of the theory, Cronbach (1991, pp. 391–392) states:

> In 1957 I obtained funds from the National Institute of Mental Health to produce, with Gleser's collaboration, a kind of handbook of measurement theory. ... "Since reliability has been studied thoroughly and is now understood," I suggested to the team, "let us devote our first few weeks to outlining that section of the handbook, to get a feel for the undertaking." We learned humility the hard way—the enterprise never got past that topic. Not until 1972 did the book appear ... that exhausted our findings on reliability reinterpreted as generalizability. Even then, we did not exhaust the topic.
>
> When we tried initially to summarize prominent, seemingly transparent, convincingly argued papers on test reliability, the messages conflicted.

To resolve these conflicts, Cronbach and his colleagues devised a rich conceptual framework and married it to analysis of random effects variance components. The net effect is "a tapestry that interweaves ideas from at least two dozen authors" (Cronbach, 1991, p. 394). In particular, Burt (1936), Hoyt (1941), Ebel (1951), and Lindquist (1953, Chap. 16) discussed ANOVA approaches to reliability. Indeed, the work by Burt and Lindquist appears to have anticipated the development of generalizability theory.

The essential features of univariate generalizability theory were largely completed with technical reports in 1960-1961. These were revised into three journal articles, each with a different first author (Cronbach et al., 1963; Gleser et al., 1965; and Rajaratnam et al., 1965).

In the mid 1960s, motivated by Harinder Nanda's studies on interbattery reliability, the Cronbach team began their development of multivariate generalizability theory, which is incorporated in their 1972 monograph. Cronbach (1976) provides some additional perspectives.

Since the Cronbach et al. (1972) monograph, a number of publications have been substantially devoted to explicating the theory at various levels of detail and complexity. For example, Brennan (1983, 1992a) provides a monograph on generalizability theory that is quite extensive but still less

detailed than Cronbach et al. (1972). Shavelson and Webb (1991) provide
an introductory monograph, and Cardinet and Tourneur (1985) provide
a monograph in French. Shorter treatments of generalizability theory are
given by Algina (1989), Brennan and Kane (1979), Crocker and Algina
(1986), Feldt and Brennan (1989), and Strube (2000). Brief introductions
are provided by Allal (1988, 1990), Brennan (1992b), Rentz (1987), and
Shavelson and Webb (1992). In addition, Brennan (2000a) treats conceptual issues, including misconceptions. Also, overviews of the theory are
incorporated in Shavelson and Webb's (1981) review of the generalizability theory literature from 1973 to 1980, and in the Shavelson et al. (1989)
coverage of additional contributions in the 1980s.

1.1   Framework of Generalizability Theory

Although classical test theory and ANOVA can be viewed as the parents
of generalizability theory, the child is both more and less than the simple
conjunction of its parents, and appreciating generalizability theory requires
an understanding of more than its lineage (see Figure 1.1). For example,
although generalizability theory liberalizes classical test theory, not all aspects of classical theory, as explicated by Feldt and Brennan (1989), are
incorporated in generalizability theory. Also, not all of ANOVA is relevant
to generalizability theory; indeed, some perspectives on ANOVA are inconsistent with generalizability theory. In addition, the ANOVA issues emphasized in generalizability theory are different from those that predominate in
many experimental design and ANOVA texts. In particular, generalizability
theory concentrates on variance components and their estimation.

Perhaps the most important aspect and unique feature of generalizability theory is its conceptual framework. Among the concepts are *universes
of admissible observations* and G (*generalizability*) studies, as well as *universes of generalization* and D (*decision*) studies. The concepts and the
methods of generalizability theory are introduced next using a hypothetical
scenario involving the measurement of writing proficiency. As illustrated by
this scenario, generalizability analyses are useful not only for understanding the relative importance of various sources of error but also for designing
efficient measurement procedures.[2]

---

[2]This hypothetical scenario is a somewhat revised version of Brennan (1992a), which
is reprinted here with permission of the publisher, the National Council on Measurement
in Education.

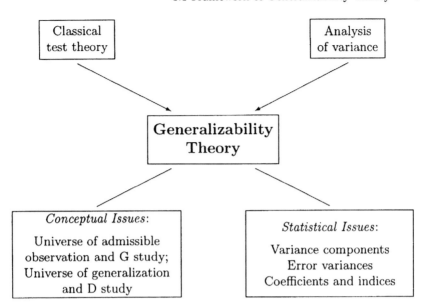

FIGURE 1.1. Parents and conceptual framework of generalizability theory.

1.1.1   Universe of Admissible Observations and G Studies

Suppose an investigator, Mary Smith, decides that she wants to construct one or more measurement procedures for evaluating writing proficiency. She might proceed as follows. First she might identify, or otherwise characterize, essay prompts that she would consider using, as well as potential raters of writing proficiency. At this point, Smith is not committing herself to actually using, in a particular measurement procedure, any specific items or raters—or, for that matter, any specific number of items or raters. She is merely characterizing the facets of measurement that might interest her or other investigators. A facet is simply a set of similar conditions of measurement. Specifically, Smith is saying that any one of the essay prompts constitutes an admissible (i.e., acceptable to her) condition of measurement for her essay-prompt facet. Similarly, any one of the raters constitutes an admissible condition of measurement for her rater facet. We say that Smith's universe of admissible observations contains an essay-prompt facet and a rater facet.

Furthermore, suppose Smith would accept as meaningful to her a pairing of any rater (r) with any prompt (t). If so, Smith's universe of admissible observations would be described as being crossed, and it would be denoted $t \times r$, where the "$\times$" is read "crossed with." Specifically, if there were N_t prompts and N_r raters in Smith's universe, then it would be described as crossed if any one of the $N_t N_r$ combinations of conditions from the two facets would be admissible for Smith. Here, it is assumed that N_t and N_r are both very large (approaching infinity, at least theoretically).

Note that it is Smith who decides which prompts and which raters constitute the universe of conditions for the prompt and rater facets, respectively. Generalizability theory does not presume that there is some particular definition of prompt and rater facets that all investigators would accept. For example, Smith might characterize the potential raters as college instructors with a Ph.D. in English, whereas another investigator might be concerned about a rater facet consisting of high school teachers of English. If so, Smith's universe of admissible observations may be of little interest to the other investigator. This does not invalidate Smith's universe, but it does suggest that other investigators need to pay careful attention to Smith's statements about facets if such investigators are to judge the relevance of Smith's universe of admissible observations for their own concerns.

In the above scenario, no explicit reference has been made to persons who respond to the essay prompts in the universe of admissible observations. However, Smith's ability to specify a meaningful universe of prompts and raters is surely, in some sense, dependent upon her ideas about a population of examinees for whom the prompts and raters would be appropriate. Without some such notion, any characterization of prompts and raters as "admissible" seems vague, at best. Even so, in generalizability theory the word *universe* is reserved for conditions of measurement (prompts and raters, in the scenario), while the word *population* is used for the objects of measurement (persons, in this scenario).

Presumably, Smith would accept as admissible to her the response of any person in the population to any prompt in the universe evaluated by any rater in the universe. If so, the population and universe of admissible observations are crossed, which is represented $p \times (t \times r)$, or simply $p \times t \times r$. For this situation, any observable score for a single essay prompt evaluated by a single rater can be represented as:

$$X_{ptr} = \mu + \nu_p + \nu_t + \nu_r + \nu_{pt} + \nu_{pr} + \nu_{tr} + \nu_{ptr}, \qquad (1.1)$$

where μ is the grand mean in the population and universe and ν designates any one of the seven uncorrelated effects, or components, for this design.[3]

The variance of the scores given by Equation 1.1, over the population of persons and the conditions in the universe of admissible observations is:

$$\sigma^2(X_{ptr}) = \sigma^2(p) + \sigma^2(t) + \sigma^2(r) + \sigma^2(pt) + \sigma^2(pr) + \sigma^2(tr) + \sigma^2(ptr). \quad (1.2)$$

That is, the total observed score variance can be decomposed into seven independent variance components. It is assumed here that the population and both facets in the universe of admissible observations are quite large

---

[3] Actually, the effect *ptr* is a residual effect involving the triple interaction and all other sources of error not explicitly represented in the universe of admissible observations.

TABLE 1.1. Expected Mean Squares and Estimators of Variance Components for $p \times t \times r$ Design

Effect(α)	$EMS(\alpha)$
p	$\sigma^2(ptr) + n_t\sigma^2(pr) + n_r\sigma^2(pt) + n_t n_r \sigma^2(p)$
t	$\sigma^2(ptr) + n_p\sigma^2(tr) + n_r\sigma^2(pt) + n_p n_r \sigma^2(t)$
r	$\sigma^2(ptr) + n_p\sigma^2(tr) + n_t\sigma^2(pr) + n_p n_t \sigma^2(r)$
pt	$\sigma^2(ptr) + n_r\sigma^2(pt)$
pr	$\sigma^2(ptr) + n_t\sigma^2(pr)$
tr	$\sigma^2(ptr) + n_p\sigma^2(tr)$
ptr	$\sigma^2(ptr)$

Effect(α)	$\hat{\sigma}^2(\alpha)$
p	$[MS(p) - MS(pt) - MS(pr) + MS(ptr)]/n_t n_r$
t	$[MS(t) - MS(pt) - MS(tr) + MS(ptr)]/n_p n_r$
r	$[MS(r) - MS(pr) - MS(tr) + MS(ptr)]/n_p n_t$
pt	$[MS(pt) - MS(ptr)]/n_r$
pr	$[MS(pr) - MS(ptr)]/n_t$
tr	$[MS(tr) - MS(ptr)]/n_p$
ptr	$MS(ptr)$

Note. α represents any one of the effects.

(approaching infinity, at least theoretically). Under these assumptions, the variance components in Equation 1.2 are called *random effects* variance components. It is important to note that they are for *single* person-prompt-rater combinations, as opposed to average scores over prompts and/or raters, which fall in the realm of D study considerations.

Now that Smith has specified her population and universe of admissible observations, she needs to collect and analyze data to estimate the variance components in Equation 1.2. To do so, Smith conducts a study in which, let us suppose, she has a sample of n_r raters use a particular scoring procedure to evaluate each of the responses by a sample of n_p persons to a sample of n_t essay prompts. Such a study is called a G (generalizability) study. The design of this particular study (i.e., the G study design) is denoted $p \times t \times r$. We say this is a two-facet design because the objects of measurement (persons) are not usually called a "facet." Given this design, the usual procedure for estimating the variance components in Equation 1.2

employs the expected mean square ($\boldsymbol{EMS}$) equations in Table 1.1. The resulting estimators of these variance components, in terms of mean squares, are also provided in Table 1.1. These so-called "ANOVA" estimators are discussed extensively in Chapter 3.

Suppose the following estimated variance components are obtained from Smith's G study.

$$\left.\begin{array}{lll} \hat{\sigma}^2(p) = .25, & \hat{\sigma}^2(t) = .06, & \hat{\sigma}^2(r) = .02, \\ \hat{\sigma}^2(pt) = .15, & \hat{\sigma}^2(pr) = .04, & \hat{\sigma}^2(tr) = .00, \\ \text{and} & \hat{\sigma}^2(ptr) = .12. & \end{array}\right\} \quad (1.3)$$

These are estimates of the actual variances (parameters) in Equation 1.2. For example, $\hat{\sigma}^2(p)$ is an estimate of the variance component $\sigma^2(p)$, which can be interpreted roughly in the following manner. Suppose that, for each person in the population, Smith could obtain the person's mean score (technically, "expected" score) over all N_t essay prompts and all N_r raters in the universe of admissible observations. The variance of these mean scores (over the population of persons) is $\sigma^2(p)$. The other "main effect" variance components for the prompt and rater facets can be interpreted in a similar manner. Note that for Smith's universe of admissible observations, the estimated variance attributable to essay prompts, $\hat{\sigma}^2(t) = .06$, is three times as large as the estimated variance attributable to raters, $\hat{\sigma}^2(r) = .02$. This suggests that prompts differ much more in average difficulty than raters differ in average stringency.

Interaction variance components are more difficult to describe verbally, but approximate statements can be made. For example, $\hat{\sigma}^2(pt)$ estimates the extent to which the relative ordering of persons differs by essay prompt, and $\hat{\sigma}^2(pr)$ estimates the extent to which persons are rank ordered differently by different raters. For the illustration considered here, it is especially important to note that $\hat{\sigma}^2(pt) = .15$ is almost four times as large as $\hat{\sigma}^2(pr) = .04$. This fact, combined with the previous observation that $\hat{\sigma}^2(t)$ is three times as large as $\hat{\sigma}^2(r)$, suggests that prompts are a considerably greater source of variability in persons' scores than are raters. The implication and importance of these facts becomes evident in subsequent sections.

1.1.2   Infinite Universe of Generalization and D Studies

The purpose of a G study is to obtain estimates of variance components associated with a universe of admissible observations. These estimates can be used to design efficient measurement procedures for operational use and to provide information for making substantive decisions about objects of measurement (usually persons) in various D (decision) studies. Broadly speaking, D studies emphasize the estimation, use, and interpretation of variance components for decision-making with well-specified measurement procedures.

Perhaps the most important D study consideration is the specification of a universe of generalization, which is the universe to which a decision-maker wants to generalize based on the results of a particular measurement procedure. Let us suppose that Smith's universe of generalization contains all the prompts and raters in her universe of admissible observations. Since both facets are assumed to be infinite, Smith's universe of generalization would be called "infinite" as well. In this scenario, then, it is assumed that Smith wants to generalize persons' scores based on the specific prompts and raters in her measurement procedure to these persons' scores for a universe of generalization that involves many other prompts and raters. In analysis of variance terminology, such a model is described as *random*, and sometimes the prompt and rater facets are said to be random, too.

The universe of generalization is closely related to replications of the measurement procedure. Let us suppose that Smith decides to design her measurement procedure such that each person will respond to n'_t essay prompts, with each response to every prompt evaluated by the same n'_r raters. Furthermore, assume that decisions about a person will be based on his or her mean score over the $n'_t n'_r$ observations associated with the person. This is a verbal description of a D study $p \times T \times R$ design. It is much like the $p \times t \times r$ design for Smith's G study, but there are some important differences.

First, the sample sizes for the D study (n'_t and n'_r) need not be the same as the sample sizes for the G study (n_t and n_r). This distinction is highlighted by the use of primes with D study sample sizes. Second, for the D study, interest focuses on *mean* scores for persons, rather than single person-prompt-rater observations that are the focus of G study estimated variance components. This emphasis on mean scores is highlighted by the use of uppercase letters for the facets in Smith's D study $p \times T \times R$ design.

Let us suppose that Smith decides that a replication of her measurement procedure would involve a *different* sample of n'_t essay prompts and a *different* sample of n'_r raters. Such measurement procedures are described as "randomly parallel." These randomly parallel replications span the entire universe of generalization, in the sense that the replications exhaust all conditions in the universe.

Universe Scores

In principle, for any person, Smith can conceive of obtaining the person's mean score for every instance of the measurement procedure in her universe of generalization. For any such person, the expected value of these mean scores is the person's *universe score*.

The variance of universe scores over all persons in the population is called *universe score variance*. It has conceptual similarities with true score variance in classical test theory.

D Study Variance Components

For Smith's D study $p \times T \times R$ design, the linear model for an observable mean score over n'_t essay prompts and n'_r raters can be represented as

$$X_{pTR} = \mu + \nu_p + \nu_T + \nu_R + \nu_{pT} + \nu_{pR} + \nu_{TR} + \nu_{pTR}. \tag{1.4}$$

The variances of the score effects in Equation 1.4 are called *D study variance components*. When it is assumed that the population and all facets in the universe of generalization are infinite, these variance components are *random effects* variance components. They can be estimated using the G study estimated variance components in Equation Set 1.3.

For example, suppose Smith wants to consider using the sample sizes $n'_t = 3$ and $n'_r = 2$ for her measurement procedure. If so, the estimated D study random effects variance components are

$$\left. \begin{array}{ccc} \hat{\sigma}^2(p) = .25, & \hat{\sigma}^2(T) = .02, & \hat{\sigma}^2(R) = .01, \\ \hat{\sigma}^2(pT) = .05, & \hat{\sigma}^2(pR) = .02, & \hat{\sigma}^2(TR) = .00, \\ \text{and} & \hat{\sigma}^2(pTR) = .02. \end{array} \right\} \tag{1.5}$$

These estimated variance components are for person *mean scores* over $n'_t = 3$ essay prompts and $n'_r = 2$ raters.

Rule. Obtaining these results is simple. Let $\hat{\sigma}^2(\alpha)$ be any one of the G study estimated variance components. Then, the estimated D study variance components $\hat{\sigma}^2(\bar{\alpha})$ are

$$\hat{\sigma}^2(\bar{\alpha}) = \begin{cases} \hat{\sigma}^2(\alpha)/n'_t & \text{if } \alpha \text{ contains } t \text{ but not } r, \\ \hat{\sigma}^2(\alpha)/n'_r & \text{if } \alpha \text{ contains } r \text{ but not } t, \text{ or} \\ \hat{\sigma}^2(\alpha)/(n'_t n'_r) & \text{if } \alpha \text{ contains both } t \text{ and } r. \end{cases} \tag{1.6}$$

The estimated variance component $\hat{\sigma}^2(p) = .25$, which is unaffected by this rule, is particularly important because it is the estimated universe score variance for the random model in this scenario. In terms of parameters, when prompts and raters are both random, universe score is defined as

$$\mu_p \equiv \mathop{E}_{T} \mathop{E}_{R} X_{pTR} = \mu + \nu_p, \tag{1.7}$$

where E stands for expected value. The variance of universe scores—that is, universe score variance—is denoted generically $\sigma^2(\tau)$, and it is simply $\sigma^2(p)$ here.

TABLE 1.2. Random Effects Variance Components that Enter $\sigma^2(\tau)$, $\sigma^2(\delta)$, and $\sigma^2(\Delta)$ for $p \times T \times R$ Design

	D Studies[a]	
	T, R Random	T Fixed
$\sigma^2(p)$	τ	τ
$\sigma^2(T) = \sigma^2(t)/n_t'$	Δ	
$\sigma^2(R) = \sigma^2(r)/n_r'$	Δ	Δ
$\sigma^2(pT) = \sigma^2(pt)/n_t'$	Δ, δ	τ
$\sigma^2(pR) = \sigma^2(pr)/n_r'$	Δ, δ	Δ, δ
$\sigma^2(TR) = \sigma^2(tr)/n_t'n_r'$	Δ	Δ
$\sigma^2(pTR) = \sigma^2(ptr)/n_t'n_r'$	Δ, δ	Δ, δ

[a] τ is universe score.

Error Variances

Given Smith's infinite universe of generalization, variance components other than $\sigma^2(p)$ contribute to one or more different types of error variance. Considered below are "absolute" and "relative" error variances.

Absolute error variance $\sigma^2(\Delta)$. Absolute error is simply the difference between a person's observed and universe score:

$$\Delta_p \equiv X_{pTR} - \mu_p. \tag{1.8}$$

For this scenario, given Equations 1.4 and 1.7,

$$\Delta_p = \nu_T + \nu_R + \nu_{pT} + \nu_{pR} + \nu_{TR} + \nu_{pTR}. \tag{1.9}$$

Consequently, the variance of the absolute errors $\sigma^2(\Delta)$ is the sum of all the variance components except $\sigma^2(p)$. This result is also provided in Table 1.2 under the column headed "T, R Random."

Given the estimated D study variance components in Equation Set 1.5, the estimate of $\sigma^2(\Delta)$ for three prompts and two raters is:

$$\hat{\sigma}^2(\Delta) = .02 + .01 + .05 + .02 + .00 + .02 = .12,$$

and its square root is $\hat{\sigma}(\Delta) = .35$, which is interpretable as an estimate of the "absolute" standard error of measurement (SEM). Consequently, with the usual caveats, $X_{pTR} \pm .35$ constitutes a 68% confidence interval for persons' universe scores.

Suppose Smith judged $\hat{\sigma}(\Delta) = .35$ to be unacceptably large for her purposes, or suppose she decided that time constraints preclude using three prompts. For either of these reasons, or other reasons, she may want to estimate $\hat{\sigma}(\Delta)$ for a number of different values of n_t' and/or n_r'. Doing so is simple. Smith merely uses the rule in Equation 1.6 to estimate the D

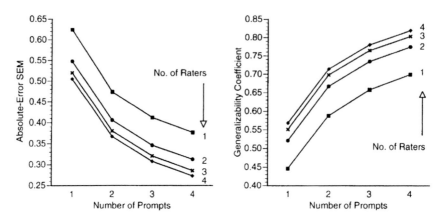

FIGURE 1.2. $\hat{\sigma}(\Delta)$ and $E\hat{\rho}^2$ for scenario with $p \times T \times R$ design.

study variance components for any pair of D study sample sizes of interest to her. Then, as indicated in Table 1.2, she sums all the estimated variance components except $\hat{\sigma}^2(p)$, and takes the square root.

The left-hand panel of Figure 1.2 illustrates results for both n'_t and n'_r ranging from one to four. It is evident from Figure 1.2 that increasing n'_t and/or n'_r leads to a decrease in $\hat{\sigma}(\Delta)$. This result is sensible, since averaging over more conditions of measurement should reduce error. Figure 1.2 also suggests that using more than three raters leads to only a very slight reduction in $\hat{\sigma}(\Delta)$. Consequently, probably it would be unnecessary to use more than three raters (and perhaps only two) for an actual measurement procedure. In addition, Figure 1.2 indicates that using additional prompts decreases $\hat{\sigma}(\Delta)$ quicker than using additional raters. This is a direct result of the fact that $\hat{\sigma}^2(t) = .06$ is larger than $\hat{\sigma}^2(r) = .02$, and $\hat{\sigma}^2(pt) = .15$ is larger than $\hat{\sigma}^2(pr) = .04$. Consequently, for this example, all other things being equal, it would seem desirable to use as many prompts as possible.

Relative error variance $\sigma^2(\delta)$. Relative error is defined as the difference between a person's observed deviation score and his or her universe deviation score:

$$\delta_p \equiv (X_{pTR} - \mu_{TR}) - (\mu_p - \mu), \tag{1.10}$$

where μ_{TR} is the expected value over persons of the observed scores X_{pTR} for the $p \times T \times R$ design. It can be shown that

$$\delta_p = \nu_{pT} + \nu_{pR} + \nu_{pTR}, \tag{1.11}$$

and the variance of these relative errors is the sum of the variance components for the three effects in Equation 1.11. This result is also given in Table 1.2, under the column headed "T, R Random." Relative error variance is similar to error variance in classical theory.

For the example introduced previously, if $n'_t = 3$ and $n'_r = 2$, then

$$\hat{\sigma}^2(\delta) = .05 + .02 + .02 = .09,$$

and its square root is $\hat{\sigma}(\delta) = .30$, which is interpretable as an estimate of the "relative" SEM. Note that this value of $\hat{\sigma}(\delta)$ is smaller than $\hat{\sigma}(\Delta) = .35$ for the same pair of sample sizes. In general, $\hat{\sigma}(\delta)$ is less than $\hat{\sigma}(\Delta)$ because, as indicated in Table 1.2, $\hat{\sigma}^2(\delta)$ involves fewer variance components than $\hat{\sigma}^2(\Delta)$. In short, relative interpretations about persons' scores are less error prone than absolute interpretations.

Coefficients

Two types of reliability-like coefficients are widely used in generalizability theory. One coefficient is called a "generalizability coefficient" and denoted $E\rho^2$. The other coefficient is an "index of dependability" that is denoted Φ.

Generalizability coefficient $E\rho^2$. A generalizability coefficient is the ratio of universe score variance to itself plus relative error variance:

$$E\rho^2 = \frac{\sigma^2(\tau)}{\sigma^2(\tau) + \sigma^2(\delta)}. \tag{1.12}$$

It is the analogue of a reliability coefficient in classical theory. For the example considered here, with $n_t' = 3$ and $n_r' = 2$,

$$E\hat{\rho}^2 = \frac{.25}{.25 + .09} = .74.$$

The right-hand panel of Figure 1.2 provides a graph of $E\hat{\rho}^2$ for values of n_t' and n_r' ranging from one to four. As observed in the discussion of SEMs, little is gained by having more than three raters, and using a relatively large number of prompts seems highly desirable.

Index of dependability Φ. An index of dependability is the ratio of universe score variance to itself plus absolute error variance:

$$\Phi = \frac{\sigma^2(\tau)}{\sigma^2(\tau) + \sigma^2(\Delta)}. \tag{1.13}$$

Φ differs from $E\rho^2$ in that Φ involves $\sigma^2(\Delta)$, whereas $E\rho^2$ involves $\sigma^2(\delta)$. Since $\sigma^2(\Delta)$ is generally larger than $\sigma^2(\delta)$, it follows that Φ is generally smaller than $E\rho^2$. The index Φ is appropriate when scores are given "absolute" interpretations, as in domain-referenced or criterion-referenced situations. For the example considered here, with $n_t' = 3$ and $n_r' = 2$,

$$\hat{\Phi} = \frac{.25}{.25 + .12} = .68.$$

1.1.3 Different Designs and/or Universes of Generalization

The previous section assumed that the D study employed a $p \times T \times R$ design and the universe of generalization was infinite, consisting of two random facets T and R. Recall that the G study also employed a fully crossed design $(p \times t \times r)$ for an infinite universe of admissible observations. In short, to this point, it has been assumed that both designs are fully crossed and the size or "extent" of both universes is essentially the same. This need not be the case, however. For example, the universe of generalization may be narrower than the universe of admissible observations. Also, the structure of the D study can be different from that employed to estimate variance components in the G study. Generalizability theory does not merely permit such differences—it effectively encourages investigators to give serious consideration to the consequences of employing different D study designs and/or different assumptions about a universe of generalization. This is illustrated below using two examples.

The $p \times T \times R$ Design with a Fixed Facet

Returning to the previously introduced scenario, suppose another investigator, Sam Jones, has access to Smith's G study estimated variance components in Equation Set 1.3. However, Jones is not interested in generalizing over essay prompts. Rather, if he were to replicate his measurement procedure, he would use different raters but the *same* prompts. If so, we would say that Jones' universe of generalization is "restricted" in that it contains a *fixed* facet T. Consequently, Jones' universe of generalization is narrower than Smith's universe of generalization. In ANOVA terminology, Jones' interest is focused on a mixed model.

Suppose, also, that Jones decides to use the same D study design structure as Smith, namely, the $p \times T \times R$ design. Under these circumstances, the last column of Table 1.2 indicates which of the random effects D study variance components need to be summed to obtain universe score variance, $\sigma^2(\tau)$, as well as $\sigma^2(\Delta)$ and $\sigma^2(\delta)$.

For example, if $n'_t = 3$ and $n'_r = 2$, then the estimated random effects D study variance components are given by Equation Set 1.5. With T fixed, the mixed model results are:

$$\hat{\sigma}^2(\tau) = \hat{\sigma}^2(p) + \hat{\sigma}^2(pT) = .25 + .05 = .30,$$

$$\begin{aligned} \hat{\sigma}^2(\Delta) &= \hat{\sigma}^2(R) + \hat{\sigma}^2(pR) + \hat{\sigma}^2(TR) + \hat{\sigma}^2(pTR) \\ &= .01 + .02 + .00 + .02 = .05, \end{aligned}$$

and

$$\hat{\sigma}^2(\delta) = \hat{\sigma}^2(pR) + \hat{\sigma}^2(pTR) = .02 + .02 = .04.$$

It is particularly important to note that, with prompts fixed, $\sigma^2(pT)$ contributes to universe score variance, not error variance. Consequently, for a restricted universe of generalization with T fixed, universe score variance is larger than it is for a universe of generalization in which both T and R are random.

Given these results, it follows from Equation 1.12 that

$$E\hat{\rho}^2 = \frac{.30}{.30 + .04} = .88.$$

Recall that, for these D study sample sizes ($n_t' = 3$ and $n_r' = 2$), when prompts were considered random, Smith obtained $E\hat{\rho}^2 = .74$. The estimated generalizability coefficient is larger when prompts are considered fixed because a universe of generalization with a fixed facet is narrower than a universe of generalization with both facets random. That is, generalizations to narrow universes are less error prone than generalizations to broader universes. It is important to note, however, this does not necessarily mean that narrow universes are to be preferred, because restricting a universe also restricts the extent to which an investigator can generalize. For example, when prompts are considered fixed, an investigator cannot logically draw inferences about what would happen if different prompts were used.

The $p \times (R{:}T)$ Design

To expand our scenario even further, consider a third investigator, Ann Hall, who decides that practical constraints preclude her from having all raters evaluate all responses of all persons to all prompts. Rather, she decides that, for each prompt, a different set of raters will evaluate persons' responses. This is a verbal description of the D study $p \times (R{:}T)$ design, where ":" is read "nested within." For this design, the total variance is the sum of five independent variance components; that is,

$$\sigma^2(X_{pTR}) = \sigma^2(p) + \sigma^2(T) + \sigma^2(R{:}T) + \sigma^2(pT) + \sigma^2(pR{:}T). \quad (1.14)$$

For a random effects model, these variance components can be estimated using Smith's estimated G study variance components, even though Smith's G study design is fully crossed, whereas Hall's D study design is partially nested. The process of doing so involves two steps.

First, the G study variance components for the $p \times (r{:}t)$ design are estimated using the results in Equation Set 1.3 for the $p \times t \times r$ design. For both designs, $\hat{\sigma}^2(p) = .25$, $\hat{\sigma}^2(t) = .06$, and $\hat{\sigma}^2(pt) = .15$. Formulas for estimating the remaining two G study variance components for the $p \times (r{:}t)$ design are:

$$\hat{\sigma}^2(r{:}t) = \hat{\sigma}^2(r) + \hat{\sigma}^2(tr) \quad (1.15)$$

and

$$\hat{\sigma}^2(pr{:}t) = \hat{\sigma}^2(pr) + \hat{\sigma}^2(ptr), \quad (1.16)$$

which give $\hat{\sigma}^2(r{:}t) = .02 + .00 = .02$, and $\hat{\sigma}^2(pr{:}t) = .04 + .12 = .16$.

Second, the rule in Equation 1.6 is applied to the estimated G study variance components for the $p \times (r{:}t)$ design. Assuming $n_t' = 3$ and $n_r' = 2$,

TABLE 1.3. Random Effects Variance Components that Enter $\sigma^2(\tau)$, $\sigma^2(\delta)$, and $\sigma^2(\Delta)$ for $p \times (R{:}T)$ Design

	D Studies[a]	
	T, R Random	T Fixed
$\sigma^2(p)$	τ	τ
$\sigma^2(T) = \sigma^2(t)/n'_t$	Δ	
$\sigma^2(R{:}T) = \sigma^2(r{:}t)/n'_r n'_t$	Δ	Δ
$\sigma^2(pT) = \sigma^2(pt)/n'_t$	Δ, δ	τ
$\sigma^2(pR{:}T) = \sigma^2(pr{:}t)/n'_r n'_t$	Δ, δ	Δ, δ

[a] τ is universe score.

the results are:

$$\hat{\sigma}^2(p) = .250, \quad \hat{\sigma}^2(T) = .020, \quad \hat{\sigma}^2(pT) = .050, \; \left.\begin{array}{c}\\\\\end{array}\right\}$$
$$\hat{\sigma}^2(R{:}T) = .003, \quad \text{and} \quad \hat{\sigma}^2(pR{:}T) = .027. \quad\quad (1.17)$$

The second column in Table 1.3 specifies how to combine the estimates in Equation Set 1.17 to obtain $\hat{\sigma}^2(\tau)$, $\hat{\sigma}^2(\Delta)$, and $\hat{\sigma}^2(\delta)$ for a universe of generalization in which both T and R are random. The third column applies when prompts are fixed. Once $\hat{\sigma}^2(\tau)$, $\hat{\sigma}^2(\Delta)$, and $\hat{\sigma}^2(\delta)$ are obtained, $E\rho^2$ and Φ can be estimated using Equations 1.12 and 1.13, respectively.

Suppose, for example, that Hall decides to generalize to a universe in which both T and R are considered random. For this universe of generalization,

$$\hat{\sigma}^2(\tau) = \hat{\sigma}^2(p) = .250,$$

$$\begin{aligned}\hat{\sigma}^2(\Delta) &= \hat{\sigma}^2(T) + \hat{\sigma}^2(pT) + \hat{\sigma}^2(R{:}T) + \hat{\sigma}^2(pR{:}T) \\ &= .020 + .050 + .003 + .027 = .100,\end{aligned}$$

and

$$\hat{\sigma}^2(\delta) = \hat{\sigma}^2(pT) + \hat{\sigma}^2(pR{:}T) = .050 + .027 = .077.$$

It follows that

$$E\hat{\rho}^2 = \frac{\hat{\sigma}^2(\tau)}{\hat{\sigma}^2(\tau) + \hat{\sigma}^2(\delta)} = \frac{.250}{.250 + .077} = .76.$$

Recall that for the $p \times T \times R$ design with T and R random, Smith obtained $E\hat{\rho}^2 = .74$, which is somewhat different from $E\hat{\rho}^2 = .76$ for the *same* universe using the $p \times (R{:}T)$ design. The difference in these two results is *not* rounding error. Rather, it is attributable to the fact that $\hat{\sigma}^2(\delta) = .090$ for the $p \times T \times R$ design is larger than $\hat{\sigma}^2(\delta) = .077$ for the $p \times (R{:}T)$ design. This illustrates that reliability, or generalizability, is affected by design structure. Recall that it has been demonstrated previously that reliability, or generalizability, is also affected by sample sizes and the size or "extent" of a universe of generalization. These results illustrate an important fact: namely, it can be very misleading to refer to *the* reliability or *the* error variance of a measurement procedure without considerable explanation and qualification.

1.1.4 Other Issues and Applications

All other things being equal, the power of generalizability theory is most likely to be realized when a G study employs a fully crossed design and a large sample of conditions for each facet in the universe of admissible observations. A large sample of conditions is beneficial because it leads to more stable estimates of G study variance components. A crossed design is advantageous because it maximizes the number of design structures that can be considered for one or more subsequent D studies. However, any design structure can be used in a G study. For example, the hypothetical scenario could have used a G study $p \times (r:t)$ design, but then an investigator could not estimate all variance components for a D study $p \times T \times R$ design.

It often happens that the distinction between a G and D study is blurred, usually because the only available data are for an operational administration of an actual measurement procedure. In this case, the procedures discussed can still be followed to estimate parameters such as error variances and generalizability coefficients, but obviously under these circumstances an investigator cannot take advantage of all aspects of generalizability theory.

In most applications of generalizability theory, examinees or persons are the objects of measurement. Occasionally, however, some other collection of conditions plays the role of objects of measurement. For example, in evaluation studies, classes are often the objects of measurement with persons and other facets being associated with the universe of generalization. It is straightforward to apply generalizability theory in such cases.

Generalizability theory has broad applicability and has been applied in numerous settings. Some real-data examples are provided in this book, but these are only a few illustrations. The theory has been applied to a vast array of educational tests and to a wide range of other types of tests, including foreign language tests (e.g., Bachman et al., 1994), personality tests (e.g., Knight et al., 1985), psychological inventories (e.g., Crowley et al., 1994), career choice instruments (e.g., Hartman et al., 1988), and cognitive ability tests (e.g., Thompson & Melancon, 1987).

Other areas of application include performance assessments in education (e.g., Linn & Burton, 1994), standard setting (e.g., Brennan, 1995b; Norcini et al., 1987), student ratings of instruction (e.g., Crooks & Kane, 1981), teaching behavior (e.g., Shavelson & Dempsey-Atwood, 1976), marketing (e.g., Finn & Kayandé, 1997), business (e.g., Marcoulides, 1998), job analyses (e.g., Webb & Shavelson, 1981), survey research (e.g., Johnson & Bell, 1985), physical education (e.g., Tobar et al., 1999), and sports (e.g., Oppliger & Spray, 1987).

Generalizability theory has also been used in numerous medical areas to study matters such as sleep disorders (e.g., Wohlgemuth et al., 1999), clinical evaluations (e.g., Boodoo & O'Sullivan, 1982), nursing (e.g., Butterfield et al., 1987), dental education (e.g., Chambers & Loos, 1997), speech perception (e.g., Demorest & Bernstein, 1993), biofeedback (e.g., Hatch et al., 1992), epidemiology (e.g., Klipstein-Grobusch et al., 1997), mental retardation (e.g., Ulrich et al., 1989), and computerized scoring of performance assessments (e.g., Clauser et al., 2000).

1.2   Overview of Book

The hypothetical scenario used in Section 1.1 is clearly idealized and incomplete, but it does highlight many of the most important and frequently used concepts and procedures in generalizability theory. The remaining chapters delve more deeply into the conceptual and statistical details and extend the theory in various ways.

This book is divided into three sets of chapters that are ordered in terms of increasing complexity. The fundamentals of univariate generalizability theory are contained in Chapters 1 to 4. The scope of these chapters is a sufficient basis for performing many generalizability analyses and for understanding much of the current literature on generalizability theory. Additional, more challenging topics in univariate theory are covered in Chapters 5 to 8. These chapters are devoted largely to statistical complexities such as the variability of estimated variance components and estimating variance components for unbalanced designs. Finally, Chapters 9 to 12 cover multivariate generalizability theory, in which each object of measurement has multiple universe scores. In these chapters, particular attention is given to tables of specifications for educational and psychological tests.

This book is intended to be both a textbook and a resource for practitioners. To serve these dual purposes, most chapters involve four components: a discussion of theory, one or more synthetic data examples, a few real-data examples, and exercises. Detailed answers to starred exercises are given in Appendix I; answers to the remaining exercises are available from the author (robert-brennan@uiowa.edu) or the publisher (www.springer-ny.com). Sometimes the four components are split over two chapters, and not all topics in all chapters are illustrated with real-data examples, but the four-way coverage predominates.

The real-data examples are intended to be instructive; they are not models for "ideal" generalizability analyses. Different applications of generalizability theory tend to involve somewhat different mixes of conceptual and statistical concerns. Consequently, a reasonable approach in one context is not necessarily appropriate in others. Most of the real-data examples are drawn from the field of educational testing, but a few are from other areas of inquiry. Note, also, that Cronbach et al. (1972) provide a number of illustrative detailed examples from psychology and education.

The synthetic data examples serve two purposes: they illustrate various results in a relatively simple manner, and they are sufficiently "small" to permit the reader to perform all computations with a hand-held calculator. In practical contexts, however, computer programs are required for performing generalizability analyses. Three computer programs are available that are specifically coordinated with the content of this book: GENOVA, urGENOVA, and mGENOVA. These are introduced at appropriate points. Using one or more of these programs, most of the computations discussed in this book can be performed for almost any data set.

The number of decimal digits used in the examples varies, depending on the context. For synthetic data examples, usually four decimal digits are reported so that readers can be assured that their computations are correct, at least to the next to the last decimal digit. Some real-data examples come from published papers in which results are reported with fewer digits. Performing computations with four, five, or even more digits is to be recommended so that rounding errors have mimimal impact on final results. However, reporting results with that many decimal digits does not mean that it is necessarily reasonable to base interpretations on the magnitude of digits far to the right of a decimal point.

Some students and practitioners who initially encounter generalizability theory are overwhelmed by the statistical issues. That is why the early chapters of this book often sidestep statistical complexities or relegate them to exercises. Other students and practitioners have little trouble with the statistical "hurdle," but the conceptual issues challenge them. In particular, in the author's experience, persons with strong statistical backgrounds are prone to view the theory simply as the application of variance components analysis to measurement issues. There is an element of truth in such a perspective, but variance components analysis is a tool used in generalizability theory—it does not encompass the theory. For example, variance components analysis per se does not tell an investigator which components contribute to which type of error, and these are central issues in generalizability theory.

1.3 Exercises

1.1* Suppose an investigator obtains the following mean squares for a G study $p \times t \times r$ design using $n_p = 100$ persons, $n_t = 5$ essay items (or tasks), and $n_r = 6$ raters.

$$MS(p) = 6.20, \quad MS(t) = 57.60, \quad MS(r) = 28.26,$$
$$MS(pt) = 1.60, \quad MS(pr) = .26, \quad MS(tr) = 8.16,$$
$$\text{and} \quad MS(ptr) = .16.$$

(a) Estimate the G study variance components assuming both t and r are infinite in the universe of admissible observations.

(b) Estimate the D study random effects variance components for a D study $p \times T \times R$ design with $n'_t = 4$, $n'_r = 2$, and with persons as the objects of measurement.

(c) For the D study design and sample sizes in (b), estimate absolute error variance, relative error variance, a generalizability coefficient, and an index of dependability.

(d) Estimate $\sigma(\Delta)$ if an investigator decides to use the D study $p \times (R{:}T)$ design with $n'_t = 3$ and $n'_r = 2$, assuming T and R are both random in the universe of generalization.

1.2* Suppose an investigator specifies that a universe of generalization consists of only two facets and both are fixed. From the perspective of generalizability theory, why is this nonsensical?

1.3 Brennan (1992a, p. 65) states that, "... the Spearman–Brown Formula does *not* apply when one generalizes over *more* than one facet." Verify this statement using the random model example of the $p \times T \times R$ design in Section 1.1.2, where $E\hat{\rho}^2 = .74$ with three prompts and two raters. Assume the number of prompts is doubled, in which case the Spearman–Brown Formula is $2\,rel/(1 + rel)$, where rel is reliability. Explain why the Spearman–Brown formula and generalizability procedures give different results.

2
Single-Facet Designs

Throughout this chapter it is assumed that the universe of admissible observations and the universe of generalization involve conditions from the same single facet, usually referred to as an item facet. Also, it is usually assumed that the population consists of persons. For a single-faceted universe, there are two designs that might be employed in a G study: the $p \times i$ or the $i{:}p$ design, where the letter p is used to index persons (or examinees), i indexes items, "$\times$" is read "crossed with," and "$:$" is read "nested within." For the $p \times i$ design, each person is administered the *same* random sample of items. For the $i{:}p$ design, each person is administered a *different* random sample of items. Similarly, there are two possible D study designs: $p \times I$ and $I{:}p$, where uppercase I is used to emphasize that D study considerations involve mean scores over sets of items.

The most common pairing is a G study $p \times i$ design with a D study $p \times I$ design. If we (verbally) neglect the distinction between i and I, then we can say that the two designs are the same. This design is the subject of the first three sections of this chapter, which systematically introduce many of the statistical issues in generalizability theory. These sections are followed by a consideration of the G study $i{:}p$ and D study $I{:}p$ designs.

Technically, the theory discussed in this chapter assumes that both the population of persons and the universe of items are infinite, which is symbolized as $N_p \to \infty$ and $N_i \to \infty$. In practice, this assumption is seldom if ever literally true, but it is a useful idealization for many studies.

2.1   G Study $p \times i$ Design

Let X_{pi} denote the score for any person in the population on any item in the universe. The expected value of a person's observed score, associated with a process in which an item is randomly selected from the universe, is

$$\mu_p \equiv \underset{i}{E} X_{pi}, \tag{2.1}$$

where the symbol E is an expectation operator, and the subscript i designates the facet over which the expectation is taken. The score μ_p can be conceptualized as an examinee's "average" or "mean" score over the *universe* of items. In a similar manner, the population mean for item i is defined as

$$\mu_i \equiv \underset{p}{E} X_{pi}, \tag{2.2}$$

and the mean over both the population and the universe is

$$\mu \equiv \underset{p}{E} \underset{i}{E} X_{pi}. \tag{2.3}$$

These mean scores are not observable because, for example, examinee responses to *all* items in the universe are never available. Nonetheless, any person-item score that *might* be observed (an observable score) can be expressed in terms of μ_p, μ_i, and μ using the linear model:

$$
\begin{aligned}
X_{pi} \;=\; &\mu && \text{(grand mean)} \\
&+ \mu_p - \mu && \text{(person effect } = \nu_p) \\
&+ \mu_i - \mu && \text{(item effect } = \nu_i) \\
&+ X_{pi} - \mu_p - \mu_i + \mu && \text{(residual effect } = \nu_{pi}),
\end{aligned}
\tag{2.4}
$$

which can be expressed more succinctly as

$$X_{pi} = \mu + \nu_p + \nu_i + \nu_{pi}. \tag{2.5}$$

Equations 2.4 and 2.5 represent the same tautologies. In each of them the observed score is decomposed into *components*, or *effects*. The only difference between the two equations is that in Equation 2.4 each effect is explicitly represented as a deviation score, while in Equation 2.5 the Greek letter ν represents an effect.[1] For example, the person effect is $\nu_p = \mu_p - \mu$, and the item effect is $\nu_i = \mu_i - \mu$. The residual effect, $\nu_{pi} = X_{pi} - \mu_p - \mu_i + \mu$, is sometimes referred to as the interaction effect. Actually, both verbal descriptions are somewhat misleading because, with a single observed score for each person-item combination, the person-item interaction effect and

---

[1] Cronbach et al. (1972) and Brennan (1992a) use $\mu\sim$ to designate a score effect. For example, they use $\mu_p \sim$ rather than ν_p.

all other residual effects are completely confounded (totally indistinguish-able).[2]

All of the effects (except μ) in Equations 2.4 and 2.5 are called *random effects* because they are associated with a process of random sampling from the population and universe. Under these assumptions, the linear model is referred to as a random effects model. The manner in which these random effects have been defined implies that

$$\mathop{E}_{p} \nu_p = \mathop{E}_{i} \nu_i = \mathop{E}_{p} \nu_{pi} = \mathop{E}_{i} \nu_{pi} = 0. \tag{2.6}$$

Equations 2.5 and 2.6 can be used directly to express mean scores in terms of score effects. For example, given the definition of μ_p in Equation 2.1, it follows that

$$
\begin{aligned}
\mu_p &= \mathop{E}_{i}(\mu + \nu_p + \nu_i + \nu_{pi}) \\
&= \mu + \nu_p + \mathop{E}_{i} \nu_i + \mathop{E}_{i} \nu_{pi} \\
&= \mu + \nu_p. \tag{2.7}
\end{aligned}
$$

Similarly,

$$\mu_i = \mu + \nu_i. \tag{2.8}$$

To this point, the words "study" and "design" have not been used. That is, the model in Equation 2.5 has been defined without explicitly specifying how any observed data are collected. Let us now assume that G study data are collected in the following manner.

- A random sample of n_p persons is selected from the population;

- an independent random sample of n_i items is selected from the universe; and

- each of the n_p persons is administered each of the n_i items, and the responses (X_{pi}) are obtained.

Technically, this is a description of the G study $p \times i$ design, with the linear model given by Equation 2.5. That is, in a sense, Equation 2.5 actually plays two roles: it characterizes the observable data for the population and universe, and it represents the actual observed data for a G study.

It is "assumed" that all effects in the model are uncorrelated. Letting a prime designate a different person or item, this means that

$$E(\nu_p \nu_{p'}) = E(\nu_i \nu_{i'}) = E(\nu_{pi} \nu_{p'i}) = E(\nu_{pi} \nu_{pi'}) = E(\nu_{pi} \nu_{p'i'}) = 0, \tag{2.9}$$

and

$$E(\nu_p \nu_i) = E(\nu_p \nu_{pi}) = E(\nu_i \nu_{pi}) = 0. \tag{2.10}$$

---

[2]For this reason, Cronbach et al. (1972) denote the residual effect as $(\mu_{pi} \sim, e)$.

The word "assumed" is in quotes because most of these zero expectations are a direct consequence of the manner in which score effects have been defined in the linear model and/or the random sampling assumptions for the $p \times i$ design (see Brennan, 1994).[3] For example, the random sampling assumptions imply that $E(\nu_p\nu_i) = (E\nu_p)(E\nu_i) = 0$.

The above development can be summarized by saying that a G study $p \times i$ design is represented by the linear model in Equation 2.5 with uncorrelated score effects. This description is adequate *provided* it is understood in the sense discussed above. Note, also, that the modeling has been specified *without* any normality assumptions, and without assuming that score effects are independent, which is a stronger assumption than uncorrelated score effects.

2.2   G Study Variance Components for $p \times i$ Design

For each score effect, or component, in Equation 2.5, there is an associated variance of the score effect, or component, which is called a *variance component*. For example, the variance component for persons is

$$\begin{aligned} \sigma^2(p) &= \underset{p}{E}(\mu_p - \underset{p}{E}\mu_p)^2 \\ &= \underset{p}{E}(\mu_p - \mu)^2 \\ &= \underset{p}{E}\nu_p^2. \end{aligned} \tag{2.11}$$

From this derivation, it is evident that $\sigma^2(p)$ might be denoted $\sigma^2(\mu_p)$ or $\sigma^2(\nu_p)$. In other words, the variance of person mean scores is identical to the variance of person score effects. Similarly, for items,

$$\sigma^2(i) = \underset{i}{E}(\mu_i - \mu)^2 = \underset{i}{E}\nu_i^2, \tag{2.12}$$

which might be denoted $\sigma^2(\mu_i)$ or $\sigma^2(\nu_i)$, indicating that the variance of item mean scores is identical to the variance of item score effects.

For the interaction of persons and items,

$$\sigma^2(pi) = \underset{p}{E}\underset{i}{E}(X_{pi} - \mu_p - \mu_i + \mu)^2 = \underset{p}{E}\underset{i}{E}\nu_{pi}^2. \tag{2.13}$$

The variance component $\sigma^2(pi)$ might be denoted $\sigma^2(\nu_{pi})$, but *not* $\sigma^2(\mu_{pi})$. Cronbach et al. (1972) denote the interaction (or residual) variance component as $\sigma^2(pi, e)$ rather than $\sigma^2(pi)$. Their notation explicitly reflects the confounding of the interaction variance component with the variance associated with other residual effects or sources of "error."[4]

---

[3]These distinctions are considered more explicitly in Section 5.4.

[4]The confounded-effects notation is not used routinely in this book, because it is awkward for multifacet designs.

These variance components provide a decomposition of the so-called "total" variance:

$$\sigma^2(X_{pi}) = \mathop{E}_{p} \mathop{E}_{i}(X_{pi} - \mu)^2 = \sigma^2(p) + \sigma^2(i) + \sigma^2(pi). \qquad (2.14)$$

The derivation of Equation 2.14 is tedious, but no assumptions are required beyond those in Section 2.1. The total variance is a variance of scores for single persons on single items. It follows that the variance components are for "single" scores, too.

These variance components might be conceptualized in the following manner. Suppose N_p and N_i were *very* large, but still finite. Under this circumstance, in theory, all items in the universe could be administered to all persons in the population. Given the resulting observed scores, values for the mean scores in Equations 2.1 to 2.3 could be obtained. Then $\sigma^2(p)$ could be computed by taking the variance, over the population of persons, of the scores μ_p. Similarly, $\sigma^2(i)$ could be computed by taking the variance, over the universe of items, of the scores μ_i. Finally, $\sigma^2(pi)$ could be computed by taking the variance, over both persons and items, of the scores $X_{pi} - \mu_p - \mu_i + \mu$. This approach to interpreting variance components is clearly a contrived scenario, but frequently it is a helpful conceptual aid.

2.2.1 Estimating Variance Components

In generalizability theory, variance components assume central importance. They are the building blocks that provide a crucial foundation for all subsequent results. To understand variance components, Equations 2.11 to 2.13 are useful, but they cannot be used directly to *estimate* variance components because these equations are expressed in terms of squares of the unknown random effects in the model. Rather, variance components are usually estimated through an analysis of variance. Table 2.1 provides ANOVA computational formulas for the $p \times i$ design. Note that α is used in Table 2.1 (and elsewhere in this book) as a generic identifier for an effect.

To this point, all reference to mean scores has implied population and/or universe mean scores. Now, however, we must clearly distinguish between these mean scores and their observed score analogues, which result from a G study. For example, the mean of the observed scores for person p is denoted $\overline{X}_p$ in Table 2.1, and it is the observed score analogue of μ_p. Similarly, $\overline{X}_i$ is the mean of the observed scores for item i, and it is analogous to μ_i. Finally, $\overline{X}$ is analogous to μ.

For the $p \times i$ design, the derivation of the so-called "ANOVA" estimators of the variance components in Table 2.1 is rather straightforward. One such derivation is outlined next. Consider, again, Equation 2.4 for the decomposition of X_{pi} into score effects, or components. Replacing population and/or universe mean scores with their observed score analogues, it is easy

TABLE 2.1. ANOVA Formulas and Notation for G Study $p \times i$ Design

Effect(α)	$df(\alpha)$	$SS(\alpha)$	$MS(\alpha)$	$\hat{\sigma}^2(\alpha)$
p	$n_p - 1$	$SS(p)$	$MS(p)$	$\hat{\sigma}^2(p) = \dfrac{MS(p) - MS(pi)}{n_i}$
i	$n_i - 1$	$SS(i)$	$MS(i)$	$\hat{\sigma}^2(i) = \dfrac{MS(i) - MS(pi)}{n_p}$
pi	$(n_p - 1)(n_i - 1)$	$SS(pi)$	$MS(pi)$	$\hat{\sigma}^2(pi) = MS(pi)$

$$SS(p) = n_i \sum_p \overline{X}_p^2 - n_p n_i \overline{X}^2$$
$$SS(i) = n_p \sum_i \overline{X}_i^2 - n_p n_i \overline{X}^2$$
$$SS(pi) = \sum_p \sum_i X_{pi}^2 - n_i \sum_p \overline{X}_p^2 - n_p \sum_i \overline{X}_i^2 + n_p n_i \overline{X}^2$$

to show that

$$X_{pi} - \overline{X} = (\overline{X}_p - \overline{X}) + (\overline{X}_i - \overline{X}) + (X_{pi} - \overline{X}_p - \overline{X}_i + \overline{X}).$$

The sum of these squared observed deviation scores is

$$\sum_p \sum_i (X_{pi} - \overline{X})^2 = n_i \sum_p (\overline{X}_p - \overline{X})^2 + n_p \sum_i (\overline{X}_i - \overline{X})^2$$
$$+ \sum_p \sum_i (X_{pi} - \overline{X}_p - \overline{X}_i + \overline{X})^2. \quad (2.15)$$

Equation 2.15 provides a decomposition of the total sums of squares for the $p \times i$ design into sums of squares attributable to persons, items, and interactions; that is,

$$\sum_p \sum_i (X_{pi} - \overline{X})^2 = SS(p) + SS(i) + SS(pi).$$

The sums of squares formulas in Equation 2.15 are basic from a definitional perspective, but for calculation purposes the formulas reported in Table 2.1 are easier to use. For example,

$$SS(p) = n_i \sum_p (\overline{X}_p - \overline{X})^2 = n_i \sum_p \overline{X}_p^2 - n_p n_i \overline{X}^2,$$

and the latter formula is computationally easier to use than the former.

Each sum of squares has a mean square associated with it, which is simply the sum of squares divided by its degrees of freedom. For example,

$$MS(p) = \frac{n_i \sum_p (\overline{X}_p - \overline{X})^2}{n_p - 1}.$$

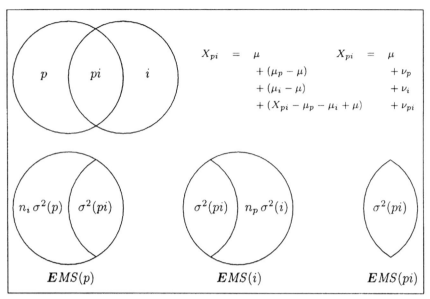

FIGURE 2.1. Venn diagrams for $p \times i$ design.

To estimate variance components we make use of well-known equations for the expected values of these mean squares ($\boldsymbol{EMS}$ equations):

$$\boldsymbol{EMS}(p) = \sigma^2(pi) + n_i \, \sigma^2(p) \tag{2.16}$$
$$\boldsymbol{EMS}(i) = \sigma^2(pi) + n_p \, \sigma^2(i) \tag{2.17}$$
$$\boldsymbol{EMS}(pi) = \sigma^2(pi). \tag{2.18}$$

Solving these equations for the variance components, and using mean squares in place of their expected values, we obtain the ANOVA estimators:

$$\hat{\sigma}^2(p) = \frac{MS(p) - MS(pi)}{n_i} \tag{2.19}$$

$$\hat{\sigma}^2(i) = \frac{MS(i) - MS(pi)}{n_p} \tag{2.20}$$

$$\hat{\sigma}^2(pi) = MS(pi). \tag{2.21}$$

For the $p \times i$ design, the equations for estimating variance components can be illustrated using a Venn diagram approach. More importantly, however, Venn diagrams provide a useful visual and conceptual aid in understanding variance components, as well as other aspects of generalizability analyses. The upper left-hand corner of Figure 2.1 provides the general form of a Venn diagram for the $p \times i$ design, followed by the linear model expressions discussed in Section 2.1. The bottom half of Figure 2.1 provides Venn diagram representations of the expected mean squares given by Equations 2.16 to 2.18. Note that circles and their intersections can be

TABLE 2.2. Synthetic Data Set No. 1 and the $p \times i$ Design

Person	Item Scores (X_{pi})												$\overline{X}_p$
	1	2	3	4	5	6	7	8	9	10	11	12	
1	1	0	1	0	0	0	0	0	0	0	0	0	.1667
2	1	1	1	0	0	1	0	0	0	0	0	0	.3333
3	1	1	1	1	1	0	0	0	0	0	0	0	.4167
4	1	1	0	1	1	0	0	1	0	0	0	0	.4167
5	1	1	1	1	1	0	1	0	0	0	0	0	.5000
6	1	1	1	0	1	1	1	0	0	0	0	0	.5000
7	1	1	1	1	1	1	1	0	0	0	0	0	.5833
8	1	1	1	1	0	1	1	1	1	1	0	0	.7500
9	1	1	1	1	1	1	1	1	1	1	1	0	.9167
10	1	1	1	1	1	1	1	1	1	1	1	1	1.0000
$\overline{X}_i$	1.0	.9	.9	.7	.7	.6	.6	.4	.3	.3	.2	.1	

$\overline{X} = .5583 \qquad \sum_p \overline{X}_p^2 = 3.7292 \qquad \sum_i \overline{X}_i^2 = 4.71 \qquad \sum_p \sum_i X_{pi}^2 = 67$

TABLE 2.3. G Study $p \times i$ Design For Synthetic Data Set No. 1

Effect(α)	$df(\alpha)$	$SS(\alpha)$	$MS(\alpha)$	$\hat{\sigma}^2(\alpha)$
p	9	7.3417	.8157	$\hat{\sigma}^2(p) = .0574$
i	11	9.6917	.8811	$\hat{\sigma}^2(i) = .0754$
pi	99	12.5583	.1269	$\hat{\sigma}^2(pi) = .1269$

interpreted as expected mean squares, while parts of circles can be interpreted as variance components, or simple functions of them. Note, however, that the actual areas in these diagrams are *not* generally proportional to the magnitudes of the variance components. Replacing parameters with estimates, it is evident from the first and third diagrams at the bottom of Figure 2.1 that $\hat{\sigma}^2(p) = [MS(p) - MS(pi)]/n_i$, as indicated in Equation 2.19.

2.2.2 Synthetic Data Example

Consider Synthetic Data Set No. 1 in Table 2.2. Using the equations in Section 2.2.1, it is easy to verify the numerical values for $\hat{\sigma}^2(p)$, $\hat{\sigma}^2(i)$, and $\hat{\sigma}^2(pi)$ in Table 2.3. Readers unfamiliar with estimating variance components are encouraged to verify these results. It is evident that the estimated variance component for persons, $\hat{\sigma}^2(p) = .0574$, is slightly less than the estimated variance for items, $\hat{\sigma}^2(i) = .0754$, but neither of these variance components is nearly as large as $\hat{\sigma}^2(pi) = .1269$. These variance components appear small in terms of their absolute values. However, since they are based on dichotomous data, none of them can be larger than 0.25.

In thinking about the magnitudes of these estimated variance components, it is helpful to remember that they provide a decomposition of the estimated total variance (see Equation 2.14) of scores (0 or 1) for single persons on single items. In particular, $\hat{\sigma}^2(i)$ and $\hat{\sigma}^2(pi)$ are *not* estimated variance components for *mean* (or total) scores over the $n_i = 12$ items in the G study. However, as shown soon, $\hat{\sigma}^2(pi)$ and sometimes $\hat{\sigma}^2(i)$ *contribute* to estimates of several types of error variance for mean scores in a D study $p \times I$ design, and their contributions to such error variances can be reduced by increasing the D study sample size for items.

In Table 2.3 numerical results involving the estimation of variance components are reported to four decimal places. In part this is done to facilitate demonstrations of certain algebraic identities. More importantly, this convention emphasizes that computations should be performed using as much numerical accuracy as possible. Generalizability analyses involve numerous computations, and premature rounding can distort estimates of variance components. However, this convention does not mean that four digits are generally required for purposes of *interpreting* estimated variance components. Indeed, as discussed extensively in Chapter 6, estimated variance components are subject to sampling variability, and when sample sizes are small, such variability is likely to be large enough to cast considerable doubt on the stability of third and fourth digits (or any digit, in extreme cases).

Concern about "significant" digits in estimates of variance components does not mean, however, that it is advisable to perform tests of statistical significance for the estimates themselves. In generalizability theory the magnitudes of estimated variance components are of central importance, rather than their statistical significance at some preassigned level. Even if an estimated variance component does not possess statistical significance, the ANOVA procedure yields an unbiased estimate. As such, it is better to use the estimate than to replace it by zero (Cronbach et al., 1972, p. 192). In particular, generalizability theory emphatically does *not* involve performing F tests on the ratios of mean squares. In generalizability theory, mean squares are used simply to estimate variance components.

2.3   D Studies for the $p \times I$ Design

To this point, discussion has focused on the model and associated variance components for a person's score on a single item (X_{pi}) in the universe of admissible observations. By contrast, if an examinee is administered a sample of n_i' items, *decisions* about the examinee will be based, surely, on his or her average (or total) score over the n_i' items, not a score on a single item. From the perspective of generalizability theory, the intent of such a decision is to generalize from the examinee's observed score to the examinee's *universe score* over all items in the *universe of generalization*.

Another perspective on the universe of generalization is based on the notion of replications of a measurement procedure (see Brennan, in press). From this perspective, multiple measurements of an examinee would consist of his or her average score on different random samples of n_i' items from the same universe. Such samples of items, or test forms, are said to be *randomly parallel*, and for the $p \times I$ design, the universe of generalization can be viewed as a universe of such randomly parallel forms.

By convention, in generalizability theory, average scores over a sample of conditions are indicated by uppercase letters. Using this notation for the D study $p \times I$ design, the linear model for the decomposition of an examinee's average score X_{pI} over n_i' items is

$$X_{pI} = \overline{X}_p = \mu + \nu_p + \nu_I + \nu_{pI}, \tag{2.22}$$

where X_{pI} and $\overline{X}_p$ mean the same thing.[5] Equation 2.22 is completely analogous to Equation 2.4, the only difference being that i (for a single item) in Equation 2.4 is replaced everywhere by I (for the mean over a set of n_i' items) in Equation 2.22.

It is particularly important to note that

$$\mu_p \equiv \mathop{E}_{I} X_{pI}, \tag{2.23}$$

which means that μ_p is defined as the expected value of X_{pI} over I in the universe of generalization. Alternatively, a person's universe score is the expected value of his or her observable mean score over all randomly parallel instances of the measurement procedure, each of which involves a different random sample of sets of conditions I. In short, the phrase "universe score" refers to the universe of generalization, not the universe of admissible observations.

The variance of the μ_p in Equation 2.23 is called *universe score variance*:

$$\sigma^2(p) = \mathop{E}_{p}(\mu_p - \mu)^2 = \mathop{E}_{p} \nu_p^2. \tag{2.24}$$

Alternatively, universe score variance is the expected value of the covariance between randomly parallel forms I and I'; that is,

$$\sigma^2(p) = E\, \sigma(X_{pI}, X_{pI'}), \tag{2.25}$$

where the covariance is taken over persons, and the expectation is taken over pairs of randomly parallel forms.

----

[5] X_{pI} is more descriptive, but $\overline{X}_p$ is simpler and much more convenient for the multifacet designs treated in later chapters.

Just as there are G study variance components associated with each of the random effects in Equation 2.5, so too there are D study variance components associated with the random effects in Equation 2.22. Definitions of these variance components are obtained by replacing i with I in Equations 2.11 through 2.13. The person variance component is unchanged by this replacement process, but the other two variance components *are* altered—namely,

$$\sigma^2(I) = \mathop{E}_{I}(\mu_I - \mu)^2 = \mathop{E}_{I} \nu_I^2 \tag{2.26}$$

and

$$\sigma^2(pI) = \mathop{E}_{p}\mathop{E}_{I}(X_{pI} - \mu_p - \mu_I + \mu)^2 = \mathop{E}_{p}\mathop{E}_{I} \nu_{pI}^2. \tag{2.27}$$

By definition, these two D study variance components apply to the population of persons and the universe of generalization. For example, $\sigma^2(I)$ is interpretable as the variance of the distribution of mean scores μ_I, where each of these means is for the population of persons and a different random sample of n_i' items. One well-known property of a distribution of mean scores for a set of uncorrelated observations is that the variance of the distribution is the variance of the individual elements divided by the sample size. It follows that

$$\sigma^2(I) = \frac{\sigma^2(i)}{n_i'}, \tag{2.28}$$

and

$$\sigma^2(pI) = \frac{\sigma^2(pi)}{n_i'}. \tag{2.29}$$

Equations 2.28 and 2.29 are also applicable when the parameters are replaced by their estimates. Therefore, the estimated G study variance components $\hat{\sigma}^2(i)$ and $\hat{\sigma}^2(pi)$ can be used to estimate $\sigma^2(I)$ and $\sigma^2(pI)$. Also, $\hat{\sigma}^2(p)$ from the G study is an estimate of $\sigma^2(p)$ for the universe of generalization. Estimates of the variance components $\sigma^2(p)$, $\sigma^2(I)$, and $\sigma^2(pI)$ are informative in that they characterize examinee performance for the universe of generalization. Moreover, combinations of them give estimates of error variances and reliability-like coefficients.

2.3.1   Error Variances

The two most frequently discussed types of error in generalizability theory are absolute error and relative error. A third type is the error associated with using the overall observed mean to estimate the mean in the population and universe.

Absolute Error Variance

Absolute error is the error involved in using an examinee's observed mean score as an estimate of his or her universe score. For person p, absolute error is defined as

$$\Delta_{pI} \equiv X_{pI} - \mu_p \tag{2.30}$$

$$= \nu_I + \nu_{pI}. \tag{2.31}$$

Absolute error is often associated with domain-referenced (criterion-referenced or content-referenced) interpretations of scores.

Since $E_I \Delta_{pI}$ is zero, the variance of Δ_{pI} for the population of persons, over randomly parallel forms, is

$$\sigma^2(\Delta) = \mathop{E}_{I}\mathop{E}_{p} \Delta_{pI}^2 = \mathop{E}_{I}\mathop{E}_{p}(\nu_I + \nu_{pI})^2.$$

Since ν_I and ν_{pI} are uncorrelated,

$$\sigma^2(\Delta) = \sigma^2(I) + \sigma^2(pI) = \frac{\sigma^2(i)}{n_i'} + \frac{\sigma^2(pi)}{n_i'}. \tag{2.32}$$

Relative Error Variance

Sometimes an investigator's interest focuses on the relative ordering of individuals with respect to their test performance, or the adequacy of the measurement procedure for making comparative decisions. In current terminology, such decisions are frequently associated with norm-referenced, or relative, interpretations of test scores. The crucial point about such interpretations is that a person's score is interpreted *not* in isolation but, rather, with respect to some measurement of group performance. The most obvious single measure of group performance is a group mean score, and the associated error is called "relative" error in generalizability theory.

Relative error for person p is defined as

$$\delta_{pI} \equiv (X_{pI} - \mathop{E}_{p} X_{pI}) - (\mu_p - \mathop{E}_{p} \mu_p) \tag{2.33}$$

$$= (X_{pI} - \mu_I) - (\mu_p - \mu). \tag{2.34}$$

In Equation 2.34, the test-form mean for the population μ_I is the reference point for a person's observed mean score X_{pI}; and the population and universe mean score μ is the reference point for the person's universe score μ_p. In other words, for relative interpretations, a person's raw score X_{pI} carries no inherent meaning. It is the person's deviation score $X_{pI} - \mu_I$ that carries the meaning, and this deviation score is to be interpreted as an estimate of the person's universe deviation score $\mu_p - \mu$.

In terms of score effects, X_{pI} is given by Equation 2.22, $\mu_I = \mu + \nu_I$, and $\mu_p = \mu + \nu_p$. It follows that

$$\delta_{pI} = [(\mu + \nu_p + \nu_I + \nu_{pI}) - (\mu + \nu_I)] - [(\mu + \nu_p) - \mu] = \nu_{pI}. \tag{2.35}$$

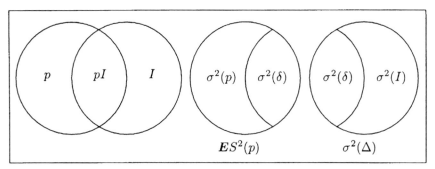

FIGURE 2.2. Venn diagrams for $p \times I$ design.

That is, relative error δ_{pI} for the $p \times I$ design is identical to the ν_{pI} effect in the linear model for X_{pI}. Since $E_p \delta_{pI}$ is zero, relative error variance is[6]

$$\sigma^2(\delta) = \underset{I}{E}\, \underset{p}{E}\, \delta_{pI}^2 = \underset{I}{E}\, \underset{p}{E}\, (\nu_{pI})^2 = \sigma^2(pI) = \frac{\sigma^2(pi)}{n_i'}. \qquad (2.36)$$

Comparing Absolute and Relative Error Variance

Relative error variance corresponds to the error variance in classical test theory, whereas absolute error variance is related to the "generic" error variance discussed by Lord and Novick (1968, pp. 177–180).

From Equations 2.32 and 2.36, it is evident that

$$\sigma^2(\Delta) = \sigma^2(\delta) + \sigma^2(I).$$

Clearly, $\sigma^2(\Delta)$ is larger than $\sigma^2(\delta)$ unless $\sigma^2(I)$ is zero. The assumption of classically parallel forms implies that μ_I is a constant for all forms, which means that $\sigma^2(I)$ is zero. It follows that the assumption of classically parallel forms does not permit a formal distinction between $\sigma^2(\delta)$ and $\sigma^2(\Delta)$. In generalizability theory, however, randomly parallel forms can (and usually do) have different means μ_I, and $\sigma^2(I)$ is generally not zero. It follows that, from the perspective of generalizability theory, any test may consist of an especially easy or difficult set of items relative to the entire universe of items. Consequently, when X_{pI} is interpreted as an estimate of μ_p, variability in μ_I *does* contribute to the error variance $\sigma^2(\Delta)$. By contrast, the definition of relative error is such that μ_I is a constant for all persons in the *deviation* scores of interest and, therefore, $\sigma^2(I)$ does not contribute to $\sigma^2(\delta)$, even though $\sigma^2(I)$ may be positive.

Figure 2.2 provides a Venn diagram perspective on the difference between $\sigma^2(\Delta)$ and $\sigma^2(\delta)$ for the $p \times I$ design. Absolute error variance $\sigma^2(\Delta)$ is

---

[6]Strictly speaking, $\sigma^2(\delta)$ in Equations 2.36 assumes that μ_I in Equation 2.34 is known, or so accurately estimated that it can be assumed known. If $\Sigma \overline{X}_p / n_p$ were used as an estimate of μ_I, then $\sigma^2(\delta)$ would increase by $[\sigma^2(p) - \sigma^2(pI)]/n_p$. This adjustment is almost always overlooked in current literature on generalizability theory.

associated with the entire I circle, whereas relative error variance $\sigma^2(\delta)$ is that part of the I circle that is contained within the p circle.

Error Variance for Estimating μ Using $\overline{X}$

The error variances $\sigma^2(\delta)$ and $\sigma^2(\Delta)$ are the most frequently discussed error variances in generalizability theory, but there are others, one of which is denoted $\sigma^2(X_{PI})$, or simply $\sigma^2(\overline{X})$. This is the error variance involved in using the mean over the sample of *both* persons and items (X_{PI} or $\overline{X}$) as an estimate of the mean over *both* the population of persons and the universe of items (μ):

$$\sigma^2(\overline{X}) \equiv \mathop{E}_{P} \mathop{E}_{I}(X_{PI} - \mu)^2. \tag{2.37}$$

It can be shown (see Exercise 2.3) that for the $p \times I$ design,

$$\begin{aligned} \sigma^2(\overline{X}) &= \sigma^2(P) + \sigma^2(I) + \sigma^2(PI) \\ &= \frac{\sigma^2(p)}{n'_p} + \frac{\sigma^2(i)}{n'_i} + \frac{\sigma^2(pi)}{n'_p n'_i}. \end{aligned} \tag{2.38}$$

Note the use of uppercase P to designate the mean score over the sample of n'_p persons.

2.3.2   Coefficients

Classically parallel forms have equal observed score variances. In generalizability theory, randomly parallel forms have no such constraint, but *expected* observed score variance $ES^2(p)$ plays a role.[7] For the $p \times I$ design,

$$ES^2(p) \equiv \mathop{E}_{I} \left[\mathop{E}_{p}(X_{pI} - \mu_I)^2 \right].$$

That is, $ES^2(p)$ is literally the expected value of the observed score variance for person mean scores, with the expectation taken over randomly parallel forms. It follows that

$$\begin{aligned} ES^2(p) &= \mathop{E}_{I} \mathop{E}_{p}(\nu_p + \nu_{pI})^2 \\ &= \sigma^2(p) + \sigma^2(pI) \\ &= \sigma^2(p) + \sigma^2(\delta). \end{aligned} \tag{2.39}$$

In terms of the Venn diagrams in Figure 2.2, expected observed score variance is represented by the entire p circle, which can be partitioned into two

---

[7]Cronbach et al. (1972) and Brennan (1992a) use $E\sigma^2(X)$ for expected observed score variance. That is, they do not explicitly designate that the variance is for persons' scores. Also, they use σ^2 rather than S^2. The latter is used here to avoid confusion with "total" variance, and because S^2 is much more convenient in multivariate generalizability theory, discussed in Chapters 9 to 12.

parts—universe score variance and relative error variance—as indicated in Equation 2.39.

Cronbach et al. (1972) define a reliability-like coefficient called a generalizability coefficient, which is denoted $E\rho^2$. A generalizability coefficient can be viewed as the ratio of universe score variance to expected observed score variance. Given the result in Equation 2.39, it follows that

$$E\rho^2 = \frac{\sigma^2(p)}{\sigma^2(p) + \sigma^2(\delta)}. \tag{2.40}$$

Technically, $E\rho^2$ is a stepped-up intraclass correlation coefficient. The notation "$E\rho^2$" introduced by Cronbach et al. (1972) is intended to imply that a generalizability coefficient is approximately equal to the expected value (over randomly parallel forms of length n_i') of the squared correlation between observed scores and universe scores. Also, $E\rho^2$ is approximately equal to the expected value of the correlation between pairs of randomly parallel forms of length n_i'.

In terms of estimates, $E\hat{\rho}^2$ for a $p \times I$ design is identical to Cronbach's (1951) coefficient alpha. For dichotomous data, $E\hat{\rho}^2$ for a $p \times I$ design is KR–20. [Note that $E\hat{\rho}^2$ is to be interpreted as $Est(E\rho^2)$.]

It is obvious from Equation 2.40 that a generalizability coefficient involves relative error variance $\sigma^2(\delta)$. Brennan and Kane (1977a,b) define a corresponding reliability-like coefficient that involves absolute error variance:

$$\Phi = \frac{\sigma^2(p)}{\sigma^2(p) + \sigma^2(\Delta)}, \tag{2.41}$$

which is called a phi coefficient or an index of dependability. Note that the denominator of Φ is *not* the variance of persons' mean scores: it is the mean-squared deviation for persons $E(\overline{X}_p - \mu)^2$.

2.3.3   Synthetic Data Example

Table 2.4 provides various D study results for the $p \times I$ design for Synthetic Data No. 1. The G study estimated variance components are given in the upper left-hand corner, followed by the D study estimated variance components for the same number of items (12) as in the G study. That is, $\hat{\sigma}^2(I) = .0754/12 = .0063$ and $\hat{\sigma}^2(pI) = .1269/12 = .0106$. Immediately below these D study estimated variance components are the estimated error variances and coefficients for $n_i' = 12$. Note that $E\hat{\rho}^2 = .844$ is identical to KR–20, since the underlying data are dichotomous. Also, $\hat{\sigma}^2(\delta) = .0106$ is identical to the usual estimate of classical error variance.

The top right-hand part of Table 2.4 provides estimated D study results for 5, 10, 15, and 20 items. For $1 \le n_i' \le 25$, the bottom left-hand figure provides $\hat{\sigma}(\Delta)$ and $\hat{\sigma}(\delta)$, which are the absolute and relative estimated standard errors of measurement—that is, the square roots of $\hat{\sigma}^2(\Delta)$ and

TABLE 2.4. D Studies for $p \times I$ Design Using G Study Variance Components for Synthetic Data Set No. 1

$\hat{\sigma}^2(\alpha)$			12	5	10	15	20
	n_i'				D Studies		
$\hat{\sigma}^2(p) = .0574$	$\hat{\sigma}^2(p)$	.0574	.0574	.0574	.0574	.0574	.0574
$\hat{\sigma}^2(i) = .0754$	$\hat{\sigma}^2(I)$	.0063		.0151	.0075	.0050	.0038
$\hat{\sigma}^2(pi) = .1269$	$\hat{\sigma}^2(pI)$	.0106		.0254	.0127	.0085	.0063
$\overline{X} = .5583$	$\hat{\sigma}^2(\delta)$	.0106		.0254	.0127	.0085	.0063
	$\hat{\sigma}^2(\Delta)$	.0169		.0405	.0202	.0135	.0101
	$\hat{\sigma}^2(\overline{X})$	.0131		.0234	.0145	.0116	.0102
	$Est[ES^2(p)]$	.0680		.0828	.0701	.0659	.0637
	$E\hat{\rho}^2$	.844		.693	.819	.871	.901
	$\widehat{\Phi}$	.773		.586	.740	.810	.850

$\hat{\sigma}^2(\delta)$. For example, the top part of Table 2.4 gives $\hat{\sigma}^2(\Delta) = .0405$ for five items, which means that the absolute error SEM is $\hat{\sigma}(\Delta) = \sqrt{.0405}$, or about .20, as reported in the SEM figure. Note that $\hat{\sigma}(\delta) \leq \hat{\sigma}(\Delta)$, and SEMs decrease as sample sizes increase. Both of these results are predictable from Equations 2.32 and 2.36.

For $1 \leq n_i' \leq 25$, the bottom right-hand figure provides $\widehat{\Phi}$ and $E\hat{\rho}^2$, with the latter denoted in the figure simply as ρ. It is evident that $E\hat{\rho}^2 > \widehat{\Phi}$, and these coefficients get larger as sample size increases. Both of these results are predictable from Equations 2.40 and 2.41.

TABLE 2.5. Real-Data Examples of Single-Facet Designs

	Dichotomous Data			Polytomous Data	
	ITBS Math Concepts	ITED Vocab	ACT Math	IWA	QUA-SAR
No. Score Cat's	2	2	2	4	5
n_p	2952	2965	3388	420	229
n_i	32	40	60	4	9
$\hat{\sigma}^2(p)$	.0292	.0372	.0319	.1655	.6624
$\hat{\sigma}^2(i)$	.0326	.0155	.0342	.0148	.1903
$\hat{\sigma}^2(pi)$	.1790	.1936	.1776	.3146	1.2341
$n_i' = n_i$					
$\hat{\sigma}^2(\delta)$	.0056	.0048	.0030	.0786	.1763
$\boldsymbol{E}\hat{\rho}^2$	.839	.885	.915	.678	.790

2.3.4 Real-Data Examples

This section provides some real-data examples of single-facet designs. The first set of examples primarily illustrates magnitudes of variance components for several different types of data. The last example considers how many blood pressure readings are needed for acceptably reliable measurement.

Dichotomous Versus Polytomous Data

Table 2.5 provides single-facet generalizability analyses for five different tests, three of which use dichotomously scored items and two of which use polytomous scored prompts or tasks:

1. *Iowa Tests of Basis Skills* (ITBS) (Hoover et al., 1993) Math Concepts test, Form K, Level 14, administered to eighth graders in Iowa— 32 items scored dichotomously;

2. *Iowa Tests of Educational Development* (ITED) (Feldt et. al., 1994) Vocabulary test, Form K, Level 17/18, administered to eleventh graders in Iowa—40 items scored dichotomously;

3. *ACT Assessment* (ACT, 1997) Mathematics test administered to an equating sample—60 items scored dichotomously;

4. *Iowa Writing Assessment* (IWA) (Hoover et al., 1994) administered to eighth graders in Iowa—two prompts evaluated by each of two raters using a four-point holistic rubric; and

5. QUASAR (see Lane et al., 1996) performance assessment tasks administered to seventh graders—each of nine tasks evaluated by a single rater using a five-point holistic rubric.

Comparing results for the various tests is not terribly informative because they are developed according to different specifications and administered to different populations. Still, there are at least two observations we can make.

- The sum of the G study single-condition variance components for the first three tests cannot exceed .25 because the data are dichotomous. No such constraint exists for IWA and QUASAR because the items (i.e., prompts or tasks) are polytomously scored.

- The estimated residual variance component is always the largest. Its impact on relative error variance is reduced by averaging over n_i' conditions of measurement, as illustrated in Table 2.5 using $n_i' = n_i$ for the various tests.

This comparison of results for different tests has been presented primarily to illustrate typical values of estimated variance components for single-facet designs with real data. No strong conclusions should be drawn. Furthermore, it can be argued that every one of the five analyses (except perhaps ITED Vocabulary) is flawed either because other random facets are involved (raters for IWA and QUASAR), or because items are explicitly classified a priori into a small number of categories (ITBS Math Concepts and ACT Math). These matters are discussed extensively in subsequent chapters. Still, analyses such as those in Table 2.5 are quite common.

Blood Pressure

Llabre et al. (1988, p. 97) apply generalizability theory to blood pressure measurements "in order to determine the number of readings needed to attain reliable estimates." Their paper provides results for a number of well-designed studies for various universes of generalization. Here, we focus on analyses in which generalization is over only one facet, replications (r) of the measurement procedure.

Each of 40 subjects (p) had their blood pressure taken three times (i.e., $n_r = 3$) using an ambulatory monitor. The design was repeated in three different locations: a laboratory, home, and work. Within a location, readings were taken on the same day. Table 2.6 provides a summary of the results reported by Llabre et al. for systolic and diastolic readings in the three settings. Note that the metric here is millimeters of mercury (mm Hg).

The results for $\widehat{\Phi}$ in Table 2.6 led Llabre et al. to conclude that

...only one reading is necessary whenever generalizations are restricted to the same day in the laboratory. At least six read-

TABLE 2.6. Llabre et al. (1988) Study of Blood Pressure

| | Estimated G Study Variance Components | | | | | |
| | Laboratory | | Home | | Work | |
Effect	Sys.	Dias.	Sys.	Dias.	Sys.	Dias.
p	125.33	64.21	143.51	49.33	150.07	57.62
r	.89	.06	6.06	1.55	$.00^a$	2.92
pr	22.57	13.66	228.84	111.41	166.82	80.29
$\hat{\sigma}(\Delta)$						
$n'_r = 1$	4.84	3.70	15.33	10.63	12.92	9.12
$n'_r = 2$	3.42	2.62	10.84	7.52	9.13	6.45
$n'_r = 3$	2.80	2.14	8.85	6.14	7.46	5.27
$n'_r = 4$	2.42	1.85	7.66	5.31	6.46	4.56
$n'_r = 5$	2.17	1.66	6.85	4.75	5.78	4.08
$n'_r = 6$	1.98	1.51	6.26	4.34	5.27	3.72
$n'_r = 10$	1.53	1.17	4.85	3.36	4.08	2.88
$\hat{\Phi}$						
$n'_r = 1$	.84	.82	.38	.30	.47	.41
$n'_r = 2$	.91	.90	.55	.47	.64	.58
$n'_r = 3$	.94	.93	.65	.57	.73	.68
$n'_r = 4$	.96	.95	.71	.64	.78	.73
$n'_r = 5$	.96	.96	.75	.69	.82	.78
$n'_r = 6$	.97	.97	.79	.72	.84	.81
$n'_r = 10$	.98	.98	.86	.81	.90	.87

^{a}Slightly negative value set to 0.

ings of systolic blood pressure are needed at home and at work, and 6 to 10 diastolic blood pressure readings may be required from work and home, respectively.

The Llabre et al. standard for their conclusions seems to be that $\hat{\Phi}$ be about .80 or greater. They might also have considered the absolute-error SEMs in Table 2.6.

Here, we are viewing the results in Table 2.6 as three separate studies— one for each location—with only one facet (replications) in the universe of generalization. Clearly, however, a more sophisticated (but considerably more complicated) analysis might explicitly represent "location" as a facet.

2.4   Nested Designs

The previous sections in this chapter have considered many issues in generalizability theory from the perspective of a G study $p \times i$ design and a

D study $p \times I$ design. Obviously, if a similar type of development were necessary for every design, then at least some aspects of the utility and "generality" of generalizability theory would be suspect. This is not the case, however. The basic concepts and procedures are appropriate for other designs, as shown most dramatically when multifacet designs are considered in subsequent chapters. Here, the basic concepts and procedures are treated in the context of a single-facet *nested* design in which each person is administered a *different* sample of the same number of items, with all items sampled from the same universe.

It is important to note that, even though the *designs* discussed in the section involve nesting, it is assumed that the population of persons is *crossed* with the universe of admissible observations and the universe of generalization.

2.4.1 Nesting in Both the G and D Studies

When the G study design is $i{:}p$, the linear model is

$$X_{pi} = \mu + (\mu_p - \mu) + (X_{pi} - \mu_p), \tag{2.42}$$

where X_{pi}, μ, and μ_p have the meanings and definitions discussed in Section 2.1.[8] Equation 2.42 can be expressed more succinctly as

$$X_{pi} = \mu + \nu_p + \nu_{i:p}, \tag{2.43}$$

where the effects are uncorrelated with zero expectations:

$$\mathop{E}_{p} \nu_p = \mathop{E}_{p} \nu_{i:p} = \mathop{E}_{i} \nu_{i:p} = 0. \tag{2.44}$$

Equation 2.43 differs from Equation 2.5 for the $p \times i$ design in two respects: Equation 2.43 does not have a distinct term for the item effect ν_i, and the residual effects in the two equations are different. These two differences both result from the fact that $\nu_{i:p}$ in the $i{:}p$ design involves the *confounding* of the ν_i and ν_{pi} effects in the $p \times i$ design. This is easily demonstrated:

$$
\begin{aligned}
\nu_{i:p} &= (X_{pi} - \mu_p) + (\mu_i - \mu_i) + (\mu - \mu) \\
&= (\mu_i - \mu) + (X_{pi} - \mu_i - \mu_p + \mu) \\
&= \nu_i + \nu_{pi}.
\end{aligned}
\tag{2.45}
$$

In the $i{:}p$ design, since each person takes a different sample of items, effects attributable solely to items are indistinguishable from interaction and other residual effects.

---

[8] In this book, the nesting operator is not used with the subscripts of X.

TABLE 2.7. ANOVA Formulas and Notation for G Study $i{:}p$ Design

Effect(α)	$df(\alpha)$	$SS(\alpha)$	$MS(\alpha)$	$\hat{\sigma}^2(\alpha)$
p	$n_p - 1$	$SS(p)$	$MS(p)$	$\hat{\sigma}^2(p) = \dfrac{MS(p) - MS(i{:}p)}{n_i}$
$i{:}p$	$n_p(n_i - 1)$	$SS(i{:}p)$	$MS(i{:}p)$	$\hat{\sigma}^2(i{:}p) = MS(i{:}p)$

$$SS(p) = n_i \sum_p \overline{X}_p^2 - n_p n_i \overline{X}^2$$
$$SS(i{:}p) = \sum_p \sum_i X_{pi}^2 - n_i \sum_p \overline{X}_p^2$$

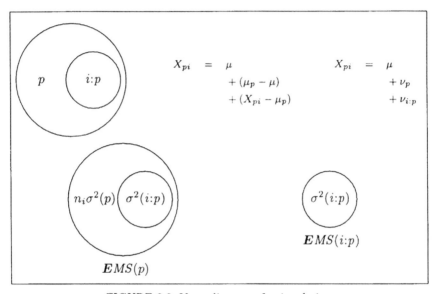

FIGURE 2.3. Venn diagrams for $i{:}p$ design.

For the G study $i{:}p$ design, there are two variance components: $\sigma^2(p)$ and $\sigma^2(i{:}p)$. The variance component $\sigma^2(p)$ is given by Equation 2.11, and

$$\sigma^2(i{:}p) = \underset{p}{E}\,\underset{i}{E}(X_{pi} - \mu_p)^2 = \underset{p}{E}\,\underset{i}{E}\,\nu_{i{:}p}^2. \tag{2.46}$$

Estimators of these variance components, in terms of mean squares, are provided in Table 2.7, and Venn diagrams representing the $i{:}p$ design and its expected mean squares are given in Figure 2.3. In these diagrams, the nesting of items within persons is represented by the inclusion of the entire item circle within the person circle.

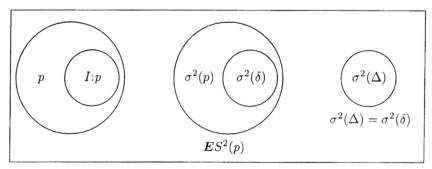

FIGURE 2.4. Venn diagrams for $I{:}p$ design.

D Studies

When the D study design also involves nesting, the design is $I{:}p$, and the linear model for an examinee's average score over n'_i items is

$$X_{pI} = \mu + \nu_p + \nu_{I{:}p}. \tag{2.47}$$

The variance of the ν_p effects is universe score variance $\sigma^2(p)$, which is the same as for the crossed design, because the universe of generalization is unchanged. The variance component associated with the $\nu_{I{:}p}$ effect is

$$\sigma^2(I{:}p) = \frac{\sigma^2(i{:}p)}{n'_i}. \tag{2.48}$$

Recall from the derivations of Δ and δ for the $p \times I$ design (Equations 2.30 and 2.33, respectively) that

$$\Delta_{pI} \equiv X_{pI} - \mu_p$$

and

$$\delta_{pI} \equiv (X_{pI} - \mathop{E}_{p} X_{pI}) - (\mu_p - \mu).$$

Both of these equations are also applicable to the $I{:}p$ random effects design, but for this design,

$$\mathop{E}_{p} X_{pI} = \mu + \mathop{E}_{p} \nu_p + \mathop{E}_{p} \nu_{I{:}p} = \mu.$$

In words, since each person takes a different randomly parallel form, taking the expectation of observed scores over the population of persons also involves taking the expectation over randomly parallel forms from the universe.

It follows that, for the $I{:}p$ design,

$$\Delta = \delta = X_{pI} - \mu_p = \nu_{I{:}p} \tag{2.49}$$

TABLE 2.8. Synthetic Data Set No. 2 and the $i:p$ Design

Person	\multicolumn{8}{c}{Item Scores $(X_{pi})^a$}	Total	$\overline{X}_p$
	1	2	3	4	5	6	7	8		
1	2	6	7	5	2	5	5	5	37	4.625
2	4	5	6	7	6	7	5	7	47	5.875
3	5	5	4	6	5	4	5	5	39	4.875
4	5	9	8	6	5	7	7	6	53	6.625
5	4	3	5	6	4	5	6	4	37	4.625
6	4	4	4	7	6	4	7	8	44	5.500
7	2	6	6	5	2	7	7	5	40	5.000
8	3	4	4	5	6	6	6	4	38	4.750
9	0	5	4	5	5	5	5	3	32	4.000
10	6	8	7	6	6	8	8	6	55	6.875

$\overline{X} = 5.275$ $\sum_p \overline{X}_p^2 = 286.0313$ $\sum_p \sum_i X_{pi}^2 = 2430$

^{a}The numbers 1, 2, ..., 8 represent different items for each person in $i:p$ design.

and

$$\sigma^2(\Delta) = \sigma^2(\delta) = \sigma^2(I:p) = \frac{\sigma^2(i:p)}{n_i'}. \qquad (2.50)$$

That is, $\sigma^2(\Delta)$ and $\sigma^2(\delta)$ are indistinguishable in the $I:p$ design, just as they are in classical theory. This is illustrated by the Venn diagrams in Figure 2.4. For the $I:p$ design, error variance for the mean is

$$\sigma^2(\overline{X}) = \sigma^2(P) + \sigma^2(I:P) = \frac{\sigma^2(p)}{n_p'} + \frac{\sigma^2(i:p)}{n_p' n_i'}. \qquad (2.51)$$

Equation 2.39 for $\boldsymbol{E}S^2(p)$ applies to the $I:p$ design, as well. Also, Equations 2.40 and 2.41 for $\boldsymbol{E}\rho^2$ and Φ, respectively, are applicable to the nested design. Moreover, since $\sigma^2(\delta) = \sigma^2(\Delta)$, it follows that $\boldsymbol{E}\rho^2 = \Phi$.

Synthetic Data Example

Consider Synthetic Data Set No. 2 in Table 2.8. These data might be viewed as examinee scores associated with their responses to eight free-response items, with each examinee taking a different set of items.[9] Using the formulas in Table 2.7 and these synthetic data, the reader can verify the values

---

[9]Careful consideration of such data usually reveals that two facets—items and raters—are confounded. Such complexities are treated in later chapters.

TABLE 2.9. G Study $i{:}p$ and D Study $I{:}p$ Designs For Synthetic Data Set No. 2

Effect(α)	$df(\alpha)$	$SS(\alpha)$	$MS(\alpha)$	$\hat{\sigma}^2(\alpha)$	D Studies		
					n_i'	4	8
p	9	62.20	6.9111	.6108	$\hat{\sigma}^2(p)$	.6108	.6108
$i{:}p$	70	141.75	2.0250	2.0250	$\hat{\sigma}^2(I{:}p)$	.5063	.2531
					$\hat{\sigma}^2(\delta) = \hat{\sigma}^2(\Delta)$	.5063	.2531
					$\hat{\sigma}^2(\overline{X})^a$	.1117	.0864
					$Est[ES^2(p)]$	1.1170	.8639
					$E\hat{\rho}^2 = \widehat{\Phi}$	.547	.707

^{a}Estimates assume $n_p' = 10$.

for the G study estimated variance components reported in Table 2.9. Since the observed scores for these synthetic data range from 0 to 9, magnitudes of these estimated variance components are many times greater than those in Table 2.3 on page 28 for Synthetic Data No. 1, where the underlying data are dichotomous. Table 2.9 also provides estimates of D study variance components, error variances, and generalizability coefficients for the D study $I{:}p$ design. Clearly, increasing n_i' leads to smaller error variances and larger coefficients, as expected.

2.4.2   Nesting in the D Study Only

The effects ν_i and ν_{pi} are confounded in the $\nu_{i{:}p}$ effect in an $i{:}p$ design. In terms of variance components, this means that

$$\sigma^2(i{:}p) = \sigma^2(i) + \sigma^2(pi),$$

where the variance components to the right of the equality are independently estimable with the $p \times i$ design. It follows from Equation 2.50 that absolute and relative error variance are:

$$\sigma^2(\Delta) = \sigma^2(\delta) = \frac{\sigma^2(i) + \sigma^2(pi)}{n_i'},$$

which can be estimated using $\hat{\sigma}^2(i)$ and $\hat{\sigma}^2(pi)$ from a crossed G study.
    Using Equation 2.51, error variance for the mean is

$$\sigma^2(\overline{X}) = \frac{\sigma^2(p)}{n_p'} + \frac{\sigma^2(i) + \sigma^2(pi)}{n_p' n_i'}. \tag{2.52}$$

Comparing this equation with the corresponding result for the $p \times I$ design in Equation 2.38, it is evident that $\sigma^2(\overline{X})$ for the nested design is smaller than for the crossed design. In this sense, the $I{:}p$ design provides more

TABLE 2.10. D Study $p \times I$ Design Based on G Study $p \times i$ Design for Synthetic Data Set No. 1

$\hat{\sigma}^2(\alpha)$			D Studies			
	n_i'	12	5	10	15	20
$\hat{\sigma}^2(p) = .0574$	$\hat{\sigma}^2(p)$	.0574	.0574	.0574	.0574	.0574
$\hat{\sigma}^2(i) = .0754$	$\hat{\sigma}^2(I{:}p)$	.0169	.0405	.0202	.0135	.0101
$\hat{\sigma}^2(pi) = .1269$						
	$\hat{\sigma}^2(\delta) = \hat{\sigma}^2(\Delta)$	.0169	.0405	.0202	.0135	.0101
	$\hat{\sigma}^2(\overline{X})^a$	.0074	.0098	.0078	.0071	.0068
	$Est[\boldsymbol{E}S^2(p)]$	.0743	.0979	.0776	.0709	.0675
	$\boldsymbol{E}\hat{\rho}^2 = \widehat{\Phi}$	.773	.586	.740	.810	.850

^{a}Estimates assume $n_p' = 10$.

dependable estimates of group mean scores than does the $p \times I$ design. This suggests that in contexts such as program evaluation where group mean scores are frequently of principal interest, an investigator is well-advised to administer different samples of items to persons, rather than the same sample of items.

Consider, for example, Table 2.10 in which the G study estimated variance components are for the $p \times i$ design based on Synthetic Data Set No. 1, and the D study results are for the $I{:}p$ design. These results can be compared with the corresponding results for the D study $p \times I$ design in Table 2.4. Note that $\hat{\sigma}^2(\delta)$ for the $I{:}p$ design equals $\hat{\sigma}^2(\Delta)$ for the $p \times I$ design, and estimated generalizability coefficients for the $I{:}p$ design are smaller than those for the $p \times I$ design.

No mention has been made of the possibility that G study estimated variance components are available for the $i{:}p$ design, but the investigator's interest is in a D study $p \times I$ design. In this circumstance, the universe score variance, $\sigma^2(\Delta)$, and Φ can be estimated, but $\sigma^2(\delta)$ and $\boldsymbol{E}\rho^2$ cannot be estimated, because $\hat{\sigma}^2(i)$ and $\hat{\sigma}^2(pi)$ are confounded in the G study estimated variance component $\hat{\sigma}^2(i{:}p)$.

2.5 Summary and Other Issues

Nested single-facet designs are relatively rare, especially when there are equal numbers of items nested within each person. However, single-facet crossed designs are very common and frequently referenced in subsequent chapters. Therefore, Table 2.11 summarizes the equations for the G study $p \times i$ design and D study $p \times I$ designs.

Many of the results in this chapter can be obtained using classical test theory. For example, the numerical values in Table 2.4 for $\boldsymbol{E}\hat{\rho}^2$ can be

TABLE 2.11. Equations for G Study $p \times i$ Design and D Study $p \times I$ Design

---

Model for G Study Design: $X_{pi} = \mu + \nu_p + \nu_i + \nu_{pi}$

$\nu_p \equiv \mu_p - \mu$ $\hat{\sigma}^2(p) = \dfrac{MS(p) - MS(pi)}{n_i}$

$\nu_i \equiv \mu_i - \mu$ $\hat{\sigma}^2(i) = \dfrac{MS(i) - MS(pi)}{n_p}$

$\nu_{pi} \equiv X_{pi} - \mu_p - \mu_i + \mu$ $\hat{\sigma}^2(pi) = MS(pi)$

Model for D Study Design: $X_{pI} = \mu + \nu_p + \nu_I + \nu_{pI}$

$$\sigma^2(I) = \frac{\sigma^2(i)}{n_i'} \qquad \sigma^2(pI) = \frac{\sigma^2(pi)}{n_i'}$$

Absolute Error: $\Delta = X_{pI} - \mu_p$

Relative Error: $\delta = (X_{pI} - \mu_I) - (\mu_p - \mu)$

Absolute Error Variance: $\sigma^2(\Delta) = \dfrac{\sigma^2(i)}{n_i'} + \dfrac{\sigma^2(pi)}{n_i'}$

Relative Error Variance: $\sigma^2(\delta) = \dfrac{\sigma^2(pi)}{n_i'}$

Exp. Obs. Score Variance: $ES^2(p) = \sigma^2(p) + \sigma^2(\delta)$

Generalizability Coefficient: $E\rho^2 = \dfrac{\sigma^2(p)}{\sigma^2(p) + \sigma^2(\delta)}$

Dependability Coefficient: $\Phi = \dfrac{\sigma^2(p)}{\sigma^2(p) + \sigma^2(\Delta)}$

---

obtained using the Spearman–Brown Prophecy Formula:

$$r' = \frac{n_i' r}{n_i + (n_i' - n_i)r},$$

where r is KR–20 for the "original" test of n_i items, and r' is the stepped-up or stepped-down reliability for the "new" test of n_i' items. The Spearman–Brown formula does not apply to many of the designs and universes considered in later chapters, however.

There are a number of issues that arise in generalizability theory over and beyond those already discussed in this chapter. A few such additional issues are briefly considered next; others are introduced in later chapters.

2.5.1 Other Indices and Coefficients

Generalizability and phi coefficients are frequently reported in generalizability analyses. Their popularity is undoubtedly related to the fact that they are reliability-like coefficients, and such coefficients have been widely used in measurement contexts since the beginning of the last century. However, other coefficients and indices are sometimes informative and perhaps even more useful in certain contexts. Perhaps the most frequently cited competitor is signal–noise ratios.

In interpreting an error variance it is often helpful to compare its magnitude directly to universe score variance. One way to do so is to form the ratio of universe score variance to error variance, which is called a signal–noise ratio. For absolute error, the signal–noise ratio and its relationships with Φ are:

$$ S/N(\Delta) = \frac{\sigma^2(p)}{\sigma^2(\Delta)} = \frac{\Phi}{1 - \Phi} \quad \text{and} \quad \Phi = \frac{S/N(\Delta)}{1 + S/N(\Delta)}. $$

Similarly, for relative error, the signal–noise ratio and its relationships with $E\rho^2$ are:

$$ S/N(\delta) = \frac{\sigma^2(p)}{\sigma^2(\delta)} = \frac{E\rho^2}{1 - E\rho^2} \quad \text{and} \quad E\rho^2 = \frac{S/N(\delta)}{1 + S/N(\delta)}. $$

As discussed by Cronbach and Gleser (1964) and Brennan and Kane (1977b), the signal–noise concept arises naturally in discussing communication systems where the signal–noise ratio compares the strength of the transmission to the strength of the interference. The signal $\sigma^2(p)$ is a function of the magnitude of the intended discriminations $\mu_p - \mu$. These intended discriminations reflect the sensitivity requirements that must be met if the measurement procedure is to achieve its intended purpose. The noise reflects the degree of precision, or the magnitude of the errors that arise in practice. If the signal is large compared to the noise, the intended discriminations are easily made. If the signal is weak compared to the noise, the intended discriminations may be completely lost.

Other indices of measurement precision are discussed by Kane (1996). In particular, he defines an error–tolerance ratio (E/T) as the root mean square of the errors of interest divided by the root mean square of the tolerances of interest. E/T will be small if errors are small relative to the tolerances, suggesting that measurements have substantial precision for the intended use. Suppose, for example, that the error root mean square is $\sigma(\Delta)$ and the tolerance root mean square is the standard deviation of universe scores $\sigma(p)$. Then, relationships between E/T and Φ are:

$$ E/T = \sqrt{\frac{1 - \Phi}{\Phi}} \quad \text{and} \quad \Phi = \frac{1}{1 + (E/T)^2}. $$

Similar relationships hold for E/T and $E\rho^2$ when the SEM is $\sigma(\delta)$ and the tolerance is $\sigma(p)$.

The notion of an error-tolerance ratio is not restricted to situations involving only square roots of variances, however. For example, for domain-referenced interpretations of scores, often interest focuses on $\mu_p - \lambda$, where λ is a cut score. Under these circumstances, the root mean square of the tolerances of interest is

$$\sqrt{E(\mu_p - \lambda)^2} = \sqrt{E[(\mu_p - \mu) + (\mu - \lambda)]^2} = \sqrt{\sigma^2(p) + (\mu - \lambda)^2},$$

the standard error of measurement of interest is

$$\sqrt{E[(X_{pI} - \lambda) - (\mu_p - \lambda)]^2} = \sqrt{E(X_{pI} - \mu_p)^2} = \sigma(\Delta),$$

and the error-tolerance ratio is

$$E/T(\lambda) = \sqrt{\frac{\sigma^2(\Delta)}{\sigma^2(p) + (\mu - \lambda)^2}}. \tag{2.53}$$

Using this error-tolerance ratio, a reliability-like coefficient is

$$\Phi(\lambda) = \cfrac{1}{1 + \cfrac{\sigma^2(\Delta)}{\sigma^2(p) + (\mu - \lambda)^2}} = \frac{\sigma^2(p) + (\mu - \lambda)^2}{\sigma^2(p) + (\mu - \lambda)^2 + \sigma^2(\Delta)}. \tag{2.54}$$

Equation 2.54 is identical to an index of dependability for domain-referenced interpretations developed by Brennan and Kane (1977a,b).

Estimating $E/T(\lambda)$ and $\Phi(\lambda)$ is slightly more complicated than it may appear, however, because $(\overline{X} - \lambda)^2$ is a biased estimator of $(\mu - \lambda)^2$. Brennan and Kane (1977a,b) showed that an unbiased estimator is

$$\text{est}(\mu - \lambda)^2 = (\overline{X} - \lambda)^2 - \hat{\sigma}^2(\overline{X}), \tag{2.55}$$

which is one reason that the error variance $\sigma^2(\overline{X})$ was introduced previously. When $\lambda = \overline{X}$, $\hat{\Phi}(\lambda) = $ KR–21, which is indicative of the fact that KR–21 involves absolute error variance, not the relative error variance of classical theory.

Table 2.12 provides estimates of the error-tolerance ratios and indices of dependability for Synthetic Data Set No. 1 when $n_i' = 12$. Note that the magnitudes of both indices depend on λ. When $\lambda = \overline{X}$, the estimate of $E/T(\lambda)$ achieves its maximum value, and the estimate of $\Phi(\lambda)$ achieves its minimum value.

2.5.2   Total Score Metric

The usual convention in generalizability theory is to report variance components and error variances in terms of the mean score metric. However,

TABLE 2.12. $\widehat{E/T}(\lambda)$ and $\widehat{\Phi}(\lambda)$ for Synthetic Data Set No. 1 with $n_i' = 12$

Variances		$\overline{X} = .5583$	.6	.7	.8
$\hat\sigma^2(p) = .0574$	$(\overline{X} - \lambda)^2$	.0000	.0017	.0201	.0584
$\hat\sigma^2(\Delta) = .0169$	$\widehat{E/T}(\lambda)$	.616	.606	.512	.406
$\hat\sigma^2(\overline{X}) = .0131$	$\widehat{\Phi}(\lambda)$	.725	.732	.792	.859

results can be expressed in terms of the total score metric, by multiplying the "mean score" variances by $(n_i')^2$. For example, for the D study results in Table 2.4 with $n_i' = 12$, the universe score variance in terms of the total score metric is

$$(n_i')^2 \, \hat\sigma^2(p) = (144)(.0574) = 8.27.$$

The error variances are

$$(n_i')^2 \, \hat\sigma^2(\delta) = (144)(.0106) = 1.53,$$

$$(n_i')^2 \, \hat\sigma^2(\Delta) = (144)(.0169) = 2.43,$$

and

$$(n_i')^2 \, \hat\sigma^2(\overline{X}) = (144)(.0131) = 1.89.$$

When interpreting universe score variance in terms of the total score metric, it is helpful to recall that universe score is defined as the expected value of observed scores over randomly parallel forms. Expressing universe score variance in terms of the total score metric simply means that the expectation is taken over the examinee's observed total scores, rather than the examinee's observed mean scores. The same type of interpretation applies to the error variances expressed in the total score metric. Also, note that for $(n_i')^2 \, \sigma^2(\overline{X})$ the observed mean score under consideration is the average of the observed *total* scores for each person in the sample. Finally, it is easy to show that $E\rho^2$ and Φ are the same for both metrics.

2.6 Exercises

2.1* Suppose that each of six persons is rated by the same three raters, resulting in the following G study data:

Person (p)	r_1	r_2	r_3
1	3	1	3
2	1	3	4
3	5	4	6
4	4	5	5
5	5	8	9
6	6	9	9

   (a) Estimate the G study variance components.

   (b) For $n'_r = 3$, estimate $\sigma^2(\delta)$, $\sigma^2(\Delta)$, $E\rho^2$, and Φ.

   (c) Verify that the average of the observed covariances equals the universe score variance estimated from the mean squares.

2.2 In the Angoff procedure for establishing a cut score, for each item in a test, each of several raters (r) provides a judgment about the probability that a minimally competent examinee would get the item correct. The final cut score is the average of these probabilities over items and raters. Brennan and Lockwood (1980) discuss a study of the Angoff procedure in which each of five raters evaluated each of 126 items, with $SS(r) = .700$, $SS(i) = 7.144$, $SS(ri) = 13.353$, and $\overline{X} = .663$. What is the standard error of $\overline{X}$?

2.3* Derive Equation 2.38 for $\sigma^2(\overline{X})$.

2.4 Using traditional formulas for KR–20 and classical error variance, $\sigma^2(E)$, verify that $E\rho^2 = $ KR–20 and $\hat{\sigma}^2(\delta) = \hat{\sigma}^2(E)$ for Synthetic Data No. 1 in Table 2.2.

2.5* The following estimated variance components are for 490 eighth-grade Iowa examinees in 1993–1994 who took Form K of Level 14 of the ITBS Math Concepts and Estimation tests (Hoover et al., 1993).

Test	n_i	$\hat{\sigma}^2(p)$	$\hat{\sigma}^2(i)$	$\hat{\sigma}^2(pi)$
Math Concepts	32	.0280	.0320	.1783
Estimation	24	.0242	.0176	.1994

Suppose it was decided to create shorter forms of both tests under the constraints that $E\rho^2 \geq .6$ for both tests, with the relative proportion of items for the two tests remaining as similar as possible. How long should the shorter tests be?

2.6   Verify that $\widehat{\Phi}(\lambda = \overline{X}) = \text{KR–21} = .725$ for Synthetic Data Set No. 1, as reported in Table 2.12.

2.7*   Consider the QUASAR data in Table 2.5. The average rating over persons and tasks was 1.43, and recall that the tasks were scored using a five-point holistic scale (1 to 5). If an important decision were based on an average rating of at least 3, how many tasks would be required to have an error–tolerance ratio no larger than .5? (Do the computations with at least three decimal digits of accuracy.)

2.8   Prove that $EMS(p) = \sigma^2(pi) + n_i\,\sigma^2(p)$.

3
Multifacet Universes of Admissible Observations and G Study Designs

Often, generalizability analyses may be viewed as two-stage processes. The goal of the first stage is to obtain estimated variance components for a G study design, given a universe of admissible observations. The second stage involves using these estimated variance components in the context of a D study design and universe of generalization to estimate quantities such as universe score variance, error variances, and coefficients. This dichotomy of stages is somewhat arbitrary at times, but it is a conceptually meaningful distinction. Therefore, this chapter treats multifacet G study designs and universes of admissible observations, while Chapters 4 and 5 treat multifacet D study design considerations and universes of generalization.

This chapter may be viewed as a treatment of analysis of variance that is tailored to *balanced* designs in generalizability theory. A balanced design has no missing data and, for any nested facet, the sample size is constant for each level of that facet. A notational system is described for characterizing G study designs and their effects. Also, algorithms, equations, and procedures are provided for expressing mean scores in terms of score effects (and vice versa), for calculating sums of squares, and for estimating variance components directly from mean squares. The algorithms and procedures are very general and apply to practically any balanced design that might be encountered in generalizability theory, ranging from single-facet G study designs to very complicated multifacet ones. In the statistical literature, the estimators of the variance components discussed here are those for the so-called "ANOVA procedure."

The topics in this chapter involve many conceptual, notational, and statistical issues that are important in generalizability theory, as presented in this book. However, some readers undoubtedly have prior familiarity with certain topics (e.g., sums of squares in Section 3.3), and computer programs such as GENOVA (see Appendix F) can be used to perform the numerical computations discussed here.

For the most part, issues are discussed from the perspective of two-facet universes and designs. This restriction is a convenience employed solely for illustrative purposes, rather than a limitation on generalizability theory or the procedures, equations, and algorithms presented here.

3.1   Two-Facet Universes and Designs

Figure 3.1 provides Venn diagrams and model equations for some possible two-facet designs. Usually the two facets are associated with the indices i and h, and p represents the objects of measurement (often persons or examinees, but not always). In standard analysis of variance terminology, effects in a design can be identified as either main effects or interaction effects. From this perspective, p is associated with a main effect. For example, in the $p \times i \times h$ design, the main effects are ν_p, ν_i, and ν_h, and all other effects (except μ) are interaction effects. More simply, we can say that p, i, and h represent main effects and pi, ph, ih, and pih represent interaction effects. For the two-facet designs in Figure 3.1, the effects are identified as follows.

Design	Main Effects	Interaction Effects
$p \times i \times h$	p, i, h	pi, ph, ih, pih
$p \times (i{:}h)$	$p, h, i{:}h$	$ph, pi{:}h$
$(i{:}p) \times h$	$p, h, i{:}p$	$ph, ih{:}p$
$i{:}(p \times h)$	$p, h, i{:}ph$	ph
$(i \times h){:}p$	$p, i{:}p, h{:}p$	$ih{:}p$
$i{:}h{:}p$	$p, h{:}p, i{:}h{:}p$	

For each of these designs, other sources of residual error e are totally confounded with the effect that contains all three indices. From some perspectives, it is unusual to use the phrase "main effect" as a description of an effect that involves nesting [e.g., the $i{:}h$ effect in the $p \times (i{:}h)$ design], but in generalizability theory it is meaningful to associate each facet with a main effect.

These notational conventions, and the linear model representations of these designs, are considered in more detail in Section 3.2. Here, these designs are first described from the perspective of Venn diagrams, and then illustrated in terms of some possible universes of admissible observations.

3.1.1   Venn Diagrams

In the Venn diagrams in Figure 3.1, each main effect is represented by a circle. Interaction effects are represented by the intersections of circles. The total number of effects in any design is the number of distinct areas in the Venn diagram.

When a main effect involves nesting, it is represented by a circle within another circle, or within the intersection of circles. For example, in the $p \times (i{:}h)$ Venn diagram, the nested nature of the $i{:}h$ main effect is represented by the i circle being within the h circle; and, in the $i{:}(p \times h)$ Venn diagram, the nested nature of the $i{:}ph$ main effect is represented by the i circle being within the intersection of the p and h circles.

3.1.2 Illustrative Designs and Universes

For the designs in Figure 3.1, the universe and population sizes are denoted N_i, N_h, and N_p. Unless otherwise noted, it is assumed that they are infinite, or large enough to be considered so. The corresponding sample sizes are denoted n_i, n_h, and n_p. Unless otherwise noted, it is assumed that conditions of i, h, and p are sampled at random. Note that the indices p, i, and h do not designate any particular facets, groups of conditions, or objects of measurement. Specifically, i does not necessarily represent items, and p does not necessarily represent persons.

Next, examples of each of the illustrative designs are briefly described, given certain universes of admissible observations. In these examples, some issues are introduced (without much explanation) to motivate discussion of related topics in subsequent sections of this and later chapters.

Items-Crossed-with-Raters in the Universe of Admissible Observations

Suppose that items (i) are crossed with raters (r) in the universe of admissible observations. This means that any one of the N_i items might be rated by any one of the N_r raters. If the population is crossed with this universe of admissible observations, then the corresponding G study design is $p \times i \times r$, in which the responses of n_p persons to n_i items are each evaluated by n_r raters. Estimated variance components for this fully crossed design can be used to estimate results for any possible two-facet design. However, for the $i \times r$ universe, an investigator might conduct a G study using any one of the other two-facet designs.

Suppose, for example, that n_p persons are administered $n_i = 12$ items, and each of $n_r = 3$ raters evaluates nonoverlapping sets of four items each. This a verbal description of the $p \times (i{:}r)$ design with $n_i = 4$. It is an entirely legitimate G study design for the universe of admissible observations that has $i \times r$. That is, even though the universe of admissible observations has crossed facets, the G study design may have nested facets. An investigator pays a price, however, for using the $p \times (i{:}r)$ design (or any design other than $p \times i \times r$) when the universe of admissible observations is crossed—namely, the estimated variance components from the $p \times (i{:}r)$ design can

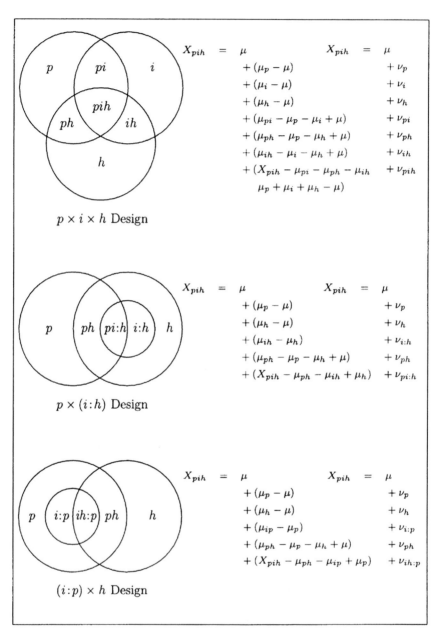

FIGURE 3.1. Venn diagrams and linear models for some two-facet designs.

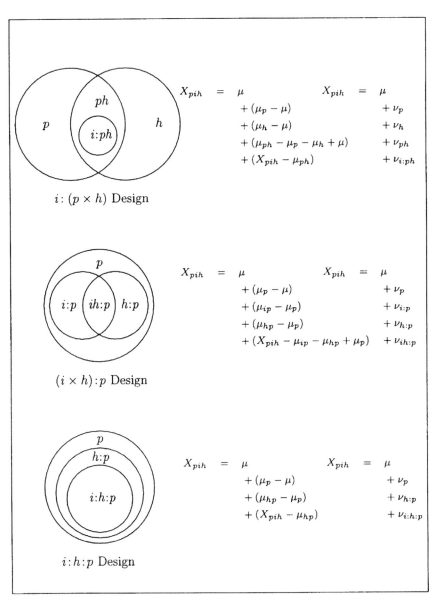

FIGURE 3.1 (continued). Venn diagrams and linear models for some two-facet designs.

be used to estimate results for only some two-facet D studies. On the other hand, the $p \times (i{:}r)$ design has a distinct advantage if rater time is at a premium.

Clearly, then, there is no universal answer to the question, "What G study design should be used?" The investigator must make this decision taking into account the nature of the universe of admissible observations, likely D study considerations (not yet specified in detail), and time and cost considerations. All things considered, however, it is generally preferable that the sample sizes for a G study design be as large as possible. Doing so helps ensure that the resulting estimated variance components will be as stable as possible.

As another example of a G study design given an $i \times r$ universe of admissible observations, suppose each person is administered a *different* sample of n_i items, but all items are evaluated by the same set of n_r raters. This is a verbal description of the $(i{:}p) \times r$ design. It has the advantage of sampling a total of $n_p n_i$ items, rather than only n_i items, but, again, the resulting estimated variance components cannot be used to estimate results for all possible two-facet D study designs.

Items-Crossed-with-Occasions in the Universe of Admissible Observations

Suppose that items (i) are crossed with occasions (o) in the universe of admissible observations. The *structure* of this universe and the one discussed above $(i \times r)$ are the same, but one of the facets is different. If the population is crossed with the $i \times o$ universe of admissible observations, then the corresponding G study design is $p \times i \times o$, in which each person responds to a set of n_i items, all of which are administered on n_o different occasions. Again, however, G study data might be collected using any one of the other two-facet designs.

For example, suppose that (a) each person responds to a set of n_i items administered on n_o occasions; (b) for each person, the occasions or times of administration are the same; and (c) for each person and each occasion, the n_i items are *different*. This is a verbal description of the $i{:} (p \times o)$ design. In conventional ANOVA terminology, this is the frequently discussed random effects two-way factorial design with replications (items) within cells. However, this ANOVA description is a potentially misleading way to characterize the $i{:} (p \times o)$ design in generalizability theory, because the i facet does not play the role of within-cell replications in the traditional ANOVA sense. Indeed, many designs frequently encountered in generalizability theory can be given a conventional ANOVA verbal description, or something close to it, but doing so can cause confusion about the nature and role of facets in generalizability theory. For this reason, such conventional ANOVA descriptions of designs are rarely used here.

As another example of a G study design for the $i \times o$ universe of admissible observations, suppose that: (a) each person is administered a set of n_i items on n_o occasions; (b) for an individual person the same items are administered on each occasion; but (c) for different persons the items are different and the occasions are different. This is a verbal description of the $(i \times o) : p$ design.

Also, the design could be $i{:}o{:}p$. For this design, (a) each person is administered different items on different occasions, and (b) for different persons the occasions (of test administration) are different. This design is sometimes called the fully nested design. As such, it is the two-facet analogue of the $i{:}p$ design discussed in Chapter 2.

Items-Nested-Within-Content Categories in the Universe of Admissible Observations

Universes of admissible observations do not always have all facets crossed. For example, many tests are best viewed as being made up of items (i) nested within content categories (h). In such cases, it is frequently reasonable to view the universe of admissible observations as $i{:}h$, with every item in the universe being associated with one and only one content category h. In addition, the number of content categories N_h is usually quite small.

For an $i{:}h$ universe of admissible observations of the type described above, a G study $p \times i \times h$ design is a logical impossibility. This design would be reasonable only if every item could be associated with every content category in the universe, which directly contradicts the previous description of the universe of admissible observations. There is frequently an additional issue involved in using this design with a nested universe of admissible observations; namely, all N_h content categories are usually represented in the design specifications, implying that $n_h = N_h < \infty$. Under these circumstances, the model is described as mixed, rather than random, and the estimated variance components are associated with a fixed facet h, with a finite universe of N_h conditions.

When the universe of admissible observations has $i{:}h$, any G study design must also have $i{:}h$. Of the designs in Figure 3.1 only two designs, other than $p \times (i{:}h)$, have this characteristic. They are the $i{:} (p \times h)$ and $i{:}h{:}p$ designs. However, when $n_h = N_h < \infty$, the $i{:}h{:}p$ design is not a likely possibility because it necessitates independent samples of content categories for each person. For the $i{:} (p \times h)$ design, each person is administered a different sample of items for each of the $n_h = N_h$ content categories.

Persons as a Facet in the Universe of Admissible Observations

Usually persons are viewed as the objects of measurement, but the theory per se does *not* dictate that persons play this role. Suppose that D study considerations ultimately will focus on class means as the objects of measurement. (For example, the reliability of class means is sometimes a topic of concern in program evaluation contexts.) In such cases, the universe of admissible observations may have persons crossed with items.

For such a universe of admissible observations, a design often considered is $(p:c) \times i$, where p, c, and i stand for persons, classes, and items, respectively. This $(p:c) \times i$ design is simply a specific instance of the generic $(i:p) \times h$ design in Figure 3.1. Again, the p, i, and h indices used in specifying the generic two-facet designs do not necessarily stand for persons, items, and some other facet, respectively.

Any group of conditions (persons, items, raters, occasions, classes, etc.) could be considered the objects of measurement. However, strictly speaking, the specification of objects of measurement is a D study consideration. Therefore, further discussion is postponed until Chapter 4.

Other Universes of Admissible Observations and G Study Designs

The foregoing discussion illustrates only a very small number of possible universes of admissible observations, G study designs, and related issues. For example, there are other possible two-facet designs, and the number of possible designs increases exponentially as the number of facets increases. Also, in theory, the number of possible facets in a universe of admissible observations is unlimited, although practical constraints usually preclude using large numbers of facets in any particular G study.

Furthermore, designs associated with nested universes are sometimes unbalanced (e.g., unequal numbers of items in content categories) and sometimes best approached by means of multivariate generalizability theory. Such complexities are treated in later chapters.

The algorithms and procedures discussed next are relevant for a very large class of balanced, univariate G study designs, no matter how many facets are involved. However, for illustrative purposes, only the $p \times i \times h$ and $p \times (i:h)$ designs are used extensively in the following sections of this chapter.

3.2   Linear Models, Score Effects, and Mean Scores

Figure 3.1 contains two expressions for the linear models associated with each of the two-facet designs introduced in Section 3.1. The first expression provides a decomposition of an observable score X_{pih} in terms of mean scores, while the second expression provides the decomposition in terms of score effects. The two expressions are simply different ways of expressing the same decomposition, which is a tautology. In this section, a notational system, algorithm, and equations are presented for expressing, relating, and interpreting effects in linear models.

The algebraic expressions for the models in Figure 3.1 do not, in and of themselves, provide a sufficient basis for analyzing G study data, however. An investigator must also specify the nature of any sampling involved. Except for Section 3.5, it is assumed in this chapter that $n < N \to \infty$ for each facet (as well as the objects of measurement "facet"), and that all sampling is random. When these assumptions are applied to equations such as those in Figure 3.1, the resulting models are referred to as *random effects* linear models.

3.2.1 Notational System for Main and Interaction Effects

In the notational system used here, any main effect can be represented in the following manner:

$$\left\{ \begin{array}{c} \text{primary} \\ \text{index} \end{array} \right\} : \left\{ \begin{array}{c} \text{first nesting} \\ \text{index (indices)} \end{array} \right\} : \left\{ \begin{array}{c} \text{second nesting} \\ \text{index (indices)} \end{array} \right\} : \ldots$$

Main effects that involve nesting are sometimes referred to as nested main effects, and those that do *not* involve nesting are sometimes called *non*nested main effects. If a main effect does not involve nesting, then it is represented by the primary effect index only. Note that a *nesting* index for a specific main effect is usually a primary index for some other main effect. For example, in the $p \times (i\!:\!h)$ design the index h is a nesting index in the nested main effect $i\!:\!h$, and h by itself is also a nonnested main effect.

For the illustrative designs in Figure 3.1 no main effect involves a "second" nesting index. However, more complicated designs do sometimes involve two and even three levels of nesting. It is for this reason that the notation for a main effect has been specified very generally. Some other authors denote a nested main effect by placing all nesting indices within parentheses, no matter what level of nesting is involved. For example, if a design involves the main effect $a\!:\!b\!:\!c$, some authors denote it $a(bc)$. Here the notation $a\!:\!b\!:\!c$ is preferred because it directly implies that levels of a are nested within levels of b which are nested within levels of c, whereas $a(bc)$ could mean that levels of a are nested within all combinations of a level of b with a level of c.

Each interaction effect can be represented as a combination of main effects in the following manner.

$$\left\{ \begin{array}{c} \text{combination of} \\ \text{primary} \\ \text{index (indices)} \end{array} \right\} : \left\{ \begin{array}{c} \text{combination of} \\ \text{first nesting} \\ \text{index (indices)} \end{array} \right\} : \left\{ \begin{array}{c} \text{combination of} \\ \text{second nesting} \\ \text{index (indices)} \end{array} \right\} : \ldots$$

The entire set of interaction effects is obtained by forming all *valid* combinations of main effects. Any such combination is valid provided no index appears as both a primary and a nesting index. Also, if an index appears multiple times in some nesting position, only the occurrence in the last nesting position is retained.

Consider, for example, the $p \times (i{:}h)$ design with the main effects p, h, and $i{:}h$. The combination of h and $i{:}h$ gives $hi{:}h$, which is not a valid interaction, because the primary index h cannot be nested within itself. As another example, consider the $(i \times h){:}p$ design with the main effects p, $i{:}p$, and $h{:}p$. The combinations $pi{:}p$ and $ph{:}p$ are not valid because p appears as a primary index *and* a nesting index in both combinations. However, the combination of $i{:}p$ and $h{:}p$ gives $ih{:}pp$, or $ih{:}p$, which is valid. Also, if $a{:}b$ and $c{:}d{:}b$ are two main effects in some design, then their combination is $ac{:}bd{:}b$, and the interaction is denoted $ac{:}d{:}b$, retaining only the last occurrence of b.

In generalizability theory, designs usually involve all possible interaction effects obtainable from the main effects; that is, any interaction effect is included in the model if it can be represented in the manner described above. Such designs are described as being *complete*. Designs such as Latin squares, in which some interactions are assumed to be zero are rarely encountered in generalizability theory. They could be accommodated, but they would have limited utility for G study purposes.

Usually, one way to determine (or verify) the total number of possible interaction effects for a design is to count the total number of distinct areas in the Venn diagram representation of the design, and then subtract the number of main effects.[1] Generating all valid combinations of main effects, as discussed above, provides all interaction effects for any complete design, without direct reference to Venn diagrams.

When the universe of admissible observations has crossed facets, Cronbach et al. (1972) typically use a sequence of confounded effects to identify an effect involving nesting. For example, for the $p \times (i{:}h)$ design, Cronbach et al. (1972) usually would denote $i{:}h$ as (i, ih). Their rationale can be viewed from the perspective of a relationship between the $p \times (i{:}h)$ design and its completely crossed counterpart $p \times i \times h$, in which the main effects and interaction effects are p, i, h, pi, ph, ih, and pih (see Figure 3.1). The main effect $i{:}h$ in the $p \times (i{:}h)$ design represents the confounding of the effects i and ih in the $p \times i \times h$ design. That is, the main effect $i{:}h$ in the $p \times (i{:}h)$ design is associated with both i and ih when the universe of admissible observations is crossed. In terms of variance components, this confounding simply means that

$$\sigma^2(i{:}h) = \sigma^2(i, ih) = \sigma^2(i) + \sigma^2(ih).$$

---

[1]For designs with more than three *crossed* facets, it is not possible to associate each effect with a distinct area in a Venn diagram when main effects are represented by circles. For four crossed facets, ellipses can be used; see, for example, Figure 9.3.

Similarly, $pi{:}h$ represents the confounding of pi and pih.[2]

Using the notation introduced above, it is relatively simple to determine which effects from a *fully crossed* design are *confounded* in an effect for a design that involves nesting; namely,

> *Confounded-Effects Rule*: The effects that are confounded are all those that involve only indices in the nested effect and that include the primary index (indices) in the nested effect.

The number of effects that are confounded in a nested effect is

$$2 \ Exp(\text{number of nesting indices}),$$

where Exp means "exponential" or "raised to the power." For example, for the effect $i{:}h{:}p$, there are $2 \ Exp(2) = 4$ confounded effects (i, ih, ip, and ihp) when the universe of admissible observations has $i \times h$, and p is crossed with both facets. A notational shorthand for this result is:

$$i{:}h{:}p \Rightarrow i, ih, ip, ihp.$$

In this book, the operator "$:$" is used to represent nested effects, because listing the sequence of confounded effects is notationally burdensome, especially for complicated designs. Even so, understanding the confounded effects notation is important, because confounded effects relate directly to the universe of admissible observations. By contrast, the nested effects notation relates most directly to the G study design.

Note, however, that a thoughtless translation of the nested effects notation to the confounded effects notation can lead to nonsensical results. For example, with a *nested* universe of admissible observations, such as $i{:}h$, it is meaningless to represent the main effect $i{:}h$ as (i, ih), because i and ih are not distinguishable in the universe of admissible observations.

3.2.2 Linear Models

Given these notational conventions, the linear model for a design can be represented, in general, as

$$X_\omega = \mu + \sum \nu_\alpha, \qquad (3.1)$$

where

$$\omega = \textit{all} \text{ indices in the design,}$$
$$\mu = \text{grand mean,}$$

---

[2]This does not mean that variance components should be estimated for a crossed design when the design actually involves nesting.

α = the index (indices), with any nesting operators, for any component in the design,

ν_α = score effect associated with α, and

$\sum \nu_\alpha$ = sum of all score effects (except μ) in the design under consideration.

Equation 3.1 provides a decomposition of the observed scores X_ω in terms of score effects ν_α. For example, in the $p \times (i{:}h)$ design, ω is simply pih, and ν_α is used generically to represent score effects associated with the components p, h, $i{:}h$, ph, or $pi{:}h$.

For the designs in Figure 3.1, the linear models resulting from specific applications of Equation 3.1 are provided to the far right of the Venn diagrams. These equations are the most concise and convenient ways to represent the linear models. However, from the perspective of generalizability theory, the linear model expressions immediately to the right of the Venn diagrams are more fundamental, because score effects are always defined in terms of mean scores in generalizability theory. To obtain these mean score decompositions of X_ω, an investigator needs to be able to express any score effect ν_α in terms of a linear combination of mean scores. An algorithm for doing so is provided later, but first we consider definitions of mean scores for the universe and population, and assumptions for random effects models.

In random effects linear models, it is assumed that $N \to \infty$ for all facets including the object of measurement "facet." For the populations and universes in Figure 3.1, this means that $N_p \to \infty$, $N_i \to \infty$, and $N_h \to \infty$. Under these circumstances, the grand mean in the population and universe is defined as

$$\mu \equiv \underset{\omega}{E} \, X_\omega, \tag{3.2}$$

where ω is the set of all indices in the design. Also, the universe (or population) mean score for any component α is defined as

$$\mu_\alpha \equiv \underset{\dot\alpha}{E} \, X_\omega, \tag{3.3}$$

where $\dot\alpha$ (note the dot over α) is the set of all indices in ω that are *not* contained in α. For example, in the $p \times (i{:}h)$ design,

$$\mu \equiv \underset{p}{E}\,\underset{i}{E}\,\underset{h}{E}\, X_{pih},$$

$$\mu_h \equiv \underset{p}{E}\,\underset{i}{E}\, X_{pih}, \text{ and}$$

$$\mu_{ih} \equiv \underset{p}{E}\, X_{pih}.$$

When the design is $p \times (i{:}h)$, there is a potential source of confusion in denoting $E_p(X_{pih})$ as μ_{ih}, rather than $\mu_{i{:}h}$. This potential confusion arises because the score effect (for the design) associated with this mean score (for the universe) is $\nu_{i:h}$. Unless a specific context dictates otherwise, in this book *mean scores* for the population and universe are represented notationally with respect to a crossed universe of admissible observations which, in turn, is crossed with the population, whereas score effects are denoted in terms of the actual design employed. For example, even if the design is $p \times (i:h)$, when the population and universe consist of crossed facets, i is *not* nested within h *in the universe* and, therefore, μ_{ih} (rather than $\mu_{i:h}$) correctly reflects the structure of the universe. However, if the universe of admissible observations had i nested within h, then $\mu_{i:h}$ would be more descriptive. These notational considerations with respect to mean scores highlight the fact that, in a specific context, a G study design should be viewed in conjunction with a specified population and universe of admissible observations.[3]

The definitions of mean scores in Equations 3.2 and 3.3, and the manner in which score effects are defined, imply that the expected value of any score effect ν_α (for any index in α) is zero:

$$E\nu_\alpha = 0. \tag{3.4}$$

If this were not so, then the linear models in Figure 3.1 would not be tautologies.

Now, given Equations 3.2 through 3.4, it follows that one can express the mean score for any component α in terms of score effects as

$$\mu_\alpha = \mu + \begin{cases} \text{sum of score effects for all components that} \\ \text{consist solely of one or more indices in } \alpha. \end{cases} \tag{3.5}$$

For example, in the $p \times (i:h)$ design,

$$\mu_{ih} = \mu + \nu_h + \nu_{i:h}, \tag{3.6}$$

which can be verified by replacing X_{pih} in Equation 3.3 with its decomposition in terms of score effects:

$$\begin{aligned} \mu_{ih} &\equiv \underset{p}{E}\, X_{pih} \\ &= \underset{p}{E}(\mu + \nu_p + \nu_h + \nu_{i:h} + \nu_{ph} + \nu_{pi:h}) \\ &= \mu + \underset{p}{E}\,\nu_p + \nu_h + \nu_{i:h} + \underset{p}{E}\,\nu_{ph} + \underset{p}{E}\,\nu_{pi:h}. \end{aligned}$$

The result in Equation 3.6 follows immediately upon recognizing that $E_p\nu_p$, $E_p\nu_{ph}$, and $E_p\nu_{pi:h}$ are all zero, because the expected value of an effect over any of its subscripts is zero.

---

[3]In this book, an observed score for single conditions in the universe of admissible observations and/or the G study is *not* denoted using the nesting operator. So, even if the design is $p \times (i:h)$, the observed score is denoted X_{pih}, not $X_{pi:h}$.

Nothing in the above discussion involves any sample size n. Rather, the linear models have been defined in terms of the structure of the G study design and the definitions of mean scores for the population and universe. However, for *random effects* linear models, it is also assumed that all sample sizes are less than their corresponding universe or population size ($n < N \to \infty$), and that G study data can be viewed as resulting from a process of independent random sampling. In addition, it is "assumed" that all effects are uncorrelated:

$$\boldsymbol{E}(\nu_\alpha \nu_{\alpha'}) = \boldsymbol{E}(\nu_\alpha \nu_\beta) = 0, \tag{3.7}$$

where ν_α and $\nu_{\alpha'}$ designate different conditions of the same effect (e.g., ν_p and $\nu_{p'}$, $p \neq p'$), and ν_α and ν_β designate different effects (e.g., ν_p and ν_h). As mentioned and illustrated in Section 2.1, the word "assumed" is actually much stronger than required in generalizability theory.[4] Finally, note that none of the foregoing statements concerning random effects models in generalizability theory involves any assumptions about distributional form (e.g., normality).

3.2.3   Expressing a Score Effect in Terms of Mean Scores

Provided next is an algorithm for converting a score effect to a linear combination of mean scores for the population and universe. This algorithm can be used to express linear models in the manner reported immediately to the right of each of the Venn diagrams in Figure 3.1.

Let α be a component with m primary indices and s nesting indices. (In this case, no distinction is made between first, second, third, etc., levels of nesting.) The score effect associated with α, namely, ν_α, is equal to:

> Step 0: μ_α;

> Step 1: *Minus* the *mean scores* for components that consist of the s nesting indices and $m-1$ of the primary indices;

> Step 2: *Plus* the *mean scores* for components that consist of the s nesting indices and $m-2$ of the primary indices;

> $\vdots$

> Step j: *Plus* (if j is even) or *Minus* (if j is odd) the *mean scores* for components that consist of the s nesting indices and $m-j$ of the primary indices.

The algorithm terminates with Step m, that is, with the mean score for the component containing only the s nesting indices. If there are no nesting

---

[4]This issue is revisited in Section 5.4.

indices in the component α, then Step m results in adding or subtracting μ.

Consider, for example, the component $pi{:}h$ in the $p \times (i{:}h)$ design. This component has a single nesting index h, and two primary indices p and i. Step 1 in the algorithm results in subtracting μ_{ph} and μ_{ih} from μ_{pih}, because both μ_{ph} and μ_{ih} involve the nesting index h and $2 - 1 = 1$ of the primary indices in pih. Step 2 results in adding μ_h to the result of Step 1 because μ_h is the mean score that involves the nesting index h and $2-2 = 0$ of the primary indices in $pi{:}h$. Therefore,

$$\nu_{pi{:}h} = \mu_{pih} - \mu_{ph} - \mu_{ih} + \mu_h.$$

In the $p \times (i{:}h)$ design with one observation per cell, μ_{pih} is indistinguishable from X_{pih} and, therefore, we usually specify $\nu_{pi{:}h}$ as

$$\nu_{pi{:}h} = X_{pih} - \mu_{ph} - \mu_{ih} + \mu_h.$$

This latter representation of $\nu_{pi{:}h}$ rather clearly indicates the "residual" nature of this effect in the $p \times (i{:}h)$ design.

3.3   Typical ANOVA Computations

This section provides a treatment of observed mean scores, score effects, sums of squares, degrees of freedom, and mean squares for complete balanced designs. Readers familiar with these topics might skim this section, paying attention primarily to notational conventions.

3.3.1   Observed Mean Scores and Score Effects

For each component α, the mean score μ_α has an observed score analogue, denoted $\overline{X}_\alpha$, which is

$$\overline{X}_\alpha = \frac{1}{\pi(\dot{\alpha})} \sum_{\dot{\alpha}} X_\omega. \tag{3.8}$$

As in Equation 3.3, $\dot{\alpha}$ (note the dot above α) refers to the set of indices in ω that are *not* included in α. For any index in $\dot{\alpha}$, the summation runs from 1 to the sample size (n) associated with the index, and

$$\pi(\dot{\alpha}) = \begin{cases} 1 \text{ if } \alpha = \omega; \text{ and, otherwise, the product} \\ \text{of the sample sizes for all indices in } \dot{\alpha}. \end{cases} \tag{3.9}$$

Consider, for example, the i component in the $p \times i \times h$ design. The observed mean score associated with a particular item is

$$\overline{X}_i = \frac{1}{n_p n_h} \sum_{p=1}^{n_p} \sum_{h=1}^{n_h} X_{pih}.$$

Similarly, for the pi component,

$$\overline{X}_{pi} = \frac{1}{n_h} \sum_{h=1}^{n_h} X_{pih}.$$

For the notational system used here, when a mean is taken over the levels of a facet, the index for that facet does *not* appear in the subscript for the mean score. Given this convention, it follows from Equation 3.8 that the estimate of the grand mean (i.e., the observed score analogue of μ) is

$$\overline{X} = \frac{1}{\pi(\omega)} \sum_{\omega} X_{\omega}, \qquad (3.10)$$

where $\pi(\omega)$ is the product of the sample sizes for all indices in the design. For example, in the $p \times i \times h$ design,

$$\overline{X} = \frac{1}{n_p n_i n_h} \sum_{p=1}^{n_p} \sum_{i=1}^{n_i} \sum_{h=1}^{n_h} X_{pih}.$$

In summary, then, we use $\overline{X}_\alpha$ to denote the observed score analogue of μ_α, and we use $\overline{X}$ to denote the observed score analogue of μ. Also, we use x_α to denote the observed score analogue of ν_α.[5] Equations 3.1, 3.5, and the algorithm in Section 3.2.3 are applicable to the observed score analogues of μ_α and ν_α. One simply replaces μ_α, ν_α, and μ with $\overline{X}_\alpha$, x_α, and $\overline{X}$, respectively. For example, in terms of observed mean scores for the $p \times i \times h$ design,

$$X_{pih} = \overline{X} + x_p + x_i + x_h + x_{pi} + x_{ph} + x_{ih} + x_{pih},$$

and application of the algorithm to the x_α terms in this equation gives

$$
\begin{aligned}
X_{pih} = \; & \overline{X} + (\overline{X}_p - \overline{X}) + (\overline{X}_i - \overline{X}) + (\overline{X}_h - \overline{X}) \\
& + (\overline{X}_{pi} - \overline{X}_p - \overline{X}_i + \overline{X}) + (\overline{X}_{ph} - \overline{X}_p - \overline{X}_h + \overline{X}) \\
& + (\overline{X}_{ih} - \overline{X}_i - \overline{X}_h + \overline{X}) \\
& + (X_{pih} - \overline{X}_{pi} - \overline{X}_{ph} - \overline{X}_{ih} + \overline{X}_p + \overline{X}_i + \overline{X}_h - \overline{X}).
\end{aligned}
$$

This familiar decomposition of X_{pih} in terms of observed mean scores is clearly analogous to the linear model decomposition in terms of μ_α mean scores in Figure 3.1.

---

[5]This is reasonably consistent with other literature in which observed deviation scores are identified with lowercase roman letters, although Brennan (1992a) uses $\overline{X}_\alpha \sim$.

3.3.2 Sums of Squares and Mean Squares

Given these notational conventions, the sum of squares for any component α in any design is

$$SS(\alpha) = \pi(\dot{\alpha}) \sum_\alpha x_\alpha^2, \qquad (3.11)$$

where the summation is taken over all indices in α. Also, the total sum of squares is

$$SS(tot) = \sum_\omega (X_\omega - \overline{X})^2, \qquad (3.12)$$

where the summation is taken over all indices in the design.

For example, for the $p \times i \times h$ design, the sum of squares for the i component is

$$SS(i) = n_p n_h \sum_{i=1}^{n_i} x_i^2 = n_p n_h \sum_{i=1}^{n_i} (\overline{X}_i - \overline{X})^2, \qquad (3.13)$$

the sum of squares for the pi component is

$$SS(pi) = n_h \sum_p \sum_i x_{pi}^2 = n_h \sum_p \sum_i (\overline{X}_{pi} - \overline{X}_p - \overline{X}_i + \overline{X})^2, \qquad (3.14)$$

and the total sum of squares is

$$SS(tot) = \sum_p \sum_i \sum_h (\overline{X}_{pih} - \overline{X})^2. \qquad (3.15)$$

The definition of $SS(\alpha)$ in Equation 3.11 can be used directly to obtain the sum of squares for any component in any complete balanced design. The process of doing so tends to be tedious, however. A simpler procedure is discussed next.

For the component α, let the "sum of squared mean scores" be

$$T(\alpha) = \pi(\dot{\alpha}) \sum_\alpha \overline{X}_\alpha^2, \qquad (3.16)$$

where the summation is taken over all the indices in α. An equivalent expression in terms of squared total scores is

$$T(\alpha) = \frac{1}{\pi(\dot{\alpha})} \sum_\alpha \left(\sum_{\dot{\alpha}} X_\omega \right)^2,$$

where the summation in parentheses is taken over all indices in ω that are not in α.

Note the distinction between $SS(\alpha)$ in Equation 3.11 which involves x_α terms, and $T(\alpha)$ in Equation 3.16 which involves $\overline{X}_\alpha$ terms. A special case of Equation 3.16 is

$$T(\mu) = \pi(\omega)\overline{X}^2, \qquad (3.17)$$

which is the "sum of squared mean scores" associated with the grand mean.

Expressing sums of squares with respect to T terms is straightforward—use the algorithm in Section 3.2.3 with ν_α, μ_α, and μ replaced by $SS(\alpha)$, $T(\alpha)$, and $T(\mu)$, respectively. For example, in the $p \times i \times h$ design, rather than using Equation 3.13 to calculate $SS(i)$, a computationally simpler expression is

$$SS(i) = T(i) - T(\mu).$$

Similarly,

$$SS(pi) = T(pi) - T(p) - T(i) + T(\mu)$$

can be used instead of Equation 3.14, and

$$SS(tot) = T(pih) - T(\mu)$$

is more efficient computationally than Equation 3.15.

Appendix A provides formulas for $T(\alpha)$ and $SS(\alpha)$ for each component in each of the illustrative designs in Figure 3.1. These tables also provide formulas for degrees of freedom $df(\alpha)$. For an effect (main effect or interaction) that is not nested, the degrees of freedom are the product of the $(n-1)$s for the indices in the effect, where n is the G study sample size associated with the index. For any nested effect, the degrees of freedom are

$$\left\{ \begin{matrix} \text{product of } (n-1)\text{s} \\ \text{for primary indices} \end{matrix} \right\} \times \left\{ \begin{matrix} \text{product of } ns \\ \text{for nesting indices} \end{matrix} \right\}.$$

Mean squares are simply

$$MS(\alpha) = \frac{SS(\alpha)}{df(\alpha)}. \tag{3.18}$$

3.3.3 Synthetic Data Examples

Table 3.1 provides Synthetic Data Set No. 3, with sample means, for the $p \times i \times o$ design with $n_p = 10$ persons, $n_i = 4$ items, and $n_o = 2$ occasions. Assume that the infinite universe of admissible observations has $i \times o$, and the data in Table 3.1 arose from a process of independent random sampling from this universe.

Clearly, this data set is not an example of an ideal G study, especially since the sample sizes are so small. However, the small sample sizes make it relatively easy to verify the usual ANOVA statistics in Table 3.3. For example, using Equation 3.16,

$$
\begin{aligned}
T(i) &= n_p n_o \sum_{i=1}^{4} \overline{X}_i^2 \\
&= (10)(2)[(4.10)^2 + (5.56)^2 + (5.80)^2 + (5.56)^2] \\
&= 2263.50,
\end{aligned}
$$

and using Equation 3.17

$$T(\mu) = n_p n_i n_o \overline{X}^2 = (10)(4)(2)(5.275)^2 = 2226.05.$$

Therefore,

$$SS(i) = T(i) - T(\mu) = 2263.50 - 2226.05 = 37.45.$$

Table 3.2 provides Synthetic Data Set No. 4, with sample means, for the $p \times (r:t)$ design with $n_p = 10$ persons and $n_t = 3$ tasks, and with each task evaluated by a different set of $n_r = 4$ raters. Even though raters are nested within tasks in the G study design, suppose raters are crossed with tasks in the universe of admissible observations. This assumption has no bearing on the computations of sums of squares, but it does affect the manner in which the sampling presumably occurred. That is, raters and tasks were both sampled independently, and $n_r = 4$ raters were randomly assigned to each of the $n_t = 3$ tasks.

The usual ANOVA statistics for Synthetic Data Set No. 4 are provided in Table 3.4. For example,

$$
\begin{aligned}
T(pt) &= n_r \sum_p \sum_t \overline{X}_{pt}^2 \\
&= 4[(5.25)^2 + (4.25)^2 + \cdots + (3.50)^2] \\
&= 2931.5000,
\end{aligned}
$$

$$
\begin{aligned}
T(p) &= n_r n_t \sum_p \overline{X}_p^2 \\
&= (4)(3)[(4.75)^2 + (5.75)^2 + \cdots + (4.00)^2] \\
&= 2800.1667,
\end{aligned}
$$

$$
\begin{aligned}
T(t) &= n_p n_r \sum_t \overline{X}_t^2 \\
&= (10)(4)[(5.50)^2 + (4.80)^2 + (3.95)^2] \\
&= 2755.7000,
\end{aligned}
$$

and

$$T(\mu) = n_p n_r n_t \overline{X}^2 = (10)(4)(3)(4.75)^2 = 2707.5000.$$

Therefore,

$$
\begin{aligned}
SS(pt) &= T(pt) - T(p) - T(t) + T(\mu) \\
&= 2931.5000 - 2800.1667 - 2755.7000 + 2707.5000 \\
&= 83.1333.
\end{aligned}
$$

TABLE 3.1. Synthetic Data Set No. 3 and the $p \times i \times o$ Design

Person	X_{pio} o_1				X_{pio} o_2				$\overline{X}_{pi}$				$\overline{X}_{po}$		$\overline{X}_p$
	i_1	i_2	i_3	i_4	i_1	i_2	i_3	i_4	i_1	i_2	i_3	i_4	o_1	o_2	
1	2	6	7	5	2	5	5	5	2.0	5.5	6.0	5.0	5.00	4.25	4.625
2	4	5	6	7	6	7	5	7	5.0	6.0	5.5	7.0	5.50	6.25	5.875
3	5	5	4	6	5	4	5	5	5.0	4.5	4.5	5.5	5.00	4.75	4.875
4	5	9	8	6	5	7	7	6	5.0	8.0	7.5	6.0	7.00	6.25	6.625
5	4	3	5	6	4	5	6	4	4.0	4.0	5.5	5.0	4.50	4.75	4.625
6	4	4	4	7	6	4	7	8	5.0	4.0	5.5	7.5	4.75	6.25	5.500
7	2	6	6	5	2	7	7	5	2.0	6.5	6.5	5.0	4.75	5.25	5.000
8	3	4	4	5	6	6	6	4	4.5	5.0	5.0	4.5	4.00	5.50	4.750
9	0	5	4	5	5	5	5	3	2.5	5.0	4.5	4.0	3.50	4.50	4.000
10	6	8	7	6	6	8	8	6	6.0	8.0	7.5	6.0	6.75	7.00	6.875
	$\overline{X}_{io}$ for $o=1$				$\overline{X}_{io}$ for $o=2$				$\overline{X}_i$				$\overline{X}_o$		$\overline{X}$
	3.5	5.5	5.5	5.8	4.7	5.8	6.1	5.3	4.10	5.65	5.80	5.55	5.075	5.475	5.275

TABLE 3.2. Synthetic Data Set No. 4 and the $p \times (r:t)$ Design

| Person | X_{prt} | | | | | | | | | | | | $\overline{X}_{pt}$ | | | $\overline{X}_p$ |
| | t_1 | | | | t_2 | | | | t_3 | | | | | | | |
	r_1	r_2	r_3	r_4	r_5	r_6	r_7	r_8	r_9	r_{10}	r_{11}	r_{12}	t_1	t_2	t_3	
1	5	6	5	5	5	3	4	5	6	7	3	3	5.25	4.25	4.75	4.750
2	9	3	7	7	7	5	5	5	7	7	5	2	6.50	5.50	5.25	5.750
3	3	4	3	3	5	3	3	5	6	5	1	6	3.25	4.00	4.50	3.917
4	7	5	5	3	3	1	4	3	5	3	3	5	5.00	2.75	4.00	3.917
5	9	2	9	7	7	7	3	7	2	7	5	3	6.75	6.00	4.25	5.667
6	3	4	3	5	3	3	6	3	4	5	1	2	3.75	3.75	3.00	3.500
7	7	3	7	7	7	5	5	7	5	5	5	4	6.00	6.00	4.75	5.583
8	5	8	5	7	7	5	5	4	3	2	1	1	6.25	5.25	1.75	4.417
9	9	9	8	8	6	6	6	5	5	8	1	1	8.50	5.75	3.75	6.000
10	4	4	4	3	3	5	6	5	5	7	1	1	3.75	4.75	3.50	4.000
	$\overline{X}_{r:t}$ for $t=1$				$\overline{X}_{r:t}$ for $t=2$				$\overline{X}_{r:t}$ for $t=3$				$\overline{X}_t$			$\overline{X}$
	6.1	4.8	5.6	5.5	5.3	4.3	4.7	4.9	4.8	5.6	2.6	2.8	5.50	4.80	3.95	4.750

TABLE 3.3. ANOVA for $p \times i \times o$ Design Using Synthetic Data Set No. 3

Effect(α)	$df(\alpha)$	$\pi(\dot{\alpha})$	$T(\alpha)$	$SS(\alpha)$	$MS(\alpha)$	$\hat{\sigma}^2(\alpha)$
p	9	8	2288.25	62.20	6.9111	.5528
i	3	20	2263.50	37.45	12.4833	.4417
o	1	40	2229.25	3.20	3.2000	.0074
pi	27	2	2382.00	56.30	2.0852	.5750
po	9	4	2303.50	12.05	1.3389	.1009
io	3	10	2274.20	7.50	2.5000	.1565
pio	27	1	2430.00	25.25	.9352	.9352
Mean(μ)	80		2226.05			
Total	79			203.95		

TABLE 3.4. ANOVA for $p \times (r\!:\!t)$ Design Using Synthetic Data Set No. 4

Effect(α)	$df(\alpha)$	$\pi(\dot{\alpha})$	$T(\alpha)$	$SS(\alpha)$	$MS(\alpha)$	$\hat{\sigma}^2(\alpha)$
p	9	12	2800.1667	92.6667	10.2963	.4731
t	2	40	2755.7000	48.2000	24.1000	.3252
$r\!:\!t$	9	10	2835.4000	79.7000	8.8556	.6475
pt	18	4	2931.5000	83.1333	4.6185	.5596
$pr\!:\!t$	81	1	3204.0000	192.8000	2.3802	2.3802
Mean(μ)		120	2707.5000			
Total	119			496.5000		

3.4   Random Effects Variance Components

This section treats definitions and interpretations of G study random effects variance components, procedures for expressing expected mean squares in terms of these variance components, procedures for estimating random effects variance components, and some other issues associated with G study random effects variance components. To avoid otherwise awkward verbal descriptions, in this section the word "facet" is used to refer to either a facet in the universe of admissible observations or the "facet" associated with the objects of measurement.

3.4.1   Defining and Interpreting Variance Components

There is a variance component associated with each of the score effects in the linear model equation for any G study design. In general, for a random effects model, the variance component for α is

$$\sigma^2(\alpha) = \sigma^2(\nu_\alpha) = \boldsymbol{E}\,\nu_\alpha^2, \tag{3.19}$$

and it is called a random effects variance component. In words, $\sigma^2(\alpha)$ is the variance, in the population and/or universe, of the score effects ν_α; or, the expected value of the square of the ν_α. For example, for the $p \times (i{:}h)$ design,

$$\sigma^2(h) = \sigma^2(\nu_h) = \boldsymbol{E}\, \nu_h^2 = E(\mu_h - \mu)^2. \tag{3.20}$$

The last result in Equation 3.20 results from expressing ν_h in terms of mean scores for the universe of admissible observations and the population. Also, since μ is a constant in Equation 3.20,

$$\sigma^2(h) = \sigma^2(\mu_h).$$

Evidently, $\sigma^2(h)$ may be interpreted either as the variance of the h-facet score effects, or as the variance of the h-facet mean scores. Indeed, whenever α is a *nonnested main effect* (i.e., α consists of a single index),

$$\sigma^2(\alpha) = \sigma^2(\nu_\alpha) = \sigma^2(\mu_\alpha).$$

When α is an *interaction effect* or a *nested main effect*,

$$\sigma^2(\alpha) = \sigma^2(\nu_\alpha) \quad \text{but} \quad \sigma^2(\alpha) \neq \sigma^2(\mu_\alpha).$$

For example, in the $p \times (i{:}h)$ design,

$$\sigma^2(ph) = \sigma^2(\nu_{ph}) = \boldsymbol{E}\, \nu_{ph}^2 = E(\mu_{ph} - \mu_p - \mu_h + \mu)^2,$$

which is *not* $\sigma^2(\mu_{ph})$. Another way to express $\sigma^2(ph)$ is

$$\begin{aligned}
\sigma^2(ph) &= \boldsymbol{E}[(\mu_{ph} - \mu) - (\mu_p - \mu) - (\mu_h - \mu)]^2 \\
&= \boldsymbol{E}[(\mu_{ph} - \mu) - \nu_p - \nu_h]^2.
\end{aligned}$$

That is, $\sigma^2(ph)$ can be viewed as the expected value of the square of $(\mu_{ph} - \mu)$ after removing the effects of p and h.

As another example, consider the variance component for the nested main effect $i{:}h$ in the $p \times (i{:}h)$ design:

$$\sigma^2(i{:}h) = \sigma^2(\nu_{i:h}) = \boldsymbol{E}\, \nu_{i:h}^2. \tag{3.21}$$

Now, if $i \times h$ in the universe of admissible observations, then Equation 3.21 can be expressed in terms of universe mean scores as:

$$\sigma^2(i{:}h) = E(\mu_{ih} - \mu_h)^2.$$

That is, $\sigma^2(i{:}h)$ is interpretable as the expected value of the squared deviation mean scores of the form $\mu_{ih} - \mu_h$. Alternatively, if $i{:}h$ in the universe of admissible observations, then Equation 3.21 can be expressed as

$$\sigma^2(i{:}h) = \underset{h}{\boldsymbol{E}} \left[\underset{i}{\boldsymbol{E}} (\mu_{i:h} - \mu_h)^2 \right],$$

using the nesting operator in $\mu_{i:h}$ to explicitly reflect the nesting in the universe. In words, when the universe of admissible observations has $i{:}h$, then $\sigma^2(i{:}h)$ is interpretable as the expected value, over the h facet, of the variance of the i-facet mean scores within a level of h.

In interpreting G study variance components, it is also helpful to view them as providing a decomposition of total variance. Specifically,

$$\sigma^2(X_\omega) = E(X_\omega - \mu)^2 = \sum \sigma^2(\alpha), \qquad (3.22)$$

where the summation is taken over all components in the design. For example, for the $p \times (i{:}h)$ design the total variance is

$$\sigma^2(X_{pih}) = \sigma^2(p) + \sigma^2(h) + \sigma^2(i{:}h) + \sigma^2(ph) + \sigma^2(pi{:}h).$$

Equation 3.22 indicates that any $\sigma^2(\alpha)$ represents that part of $\sigma^2(X_\omega)$ that is uniquely attributable to the component α. Furthermore, since X_ω is the observed score associated with a single condition of each facet, the variance components are for single conditions, too. ·

3.4.2 Expected Mean Squares

The sum of squares for a component divided by its degrees of freedom is a mean square, and its expected value is called an *expected* mean square. Consider, for example, the sum of squares for the i component in the $p \times i \times h$ design, which is given in Equation 3.13. The mean square for the i component is

$$MS(i) = \frac{n_p n_h}{n_i - 1} \sum_i (\overline{X}_i - \overline{X})^2,$$

and its expected value is

$$EMS(i) = \frac{n_p n_h}{n_i - 1} E\left[\sum_i (\overline{X}_i - \overline{X})^2\right]. \qquad (3.23)$$

Now,

$$\overline{X}_i = \frac{1}{n_p n_h} \sum_p \sum_h X_{pih}$$

$$= \mu + \frac{\sum_p \nu_p}{n_p} + \nu_i + \frac{\sum_h \nu_h}{n_h} + \frac{\sum_p \nu_{pi}}{n_p}$$

$$+ \frac{\sum_p \sum_h \nu_{ph}}{n_p n_h} + \frac{\sum_h \nu_{ih}}{n_h} + \frac{\sum_p \sum_h \nu_{pih}}{n_p n_h},$$

and

$$\overline{X} = \frac{1}{n_p n_i n_h} \sum_p \sum_i \sum_h X_{pih}$$

$$= \mu + \frac{\sum_p \nu_p}{n_p} + \frac{\sum_i \nu_i}{n_i} + \frac{\sum_h \nu_h}{n_h} + \frac{\sum_p \sum_i \nu_{pi}}{n_p n_i}$$
$$+ \frac{\sum_p \sum_h \nu_{ph}}{n_p n_h} + \frac{\sum_i \sum_h \nu_{ih}}{n_i n_h} + \frac{\sum_p \sum_i \sum_h \nu_{pih}}{n_p n_i n_h}.$$

Replacing these two expressions in Equation 3.23, it can be shown that

$$EMS(i) = \sigma^2(pih) + n_p \sigma^2(ih) + n_h \sigma^2(pi) + n_p n_h \sigma^2(i).$$

The end result is relatively simple because of uncorrelated effects (recall Equation 3.7), but the process involves a substantial amount of algebra, so much that it would be very impractical to derive EMS equations for every component in every G study design that might be encountered.

It is fortunate, therefore, that simpler procedures are available. In particular, the notational system used in this book makes it easy to express the expected mean squares for each component in a random effects model. In general, for any component β,

$$EMS(\beta) = \sum \pi(\dot{\alpha}) \, \sigma^2(\alpha), \qquad (3.24)$$

where the summation is taken over all α that contain at least *all of the indices* in β, $\pi(\dot{\alpha})$ is defined in Equation 3.9, and $\sigma^2(\alpha)$ is the random effects variance component for α. Based on Equation 3.23, Appendix B reports the expected mean squares for every component in each of the illustrative designs in Figure 3.1. Also, these expected mean squares can be obtained using procedures originally reported by Cornfield and Tukey (1956). Indeed, Equation 3.23 is nothing more than a formula that summarizes one special case of the Cornfield and Tukey (1956) procedures.

In Chapter 2 for single-facet designs, it was shown that expected mean squares, and the random effects variance components entering each of them, can be represented in terms of Venn diagrams. Venn diagram representations for the $p \times i \times h$ and $p \times (i{:}h)$ designs are provided in Figures 3.2 and 3.3, respectively.

Estimating variance components involves substituting $MS(\beta)$ for $EMS(\beta)$ and $\hat{\sigma}^2(\alpha)$ for $\sigma^2(\alpha)$ in Equation Set 3.24; that is,

$$MS(\beta) = \sum \pi(\dot{\alpha}) \, \hat{\sigma}^2(\alpha). \qquad (3.25)$$

Since these are simultaneous linear equations with as many unknowns as equations, they have a unique solution. In general, estimating variance components through "equating" mean squares to their expected values is called the "ANOVA procedure," which gives best quadratic unbiased estimates (BQUE) without any normality assumptions (see Searle et al., 1992).

A number of specific procedures might be used to solve for the $\hat{\sigma}^2(\alpha)$, including

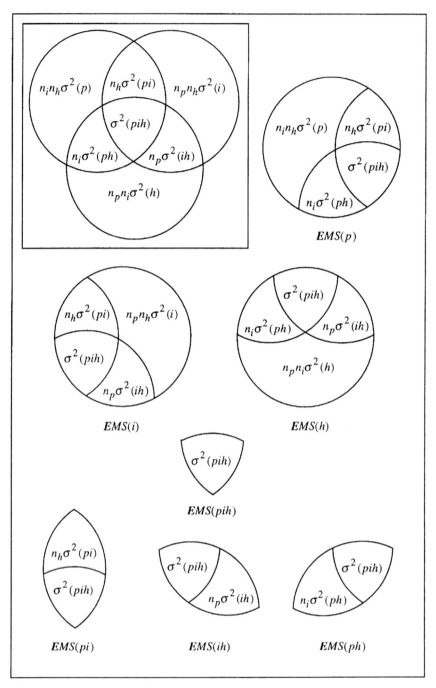

FIGURE 3.2. Venn diagram representation of expected mean squares for the $p \times i \times h$ random effects design.

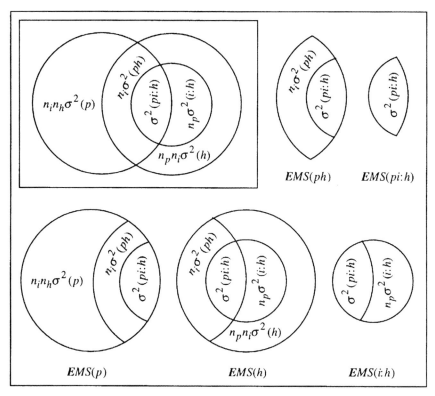

FIGURE 3.3. Venn diagram representation of expected mean squares for the $p \times (i{:}h)$ random effects design.

- an algebraic procedure discussed by Cronbach et al. (1972) that involves explicit use of the EMS equations,

- an algorithm that works directly with mean squares, and

- a matrix procedure described in Appendix C.

All three procedures give the same results provided no negative estimates occur. Negative estimates are discussed in Section 3.4.6. The ANOVA procedure, as well as other estimation procedures, for both balanced and unbalanced designs are discussed in more detail later in Section 7.3.1.

3.4.3 Estimating Variance Components Using the EMS Procedure

Cronbach et al. (1972) discuss a procedure for estimating variance components that is called the EMS procedure here. It involves replacing parameters with estimators in the EMS equations, and then solving them "in reverse," in the sense illustrated next. Consider the synthetic data example of the $p \times (r{:}t)$ design in Table 3.4. Using Equation 3.25, the mean squares expressed in terms of estimators of the variance components are:

$$
\begin{aligned}
MS(p) &= 10.2963 = \hat{\sigma}^2(pr{:}t) + 4\,\hat{\sigma}^2(pt) + 12\,\hat{\sigma}^2(p) && (3.26) \\
MS(t) &= 24.1000 = \hat{\sigma}^2(pr{:}t) + 4\,\hat{\sigma}^2(pt) + 10\,\hat{\sigma}^2(r{:}t) + 40\,\hat{\sigma}^2(t) \\
MS(r{:}t) &= 8.8556 = \hat{\sigma}^2(pr{:}t) + 10\,\hat{\sigma}^2(r{:}t) \\
MS(pt) &= 4.6185 = \hat{\sigma}^2(pr{:}t) + 4\,\hat{\sigma}^2(pt) && (3.27) \\
MS(pr{:}t) &= 2.3802 = \hat{\sigma}^2(pr{:}t).
\end{aligned}
$$

Starting at the bottom and replacing 2.3802 for $\hat{\sigma}^2(pr{:}t)$ in Equation 3.27 gives

$$
\hat{\sigma}^2(pt) = \frac{4.6185 - 2.3802}{4} = .5596.
$$

Also, replacing 2.3802 for $\hat{\sigma}^2(pr{:}t)$ and .5596 for $\hat{\sigma}^2(pt)$ in Equation 3.26 gives

$$
\hat{\sigma}^2(p) = \frac{10.2963 - 2.3802 - 4(.5596)}{12} = .4731.
$$

The other $\hat{\sigma}^2(\alpha)$ can be obtained similarly.

3.4.4   Estimating Variance Components Directly from Mean Squares

For simple designs, the Cronbach et al. *EMS* procedure for estimating random effects variance components is not too tedious, but it does become burdensome for more complicated designs. Frequently, therefore, the following algorithm provides a simpler procedure in that it precludes the need to determine expected mean squares.

Let α be any component consisting of t indices. Also, identify all components that consist of the t indices in α and *exactly one* additional index; and call the set of "additional" indices $\mathcal{A}$.[6] In this case, it does *not* matter whether an index is nested. In general, for the random effects model,

$$
\hat{\sigma}^2(\alpha) = \frac{1}{\pi(\dot{\alpha})} \left[\begin{array}{c} \text{some combination} \\ \text{of mean squares} \end{array} \right], \qquad (3.28)
$$

where $\pi(\dot{\alpha})$ has been defined in Equation 3.9, and the appropriate combination of mean squares is obtained by

Step 0: $MS(\alpha)$;

---

[6]The "additional" indices in $\mathcal{A}$ plus the indices in α will *not* always constitute all indices in the design.

Step 1: *Minus* the *mean squares* for all components that consist of the t indices in α and exactly *one* of the indices in $\mathcal{A}$;

Step 2: *Plus* the *mean squares* for all components that consist of the t indices in α and any *two* of the indices in $\mathcal{A}$;

$\vdots$

Step j: *Plus* (if j is even) or *Minus* (if j is odd) the *mean squares* for all components that consist of the t indices in α and any j of the indices in $\mathcal{A}$.

For the estimated variance component that contains all indices in the design, the algorithm terminates at Step 0; that is, the variance component is estimated by its mean square. Otherwise, the algorithm terminates when a step results in one mean square added or subtracted. Except in quite complicated designs, it is rare that more than two steps are required. Appendix B provides estimators of variance components, in terms of mean squares, for each component in each of the illustrative designs in Figure 3.1.

Consider, for example, the $p \times i \times h$ design. For $\hat{\sigma}^2(p)$ in this design $\pi(\dot{\alpha}) = n_i n_h$, and the set $\mathcal{A}$ consists of the indices i and h because both pi and ph consist of p and exactly one additional index. Therefore, Step 1 results in subtracting $MS(pi)$ and $MS(ph)$ from $MS(p)$, Step 2 results in adding $MS(pih)$, and the algorithm terminates. The resulting estimator is:

$$\hat{\sigma}^2(p) = \frac{MS(p) - MS(pi) - MS(ph) + MS(pih)}{n_i n_h}.$$

For the variance component $\hat{\sigma}^2(pi)$, $\pi(\dot{\alpha}) = n_h$; and $\mathcal{A}$ consists of the index h, only, because pih is the only component that contains the two indices in $\alpha = pi$ and exactly one additional index h. Therefore, Step 1 results in subtracting $MS(pih)$ from $MS(pi)$, and the algorithm terminates with

$$\hat{\sigma}^2(pi) = \frac{MS(pi) - MS(pih)}{n_h}.$$

For $\hat{\sigma}^2(pih)$, $\pi(\dot{\alpha}) = 1$, and $\mathcal{A}$ consists of no indices, because $\alpha = pih$ contains all the indices in the design; hence,

$$\hat{\sigma}^2(pih) = MS(pih).$$

The algorithm has been expressed in terms of estimated random effects variance components and mean squares; the same basic algorithm applies to random effects variance components (i.e., the parameters) and expected mean squares. Partial understanding of this algorithm can be obtained through consideration of Venn diagrams. For example, Figures 3.4 and 3.5 provide Venn diagram illustrations of which expected mean squares are added and subtracted to obtain certain variance components for the $p \times i \times h$ and $p \times (i:h)$ designs.

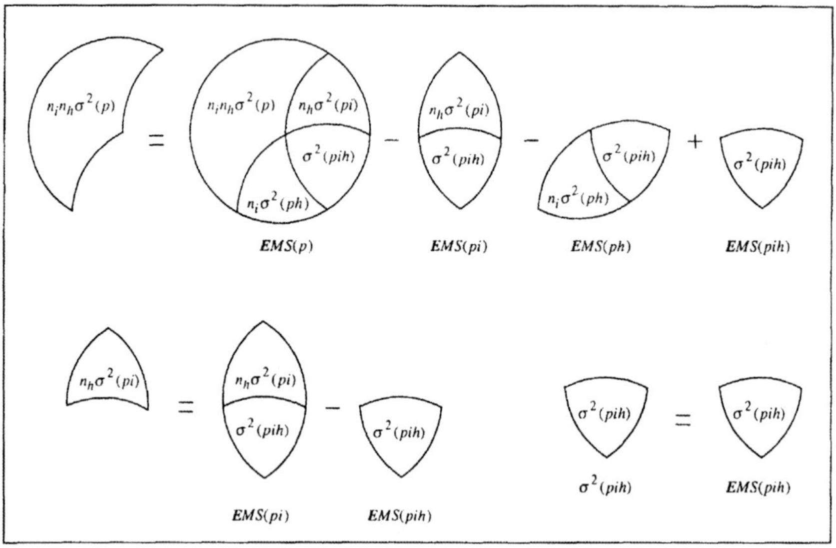

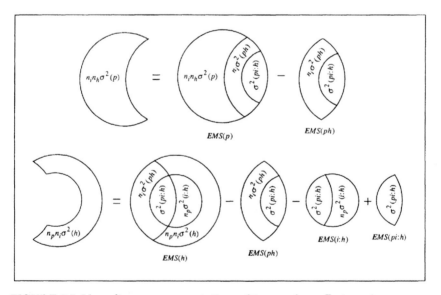

FIGURE 3.4. Venn diagram representations of three random effects variance components in terms of expected mean squares for the $p \times i \times h$ random effects design.

FIGURE 3.5. Venn diagram representations of two random effects variance components in terms of expected mean squares for the $p \times (i{:}h)$ random effects design.

3.4.5 Synthetic Data Examples

Using the algorithm in the previous section, the reader can verify the values of the estimated variance components reported in Tables 3.3 and 3.4. Alternatively, these estimates can be obtained using the Cronbach et al. *EMS* procedure or the matrix procedure in Appendix C.

Synthetic Data Set No. 3.

Consider the estimated variance components for the $p \times i \times o$ design in Table 3.3 on page 74 based on Synthetic Data Set No. 3, and recall that the observed scores have values between 0 and 9 (see Table 3.1 on page 72). To understand these estimated variance components requires considering their definitions, given the $i \times o$ universe of admissible observations. For example, $\hat{\sigma}^2(p) = .5528$ is the estimate of the variance of person mean scores, where each mean score is over all items and occasions in the universe. As such $\hat{\sigma}^2(p)$ is *not* the same thing as the variance of the *observed* mean scores for persons in Table 3.1, although the large value of $\hat{\sigma}^2(p)$ (relative to most of the other estimated variance components) is influenced by the rather substantial variability in persons' observed mean scores. By contrast, the relatively small value for $\hat{\sigma}^2(o) = .0074$ is influenced by the similarity in observed mean scores for occasions. It is also evident from Table 3.3 that $\hat{\sigma}^2(p)$ and $\hat{\sigma}^2(i)$ have similar and relatively large values, suggesting that variability attributable to persons and items is about equal and substantial.

By far, the largest estimated variance component in Table 3.3 is clearly $\hat{\sigma}^2(pio) = .9352$, which incorporates variability attributable to the three-way interaction as well as any other unexplained sources of variation. The large magnitude of $\hat{\sigma}^2(pio)$ is not too surprising, since the largest estimated variance component is often the one associated with the highest order interaction. Moreover, $\hat{\sigma}^2(pi) = .5750$ is noticeably large relative to $\hat{\sigma}^2(po)$ and $\hat{\sigma}^2(io)$, which suggests that person–item interaction effects are substantial and need to be taken into account in designing a measurement procedure, as discussed in Chapter 4.

Synthetic Data Set No. 4.

Consider the estimated variance components for the $p \times (r{:}t)$ design in Table 3.4 on page 74, and the associated raw data and mean scores in Table 3.2 on page 73. Suppose the universe of admissible observations has $r \times t$, not $r{:}t$. Then $\hat{\sigma}^2(r{:}t) = .6475$ is an estimate of $\sigma^2(r) + \sigma^2(rt)$, and $\hat{\sigma}^2(pr{:}t) = 2.3802$ is an estimate of $\sigma^2(pr) + \sigma^2(prt)$. These two estimated variance components are the largest, which is partly influenced by the fact that each of them is an estimate of the sum of two variance components for the universe.

Note also that $\hat{\sigma}^2(t) = .3252$, which suggests that the task means (over the population of persons and universe of raters) are somewhat different.

TABLE 3.5. Illustration of Negative Estimates of Variance Components

Effect(α)	$df(\alpha)$	$T(\alpha)$	$SS(\alpha)$	$MS(\alpha)$	$\hat{\sigma}^2(\alpha)$ from EMS	Alg.
p	9	2288.25	62.20	6.9111	.6751	.6965
o	1	2229.25	3.20	3.2000	−.1073	−.1030
$i{:}o$	6	2274.20	44.95	7.4917	.5981	.5981
po	9	2303.50	12.05	1.3389	−.0428	−.0428
$pi{:}o$	54	2430.00	81.55	1.5102	1.5102	1.5102
Mean(μ)		2226.05				

To put it another way, there is some evidence of systematic task effects. Also, $\hat{\sigma}^2(pt) = .5596$, which suggests that the rank order of persons varies by task. Finally, the similarity in the magnitudes of $\hat{\sigma}^2(p) = .4731$ and $\hat{\sigma}^2(pt) = .5596$ suggests that the variability of persons' scores (over the infinite universe of tasks and raters) is about as large as the variability attributable to different rank ordering of persons by task. As discussed in Chapter 4, the rather large magnitude of $\hat{\sigma}^2(pt)$ needs to be taken into account in designing a measurement procedure.

3.4.6   Negative Estimates of Variance Components

Estimated variance components are subject to sampling variability, and the smaller the sample sizes, the more likely it is that estimates will vary. (This subject of sampling variability is treated extensively in Chapter 6.) One possible consequence of sampling variability of estimated variance components is that one or more *estimated* variance components may be negative even though, by definition, variance components are nonnegative.

Consider, for example, Table 3.5 which provides ANOVA results and estimated variance components using Synthetic Data Set No. 3 in Table 3.1, assuming that the items administered on the two occasions were different. Recall that these data were analyzed previously according to the $p \times i \times o$ design. Here, for illustrative purposes *only*, we are simply analyzing these data *as if* they were associated with a $p \times (i{:}o)$ design.

The column in Table 3.5 headed "EMS" reports the estimated variance components using the Cronbach et al. (1972) procedure. Note, in particular, that whenever an estimated variance component is negative, it is replaced by zero *everywhere* it occurs in all the EMS equations. (Actual values of the negative estimates are reported in Table 3.5 so that the reader can verify the computations.) This is the procedure suggested by Cronbach et al. (1972) for dealing with negative estimates.

Alternatively, one can use the algorithm in Section 3.4.4 and simply set all negative estimates to zero. The estimated variance components that result from this procedure are also provided in Table 3.5. (Again, the actual values of the negative estimates are reported for verification purposes.) These results are necessarily identical to those obtained using the matrix procedure in Appendix C.

In examining the sets of estimated variance components, it is evident that $\hat{\sigma}^2(i{:}o)$ and $\hat{\sigma}^2(pi{:}o)$ are identical for the *EMS* procedure and the algorithm, but $\hat{\sigma}^2(p)$ is slightly different, and the actual magnitudes of the negative estimates of $\sigma^2(o)$ are different. The reason for these differences is that the zero-replacement aspect of the *EMS* procedure has an impact on not only those estimates of variance components that turn out to be negative, but also (usually) other estimates of variance components, as well. For example, since $EMS(p)$ involves $\sigma^2(po)$, setting $\hat{\sigma}^2(po)$ to zero influences the estimate of $\sigma^2(p)$. By contrast, using the algorithm, $\hat{\sigma}^2(p)$ is expressed directly as a function of mean squares and sample sizes, and these do not change when some other estimated variance component is negative.

It follows that, using the *EMS* procedure, when one or more estimates is (are) negative, other estimated variance components may be biased. This does not occur if the algorithm or matrix procedure is used. Therefore, from the perspective of unbiased estimates being desirable, the algorithm or matrix procedure seems preferable in the presence of negative estimates. (Note that setting a negative estimate to zero results in the estimate being biased no matter which procedure is used.) However, the *EMS* procedure might be preferable if there were a substantive, theory-based reason to believe that the variance component (i.e., parameter) associated with the negative estimate is indeed zero. From a practical perspective, choosing among the procedures is seldom a crucial issue, because they almost always provide very similar results, even when there are several negative estimates.

Negative estimates tend to arise when sample sizes are small, especially when the design involves a large number of effects. Obvious remedies include, of course, using larger sample sizes. Also, the possibility of obtaining negative estimates can be avoided completely using Bayesian procedures (see, e.g., Box & Tiao, 1973) and other procedures discussed briefly in Section 7.3.1, but doing so usually necessitates distributional-form assumptions and substantial computational complexities.

3.5   Variance Components for Other Models

For G study purposes it is frequently best to report variance components for the random effects model. As discussed in Chapter 4, these variance components are easily used for a variety of D studies, including D studies in which one or more facets is (are) fixed in the universe of generalization. However, it sometimes occurs that the relationship between a G study and a universe of admissible observations is such that estimated random effects variance components would clearly misrepresent this universe. For example, when items are nested within content categories in a universe of admissible observations, and the G study design incorporates all content categories, estimated random effects variance components are inappropriate. Therefore, this section considers procedures for estimating G study variance components for any model.

Throughout this section an uppercase M is used as a general designator for a model. For example, if $n = N$ for some facets and $N \to \infty$ for all other facets, then the model M is the mixed model; if $N \to \infty$ for all facets, then M is the random model; and if $n < N < \infty$ for some facet, then the model involves sampling from a finite universe of conditions for that facet. When the model is mixed, we usually replace the letter M with the uppercase letter(s) for the specific facet(s) that is (are) fixed. Throughout this section the word "facet" is used to refer to either a facet in the universe of admissible observations, or the "facet" associated with the object of measurement.

Reading of the rest of this section may be postponed until Chapter 5, because Chapter 4 involves using only random effects G study estimated variance components.

3.5.1   Model Restrictions and Definitions of Variance Components

Whether or not the model is random, equations like those in Figure 3.1 are still applicable, without *necessarily* making any notational changes. What changes are the definitions of mean scores, the assumptions about model effects, and the definitions of the variance components. These issues are treated in detail by Brennan (1994). Here, some relevant issues are summarized based on the $p \times (i:h)$ design with $N_p \to \infty$, $N_i \to \infty$, and $n_h = N_h < \infty$ (i.e., the mixed model with the h facet fixed).

This mixed model is frequently a good characterization of a measurement procedure associated with a table of specifications, in which the content categories play the role of levels of the h facet. Specifically, in the universe of admissible observations, an infinite universe of items is nested within each of the N_h content categories, and all content categories are represented in the $p \times (i:h)$ design (i.e., $n_h = N_h < \infty$). Of course, tests frequently consist of *un*equal numbers of items within content categories. For such tests, a design characterizing the data would be described as *unbalanced*. Unbalanced designs with mixed models are best treated using multivariate generalizability theory, as introduced in Section 9.1 and discussed more fully in Chapter 11. Here, to avoid such complexities, it is assumed that the number of items is a constant in each content category of any form of a test.

When $n_h = N_h$ the following means scores are defined in the same way as they are in the random model.

$$\mu_h \equiv \mathop{E}_p \mathop{E}_i X_{pih}, \quad \mu_{ph} \equiv \mathop{E}_i X_{pih}, \quad \text{and} \quad \mu_{i:h} \equiv \mathop{E}_p X_{pih}.$$

Note that we use the nesting operator in denoting $\mu_{i:h}$ because the universe of admissible observations has i nested within h. However, an expectation is not defined over the finite number of levels of the h facet (N_h). Consequently, the definitions of μ and μ_p for the mixed model are different from their definitions for the random model. For the mixed model,

$$\mu \equiv \frac{1}{n_h} \sum_h \left[\mathop{E}_p \mathop{E}_i X_{pih} \right] \tag{3.29}$$

and

$$\mu_p \equiv \frac{1}{n_h} \sum_h \left[\mathop{E}_i X_{pih} \right], \tag{3.30}$$

where the summation runs from 1 to $n_h = N_h$.

It is important to recognize that these mean scores are defined for the universe of admissible observations and the population of persons. As such, they are defined independent of (and logically prior to) specifying the linear model for the $p \times (i:h)$ G study design (or any G study design, for that matter). Furthermore, the score effects in the linear model for the $p \times (i:h)$ design are defined *in terms of* these mean scores (see Section 3.2.3), and the linear model is, by definition, a tautology. It follows from the above definitions that

$$\mathop{E} \nu_p = \mathop{E} \nu_{i:h} = \mathop{E} \nu_{pi:h} = 0$$

and

$$\sum_h \nu_h = \sum_h \nu_{ph} = 0. \tag{3.31}$$

The zero sums in Equation 3.31 are an *integral* part of the mixed model in generalizability theory. We refer to them as "restrictions" on the model to distinguish them from additional constraints that are *sometimes* imposed when the linear model is specified. The perspective on mixed models in generalizability theory is similar to that of Scheffé (1959).

Note, in particular, that *no* restrictions are employed in the *general* linear model (see, e.g., Searle, 1971, Chap. 9). Consequently, there are a number of procedures for estimating variance components (and computer programs for doing so) that do not employ these restrictions, and such procedures are not applicable in generalizability theory. To put it bluntly, generalizability theory with fixed facets is not isomorphic with other variance components perspectives on the so-called "general mixed model."

In addition to model restrictions, the mixed model necessitates different definitions of variance components. Table 3.6 provides these definitions for

TABLE 3.6. Variance Components for $p \times (i{:}h)$ Design with h Fixed

Effect	Definition of $\sigma^2(\alpha\|H)$	$EMS(\alpha\|H)$
p	$\boldsymbol{E}\,\nu_p^2$	$\sigma^2(pi{:}h\|H) + n_i n_h\,\sigma^2(p\|H)$
h	$\sum_h \nu_h^2/(n_h - 1)$	$\sigma^2(pi{:}h\|H) + n_i\,\sigma^2(ph\|H)$
		$+ n_p\,\sigma^2(i{:}h\|H) + n_p n_i\,\sigma^2(h\|H)$
$i{:}h$	$\sum_h (\boldsymbol{E}\,\nu_{i{:}h}^2)/n_h$	$\sigma^2(pi{:}h\|H) + n_p\,\sigma^2(i{:}h\|H)$
ph	$\sum_h (\boldsymbol{E}\,\nu_{ph}^2)/(n_h - 1)$	$\sigma^2(pi{:}h\|H) + n_i\,\sigma^2(ph\|H)$
$pi{:}h$	$\sum_h (\boldsymbol{E}\,\nu_{pi{:}h}^2)/n_h$	$\sigma^2(pi{:}h\|H)$

the $p \times (i{:}h)$ design. They are denoted generically as $\sigma^2(\alpha|H)$ to emphasize that the h facet is fixed and to distinguish them from $\sigma^2(\alpha)$ for the random model. The definitions of $\sigma^2(\alpha|H)$ in Table 3.6 are consistent with those used by Cornfield and Tukey (1956) and Cronbach et al. (1972), but these are not always the definitions used by other authors in other contexts. Brennan (1994) discusses this issue in more detail. Note in particular that, for $\sigma^2(\alpha|H)$ in Table 3.6, a divisor of n_h is used only if α includes h as a nesting index (an index *after* the nesting operator ":"), and a divisor of $(n_h - 1)$ is used if α includes h as a primary index (an index *before* the nesting operator). This convention is consistent with the use of n_h or $(n_h - 1)$ in the degrees of freedom for α.

The interpretation of variance components for the $p \times (i{:}h)$ mixed model is similar to the interpretation for the random effects model. The primary difference is that the concept of an expectation over h is replaced by the concept of an average over the levels of h. For example, $\sigma^2(p|H)$ is the variance of the person mean scores, where each mean is the *average* over the n_h levels of h, of the expected value over the infinite universe of i within h (see Equation 3.30). Also, $\sigma^2(h|H)$ is the variance of the mean scores for the n_h levels of h, and $\sigma^2(i{:}h|H)$ is the average variance of the i-facet mean scores within a level of h.

In much of the traditional statistical literature, $\sigma^2(h|H)$ would be called a quadratic form, rather than a variance component, because $\sigma^2(h|H)$ is associated with fixed effects only. In current literature on generalizability theory, however, this distinction is seldom drawn, because $\sigma^2(h|H)$ has the form of a variance for the levels of a facet in the universe. Here, we usually call $\sigma^2(h|H)$ a variance component although, in subsequent chapters, the term "quadratic form" is used in certain contexts. Note that, for the $p \times (i{:}h)$ design with $n_h = N_h < \infty$, the only quadratic form is $\sigma^2(h|H)$. In particular, $\sigma^2(ph|H)$ is *not* a quadratic form, even though it involves the h facet, because the ν_{ph} effects are random effects. An effect is random if *any* of its subscripts is associated with a facet for which $N \to \infty$.

3.5.2 Expected Mean Squares

For any model M, the expected mean square for the component β is

$$EMS(\beta) = \sum k(\alpha|\beta)\, \pi(\dot{\alpha})\, \sigma^2(\alpha|M), \qquad (3.32)$$

where the summation is taken over all α that contain at least *all of the indices* in β, $\pi(\dot{\alpha})$ is given by Equation 3.9, $\sigma^2(\alpha|M)$ is the variance component for the model M, and

$$k(\alpha|\beta) = \begin{cases} \text{the product of the terms } (1 - n/N) \text{ for} \\ \text{all } primary \text{ indices in } \alpha \text{ that are not in } \beta \\ [k(\alpha|\beta) = 1 \text{ if } \alpha = \beta]. \end{cases} \qquad (3.33)$$

Equation 3.24 for the random effects model is simply a special case of Equation 3.32 in which $k(\alpha|\beta) = 1$ for all facets, and the $\sigma^2(\alpha|M)$ are the random effects variance components, $\sigma^2(\alpha)$. Another special case is the mixed model with h fixed, with the EMS equations given in Table 3.6. (To say that the h facet is fixed means that the conditions of h in the G study exhaust those in the universe of admissible observations, which implies that $n_h = N_h < \infty$ and, consequently, $1 - n_h/N_h = 0$.) Equation 3.32 applies to the general case of sampling from a finite universe and/or population. Such applications are rare in generalizability theory, although Cronbach et al. (1997) provide a hypothetical example.

3.5.3 Obtaining $\sigma^2(\alpha|M)$ from $\sigma^2(\alpha)$

The simultaneous EMS equations can be solved for the variance components, with numerical results obtained by replacing parameters with estimates.[7] A procedure that gives the same results, but is often simpler, is discussed next. This procedure has the additional important advantage of revealing relationships between variance components for different models.

Given the variance components for a random model $\sigma^2(\alpha)$, the variance components for any model $\sigma^2(\alpha|M)$ are:

$$\sigma^2(\alpha|M) = \sum \frac{\sigma^2(\beta)}{\Pi(\beta|\alpha)}, \qquad (3.34)$$

where the summation is taken over all β that contain at least *all of the indices* in α, and

$$\Pi(\beta|\alpha) = \begin{cases} \text{the product of the universe sizes } (N) \\ \text{for all indices in } \beta \text{ except those in } \alpha \\ [\Pi(\beta|\alpha) = 1 \text{ if } \alpha = \beta]. \end{cases} \qquad (3.35)$$

---

[7]The equations could be solved "in reverse" with negative estimates (if present) replaced by zero everywhere they occur, as in the Cronbach et al. EMS procedure for random models. This possibility is ignored here.

Here, no distinction is made between nested and nonnested indices.

Replacing parameters with estimates, Equation 3.34 can be used to estimate variance components for any model M. If one or more of the $\hat{\sigma}^2(\alpha)$ is (are) negative, the negative value(s) should be *retained* in applying Equation 3.34. Even if some of the resulting $\hat{\sigma}^2(\alpha|M)$ are negative, they will be identical to the matrix-based results in Appendix C, which are the ANOVA procedure results.[8]

For example, for the random effects $p \times (i{:}h)$ design with H fixed, Equation 3.34 gives

$$\sigma^2(p|H) = \sigma^2(p) + \frac{\sigma^2(ph)}{N_h} + \frac{\sigma^2(pi{:}h)}{N_i N_h}$$

$$= \sigma^2(p) + \frac{\sigma^2(ph)}{n_h},$$

because $N_h = n_h$ and $N_i \to \infty$. This result is theoretically justified by careful consideration of the **EMS** equations for the model M and the random effects model. That is,

$$\textbf{EMS}(p) = \sigma^2(pi{:}h) + n_i \sigma^2(ph) + n_i n_h \sigma^2(p),$$

and for the mixed model with h fixed

$$\textbf{EMS}(p) = \sigma^2(pi{:}h|H) + n_i n_h \sigma^2(p|H).$$

Since $\textbf{EMS}(pi{:}h) = \sigma^2(pi{:}h) = \sigma^2(pi{:}h|H)$, it follows that

$$\sigma^2(p|H) = \sigma^2(p) + \frac{\sigma^2(ph)}{n_h},$$

which is the result obtained using Equation 3.34.

3.5.4   Example: APL Program

The Adult Performance Level (APL) Program was designed to measure functional competency in five content areas: community resources (CR), occupational knowledge (OK), health (H), and government and law (GL).[9] That is, the universe of admissible observations can be viewed as having items nested with $n_h = N_h = 5$ content areas.

The APL Survey instrument has eight dichotomously scored items nested within each of the categories. That is, the design of the Survey instrument is $p \times (i{:}h)$ with $n_h = N_h = 5$ fixed categories and $n_i = 8 < N_i \to \infty$.

---

[8]If any of the resulting $\hat{\sigma}^2(\alpha|M)$ are negative, usually they are set to zero for both reporting and D study purposes.

[9]This program, which was developed by ACT (1976), is no longer in active use.

TABLE 3.7. G Study for APL Survey $p \times (i{:}h)$ Design for Southern Region

Effect(α)	$df(\alpha)$	$\pi(\dot{\alpha})$	$SS(\alpha)$	$MS(\alpha)$	$\hat{\sigma}^2(\alpha\|H)$
p	607	40	1014.4553	1.6713	.0378
h	4	4864	88.2773	22.0693	.0013[a]
$i{:}h$	35	608	556.0082	15.8860	.0259
ph	2428	8	484.3227	.1995	.0051
$pi{:}h$	21245	1	3375.2418	.1589	.1589

[a]Strictly speaking, $\hat{\sigma}^2(h\|H)$ is a quadratic form.

This is essentially a description of a simple table of specifications model. It is simple in two senses: there is only one set of fixed categories, and the $p \times (i{:}h)$ design is balanced (i.e., equal numbers of items per content area). More complicated table of specification models are considered later in Chapters 9 to 11.

In 1977, the APL Survey was administered to examinees in four regions of the United States: northeast, north central, south, and west. In Section 5.2, an analysis is provided that involves all four regions. Here, attention is directed to the southern region only. Specifically, Table 3.7 reports results for $n_p = 608$ southern examinees.

The EMS equations in Table 3.6 can be used to estimate $\sigma^2(p\|H)$. Specifically, replacing parameters with estimates, it is evident that

$$\hat{\sigma}^2(p|H) = \frac{MS(p) - MS(pi{:}h)}{n_i n_h} = \frac{1.6713 - .1589}{40} = .0378.$$

Alternatively, it is easy to verify that the estimated random effects variance components for p and ph are $\hat{\sigma}^2(p) = .0368$ and $\hat{\sigma}^2(ph) = .0051$, respectively, and using Equation 3.34,

$$\hat{\sigma}^2(p|H) = \hat{\sigma}^2(p) + \frac{\hat{\sigma}^2(ph)}{n_h} = .0368 + \frac{.0051}{5} = .0378.$$

In Table 3.7, $\hat{\sigma}^2(p\|H)$ is somewhat larger than $\hat{\sigma}^2(i{:}h\|H)$, suggesting that variability in person means scores (over the universe of items within all content categories) is greater than the variability in item mean scores (over the population of persons) within a content area. The fact that $\hat{\sigma}^2(pi{:}h\|H)$ is quite large relative to the other estimates suggests that there is substantial variability attributable to interactions of persons and items within content areas and/or residual errors.

The observed means over persons and items in the five content categories were

CR	OK	CE	H	GL
.75	.64	.63	.67	.57

Their similarity partly explains the relatively small value for the quadratic form $\hat{\sigma}^2(h|H) = .0013$. Note that $\hat{\sigma}^2(ph|H) = .0051$ is quite small, too, suggesting that interactions between persons and the fixed content categories are not very large. Therefore, an examinee who scores relatively high in one content area is likely to score high in another.

3.6 Exercises

3.1* An arithmetic test with 10 addition and 10 subtraction items is administered to 25 children in each of three classrooms. Provide the symbolic representation (i.e., sequence of letters with "×"s and ":"s, as required), Venn diagram, and linear model for this design.

3.2* A group of 30 students is administered a reading comprehension test consisting of three passages. Each passage contains a set of items asking factual questions about the content of the passage and a set of items requiring inferences. Provide the symbolic representation, Venn diagram, and linear model for this design. Describe the likely universe of admissible observations.

3.3 An investigator has 120 items and randomly splits them into three sets of items. Then, she randomly splits a group of 300 students into three subgroups, and administers the first set of items to the first subgroup, the second set to the second subgroup, and the third set to the third subgroup. What is a common name for this type of design? Provide the symbolic representation, Venn diagram, and linear model for this design. Viewing the data from this design as a G study, what are the variance components, and how would you estimate them?

3.4* Using the algorithm in Section 3.4.4, determine the estimators of $\sigma^2(p)$ and $\sigma^2(h)$ for the random effects $p \times (i\!:\!h)$ design.

3.5* Gillmore et al. (1978) discuss a study in which 14 students (p) are nested within each of two classes (c), which are in turn nested within each of 42 teachers (t). All students responded to 11 items (i) in an instructional assessment questionnaire, in which each item had six response categories scored 0 to 5. Provide the linear model and Venn diagram for the design. Suppose the grand mean was $\overline{X} = 2.5$ and

$$\sum_t \overline{X}_t^2 = 269.4830 \qquad \sum_c \sum_t \overline{X}_{ct}^2 = 545.9125$$

$$\sum_p \sum_c \sum_t \overline{X}_{pct}^2 = 8229.2590 \qquad \sum_i \overline{X}_i^2 = 69.0028$$

$$\sum_t \sum_i \overline{X}_{ti}^2 = 3003.0981 \qquad \sum_c \sum_i \sum_t \overline{X}_{cit}^2 = 6111.2862$$

$$\sum_p \sum_i \sum_c \sum_t X_{pict}^2 = 97506.4529.$$

Provide the degrees of freedom, T terms, sums of squares, mean squares, and estimated random effects variance components.

3.6 Consider the following scores for 10 students (p) on two science performance tasks (t) with all responses scored by the same three raters (r).

p	t_1			t_2		
	r_1	r_2	r_3	r_1	r_2	r_3
1	3	2	1	4	3	1
2	3	1	1	4	1	1
3	3	3	3	4	2	2
4	3	2	1	4	2	1
5	3	2	2	4	2	2
6	3	3	2	4	4	2
7	3	3	2	4	4	3
8	3	1	0	4	1	0
9	3	4	1	4	4	1
10	3	2	2	4	3	3

Provide the degrees of freedom, T terms, sums of squares, mean squares, and estimated random effects variance components.

3.7* Consider the $p \times (i{:}h)$ design for the model M with $n_p < N_p < \infty$, $n_i < N_i < \infty$, and $n_h < N_h < \infty$.

(a) What are the expected mean squares for p, ph, and $pi{:}h$?

(b) What is $\sigma^2(ph|M)$ in terms of random effects variance components?

(c) Verify (b) given (a).

3.8 Consider, again, Synthetic Data Set No. 1 in Table 2.2, and suppose that persons 1 to 5 are in one class, and persons 6 to 10 are in a second class.

(a) What is the G study design?

(b) What are the degrees of freedom, T terms, sums of squares, mean squares, and G study estimated random effects variance components?

(c) What are the G study estimated variance components under the assumption that items are fixed—i.e., $n_i = N_i = 12$?

(d) What are the G study estimated variance components under the assumptions that $n_c = 2 < N_c = 10$, $n_p = 5 < N_p = 30$, and $n_i = 12 < N_i \to \infty$?

4

Multifacet Universes of Generalization and D Study Designs

In this chapter it is assumed that all facets in the universe of admissible observations can be considered infinite in size and, consequently, the G study variance components have been estimated for a *random* model. Given these assumptions, this chapter provides procedures, equations, and rules for estimating universe score variance, error variances, generalizability coefficients, and other quantities for random and mixed model D studies with any number of facets. The random model equations and rules are quite general. The mixed model procedures are somewhat simplified; more general procedures are treated in Chapter 5.

The substance of this chapter is at the very center of generalizability theory. In particular, considering multifacet universes of generalization forces an investigator to draw careful distinctions among which facets are fixed and which are random for a measurement procedure. Such decisions affect both universe score variance and error variances. Also, the investigator must consider D study sample sizes and D study design structure, both of which affect error variances.

Some of the issues covered in this chapter overlap those in Chapter 2 for single-facet designs. To preclude frequent cross-referencing back to Chapter 2, there is some duplication of equations.

4.1   Random Models and Infinite Universes of Generalization

Linear models for multifacet D study designs can be specified using the procedures discussed in Chapter 3 for G study designs, except that for D study designs an uppercase index is used if the index is associated with a facet in the universe of generalization. This notational convention emphasizes that D study linear models are for the decomposition of observed *mean* scores for the objects of measurement over sets of conditions (not single conditions) in the universe of generalization. Unless otherwise noted, persons are the objects of measurement in this chapter.

Consider, for example, the $p \times I \times H$ design. The linear model for the decomposition of a person's observed mean scores over n'_i and n'_h conditions is

$$X_{pIH} = \mu + \nu_p + \nu_I + \nu_H + \nu_{pI} + \nu_{pH} + \nu_{IH} + \mu_{pIH}, \qquad (4.1)$$

and the linear model associated with the $p \times (I{:}H)$ design is

$$X_{pIH} = \mu + \nu_p + \nu_H + \nu_{I:H} + \nu_{pH} + \nu_{pI:H}. \qquad (4.2)$$

As in Chapter 2, we frequently use $\overline{X}_p$ as an abbreviation for the mean for a person over all sampled conditions of all facets, such as X_{pIH} in Equations 4.1 and 4.2.

Each of the effects in linear models such as those in Equations 4.1 and 4.2 can be expressed also in terms of a linear combination of population and/or universe (of generalization) mean scores that define the effect. To do so, one simply uses the algorithm in Section 3.2.3, replacing any lowercase index with an uppercase index if the index is associated with a facet in the universe of generalization. So, for example, in terms of mean scores, the ν_{pH} effect in both Equations 4.1 and 4.2 is

$$\nu_{pH} = \mu_{pH} - \mu_p - \mu_H + \mu.$$

In short, linear models for persons' observed mean scores can be expressed explicitly in terms of linear combinations of mean scores or, more compactly, in terms of score effects. In this sense, linear models for D study designs are completely analogous to those discussed in Chapter 3 for G study designs.

Since we are assuming in this section that all effects are random (except μ), the linear models discussed here are *random effects* linear models with definitions and assumptions like those discussed in Sections 3.2 and 3.4. The only difference is that these definitions and assumptions now relate to means over *sets* of conditions in the universe of generalization rather than single conditions in the universe of admissible observations. In particular, the manner in which mean scores are defined implies that the expected value of any score effect is zero. For example, for both the $p \times I \times H$ and $p \times (I\!:\!H)$ designs,

$$\mu_H \equiv \mathop{E}_{p}\mathop{E}_{I} X_{pIH} \qquad \text{and} \qquad \mu \equiv \mathop{E}_{p}\mathop{E}_{I}\mathop{E}_{H} X_{pIH}.$$

It follows that

$$\mathop{E}_{H}\nu_H \equiv \mathop{E}_{H}(\mu_H - \mu) = \mathop{E}_{H}\mu_H - \mu = 0.$$

Also, all effects are assumed to be uncorrelated. This is actually a stronger statement than necessary in that many pairs of effects are necessarily un-correlated by the manner in which they are defined in generalizability (see Brennan, 1994; Exercise 4.2).

A particular D study design is associated with a measurement procedure and a universe of generalization. Suppose, for example, that $n'_i = 5$, $n'_h = 2$, and the D study design is $p \times I \times H$. In this case, any single instance of the measurement procedure would result in persons' observed mean scores associated with a random sample of $n'_i = 5$ conditions from the i facet and an independent random sample of $n'_h = 2$ conditions from the h facet, and each person's observed mean score would be based on the average score over the same 10 pairs of sampled conditions. Another instance of the same measurement procedure would simply involve a *different* random sample of size $n'_i = 5$ and $n'_h = 2$ conditions, using the same design. The set of all such randomly parallel instances of the measurement procedure is analogous to the set of classically parallel forms in classical theory. The set of all randomly parallel forms exhausts all conditions in the universe of generalization.

A particular D study design reflects the manner in which a measure-ment procedure is structured, but not necessarily the manner in which the universe of generalization is structured. Especially for random models, it is quite common for a D study design to involve nesting of one or more facets, although all facets are crossed in the universe of generalization. For example, when items are nested within raters in a D study random effects design, it is often the case that an investigator would allow any rater to evaluate any item, at least in principle. If so, the item and rater facets are crossed in the universe of generalization $(I \times H)$, and presumably the investigator has chosen a nested design for practical reasons, or in order to make the design efficient, in some sense.

4.1.1 Universe Score and Its Variance

Let $\mathcal{R}$ be the set of random facets in the universe of generalization. For random models, $\mathcal{R}$ exhausts the facets in the universe of generalization, and the universe score for a person p is defined as

$$\mu_p \equiv \mathop{E}_{\mathcal{R}} X_{p\mathcal{R}}. \tag{4.3}$$

For example, when the facets are I and H, the universe score for person p is

$$\mu_p \equiv \mathop{E}_{I} \mathop{E}_{H} X_{pIH}.$$

That is, a person's universe (of generalization) score is his or her expected score over I and H in the universe of generalization. Alternatively, a person's universe score is the expected value of his or her observable mean score over all randomly parallel instances of the measurement procedure, each of which involves a different random sample of sets of conditions I and H.

By convention, the universe of admissible observations is viewed as a universe of *single* conditions of facets (e.g., i and h), while the universe of generalization is usually viewed as a universe consisting of *sets* of conditions of facets (e.g., I and H). When the two universes are the same, *except* for this distinction, it is clear that

$$\mu_p \equiv \mathop{E}_{I} \mathop{E}_{H} X_{pIH} = \mathop{E}_{i} \mathop{E}_{h} X_{pih}.$$

However, it is misleading to think about universe score as an expected value over conditions in the universe of admissible observations, because universe score is defined with respect to the universe of generalization. Universe score is interpretable as an expected value over instances of a measurement procedure, and such instances are not defined until the investigator considers the universe of generalization and D study issues.

The variance of the universe scores defined in Equation 4.3 is universe score variance:

$$\sigma^2(p) \equiv \mathop{E}_{p} \left(\mu_p - \mathop{E}_{p} \mu_p \right)^2 = \mathop{E}_{p} \nu_p^2. \tag{4.4}$$

For random models, then, a trivial rule for identifying universe score variance when persons are the objects of measurement is:

Rule 4.1.1: $\sigma^2(p)$ is universe score variance.

Universe score variance, as defined in Equation 4.4, also equals the expected value of the covariance of person mean scores over pairs of randomly parallel instances of the measurement procedure, say, $\mathcal{R}$ and $\mathcal{R}'$; that is,

$$\sigma^2(p) = \mathop{E} \sigma \left(X_{p\mathcal{R}}, X_{p\mathcal{R}'} \right), \tag{4.5}$$

where the covariance is taken over persons, and the expectation is taken over pairs of randomly parallel forms. For example, consider the $p \times I \times H$ design. Let I and I' represent different random samples of size n_i' from one facet, and let H and H' represent different random samples of size n_h' from the other facet. Then, $\mathcal{R}$ includes these two facets (i.e., I/I' and H/H'), and

$$\mathop{E} \sigma \left(X_{pIH}, X_{pI'H'} \right) \;=\; \mathop{E} \left[\mathop{E}_{p} \left(X_{pIH} - \mu_{IH} \right) \left(X_{pI'H'} - \mu_{I'H'} \right) \right]$$

TABLE 4.1. Random Effects Variance Components that Enter $\sigma^2(\tau)$, $\sigma^2(\delta)$, and $\sigma^2(\Delta)$ for Random and Mixed Model D Studies for the $p \times I \times H$ Design

	D Studies		
$\sigma^2(\bar{\alpha})$	I, H Random	H Fixed[a]	I Fixed[a]
$\sigma^2(p)$	τ	τ	τ
$\sigma^2(I) = \sigma^2(i)/n_i'$	Δ	Δ	
$\sigma^2(H) = \sigma^2(h)/n_h'$	Δ		Δ
$\sigma^2(pI) = \sigma^2(pi)/n_i'$	Δ, δ	Δ, δ	τ
$\sigma^2(pH) = \sigma^2(ph)/n_h'$	Δ, δ	τ	Δ, δ
$\sigma^2(IH) = \sigma^2(ih)/n_i'n_h'$	Δ	Δ	Δ
$\sigma^2(pIH) = \sigma^2(pih)/n_i'n_h'$	Δ, δ	Δ, δ	Δ, δ

[a] Discussed in Section 4.3.

TABLE 4.2. Random Effects Variance Components that Enter $\sigma^2(\tau)$, $\sigma^2(\delta)$, and $\sigma^2(\Delta)$ for Random and Mixed Model D Studies for the $p \times (I{:}H)$ Design

	D Studies	
$\sigma^2(\bar{\alpha})$	I, H Random	H Fixed[a]
$\sigma^2(p)$	τ	τ
$\sigma^2(H) = \sigma^2(h)/n_h'$	Δ	
$\sigma^2(I{:}H) = \sigma^2(i{:}h)/n_i'n_h'$	Δ	Δ
$\sigma^2(pH) = \sigma^2(ph)/n_h'$	Δ, δ	τ
$\sigma^2(pI{:}H) = \sigma^2(pi{:}h)/n_i'n_h'$	Δ, δ	Δ, δ

[a] Discussed in Section 4.3.

$$= E\left[\underset{p}{E} \left(\nu_p + \nu_{pI} + \nu_{pH} + \nu_{pIH}\right) \times \right.$$
$$\left. \left(\nu_p + \nu_{pI'} + \nu_{pH'} + \nu_{pI'H'}\right)\right].$$

After forming all the cross-products and taking expectations of them, all that remains is $E_p\nu_p^2 = \sigma^2(p)$, the universe score variance.

It is especially important to note that universe scores and universe score variance are defined for the universe of generalization and do not depend on D study sample sizes or the structure of the D study design [e.g., $p \times I \times H$ versus $p \times (I{:}H)$]. By contrast, as discussed soon, error variances do depend on both D study sample sizes and design structure.

4.1.2 D Study Variance Components

In Chapter 3 the notation $\sigma^2(\alpha)$ was introduced to designate variance components associated with random effects in linear models for G study designs. For D study designs, the notation $\sigma^2(\bar{\alpha})$ is used to emphasize that D study variance components are associated with *means* for sets of sampled conditions.[1] For random models, obtaining the $\sigma^2(\bar{\alpha})$ from the $\sigma^2(\alpha)$ is very simple:

$$\sigma^2(\bar{\alpha}) = \frac{\sigma^2(\alpha)}{d(\bar{\alpha})}, \tag{4.6}$$

where

$$d(\bar{\alpha}) = \begin{cases} 1 \text{ if } \bar{\alpha} = p, \text{ and, otherwise, the product} \\ \text{of the D study sample sizes } (n') \text{ for all} \\ \text{indices in } \bar{\alpha} \text{ except } p. \end{cases} \tag{4.7}$$

Tables 4.1 and 4.2 provide equations for the $\sigma^2(\bar{\alpha})$ in terms of the $\sigma^2(\alpha)$ for the $p \times I \times H$ and $p \times (I{:}H)$ designs, respectively. Consider, for example, $\sigma^2(pH)$ for either design. This variance component is the expected value of the square of the ν_{pH} effects in the linear model for each design, where ν_{pH} is the average of the ν_{ph} effects over n'_h conditions. Assuming score effects are uncorrelated, it follows that

$$\sigma^2(pH) \equiv E\nu^2_{pH} = E\left[\frac{\sum_h \nu_{ph}}{n'_h}\right]^2 = \frac{\sigma^2(ph)}{n'_h}. \tag{4.8}$$

This sequence of equalities essentially states the well-known result that the variance of a mean for a set of uncorrelated observations is the variance of the individual elements divided by the sample size (see Exercise 4.1). No normality assumptions are required to derive this result.

4.1.3 Error Variances

The two most frequently discussed types of error in generalizability theory are absolute error and relative error. A third type is the error associated with using the overall observed mean to estimate the mean in the population and universe. All three of these types of errors were introduced in Chapter 2 for single-facet universes and designs. They are extended here to multifacet universes and designs for random models.

Absolute Error Variance

Absolute error for a person p is defined as

$$\Delta_{p\mathcal{R}} \equiv X_{p\mathcal{R}} - \mu_p, \tag{4.9}$$

---

[1] As discussed in Chapter 2, in generalizability theory, the convention is to use the mean-score metric, although the total-score metric could be used. This issue is discussed more extensively in Section 7.2.5.

which is often abbreviated

$$\Delta_p \equiv \overline{X}_p - \mu_p. \tag{4.10}$$

It is the error involved in using an examinee's observed mean score as an estimate of his or her universe score. Consider, for example the $p \times I \times H$ design in Equation 4.1. Since $\mu_p = \mu + \nu_p$, it follows from the definition in Equation 4.9 that

$$\Delta_p = \nu_I + \nu_H + \nu_{pI} + \nu_{pH} + \nu_{IH} + \nu_{pIH}. \tag{4.11}$$

The expected value of Δ_p over the infinite facets in $\mathcal{R}$ is zero; that is,

$$\underset{\mathcal{R}}{E} \, \Delta_{p\mathcal{R}} = 0. \tag{4.12}$$

Clearly, for example, the expected value, over I and H, of Δ_p in Equation 4.11 is zero.

Absolute error variance is defined as

$$\sigma^2(\Delta) \equiv \underset{p}{E} \left[\underset{\mathcal{R}}{E} \left(\Delta_{p\mathcal{R}} - \underset{\mathcal{R}}{E} \Delta_{p\mathcal{R}} \right)^2 \right] = \underset{p}{E} \left[\underset{\mathcal{R}}{E} \Delta_{p\mathcal{R}}^2 \right]. \tag{4.13}$$

The term in square brackets is the error variance for a person p, called conditional error variance, which is discussed in detail in Section 5.4. Here, we simply note that the overall Δ-type error variance in Equation 4.13 is interpretable as the expected value (over persons) of conditional error variance.[2] For example, the overall error variance associated with Δ_p in Equation 4.11 is

$$
\begin{aligned}
\sigma^2(\Delta) &= \sigma^2(I) + \sigma^2(H) + \sigma^2(pI) + \sigma^2(pH) + \sigma^2(IH) + \sigma^2(pIH) \\
&= \frac{\sigma^2(i)}{n_i'} + \frac{\sigma^2(h)}{n_h'} + \frac{\sigma^2(pi)}{n_i'} + \frac{\sigma^2(ph)}{n_h'} + \frac{\sigma^2(ih)}{n_i' n_h'} + \frac{\sigma^2(pih)}{n_i' n_h'}.
\end{aligned}
$$

In terms of the notational system used here, it is easy to determine which of the $\sigma^2(\bar{\alpha})$ enter $\sigma^2(\Delta)$:

Rule 4.1.2: $\sigma^2(\Delta)$ is the sum of all $\sigma^2(\bar{\alpha})$ except $\sigma^2(p)$.

The second columns of Tables 4.1 and 4.2 report which variance components enter $\sigma^2(\Delta)$ for the random effects $p \times I \times H$ and $p \times (I : H)$ designs, respectively.

---

[2]Note that $\boldsymbol{E}_{\mathcal{R}}$ is *within* the square brackets in the definition of $\sigma^2(\Delta)$ in Equation 4.13 in order to express that definition as the expected value (over the population) of conditional error variance. Overall Δ-type error variance would be unchanged, however, if the positions of the $\boldsymbol{E}_p$ and $\boldsymbol{E}_{\mathcal{R}}$ operators were interchanged.

Relative Error Variance

Relative error for a person p is defined as

$$\delta_{pR} \equiv (X_{pR} - \underset{p}{E} X_{pR}) - (\mu_p - \mu), \qquad (4.14)$$

which is often abbreviated

$$\delta_p \equiv (\overline{X}_p - \underset{p}{E} \overline{X}_p) - (\mu_p - \mu). \qquad (4.15)$$

As such, δ_p is the error associated with using an examinee's observable deviation score as an estimate of his or her universe deviation score. For the $p \times I \times H$ design,

$$\underset{p}{E} \overline{X}_p = \mu + \nu_I + \nu_H + \nu_{IH} = \mu_{IH},$$

and, therefore,

$$\begin{aligned}
\delta_p &= (\overline{X}_p - \mu_{IH}) - (\mu_p - \mu) \\
&= \nu_{pI} + \nu_{pH} + \nu_{pIH}. \qquad (4.16)
\end{aligned}$$

The expected value of δ_p over the population and/or over the infinite facets in R is zero; that is,

$$\underset{p}{E} \delta_{pR} = \underset{R}{E} \delta_{pR} = 0. \qquad (4.17)$$

Clearly, for example, for δ_p in Equation 4.16, the expected value over p is zero, and the expected value over I and H is also zero.

Relative error variance is defined as

$$\sigma^2(\delta) \equiv \underset{R}{E} \left[\underset{p}{E} \left(\delta_{pR} - \underset{p}{E} \delta_{pR} \right)^2 \right] = \underset{R}{E} \left[\underset{p}{E} \delta_{pR}^2 \right]. \qquad (4.18)$$

Cronbach et al. (1972) designated relative error variance as $E\sigma^2(\delta)$, where the expectation E is taken over R. Although their notation is not used here, the definition in Equation 4.18 is consistent with the Cronbach et al. notation in that E_R is outside the square brackets. We return to this matter in the next section when we discuss expected observed score variance.

For the $p \times I \times H$ design, δ_p is given by Equation 4.16, and its variance is

$$\begin{aligned}
\sigma^2(\delta) &= \sigma^2(pI) + \sigma^2(pH) + \sigma^2(pIH) \\
&= \frac{\sigma^2(pi)}{n_i'} + \frac{\sigma^2(ph)}{n_h'} + \frac{\sigma^2(pih)}{n_i' n_h'}.
\end{aligned}$$

The following rule can be used to obtain $\sigma^2(\delta)$ for any random effects design.

Rule 4.1.3: $\sigma^2(\delta)$ is the sum of all $\sigma^2(\bar{\alpha})$ such that $\bar{\alpha}$ includes p and at least one other index.

In other words, $\sigma^2(\delta)$ is the sum of all variance components that involve interactions of p with facets in the universe of generalization. The second columns of Tables 4.1 and 4.2 report which variance components enter $\sigma^2(\delta)$ for the random effects $p \times I \times H$ and $p \times (I:H)$ designs, respectively.

Comparing Absolute and Relative Error Variance

Absolute error variance is always at least as large as relative error variance. For example, for the $p \times I \times H$ design

$$\sigma^2(\Delta) = \sigma^2(\delta) + \sigma^2(I) + \sigma^2(H) + \sigma^2(IH),$$

and for the $p \times (I:H)$ design

$$\sigma^2(\Delta) = \sigma^2(\delta) + \sigma^2(H) + \sigma^2(I:H).$$

Assuming the universe of generalization has $I \times H$, the difference between absolute and relative error variance for both designs is

$$\sigma^2(\mu_{IH}) = \mathop{E}_{I} \mathop{E}_{H} (\mu_{IH} - \mu)^2,$$

because

$$\mu_{IH} - \mu = \nu_I + \nu_H + \nu_{IH} = \nu_H + \nu_{I:H}.$$

Absolute error variance takes into account the potential differences among overall means (μ_{IH}) for randomly parallel instances of a measurement procedure. If the μ_{IH} are all equal, then $\sigma^2(\delta)$ and $\sigma^2(\Delta)$ are indistinguishable. The same kind of statement was made in Chapter 2 when classically parallel forms were discussed for single-facet universes.

The structure of a D study design influences the magnitude of error variances. Suppose the universe of generalization is $I \times H$, and consider the $p \times I \times H$ and $p \times (I:H)$ design structures with identical sample sizes. For the $p \times I \times H$ design,

$$\sigma^2(\delta) = \frac{\sigma^2(ph)}{n'_h} + \frac{\sigma^2(pi)}{n'_i} + \frac{\sigma^2(pih)}{n'_i n'_h}.$$

By contrast, for the $p \times (I:H)$ design,

$$\begin{aligned}
\sigma^2(\delta) &= \frac{\sigma^2(ph)}{n'_h} + \frac{\sigma^2(pi:h)}{n'_i n'_h} \\
&= \frac{\sigma^2(ph)}{n'_h} + \frac{\sigma^2(pi)}{n'_i n'_h} + \frac{\sigma^2(pih)}{n'_i n'_h},
\end{aligned}$$

since $\sigma^2(pi:h) = \sigma^2(pi) + \sigma^2(pih)$, as discussed in Section 3.2. Note that $\sigma^2(pi)$ is divided by n'_i when the design is $p \times I \times H$, but it is divided by

$n'_i n'_h$ when the design is $p \times (I{:}H)$. Therefore, all other things being equal, $\sigma^2(\delta)$ is smaller for the $p \times (I{:}H)$ design than for the $p \times I \times H$ design. A similar statement applies to $\sigma^2(\Delta)$. In this sense, for a fixed number of observations per person, the $p \times (I{:}H)$ design is more efficient than the $p \times I \times H$ design. This result is associated with the fact that, for the $p \times (I{:}H)$ design, the total number of sampled items is $n'_i n'_h$, whereas only n'_i items are involved in the $p \times I \times H$ design—sampling more conditions of facets tends to reduce error.

Error Variance for Estimating μ Using $\overline{X}$

Absolute error and relative error are, by far, the most frequently discussed and used, but other types of error may be of interest. For example, the error variance associated with using the observed grand mean $\overline{X}$ as an estimate of the mean in the population and universe μ is

$$\sigma^2(\overline{X}) \equiv \boldsymbol{E}(\overline{X} - \mu)^2 = \underset{\mathcal{R}}{\boldsymbol{E}}\,\underset{P}{\boldsymbol{E}}(X_{P\mathcal{R}} - \mu)^2, \qquad (4.19)$$

where $X_{P\mathcal{R}} = \overline{X}$ is the D study mean over the observed levels of *all* facets, including persons. That is why the person subscript in Equation 4.19 is an uppercase P. For random models, $\sigma^2(\overline{X})$ can be obtained by first dividing each $\sigma^2(\bar{\alpha})$ by n'_p if $\bar{\alpha}$ includes p, and then summing these with all other $\sigma^2(\bar{\alpha})$. This verbal recipe is equivalent to the formula:

$$\sigma^2(\overline{X}) = \frac{\sigma^2(p) + \sigma^2(\delta)}{n'_p} + \left[\sigma^2(\Delta) - \sigma^2(\delta)\right]. \qquad (4.20)$$

Note, in particular, that $\sigma^2(\overline{X})$ includes the universe score variance $\sigma^2(p)$. Alternatively, for random models, $\sigma^2(\overline{X})$ can be obtained by first dividing each of the $\sigma^2(\alpha)$ by the product of the D study sample sizes for indices in α, and then summing (see Exercise 4.3).

4.1.4   Coefficients and Signal–Noise Ratios

Traditionally, two coefficients predominate in generalizability theory, both of which have a range of zero to one. A generalizability coefficient is very much like a reliability coefficient in classical theory. It employs relative error variance. A phi coefficient employs absolute error variance.

The basic equation for a generalizability coefficient for multifacet random designs is the same as that for single-facet random designs:

$$\boldsymbol{E}\rho^2 = \frac{\sigma^2(p)}{\sigma^2(p) + \sigma^2(\delta)}. \qquad (4.21)$$

Of course, the variance components that enter $\sigma^2(\delta)$ for multifacet designs are different from those that enter $\sigma^2(\delta)$ for single-facet designs. This fact has consequences, many of which are illustrated in examples later in this chapter. The symbolism $E\rho^2$ used to represent a generalizability coefficient highlights that it is interpretable as an approximation to the squared correlation between universe scores and observed scores.

The denominator of $E\rho^2$ in Equation 4.21 is sometimes called expected observed score variance and defined as

$$ES^2(p) \equiv \underset{\mathcal{R}}{E}\left[\underset{p}{E}\left(X_{p\mathcal{R}} - \underset{p}{E}\, X_{p\mathcal{R}} \right)^2 \right] = \sigma^2(p) + \sigma^2(\delta), \qquad (4.22)$$

where $\overline{X}_p = X_{p\mathcal{R}}$, and the E in $ES^2(p)$ means that the expectation is taken over the set of random facets, $\mathcal{R}$. For both this definition and the definition of $\sigma^2(\delta)$ in Equation 4.18, the expectation over $\mathcal{R}$ is taken last. $ES^2(p)$ is interpretable as the expected value over randomly parallel instances of the measurement procedure (i.e., $\mathcal{R}$) of the *observed* variance for the population. Given Equation 4.22 and *Rule* 4.1.3, it is evident that:

Rule 4.1.4: $ES^2(p)$ is the sum of all $\sigma^2(\bar{\alpha})$ that include p.

The basic equation for an index of dependability, or phi coefficient, is the same for multifacet designs as it is for single-facet designs:

$$\Phi = \frac{\sigma^2(p)}{\sigma^2(p) + \sigma^2(\Delta)}. \qquad (4.23)$$

Of course, the variance components that enter $\sigma^2(\Delta)$ for multifacet designs are different from those that enter $\sigma^2(\Delta)$ for single-facet designs. A phi coefficient differs from a generalizability coefficient only in that Φ involves absolute error variance rather than the relative error variance in $E\rho^2$. Among other things, this difference means that the denominator of Φ is *not* the observed variance for persons; it is the mean squared deviation $E(\overline{X}_p - \mu)^2$ for persons.

The information in the coefficients $E\rho^2$ and Φ can be expressed as the signal–noise ratios

$$S/N(\delta) = \frac{\sigma^2(p)}{\sigma^2(\delta)} \qquad (4.24)$$

and

$$S/N(\Delta) = \frac{\sigma^2(p)}{\sigma^2(\Delta)}, \qquad (4.25)$$

respectively, which have limits ranging from zero to infinity. Relationships between $E\rho^2$ and $S/N(\delta)$ are:

$$E\rho^2 = \frac{S/N(\delta)}{1 + S/N(\delta)} \quad \text{and} \quad S/N(\delta) = \frac{E\rho^2}{1 - E\rho^2}.$$

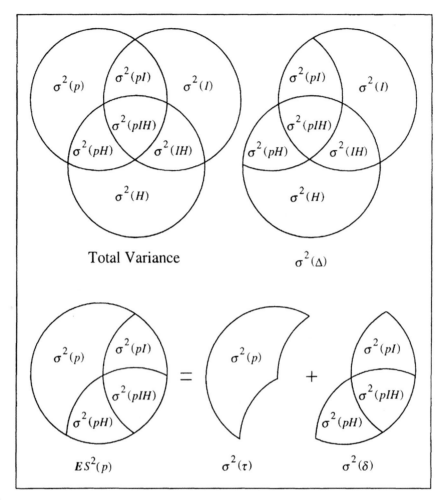

FIGURE 4.1. Venn diagram representation of variance components for the $p \times I \times H$ design with both I and H random.

Corresponding relationships between Φ and $S/N(\Delta)$ are:

$$\Phi = \frac{S/N(\Delta)}{1 + S/N(\Delta)} \quad \text{and} \quad S/N(\Delta) = \frac{\Phi}{1 - \Phi}.$$

Kane (1996) discusses coefficients, signal–noise ratios, and a number of other measures of precision that might be applied in generalizability analyses. In particular, the error–tolerance ratios discussed in Section 2.5.1 for single-facet designs also apply to multifacet designs.

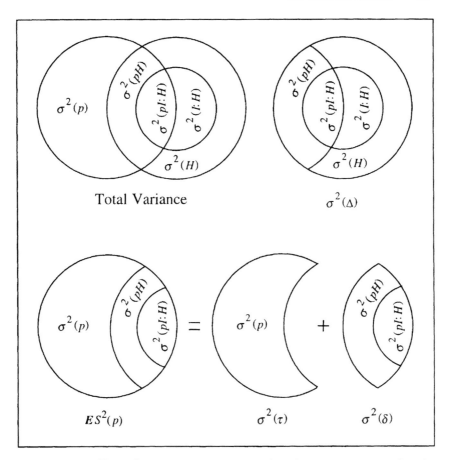

FIGURE 4.2. Venn diagram representation of variance components for the $p \times (I:H)$ design with both I and H random.

4.1.5 Venn Diagrams

Figure 4.1 provides Venn diagram representations of variance components for the $p \times I \times H$ design, and Figure 4.2 provides the same type of information for the $p \times (I:H)$ design. For each figure, the full Venn diagram in the upper left-hand corner represents the so-called "total variance":[3]

$$E(\overline{X}_p - \mu)^2 = \underset{\mathcal{R}}{E}\, \underset{p}{E}\, (X_{p\mathcal{R}} - \mu)^2$$

---

[3]Much of the generalizability theory literature uses the phrase "total variance," but it is not total *variance* for persons' mean scores; rather, it is a mean-squared deviation for persons.

and its decomposition. Note that total variance is not the same as $\sigma^2(\overline{X})$ in Equation 4.19 or $ES^2(p)$ in Equation 4.22. The size of any particular area in these Venn diagrams does not reflect anything about its magnitude.

The upper right-hand corner of each figure represents $\sigma^2(\Delta)$ as the full Venn diagram excluding the universe score variance $\sigma^2(p)$. The bottom half of the two figures illustrates that the entire p circle can be associated with expected observed score variance, which can be partitioned into universe score variance [designated $\sigma^2(\tau)$ at the bottom of the figures] and relative error variance. Comparing the Venn diagram representations of $\sigma^2(\delta)$ and $\sigma^2(\Delta)$ gives a visual perspective on the additional variance components that enter $\sigma^2(\Delta)$, namely, all variance components outside the p circle. These additional variance components represent contributions to total variance that are constant for all persons.

In comparing the Venn diagrams for the $p \times I \times H$ and $p \times (I{:}H)$ designs, note that

$$\sigma^2(I{:}H) = \sigma^2(I) + \sigma^2(IH)$$

and

$$\sigma^2(pI{:}H) = \sigma^2(pI) + \sigma^2(pIH).$$

Therefore, the areas associated with $\sigma^2(I{:}H)$ and $\sigma^2(pI{:}H)$ in Figure 4.2 represent the combination of two areas in Figure 4.1.

4.1.6 Rules and Equations for Any Object of Measurement

To this point in this chapter it has been assumed that persons are the objects of measurement. In principle, however, any facet can play that role. Cardinet et al. (1976) refer to this fact as the "symmetry" of generalizability theory. A common example occurs when students are nested within classrooms, and classes are the objects of measurement, that is, the units about which decisions are made. No matter what objects of measurement are involved, the equations and rules discussed in this chapter still apply provided the index p is replaced by the index representing the objects of measurement.

Letting τ be a generic index designating the objects of measurement, Table 4.3 summarizes the notation, rules, and most frequently used equations for random model D studies. The equation numbers in Table 4.3 correspond to those in the text. The only differences in the equations are that those in Table 4.3 are expressed with respect to τ, not p, and $d(\bar\alpha|\tau)$ is used in Table 4.3, rather than $d(\bar\alpha)$.

These rules and equations assume that the objects of measurement are *not* nested within some other facet. Generalizability theory can treat such nested, or stratified, objects of measurement, but to do so requires special considerations (see Section 5.2).

In practical applications, of course, the $\sigma^2(\alpha)$ in Table 4.3 are replaced with estimates, which are usually those discussed in Chapter 3. In doing

TABLE 4.3. Rules and Equations for Random Model D Studies

Let τ = objects of measurement.
Let $\mathcal{R}$ = set of all random facets in universe of generalization.

D study variance components: $\sigma^2(\bar{\alpha}) = \sigma^2(\alpha)/d(\bar{\alpha}|\tau)$ (4.6')

$$\text{where } d(\bar{\alpha}|\tau) = \begin{cases} 1 \text{ if } \bar{\alpha} = \tau, \text{ and, otherwise, the product} \\ \text{of the D study sample sizes } (n') \text{ for all} \\ \text{indices in } \bar{\alpha} \text{ except } \tau. \end{cases}$$ (4.7')

Rule 4.1.1: $\sigma^2(\tau)$ is universe score variance.
Rule 4.1.2: $\sigma^2(\Delta)$ is the sum of all $\sigma^2(\bar{\alpha})$ except $\sigma^2(\tau)$.
Rule 4.1.3: $\sigma^2(\delta)$ is the sum of all $\sigma^2(\bar{\alpha})$ such that $\bar{\alpha}$
        includes τ and at least one other index.

$$\sigma^2(\overline{X}) = \frac{\sigma^2(\tau) + \sigma^2(\delta)}{n'_\tau} + [\sigma^2(\Delta) - \sigma^2(\delta)].$$ (4.20')

$$E\rho^2 = \frac{\sigma^2(\tau)}{\sigma^2(\tau) + \sigma^2(\delta)}.$$ (4.21')

$$\Phi = \frac{\sigma^2(\tau)}{\sigma^2(\tau) + \sigma^2(\Delta)}.$$ (4.23')

---

so, it is effectively assumed that $N_\tau = N'_\tau \to \infty$ for the objects of measurement. Even when this assumption is clearly false, estimates of $\sigma^2(\tau)$, $\sigma^2(\delta)$, and $\sigma^2(\Delta)$ are not likely to be much affected when the sample size n'_τ is even moderately large.[4]

4.1.7 D Study Design Structures Different from the G Study

Equation 4.7 for determining D study variance components assumes that the structures of the G and D studies are the same. Suppose, however, that the G study uses the $p \times i \times h$ design, but the D study uses the $p \times (I:H)$ design. In this case, before using Equation 4.7 to obtain the $d(\bar{\alpha})$, the G study variance components for the $p \times (i:h)$ design need to be represented in terms of those for the $p \times i \times h$ design. Doing so gives

$$\sigma^2(i{:}h) = \sigma^2(i) + \sigma^2(ih)$$

and

$$\sigma^2(pi{:}h) = \sigma^2(pi) + \sigma^2(pih).$$

---

[4]This is a very much different issue from what happens when a universe-of-generalization facet is fixed in the D study, as discussed later in Section 4.3.

These equalities are basically examples of the rule discussed in Section 3.2 for determining which effects in a *fully crossed* design are *confounded* in an effect from a design that involves nesting:

> *Confounded-Effects Rule*: The effects that are confounded are all those that involve only indices in the nested effect and that include the primary index (indices) in the nested effect.

When *both* the G and D study designs involve nesting, these steps can be used:

1. use the confounded-effects rule to express each effect in the G study design in terms of effects from the fully crossed design;

2. use the confounded-effects rule to express each effect in the D study design (single-conditions-version) in terms of effects from the fully crossed design; and

3. use 1 and 2 to determine, if possible,[5] which effects from the G study design are confounded in effects for the D study design (single-conditions-version).

For example, suppose the G study uses the $p \times (i{:}h)$ design, but the D study employs the $I{:}H{:}p$ design, and we want to know what effects from the $p \times (i{:}h)$ design are confounded in the $i{:}h{:}p$ effect. For the $i{:}h{:}p$ design,

$$i{:}h{:}p \Rightarrow (i, ih, pi, pih);$$

and for the $p \times (i{:}h)$ design,

$$i{:}h \Rightarrow (i, ih) \qquad \text{and} \qquad pi{:}h \Rightarrow (pi, pih).$$

It follows that $\sigma^2(i{:}h{:}p) = \sigma^2(i{:}h) + \sigma^2(pi{:}h)$.

4.2   Random Model Examples

In this section, both synthetic and real-data examples are used to illustrate random model rules and procedures for D studies. These examples highlight some important differences between classical results and those in generalizability theory, and they demonstrate a few different ways of depicting results numerically and graphically. The exercises at the end of this chapter provide additional instructive examples.

4.2.1 $p \times I \times O$ Design with Synthetic Data Set No. 3

Table 4.4 reports results for D studies employing the $p \times I \times O$ design, based on the G study estimated variance components for Synthetic Data

---

[5]It is not always possible to express effects for every nested design in terms of effects for another nested design (see Exercise 4.4).

TABLE 4.4. Random Effects D Study $p \times I \times O$ Designs For Synthetic Data 3

$\hat{\sigma}^2(\alpha)$		D Studies					
	n'_o	1	1	1	2	2	2
	n'_i	4	8	16	4	8	16
$\hat{\sigma}^2(p) = .5528$	$\hat{\sigma}^2(p)$	.553	.553	.553	.553	.553	.553
$\hat{\sigma}^2(i) = .4417$	$\hat{\sigma}^2(I)$	.110	.055	.028	.110	.055	.028
$\hat{\sigma}^2(o) = .0074$	$\hat{\sigma}^2(O)$	.007	.007	.007	.004	.004	.004
$\hat{\sigma}^2(pi) = .5750$	$\hat{\sigma}^2(pI)$	.144	.072	.036	.144	.072	.036
$\hat{\sigma}^2(po) = .1009$	$\hat{\sigma}^2(pO)$	.101	.101	.101	.050	.050	.050
$\hat{\sigma}^2(io) = .1565$	$\hat{\sigma}^2(IO)$	.039	.020	.010	.020	.010	.005
$\hat{\sigma}^2(pio) = .9352$	$\hat{\sigma}^2(pIO)$	.234	.117	.058	.117	.058	.029
	$\hat{\sigma}^2(\tau)$	.55	.55	.55	.55	.55	.55
	$\hat{\sigma}^2(\delta)$	.48	.29	.20	.31	.18	.12
	$\hat{\sigma}^2(\Delta)$	.64	.37	.24	.45	.25	.15
	$E\hat{\rho}^2$	.54	.66	.74	.64	.75	.83
	$\widehat{\Phi}$	.47	.60	.70	.55	.69	.78

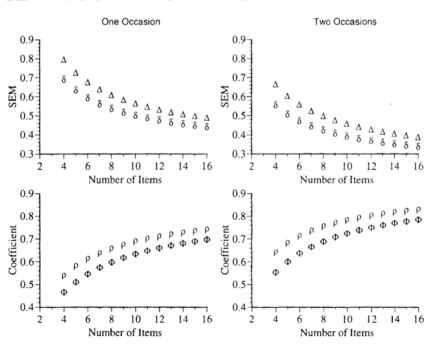

Set No. 3 in Table 3.3 on page 74. At the top of Table 4.4, the first three D studies are for one occasion, and the next three are for two occasions. For both sets of D studies, $n'_i = 4, 8$, and 16. These results can be obtained using the equations and rules summarized in Table 4.3 on page 109.

The figures at the bottom of Table 4.4 provide graphs of $\hat{\sigma}(\delta)$, $\hat{\sigma}(\Delta)$, $\boldsymbol{E}\hat{\rho}^2$, and $\hat{\Phi}$ for one and two occasions with $4 \leq n'_i \leq 16$. Note that the first two figures report estimated standard errors of measurement (i.e., the square roots of the estimated error variances), rather than estimated error variances. Also, the symbol ρ means $\boldsymbol{E}\hat{\rho}^2$, *not* its square root.[6]

These numerical and graphical results illustrate that

- relative error variance is smaller than absolute error variance;

- a generalizability coefficient is larger than an index of dependability; and

- as sample sizes increase, universe score variance is unchanged, error variances decrease, and coefficients increase.

An investigator might reasonably ask questions about which variance components contribute the most to error. Clearly, the answer depends upon the sample sizes, but $\hat{\sigma}^2(pIO)$ is generally large. Also, with only one or two occasions, $\hat{\sigma}^2(pO)$ tends to be relatively large suggesting that the rank orderings of persons' mean scores over one or two randomly selected occasions is generally different from the rank orderings of the same persons' mean scores for another one or two randomly selected occasions.

Apparent Similarities with Some Traditional Reliability Coefficients

The values of $\boldsymbol{E}\hat{\rho}^2$ for two occasions may appear conceptually similar to test–retest reliability coefficients—sometimes called coefficients of stability—but this apparent similarity is misleading for two reasons. For the $p \times I \times O$ random effects design, generalization is over both sets of items and sets of occasions, whereas a test–retest correlation keeps items fixed in the sense discussed later in Section 4.3. Also, the generalizability coefficients for two occasions are for decisions involving mean scores over two administrations of the same test. By contrast, a test–retest correlation provides an estimate of reliability for scores based on a *single* administration of a test.

The values of $\boldsymbol{E}\hat{\rho}^2$ for one occasion may appear conceptually similar to internal consistency reliability coefficients, but again there is an important difference. Even with $n'_o = 1$ occasion, results for the $p \times I \times O$ random effects design are for generalization *over* occasions. By contrast, for internal consistency coefficients, occasion is generally a fixed "hidden" facet, as

---

[6] Using ρ^2 or $\boldsymbol{E}\rho^2$ renders a graph too confusing.

discussed more fully later in Sections 4.4.3 and 5.1.4. If generalization is intended to a broader set of occasions than the single occasion on which data are collected, then the random model coefficients with $n'_o = 1$ are more appropriate than traditional measures of internal consistency. Note, however, that *estimating* the random model coefficients requires a G study with $n_o \geq 2$, even when the D study uses $n'_o = 1$.

Inapplicability of Spearman–Brown Formula

One of the principal results from classical test theory is the Spearman–Brown formula for predicting reliability for forms of different lengths. Under the assumptions of classical theory, reliability always increases as test length increases. This is not necessarily true in generalizability theory when there are two or more random facets.

Consider, for example, the D study in Table 4.4 for $n'_o = 2$ and $n'_i = 4$, which gives $E\hat{\rho}^2 = .64$. For a test twice as long, the Spearman–Brown formula predicts that

$$E\hat{\rho}^2 = \frac{2(.64)}{1 + .64} = .78.$$

By contrast, the D study with $n'_o = 2$ and $n'_i = 8$ gives $E\hat{\rho}^2 = .75$. This discrepancy is not attributable to rounding error. Using the Spearman–Brown formula has the effect of dividing *all* variance components that enter $\hat{\sigma}^2(\delta)$ by the same constant. For this example, then, $\hat{\sigma}^2(pI)$, $\hat{\sigma}^2(pO)$, and $\hat{\sigma}^2(pIO)$ are all divided by two, whereas generalizability theory leaves $\hat{\sigma}^2(pO)$ unchanged for any number of items. It follows that the Spearman–Brown formula predicts a larger value for reliability (and a smaller relative error variance) than that obtained using generalizability theory. Except in trivial cases, the Spearman–Brown formula does not apply when generalization is over more than one facet.

4.2.2 $p \times (I{:}O)$ Design with Synthetic Data Set No. 3

The D study results in Table 4.5 are based on the G study estimated random effects variance components for Synthetic Data Set No. 3. For Table 4.5, however, the design structure is $p \times (I{:}O)$, not $p \times I \times O$. Therefore, before applying the rules and equations for random model D studies, the G study $p \times i \times o$ variance components need to be converted to variance components for the $p \times (i{:}o)$ design in the manner discussed in Section 4.1.7 and illustrated in Table 4.5.

The D study results in Table 4.5 illustrate the theoretical arguments in Section 4.1.7. For example, for the random model $p \times I \times O$ design with $n'_o = 2$ and $n'_i = 4$, Table 4.4 reports that

$$\hat{\sigma}^2(\delta) = .31 \quad \text{and} \quad E\hat{\rho}^2 = .64.$$

TABLE 4.5. Random Effects D Study $p \times (I:O)$ Designs For Synthetic Data 3

$\hat{\sigma}^2(\alpha)$ for $p \times i \times o$	$\hat{\sigma}^2(\alpha)$ for $p \times (i:o)$	D Studies				
		n_o'	1	1	2	2
		n_i'	4	8	4	8
$\hat{\sigma}^2(p) = .5528$	$\hat{\sigma}^2(p) = .5528$	$\hat{\sigma}^2(p)$	.553	.553	.553	.553
$\hat{\sigma}^2(o) = .0074$	$\hat{\sigma}^2(o) = .0074$	$\hat{\sigma}^2(O)$	.007	.007	.004	.004
$\left.\begin{array}{l}\hat{\sigma}^2(i) = .4417 \\ \hat{\sigma}^2(io) = .1565\end{array}\right\}$	$\hat{\sigma}^2(i:o) = .5982$	$\hat{\sigma}^2(I:O)$	.150	.075	.075	.037
$\hat{\sigma}^2(po) = .1009$	$\hat{\sigma}^2(po) = .1009$	$\hat{\sigma}^2(pO)$	.101	.101	.051	.051
$\left.\begin{array}{l}\hat{\sigma}^2(pi) = .5750 \\ \hat{\sigma}^2(pio) = .9352\end{array}\right\}$	$\hat{\sigma}^2(pi:o) = 1.5102$	$\hat{\sigma}^2(pI:O)$	.378	.189	.189	.094
		$\hat{\sigma}^2(\tau)$	.55	.55	.55	.55
		$\hat{\sigma}^2(\delta)$	.48	.29	.24	.15
		$\hat{\sigma}^2(\Delta)$	.64	.37	.32	.19
		$E\hat{\rho}^2$	.54	.66	.70	.79
		$\hat{\Phi}$	.47	.60	.63	.75

By contrast, for the same sample sizes, the random model $p \times (I:O)$ design in Table 4.5 gives

$$\hat{\sigma}^2(\delta) = .24 \quad \text{and} \quad E\hat{\rho}^2 = .70.$$

That is, all other things being equal, the nested design leads to smaller error variances and, hence, larger coefficients.

Coefficients of Equivalence and Stability

The two D studies with $n_o' = 1$ in Table 4.5 for the $p \times (I:O)$ design give identical results to the corresponding D studies in Table 4.4 for the $p \times I \times O$ design. When n_o' is a *single* randomly selected occasion, there is no functional difference between the two designs, and occasion is a "hidden" random facet for both of them. (Hidden facets are discussed more fully later in Section 5.1.4.)

Data collected using a G study $p \times (i:o)$ design with $n_o = 2$ administrations are often used to estimate reliability in the classical sense of "equivalence and stability." Such a reliability coefficient is estimated by the correlation between persons' scores for the two administrations. This correlation involves neither the same items nor the same occasions, and it is an estimate of reliability for a single administration. Therefore, it corresponds conceptually with a generalizability coefficient for random effects $p \times (I:O)$ (or $p \times I \times O$) design with $n_o = 1$. Indeed, the expected value of the numerator of the correlation (i.e., the covariance) is the universe score variance, as discussed in Section 4.1.1. There is a difference in the denominators, however, because a correlation uses the geometric mean of the two variances, whereas a generalizability coefficient uses the expected observed score variance (i.e., an arithmetic mean).

4.2.3 $p \times (R:T)$ Design with Synthetic Data Set No. 4

Table 4.6 reports results for six D studies using the $p \times (R:T)$ design, based on the G study estimated variance components in Table 3.4 for the $p \times (r:t)$ design and Synthetic Data Set No. 4. For this D study design, it is important to note that since raters are nested within tasks, examinee responses to each of the n'_t tasks are evaluated by a *different* set of n'_r raters, and the total number of ratings for each examinee is $n'_t n'_r$.[7]

Table 4.6 reports results for all the possible random effects D studies such that $n'_t \leq 6$ and $n'_t n'_r$ is as close as possible to 12 without exceeding it. At the bottom of the table are "box-bar" plots that graphically summarize statistics for the various D studies in the manner illustrated in Figure 4.3.[8] In these plots, the height of the solid box represents estimated universe score variance, and the magnitudes of $\hat{\sigma}^2(\delta)$ and $\hat{\sigma}^2(\Delta)$ are represented by the lengths of the bars below the box.

An investigator might examine the D study results in Table 4.6 if time and/or cost constraints required using no more than six tasks and no more than 12 ratings per examinee. Clearly, both $\hat{\sigma}^2(\Delta)$ and $\hat{\sigma}^2(\delta)$ change even when the total number of ratings remains constant at 12. Also, using more tasks leads to greater reduction in error variance than using more raters per task. For example, when $n'_t = 2$ and $n'_r = 6$, $\hat{\sigma}^2(\Delta) = .70$; whereas when $n'_t = 6$ and $n'_r = 2$, $\hat{\sigma}^2(\Delta) = .40$. Under the constraint that $n'_t n'_r \leq 12$, the D studies in Table 4.6 indicate that error variance is minimized by using six tasks with two raters per task. However, the investigator might also observe that error variances for five tasks are not much larger than those for six tasks, and the results for five tasks require a total of only 10 ratings per examinee, as opposed to 12. This might lead the investigator to use only five tasks.

Table 4.6 also reports that:

- $E\hat{\rho}^2 = .50$ for two tasks with six raters per task—a total of 12 ratings per examinee; while

---

[7]This does not necessarily mean that the universe of generalization has $R:T$; it is certainly possible that the universe has $R \times T$.

[8]The generic Venn diagram in Figure 4.3 does not represent each of the individual D study variance components. Rather, it depicts the relationships among universe score variance, relative error variance, and absolute error variance, with $\sigma^2(\mathcal{R}) = \sigma^2(\Delta) - \sigma^2(\delta)$ being the set of all D study variance components that include indices for random facets only. For the particular box-bar plot in Figure 4.3, $E\rho^2 = 32/40 = .80$ and $\Phi = 32/52 = .62$.

TABLE 4.6. Random Effects D Study $p \times (R{:}T)$ Designs For Synthetic Data 4

$\hat{\sigma}^2(\alpha)$		D Studies					
	n'_t	1	2	3	4	5	6
	n'_r	12	6	4	3	2	2
	$n'_t n'_r$	12	12	12	12	10	12
$\hat{\sigma}^2(p) = .4731$	$\hat{\sigma}^2(p)$	.473	.473	.473	.473	.473	.473
$\hat{\sigma}^2(t) = .3252$	$\hat{\sigma}^2(T)$	.325	.163	.108	.081	.065	.054
$\hat{\sigma}^2(r{:}t) = .6475$	$\hat{\sigma}^2(R{:}T)$	.054	.054	.054	.054	.065	.054
$\hat{\sigma}^2(pt) = .5596$	$\hat{\sigma}^2(pT)$	.560	.280	.187	.140	.112	.093
$\hat{\sigma}^2(pr{:}t) = 2.3802$	$\hat{\sigma}^2(pR{:}T)$	.198	.198	.198	.198	.238	.198
	$\hat{\sigma}^2(\tau)$	.47	.47	.47	.47	.47	.47
	$\hat{\sigma}^2(\delta)$	.76	.48	.39	.34	.35	.29
	$\hat{\sigma}^2(\Delta)$	1.14	.69	.55	.47	.48	.40
	$E\hat{\rho}^2$	.38	.50	.55	.58	.58	.62
	$\widehat{\Phi}$	.29	.41	.46	.50	.50	.54
	$\widehat{S/N}(\delta)$	.62	.98	1.23	1.40	1.35	1.63
	$\widehat{S/N}(\Delta)$	.42	.68	.87	1.00	.99	1.19

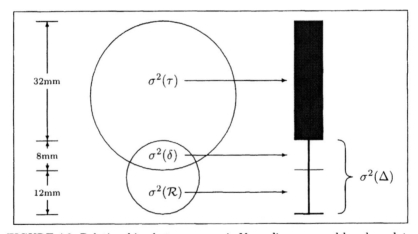

FIGURE 4.3. Relationships between generic Venn diagrams and box-bar plots.

- $E\hat{\rho}^2 = .58$ for five tasks with two raters per task—a total of 10 ratings per examinee.

In other words, for these sample sizes, *decreasing* the total number of ratings per examinee leads to an *increase* in reliability. Such a result cannot occur in any single application of classical theory with its undifferentiated error term, but it can arise in generalizability theory when there is more than one source of random error.

This paradoxical result (from the perspective of classical theory) is a direct consequence of the fact that $\hat{\sigma}^2(pt)$ is relatively large, and its influence on $\hat{\sigma}^2(\delta)$ decreases as the number of tasks increases, no matter what changes occur for n_r'. In this particular example, the slight *increase* in $\hat{\sigma}^2(pR{:}T)$ as $n_t'n_r'$ goes from 12 to 10 does not offset the sizeable *decrease* in $\hat{\sigma}^2(pT)$ as n_t' goes from two to five.

There are several lessons to be learned from this example and the previous ones in this chapter. First, generalizability theory allows for, and indeed encourages, a multidimensional perspective on error variances. Second, to understand error variances, one needs to consider both the variance components that enter them and D study sample sizes. Third, conventional wisdom based on classical theory and single-faceted universes frequently does not apply in generalizability theory. This does not mean that classical theory is wrong and generalizability theory is right in a universal sense, but it does mean that a single application of classical theory cannot differentiate among the contributions of multiple sources of error.[9]

4.2.4 Performance Assessments

Generalizability theory is particularly well suited to evaluating assessments that are based on ratings of human performance. The use of generalizability theory in performance assessments in educational measurement has been studied by Brennan (1996b, 2000b), Brennan and Johnson (1995), Cronbach et al. (1997), and Kane et al. (1999), among others.

Shavelson et al. (1993, p. 222) provide an instructive example of a performance assessment program in science called the California Assessment Program (CAP). They state that:

> Students were posed five independent tasks. More specifically, students rotated through a series of five self-contained stations at timed intervals (about 15 mins.). At one station, students

---

[9]Gulliksen (1950, pp. 211–214) discusses reliability of essay scores from a classical perspective, and Hoover and Bray (1995) outline how classical theory can be used to examine multiple sources of error in essay examinations. Isolating the sources requires multiple analyses of different data sets or different subsets of the same data set. In this sense the process is somewhat disjointed and not nearly as integrated and "rich" as the approach in generalizability theory.

were asked to complete a problem solving task (determine which of these materials may serve as a conductor). At the next station, students were asked to develop a classification system for leaves and then to explain any adjustments necessary to include a new mystery leaf in the system. At yet another, students were asked to conduct tests with rocks and then use the results to determine the identity of an unknown rock. At the fourth station, students were asked to estimate and measure various characteristics of water (e.g., temperature, volume). And at the fifth station, students were asked to conduct a series of tests on samples of lake water to discover why fish are dying (e.g., is the water too acidic?). At each station, students were provided with the necessary materials and asked to respond to a series of questions in a specified format (e.g., fill in a table).

A predetermined scoring rubric developed by teams of teachers in California was used to evaluate the quality of students' written responses . . . to each of the tasks. Each rubric was used to score performance on a scale from 0 to 4 (0 = no attempt, 1 = serious flaws, 2 = satisfactory, 3 = competent, 4 = outstanding). All tasks were scored by three raters.

For the CAP, the G study design is $p \times t \times r$ with $n_t = 5$ tasks and $n_r = 3$ raters. We assume here that tasks and raters are both random. Table 4.7 reports the G study estimated variance components, along with D studies for various sample sizes.

The G study estimated variance component for persons $\hat{\sigma}^2(p) = .298$ is relatively large, but the estimated variance component for the pt interactions $\hat{\sigma}^2(pt) = .493$ is even larger. By contrast, $\hat{\sigma}^2(r)$, $\hat{\sigma}^2(pr)$, and $\hat{\sigma}^2(rt)$ are all close to zero, which suggests that the rater facet does not contribute much to variability in observed scores. Also, since $\hat{\sigma}^2(t) = .003$, it appears that tasks are quite similar in average difficulty.

The CAP G study results are typical of generalizability results for many programs that involve performance assessments; that is, $\hat{\sigma}^2(pt)$ tends to be relatively large, the rater facet often contributes relatively little variance, and $\hat{\sigma}^2(t)$ is often small. As discussed by Brennan (1996b), other published studies using the $p \times t \times r$ design that give similar results include science and math assessments discussed by Shavelson et al. (1993) and tests of listening and writing discussed by Brennan et al. (1995).

The D study results at the top of Table 4.7 illustrate that using three raters gives very little improvement in measurement precision over that obtained using two raters, and the figures at the bottom suggest that a single rater may be adequate for many purposes. However, it is very clear that the number of tasks has a substantial impact on error variances and coefficients, which is to be expected since $\hat{\sigma}^2(pt)$ is quite large.

TABLE 4.7. Random Effects D Study $p \times T \times R$ Designs For CAP Data

		D Studies					
	n_t'	5	5	5	10	10	10
$\hat{\sigma}^2(\alpha)$	n_r'	1	2	3	1	2	3
$\hat{\sigma}^2(p) = .298$	$\hat{\sigma}^2(p)$	.298	.298	.298	.298	.298	.298
$\hat{\sigma}^2(t) = .092$	$\hat{\sigma}^2(T)$	.018	.018	.018	.009	.009	.009
$\hat{\sigma}^2(r) = .003$	$\hat{\sigma}^2(R)$	.003	.002	.001	.003	.002	.001
$\hat{\sigma}^2(pt) = .493$	$\hat{\sigma}^2(pT)$	.099	.099	.099	.049	.049	.049
$\hat{\sigma}^2(pr) = .000$	$\hat{\sigma}^2(pR)$	.000	.000	.000	.000	.000	.000
$\hat{\sigma}^2(tr) = .002$	$\hat{\sigma}^2(TR)$	.000	.000	.000	.000	.000	.000
$\hat{\sigma}^2(ptr) = .148$	$\hat{\sigma}^2(pTR)$	.030	.015	.010	.015	.007	.005
	$\hat{\sigma}^2(\tau)$	.30	.30	.30	.30	.30	.30
	$\hat{\sigma}^2(\delta)$	.13	.11	.11	.06	.06	.05
	$\hat{\sigma}^2(\Delta)$	.15	.13	.13	.08	.07	.06
	$E\hat{\rho}^2$	.70	.72	.73	.82	.84	.85
	$\hat{\Phi}$	.67	.69	.70	.80	.82	.82

Note. G study estimated variance components were provided by Xiaohong Gao.

The Shavelson et al. (1993) study does not include all potentially relevant facets. For example, the occasion on which tasks were administered was not varied. Therefore, their study does not permit generalizing to different testing occasions. This limitation may be quite important in that the literature contains some evidence that the relatively large value of $\sigma^2(pt)$ is sometimes better characterized as $\sigma^2(pto)$ (see Ruiz-Primo et al., 1993; and Webb et al., 2000).

In many performance assessments, there are other facets that may be important but are frequently overlooked, including the following.

- Rating occasions: The occasion on which ratings are obtained is often different from the occasion on which task responses are obtained. This happens, for example, when examinee responses are centrally scored. That is, there are often two occasion facets: testing occasion and rating occasion. There is evidence that the rating occasion facet can contribute significantly to error (see Wainer, 1993).

- Scoring rubrics: In almost all performance assessments, only one scoring rubric is used. However, there are often other rubrics that could have been used.

- Modes of testing: Most performance assessments involve only one method or mode of testing (e.g., actual performance observed by raters, videotaped performance, performance results recorded in a notebook, etc.). In the absence of evidence to the contrary, there is no compelling reason to believe that scores would be invariant over modes of testing.

It is not always the case that these other facets (especially rubrics and modes of testing) are best viewed as random. An investigator might view one or more of them as fixed, using the methodology discussed in Section 4.3.

4.3   Simplified Procedures for Mixed Models

Thus far in this chapter it has been assumed that all facets in both the universe of admissible observations and the universe of generalization are infinite in size. Frequently, however, an investigator wants to consider a *restricted* universe of generalization, that is, a universe of generalization that involves only a subset of the conditions in the universe of admissible observations.[10] In such cases, the procedures discussed in this section can be used provided certain conditions prevail:

---

[10] A more elegant way to conceptualize a restricted universe is considered in Section 4.5.

1. the estimated G study variance components are for a random model, and

2. each facet in the D study is either random or fixed, in the sense discussed next.

If these conditions do not apply, then the procedures in Section 5.1 can be used.

Recall that n represents the G study sample size for a facet, and N represents the size of the facet in the universe of admissible observations. Similarly, n' represents the D study sample size for a facet, and N' represents the size of the facet in the universe of generalization. We say that a facet is *fixed* in the D study when $n' = N' < \infty$. This definition involves two considerations: the number of conditions for the facet in the universe of generalization is finite ($N' < \infty$), and the D study sample size equals the number of conditions for the facet in the universe of generalization ($n' = N'$). Furthermore, since the conditions in the D study must be contained in the universe of generalization, it follows that the conditions of the fixed facet in the D study must be the *same conditions* as those in the universe of generalization.

In generalizability theory it is always assumed that, for at least one facet, generalization is intended to a larger universe of conditions than those actually represented in the D study. If that were not the case, there would be no errors of measurement, by definition! In this section, it is assumed that for every facet $n' < N' \to \infty$ or $n' = N' < \infty$, and for at least one facet $n' < N' \to \infty$. In traditional ANOVA terminology, therefore, this section treats *mixed* models.

4.3.1 Rules

For mixed models, a simple set of equations and rules can be used to obtain universe score variance, error variances, and coefficients. These equations and rules are based on using the same $\sigma^2(\bar{\alpha})$ as those discussed in Section 4.1 for random models. That is, here we use D study *random* effects variance components to obtain results for *mixed* models. This apparent incongruity means that these simplified procedures have limitations. However, these simplified procedures are quite adequate for much practical work, and they provide an instructive basis for comparing error variances and coefficients for random and mixed models.

The notation and equations summarized in Table 4.3 on page 109 still apply, but the rules in Table 4.3 need to be altered for mixed models. For mixed models, an observed mean score might be represented as $X_{p\mathcal{R}\mathcal{F}}$, where $\mathcal{F}$ is the set of fixed facets in the universe of generalization. Recognizing that $\mathcal{R}$ contains only random facets, the rules for mixed models are:

Rule 4.3.1: $\sigma^2(\tau)$ is the sum of all $\sigma^2(\bar{\alpha})$ such that $\bar{\alpha}$ includes τ and does *not* include any index in $\mathcal{R}$;

Rule 4.3.2: $\sigma^2(\Delta)$ is the sum of all $\sigma^2(\bar{\alpha})$ such that $\bar{\alpha}$ includes at least one of the indices in $\mathcal{R}$; and

Rule 4.3.3: $\sigma^2(\delta)$ is the sum of all $\sigma^2(\bar{\alpha})$ such that $\bar{\alpha}$ includes τ *and* at least one of the indices in $\mathcal{R}$.

Tables 4.1 and 4.2 on page 99 illustrate these rules for mixed models with the $p \times I \times H$ and $p \times (I:H)$ designs, respectively.

To get a somewhat intuitive notion of why Rules 4.3.1 to 4.3.3 work, it is helpful to consider the expected mean squares for different models (see Section 3.5). For example, for the $p \times (I:H)$ design,

$$
\begin{aligned}
\boldsymbol{E}MS(p) &= \sigma^2(pi{:}h) + n_i\sigma^2(ph) + n_in_h\sigma^2(p) \\
&= \sigma^2(pi{:}h|H) + n_in_h\sigma^2(p|H),
\end{aligned}
$$

where the first equation is for the random model, and the second is for the mixed model with H fixed. Since $\boldsymbol{E}MS(pi{:}h) = \sigma^2(pi{:}h) = \sigma^2(pi{:}h|H)$, it follows that

$$
\sigma^2(p|H) = \sigma^2(p) + \frac{\sigma^2(ph)}{n_h},
$$

where $\sigma^2(p|H)$ is universe score variance, $\sigma^2(\tau)$, for the mixed model when $n'_h = n_h$. This result is identical to that obtained using Rule 4.3.1.

4.3.2   Venn Diagrams

Visual perspectives on the $p \times I \times H$ and $p \times (I:H)$ designs with H fixed are provided by the Venn diagrams in Figure 4.4 and Figure 4.5, respectively. Figure 4.4 for the $p \times I \times H$ design with H fixed can be compared with Figure 4.1 for the same design with both H and I random. Note that, with H fixed, $\sigma^2(\Delta)$ does not include $\sigma^2(H)$ because every instance of the measurement procedure would contain the same n'_h conditions. Also, note that the expected observed score variance is the same whether or not H is fixed. What changes is the manner in which $\boldsymbol{E}S^2(p)$ is decomposed into $\sigma^2(\tau)$ and $\sigma^2(\delta)$. Specifically, when H is fixed, $\sigma^2(pH)$ contributes to universe score variance rather than $\sigma^2(\delta)$. In short, restricting the universe of generalization leads to an increase in universe score variance and a decrease in relative error variance, as compared to the random model results. The same conclusions are evident in comparing Figure 4.5 for the $p \times (I:H)$ design with H fixed with Figure 4.2 for the same design with both H and I random.

It is evident from Tables 4.1 and 4.2, and from Figures 4.4 and 4.5, that universe score variance for both the $p \times I \times H$ and $p \times (I:H)$ designs is

$$
\sigma^2(\tau) = \sigma^2(p) + \sigma^2(pH)
$$

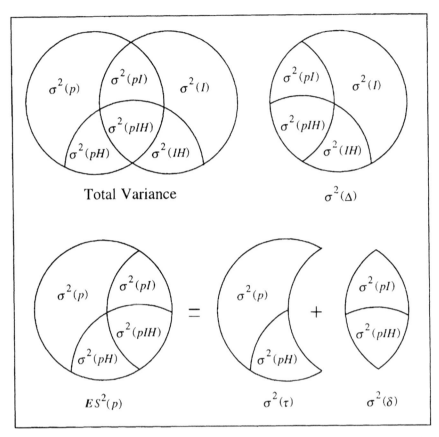

FIGURE 4.4. Venn diagram representation of variance components for the $p \times I \times H$ design with I random and H fixed.

when H is fixed. In general, for a specified universe of generalization, universe score variance is unaffected by the structure of the D study. However, error variances *are* affected. For example, suppose the universe of generalization has $I \times H$. If the D study design has H fixed, then for the $p \times I \times H$ design,

$$\sigma^2(\delta) = \frac{\sigma^2(pi)}{n_i'} + \frac{\sigma^2(pih)}{n_i'n_h'};$$

and for the $p \times (I:H)$ design,

$$\sigma^2(\delta) = \frac{\sigma^2(pi:h)}{n_i'n_h'} = \frac{\sigma^2(pi) + \sigma^2(pih)}{n_i'n_h'} = \frac{\sigma^2(pi)}{n_i'n_h'} + \frac{\sigma^2(pih)}{n_i'n_h'}.$$

Clearly, $\sigma^2(\delta)$ is smaller for the $p \times (I:H)$ design with H fixed. A similar statement holds for $\sigma^2(\Delta)$. Recall from Section 4.1 that the same conclusions hold for random models. So, in general, all other things being equal, nested designs lead to smaller error variances than crossed designs.

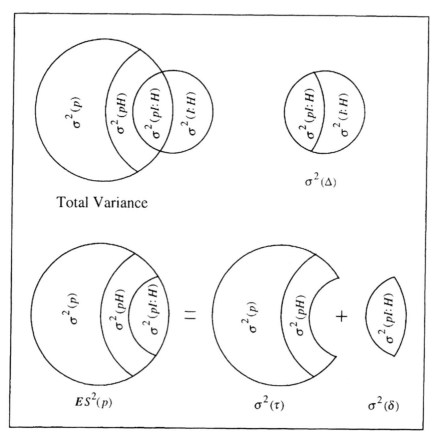

FIGURE 4.5. Venn diagram representation of variance components for the $p \times (I:H)$ design with I random and H fixed.

The simplified procedures for mixed models based on Rules 4.3.1 to 4.3.3 are very useful, but they have limitations. In particular, they use random effects D study variance components, rather than the D study components for the actual mixed model. This deficiency is remedied by the procedures in Section 5.1. Also, when these rules are used in practice, with parameters replaced by estimates, strictly speaking it must be assumed that the G and D studies contain the same levels of the fixed facet(s), not simply the same number of levels. Under these circumstances, the simplified procedures and those in Section 5.1 give the *same* unbiased estimates of universe score variance, error variances, and coefficients. If the G and D studies do not contain the same conditions, then the mixed model estimates are approximations only.

4.4 Mixed Model Examples

In this section, both synthetic and real-data examples are used to illustrate the simplified mixed model procedures in Section 4.3. Many aspects of these examples focus on differences between random and mixed model results. Particular emphasis is placed on illustrating how some classical reliability statistics are interpretable from the perspective of random and/or mixed model results in generalizability theory, how the reliability of class means is easily accommodated in generalizability theory, and how generalizability theory provides an elegant explanation of the reliability–validity paradox. The exercises at the end of this chapter provide additional instructive examples.

4.4.1 $p \times I \times O$ Design with Items Fixed

Section 4.2.1 treated random model D studies for the $p \times I \times O$ design, based on the following G study estimated variance components for Synthetic Data Set No. 3.

$$\hat{\sigma}^2(p) = .5528 \qquad \hat{\sigma}^2(i) = .4417 \qquad \hat{\sigma}^2(o) = .0074$$
$$\hat{\sigma}^2(pi) = .5750 \qquad \hat{\sigma}^2(po) = .1009 \qquad \hat{\sigma}^2(io) = .1565$$
$$\hat{\sigma}^2(pio) = .9352.$$

Now let us consider the $p \times I \times O$ design for sample sizes of $n'_o = 2$ and $n'_i = 4$ with items fixed. Using Rules 4.3.1 to 4.3.3 on page 122 and the equations in Table 4.3 on page 109, it is easy to verify the mixed model results in the first row of the following table.

Model	$\hat{\sigma}^2(\tau)$	$\hat{\sigma}^2(\delta)$	$Est[\boldsymbol{ES}^2(p)]$	$\hat{\sigma}^2(\Delta)$	$E\hat{\rho}^2$	$\hat{\Phi}$
I fixed	.70	.17	.86	.19	.81	.79
I, O random	.55	.31	.86	.45	.64	.55

These mixed model results in the first row can be compared with the previously discussed random model results in the second row. (For both rows, $n'_o = 2$ and $n'_i = 4$.) Clearly, fixing items leads to decreased error variances and increased coefficients, as theory dictates. Note, also, that expected observed score variance is the same for both models. For a given design structure and sample sizes, fixing a facet affects which variance components contribute to $\hat{\sigma}^2(\tau)$ and $\hat{\sigma}^2(\delta)$, but it does not change their sum.

4.4.2 $p \times (R{:}T)$ Design with Tasks Fixed

Section 4.2.3 considered the $p \times (R{:}T)$ design for Synthetic Data Set No. 4 under random model assumptions. Suppose, however, that an investigator also wants to consider a mixed model in which generalization is over raters,

TABLE 4.8. D Studies for $p \times (R\!:\!T)$ Design for Synthetic Data Set No. 4 Under Random Model and Model with Tasks Fixed

		D Studies			
	n'_t	3	3	3	3
	n'_r	1	2	3	4
$\hat{\sigma}^2(\alpha)$	$n'_t n'_r$	3	6	9	12
$\hat{\sigma}^2(p) = .4731$	$\hat{\sigma}^2(p)$	.473	.473	.473	.473
$\hat{\sigma}^2(t) = .3252$	$\hat{\sigma}^2(T)$	.108	.108	.108	.108
$\hat{\sigma}^2(r\!:\!t) = .6475$	$\hat{\sigma}^2(R\!:\!T)$	.216	.108	.072	.054
$\hat{\sigma}^2(pt) = .5596$	$\hat{\sigma}^2(pT)$	.187	.187	.187	.187
$\hat{\sigma}^2(pr\!:\!t) = 2.3802$	$\hat{\sigma}^2(pR\!:\!T)$	.793	.397	.265	.198

	Model	Ran	T_f	Ran	T_f	Ran	T_f	Ran	T_f
	$\hat{\sigma}^2(\tau)$	.47	.66	.47	.66	.47	.66	.47	.66
	$\hat{\sigma}^2(\delta)$	.98	.79	.58	.40	.45	.26	.39	.20
	$\hat{\sigma}^2(\Delta)$	1.30	1.01	.80	.51	.63	.34	.55	.25
	$Est[ES^2(p)]$	1.45	1.45	1.06	1.06	.92	.92	.86	.86
	$E\hat{\rho}^2$	.33	.45	.45	.62	.51	.71	.55	.77
	$\widehat{\Phi}$	.27	.40	.37	.57	.43	.66	.46	.72

Note. In the lower half of the table, the first entries are for a random model (*Ran*), and the second entries are for a model with tasks fixed (T_f).

but not tasks. That is, the investigator is also interested in a universe of generalization in which the same tasks would be used for every instance of the measurement procedure. Table 4.8 provides D study results for both the random and mixed models under the assumption that $n'_t = 3$, which is the same number of tasks as those in the synthetic data.

For the mixed model results with T fixed and only one random facet (R), as the number of raters increases, error variances decrease and coefficients increase. By comparison with the random model, under the mixed model:

- universe score variance is larger,

- error variances are smaller,

- expected observed score variance is unchanged, and

- coefficients are larger.

These differences are predictable from theory. The principal explanation is that, for the random model, $\hat{\sigma}^2(pT)$ is part of error variances, whereas $\hat{\sigma}^2(pT)$ contributes to universe score variance for the mixed model with T fixed.

The D study results for $n'_r = 1$ in Table 4.8 require careful consideration. These results are estimates for a measurement procedure having $n'_t = 3$ tasks with each one evaluated by a *single* different rater. Clearly, if such data were analyzed, the task and rater facets would be completely confounded, and it would be impossible to treat raters as random and tasks as fixed. Doing so is possible here because a G study was conducted that involved multiple raters and multiple tasks, which permits us to untangle the contributions of the task and rater facets.

4.4.3   Perspectives on Traditional Reliability Coefficients

In classical theory, three types of reliability coefficients are frequently discussed:

- *Stability and equivalence*: correlation for scores for different forms of a test administered on different occasions;

- *Stability* or *test–retest*: correlation of scores for the same form of a test administered on different occasions; and

- *Internal consistency*: Cronbach's alpha or KR-20 based on a single administration of a test form.

Table 4.9 provides a generalizability theory perspective on these coefficients. Internal consistency coefficients are discussed more fully in Section 5.1.4, but the theory discussed thus far is a sufficient basis for the most important distinctions among these coefficients.

Relationships Among Coefficients

These three traditional coefficients are estimates of reliability for a single test form, that is, a form administered on a single occasion. That is why the D study sample size for occasions is $n'_o = 1$ for all coefficients, even though the data collection design (G study) may involve two administrations. To emphasize that $n'_o = 1$, a lowercase o, rather than O, is used in the D study notation in Table 4.9. Second, the coefficients differ with respect to which facets are fixed and which are random in the universes of generalization. For a coefficient of stability and equivalence, both occasion and items are random; for a coefficient of stability, items are fixed; and for an internal consistency coefficient there is a single fixed occasion.

The denominators (expected observed score variance) of the coefficients in Equations 4.26 to 4.28 are all the same. So, differences in their numerators (universe score variance) will dictate relationships among them.

TABLE 4.9. Some Classical Reliability Coefficients from the Perspective of Generalizability Theory

---

Stability and Equivalence: Correlation of scores for different forms of a test administered on two different occasions.

G Study: $p \times (i{:}o)$ $n_o = 2$

D Study: $p \times (I{:}o)$ $n'_o = 1$ (random); I random

$$E\rho^2_{se} = \frac{\sigma^2(p)}{\sigma^2(p) + [\sigma^2(po) + \sigma^2(pI{:}o)]}$$

$$= \frac{\sigma^2(p)}{\sigma^2(p) + [\sigma^2(po) + \sigma^2(pI) + \sigma^2(pIo)]}. \quad (4.26)$$

Stability: Correlation of scores for the same form of a test administered on two different occasions.

G Study: $p \times i \times o$ $n_o = 2$

D Study: $p \times I \times o$ $n'_o = 1$ (random); I fixed

$$E\rho^2_s = \frac{\sigma^2(p) + \sigma^2(pI)}{\sigma^2(p) + \sigma^2(pI) + [\sigma^2(po) + \sigma^2(pIo)]}. \quad (4.27)$$

Internal Consistency (Equivalence): Cronbach's alpha or KR-20 based on a single administration of a test form.

Note. To conceptualize an internal consistency estimate of reliability and relate it to other coefficients, it is helpful to view the universe of admissible observations as having an occasion facet, with a G study involving the collection of data on two or more occasions.

G Study: $p \times i \times o$ or $p \times (i{:}o)$ $n_o \geq 2$

D Study: $p \times I \times o$ or $p \times (I{:}o)$ $n'_o = 1$ (fixed); I random

$$E\rho^2_{ic} = \frac{\sigma^2(p) + \sigma^2(po)}{\sigma^2(p) + \sigma^2(po) + [\sigma^2(pI) + \sigma^2(pIo)]}. \quad (4.28)$$

Clearly, a coefficient of equivalence and stability is likely to be smaller than either of the other two. A stability coefficient will be smaller than an internal consistency coefficient if $\sigma^2(pI) < \sigma^2(po)$, which is highly likely whenever n_i' is large. These conclusions are in accord with experience; that is, almost always,

$$E\hat{\rho}_{es}^2 < E\hat{\rho}_s^2 < E\hat{\rho}_{ic}^2.$$

Similarly, almost always,

$$\hat{\sigma}^2(\delta_{es}) > \hat{\sigma}^2(\delta_s) > \hat{\sigma}^2(\delta_{ic}).$$

Interrater Reliability

Interrater coefficients are another frequently discussed type of reliability. Actually, there are at least two interrater coefficients, both of which are often misunderstood:

- *Standardized:* correlation between the scores assigned by the same two raters to student responses to the *same* task; and

- *Nonstandardized:* correlation between the scores assigned by the same two raters to student responses to *different* tasks.

The magnitudes of standardized coefficients are often quite high, while nonstandardized coefficients tend to be small. As discussed by Brennan (2000b), these results are predictable from a careful consideration of the D study designs, sample sizes, and universes of generalization that are implicit in these two coefficients.

In the terminology of generalizability theory, the standardized coefficient uses the G study $p \times t \times r$ design with $n_r = 2$ raters evaluating the same $n_t = 1$ task. For the D study, the design is $p \times t \times r$ with a single *random* rater and a single *fixed* task. It follows that the standardized coefficient is

$$E\rho^2 = \frac{\sigma^2(p) + \sigma^2(pt)}{\sigma^2(p) + \sigma^2(pt) + [\sigma^2(pr) + \sigma^2(ptr)]}. \tag{4.29}$$

For this coefficient, $n_t' = 1$ because only one task is involved in the correlation, and $n_r' = 1$ because a correlation between two raters gives an estimate of reliability for a single rater.

The nonstandardized coefficient has tasks nested within persons, which means that it effectively uses the G study $(t{:}p) \times r$ design with $n_r = 2$ raters evaluating a different task for each person. For the D study, the design is $(t{:}p) \times r$ with a single *random* rater and a single *random* task. It follows that the nonstandardized coefficient is

$$E\rho^2 = \frac{\sigma^2(p)}{\sigma^2(p) + [\sigma^2(t{:}p) + \sigma^2(pr) + \sigma^2(tr{:}p)]}, \tag{4.30}$$

where $\sigma^2(t{:}p) = \sigma^2(t) + \sigma^2(pt)$ and $\sigma^2(tr{:}p) = \sigma^2(tr) + \sigma^2(ptr)$.

The nonstandardized coefficient in Equation 4.30 is smaller than the standardized coefficient in Equation 4.29 for two reasons: universe score variance for the nonstandardized coefficient is smaller because it does not contain $\sigma^2(pt)$; relative error variance for the nonstandardized coefficient is larger by $\sigma^2(t) + \sigma^2(pt) + \sigma^2(tr)$.

The standardized and nonstandardized interrater reliability coefficients may be of value in evaluating the extent to which raters are functioning as intended, but neither coefficient characterizes the reliability of student scores based on *two* ratings. To do so, variance components containing r need to be divided by $n'_r = 2$. Also, in most cases, investigators want to generalize over tasks, which means that $\sigma^2(pt)$ should be part of error variance, not universe score variance. Finally, often student scores are based on more than one task. If so, then variance components containing t should be divided by n'_t.

4.4.4   Generalizability of Class Means

Usually persons are the objects of measurement, but aggregates of persons, such as classes, are sometimes the entities about which decisions are made. Kane and Brennan (1977) provide an extensive treatment of the generalizability of class means. Using the procedures, rules, and equations that have been discussed in this chapter, the reader can verify that coefficients for the $(P{:}c) \times I$ design with classes as the objects of measurement are as follows.

- For generalizing over persons and items (P and I random),

$$E\rho^2(P,I) = \frac{\sigma^2(c)}{\sigma^2(c) + [\sigma^2(cI) + \sigma^2(P{:}c) + \sigma^2(PI{:}c)]}. \qquad (4.31)$$

- For generalizing over items, only (I random; P fixed),

$$E\rho^2(I) = \frac{\sigma^2(c) + \sigma^2(P{:}c)}{\sigma^2(c) + \sigma^2(P{:}c) + [\sigma^2(cI) + \sigma^2(PI{:}c)]}. \qquad (4.32)$$

- For generalizing over persons, only (P random; I fixed),

$$E\rho^2(P) = \Phi(P) = \frac{\sigma^2(c) + \sigma^2(cI)}{\sigma^2(c) + \sigma^2(cI) + [\sigma^2(P{:}c) + \sigma^2(PI{:}c)]}. \qquad (4.33)$$

In each of the coefficients, relative error variances are in square brackets. For $\Phi(P,I)$ and $\Phi(I)$, the variance component $\sigma^2(I)$ contributes to absolute error variance and is added to the denominators of Equations 4.31 and 4.32, respectively.

Kane et al. (1976) studied the generalizability of class means in the context of student evaluations of teaching. One of the questionnaires they used

TABLE 4.10. Random Effects D Studies for the $(P:c) \times I$ Design for Kane et al. (1976) Data with $n_i' = 8$ Attribute Items

n_p'	Model	$\hat{\sigma}^2(\tau)$	$\hat{\sigma}^2(\delta)$	$\hat{\sigma}^2(\Delta)$	$E\hat{\rho}^2$	$\hat{\Phi}$
10	P, I random	.030	.027	.043	.53	.41
10	P random; I fixed	.036	.021	.021	.64	.64
10	I random; P fixed	.047	.010	.026	.83	.64
20	P, I random	.030	.017	.033	.65	.48
20	P random; I fixed	.036	.010	.010	.78	.78
20	I random; P fixed	.039	.008	.024	.83	.61

was administered in physics courses at the University of Illinois in 1972. "Fifteen classes that had twenty or more students were randomly selected, with the restriction that only one section taught by each instructor was included in the sample" (Kane et al., 1976, p. 177).

The questionnaire contained a set of eight items concerning attributes (e.g., ability to answer questions). For this item set, the G study estimated variance components were:

$$\hat{\sigma}^2(c) = .03, \quad \hat{\sigma}^2(p:c) = .17, \quad \hat{\sigma}^2(i) = .13,$$

$$\hat{\sigma}^2(ci) = .05, \quad \text{and} \quad \hat{\sigma}^2(pi:c) = .28.$$

Table 4.10 provides results for error variances and coefficients for random and mixed D studies with $n_i' = 8$ items and various numbers of persons within classes.

In terms of generalizability coefficients, there is a consistent relationship among the three models; namely, fixing persons results in larger coefficients than fixing items, and both mixed models give larger coefficients than the random model. In terms of phi coefficients, however, for $n_p' > 10$, fixing persons gives *smaller* coefficients than fixing items. Numerous considerations may be involved in deciding which coefficients are most appropriate. For example, if the attribute items are used to make comparative decisions about instructors, then generalizability coefficients would seem sensible. On the other hand, if decisions about instructors are based on a specific score for the attribute items, then phi coefficients are probably more reasonable.

Deciding which model(s) is (are) appropriate is often a more difficult matter. If the attribute items are viewed as the only such items that are of interest, then treating them as fixed is sensible. On the other hand, if they are viewed as only one potential set of attribute items that might be of interest, then it seems more reasonable to treat them as random. Similarly, if the students in the particular classes are viewed as the only ones of interest, then they might be considered fixed. But, if these students are viewed as a sample of the potential students who might have taken the classes, then treating students as random seems sensible.

In this author's opinion, treating students as fixed often involves somewhat suspect logic, although treating them as random has problems, too. If students are truly fixed, then strictly speaking they are the only students of interest. Also, since the modeling here is for balanced designs, treating students as fixed means that the number of students in each class is truly equal.[11] On the other hand, random model results effectively assume that students are randomly assigned to classes and the population size for students is infinite. Neither of these conditions is likely to be literally true.

In practice, it is often advisable to provide D study results for multiple models, as has been done in Table 4.10, even if some results require cautious interpretations. Doing so provides relevant information for different purposes, and for investigators who have different perspectives on the measurement procedure.

4.4.5 A Perspective on Validity

In an extensive consideration of validity from the perspective of generalizability theory, Kane (1982) argues that a restricted universe of generalization (what he calls a universe of allowable observations) for a standardized measurement procedure can be conceptualized as a reliability-defining universe, while the broader universe of generalization can be considered a validity-defining universe. Doing so provides an elegant explanation of the reliability–validity paradox, whereby attempts to increase reliability through standardization (i.e., fixing facets) can actually lead to a decrease in some measures of validity (Lord & Novick, 1968, p. 334).

---

[11]If students are randomly discarded to obtain an equal number per class, then it is difficult to make a convincing argument that students are fixed.

Let $\mathcal{F}$ be the set of fixed facets in the *restricted* universe, and recall that $\mathcal{R}$ designates the set of random facets. For simplicity, assume that $\mathcal{R}$ and $\mathcal{F}$ contain only one facet each, and the D study design is fully crossed with observed mean scores:

$$X_{p\mathcal{R}\mathcal{F}} = \mu + \nu_p + \nu_{\mathcal{R}} + \nu_{\mathcal{F}} + \nu_{p\mathcal{R}} + \nu_{p\mathcal{F}} + \nu_{\mathcal{R}\mathcal{F}} + \nu_{p\mathcal{R}\mathcal{F}}.$$

Universe scores for the restricted universe of generalization are

$$\mu_{p\mathcal{F}} = \mathop{E}_{\mathcal{R}} X_{p\mathcal{R}\mathcal{F}},$$

whereas universe scores for the more broadly defined universe are

$$\mu_p = \mathop{E}_{\mathcal{R}} \mathop{E}_{\mathcal{F}} X_{p\mathcal{R}\mathcal{F}}.$$

That is, in the unrestricted universe, the facet in $\mathcal{F}$ is treated as random. For example, an investigator might consider tasks to be fixed (i.e., standardized) in the sense that another form of the measurement procedure would use the same tasks, but the same investigator might have an interest in the extent to which scores for the fixed tasks generalize to a broader universe of tasks.

In Kane's terminology, inferences from $X_{p\mathcal{R}\mathcal{F}}$ to $\mu_{p\mathcal{F}}$ are in the realm of reliability, while inferences from $X_{p\mathcal{R}\mathcal{F}}$ to μ_p relate to validity. It follows that a squared validity coefficient is

$$E\rho^2(X_{p\mathcal{R}\mathcal{F}}, \mu_p) = \frac{\sigma^2(p)}{\sigma^2(p) + [\sigma^2(p\mathcal{R}) + \sigma^2(p\mathcal{F}) + \sigma^2(p\mathcal{R}\mathcal{F})]}, \tag{4.34}$$

and a reliability coefficient for the restricted universe of generalization is

$$E\rho^2(X_{p\mathcal{R}\mathcal{F}}, \mu_{p\mathcal{F}}) = \frac{\sigma^2(p) + \sigma^2(p\mathcal{F})}{\sigma^2(p) + \sigma^2(p\mathcal{F}) + [\sigma^2(p\mathcal{R}) + \sigma^2(p\mathcal{R}\mathcal{F})]}. \tag{4.35}$$

Furthermore, the squared validity coefficient corrected for attenuation is

$$E\rho^2(\mu_{p\mathcal{F}}, \mu_p) = \frac{E\rho^2(X_{p\mathcal{R}\mathcal{F}}, \mu_p)}{E\rho^2(X_{p\mathcal{R}\mathcal{F}}, \mu_{p\mathcal{F}})} = \frac{\sigma^2(p)}{\sigma^2(p) + \sigma^2(p\mathcal{F})}. \tag{4.36}$$

This squared disattenuated validity coefficient is the squared correlation between scores for a perfectly reliable standardized measurement procedure ($\mu_{p\mathcal{F}}$) and universe scores for the more broadly defined universe of generalization (μ_p). As such, Equation 4.36 relates the restricted universe to the broader universe of generalization.

The three coefficients in Equations 4.34 through 4.36 have a simple relationship:

$$E\rho^2(X_{p\mathcal{R}\mathcal{F}}, \mu_p) = E\rho^2(X_{p\mathcal{R}\mathcal{F}}, \mu_{p\mathcal{F}}) \times E\rho^2(\mu_{p\mathcal{F}}, \mu_p). \tag{4.37}$$

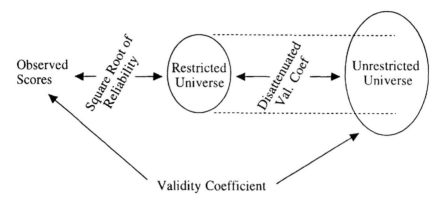

FIGURE 4.6. Relationships between universes and among coefficients involved in the reliability–validity paradox.

That is, the dependability of inferences from observed scores to μ_p can be factored into two parts. The first factor is a reliability coefficient and represents the dependability of inferences from observed scores to $\mu_{p\mathcal{F}}$. The second factor is a squared disattenuated validity coefficient and represents the dependability of inferences from $\mu_{p\mathcal{F}}$ to μ_p (Kane, 1982, p. 145).

The reliability–validity paradox arises because of the differential role played by $\sigma^2(p\mathcal{F})$ in the coefficients in Equation 4.37. Specifically, as $\sigma^2(p\mathcal{F})$ gets larger,

- reliability *increases* because $\sigma^2(p\mathcal{F})$ contributes to universe score variance for the restricted universe, and

- validity *decreases* because $\sigma^2(p\mathcal{F})$ contributes to relative error variance for the broader universe of generalization.

Note also that, as $\sigma^2(p\mathcal{F})$ gets larger, the squared disattenuated validity coefficient decreases, which weakens inferences from the restricted universe to the more broadly defined universe of generalization.

Figure 4.6 depicts relationships between universes and among coefficients that are involved in the reliability–validity paradox. In this figure, the square root of reliability is used for consistency with the usual definition of a validity coefficient as a correlation, as opposed to a squared correlation. In this sense, Figure 4.6 depicts the square root of Equation 4.37.

The discussion of interrater reliability in Section 4.4.3 provides an example of Equation 4.37 and the reliability–validity paradox. In terms of that discussion, $\mathcal{R}$ is r, $\mathcal{F}$ is t, interrater reliability is

$$E\rho^2(X_{ptr}, \mu_{pt}) = \frac{\sigma^2(p) + \sigma^2(pt)}{\sigma^2(p) + \sigma^2(pt) + [\sigma^2(pr) + \sigma^2(ptr)]},$$

the squared validity coefficient is

$$E\rho^2(X_{ptr}, \mu_p) = \frac{\sigma^2(p)}{\sigma^2(p) + [\sigma^2(pr) + \sigma^2(pt) + \sigma^2(ptr)]}, \tag{4.38}$$

and the squared disattenuated validity coefficient is

$$E\rho^2(\mu_{pt}, \mu_p) = \frac{\sigma^2(p)}{\sigma^2(p) + \sigma^2(pt)}. \tag{4.39}$$

So, as $\sigma^2(pt)$ gets larger, interrater reliability will increase, and squared validity as defined by Equation 4.38 will decrease. Also, as $\sigma^2(pt)$ gets larger, the link between the restricted universe and the broader one, as defined by Equation 4.39, will become weaker.[12]

From Kane's perspective on the reliability–validity paradox, it makes sense to call Equation 4.38 a squared validity coefficient. Clearly, however, Equation 4.38 is a generalizability coefficient for a universe of generalization in which both raters and tasks are random. This terminological ambiguity illustrates that generalizability theory "blurs" arbitrary distinctions between reliability and validity (Cronbach et al., 1972, p. 380; Brennan, in press) and forces an investigator to concentrate on the intended inferences, whatever terms are used to characterize them.

In this discussion of the reliability–validity paradox, it was assumed that $\mathcal{R}$ and $\mathcal{F}$ contain only one facet each. The theory discussed here has no such restriction, however, although equations become more complicated when $\mathcal{R}$ and/or $\mathcal{F}$ contain multiple facets. Also, it is entirely possible that the more broadly defined universe could have one or more fixed facets *provided* they are also fixed in the restricted universe. That is, there could be two sets of fixed facets: those that are fixed in both universes and those that are fixed in the restricted universe only.

4.5 Summary and Other Issues

A great deal of the essence of generalizability theory is captured by addressing one central question, "What constitutes a replication of a measurement procedure?" Answering this question requires particular attention to which facets are fixed and which are random in an investigator-defined multi-faceted universe of generalization. Generalizability theory does not provide answers to questions about which facets *should* be fixed and which should be random. Such answers must be sought in the investigator's substantive theory and intended interpretations. But, generalizability theory *does*

---

[12]The squared disattenuated validity coefficient in Equation 4.39 is essentially what Gulliksen (1950, p. 214) calls "content" reliability. It is also called "score" reliability in some of the performance assessment literature (see Dunbar et al., 1991).

provide powerful tools for determining the measurement consequences of decisions about which facets are fixed and which are random.

From the perspective of generalizability theory, defining a measurement procedure requires explicit decisions about three issues:

1. which facets are fixed and which are random in the universe of generalization,

2. D study sample sizes, and

3. D study design structure.

The first issue influences universe score variance, but all three issues influence error variances. In this sense, understanding error variance is a more challenging undertaking than understanding universe score variance. In particular, all three issues need to be taken into account in assessing the influence of particular facets on error variance.

In an attempt to simplify interpretations, G study estimated variance components are sometimes converted to percents. For example, for the class means example in Section 4.4.4, the sum of the estimated variance components is .66, and the percents are given below.

Effect	c	$p{:}c$	i	ci	$pi{:}c$
$\hat{\sigma}^2(\alpha)$	.03	.17	.13	.05	.28
Percents	4.55	25.76	19.70	7.58	42.42

The virtue of these percents is that they are scale independent. However, they are easily misinterpreted as indicators of the "importance" of each effect and/or facet. Such interpretations generally require attending to particular statistics, such as error variances, and the kinds of D study issues mentioned in the previous paragraph. For the class means example, Table 4.10 clearly indicates that judgments about the contribution of various facets to measurement precision depend heavily on such D study issues.

One of the distinct advantages of the procedures discussed in this chapter is that they do not require much computational effort once G study variance components have been estimated. A hand calculator is often sufficient. However, the computer program GENOVA (see Appendix F) can be used to perform all computations that have been discussed here.

4.6   Exercises

4.1* Assuming effects are uncorrelated, prove the last equality in Equation 4.8; namely,

$$E\left[\frac{\sum_h \nu_{ph}}{n'_h}\right]^2 = \frac{\sigma^2(ph)}{n'_h}.$$

4.2   For the $p \times I \times H$ design, show that the manner in which score effects are defined in generalizability theory is a sufficient basis for the uncorrelated-effects result $\sigma(\nu_{pH}, \nu_H) = 0$, where the covariance is taken over p and H.

4.3   For the $p \times (I:H)$ random effects design, starting with the definition of $\sigma^2(\overline{X})$ in Equation 4.19, show that

$$\sigma^2(\overline{X}) = \frac{\sigma^2(p)}{n'_p} + \frac{\sigma^2(h)}{n'_h} + \frac{\sigma^2(i{:}h)}{n'_i n'_h} + \frac{\sigma^2(ph)}{n'_p n'_h} + \frac{\sigma^2(pi{:}h)}{n'_p n'_i n'_h},$$

and that Equation 4.20 applies.

4.4* Demonstrate that it is generally not possible to use the variance components for a G study $p \times (i{:}h)$ design to estimate results for a D study $(I \times H){:}p$ design. Show that it is possible to do so if it is assumed that $\sigma^2(ih{:}p) = 0$.

4.5   Consider, again, the $p \times t \times r$ example in Exercise 3.6.

(a) Provide estimates of D study variance components, universe score variance, error variance, and coefficients for the random model with sample sizes of $n'_t = 3$ and $n'_r = 2$, and with sample sizes of $n'_t = 6$ and $n'_r = 1$.

(b) For both D studies the total number of observations per person is $n'_t n'_r = 6$. Why, then, are the error variances for the second D study larger/smaller than those for the first?

(c) Estimate the variance components for the $p \times i$ design for the undifferentiated data set with two tasks and three raters, and determine Cronbach's alpha for these data.

(d) Why is Cronbach's alpha smaller/larger than the generalizability coefficient in (a) when $n'_t = 3$ and $n'_r = 2$?

4.6* Consider the G study results in Table 3.4 on page 74 for the random effects $p \times (r{:}t)$ design and Synthetic Data Set No. 4. Under a random model, what are the D study estimated variance components, universe score variance, error variances and coefficients for the sample sizes:

(a) $n'_r = 1$ and $n'_i = 12$, and

(b) $n'_r = 2$ and $n'_i = 6$?

Provide box-bar plots, as well. What is the similarity between these results and the corresponding results in Table 4.6 for the $p \times (R{:}T)$ design?

4.7* Recall the Gillmore et al. (1978) example of the $(p{:}c{:}t) \times i$ design in Exercise 3.5. Suppose teachers are the objects of measurement. Under random model assumptions, if $n'_p = 20$ and $n'_i = 11$, what is the minimum number of classes required for the signal–noise ratio $\widehat{S/N}(\delta)$ to be at least 2?

4.8 For the CAP example in Table 4.7, suppose the task facet were fixed with $n'_t = 5$ and the rater facet were random with $n'_r = 1$. What are the estimated universe score variance, error variances, and coefficients for the $p \times T \times R$ design and for the $p \times (R{:}T)$ design? Report results to three decimal places. Why are these results so different from those for the random model?

4.9* For the random and mixed models in Table 4.8, what is the estimated standard error of the mean $\hat{\sigma}(\overline{X})$, assuming $n'_p = 10$, $n'_t = 3$, and $n'_r = 2$?

4.10* Many writing assessments involve exercises or tasks in different genres (e.g., narrative, informational, informative, persuasive, etc.). Suppose G study estimated random effects variance components are available for the $p \times (r{:}t{:}g)$ design, where p stands for students, r stands for raters, t stands for exercises or tasks, and g stands for genre.

   (a) Provide a formula for an estimated generalizability coefficient based on a D study design in which genre is fixed at $n'_g = 2$, there is only one exercise for each genre, and there is only one rating of each exercise.

   (b) Suppose data were available for this D study design and Cronbach's alpha were computed for the four ratings. Estimate alpha based on the G study estimated random effects variance components.

   (c) Why is alpha larger/smaller than the generalizability coefficient in (a)?

4.11 Assuming items are fixed and persons are random, provide an expression for $\boldsymbol{E}\hat{\rho}^2$ for class means in terms of mean squares, when G and D study sample sizes are the same.

4.12 In Table 4.10, $\boldsymbol{E}\hat{\rho}^2$ for class means with items random and persons fixed is nearly constant for all values of n'_p. Explain this result in terms of signal–noise ratios.

4.13 Recall the blood pressure example in Section 2.3.4 from Llabre et al. (1988). In the same paper they describe a study in which each of 40 persons had their blood pressure taken in a laboratory using a mercury sphygmomanometer. Measurements were taken on two consecutive days (d), with three replications (r) collected on each day. The resulting G study estimated variance components for the $p \times (r\!:\!d)$ design were:

Effect	Systolic	Diastolic
p	83.41	36.49
d	.73	0^a
$r\!:\!d$	.30	0^a
pd	24.91	21.69
$pr\!:\!d$	9.91	7.99

[a] Slightly negative estimate replaced by 0.

(a) Assuming an investigator wants to generalize over both days and replications within day, which of these two facets is likely to contribute more to errors in blood pressure measurements?

(b) Assuming a random model with $n'_d = 2$, what is the minimum number of replications required for systolic blood pressure to have $\widehat{\Phi} \geq .8$? In this case, what is $\hat{\sigma}(\Delta)$?

(c) Assuming a random model with $n'_d = 2$, what is the minimum number of replications required for diastolic blood pressure to have $\widehat{\Phi} \geq .8$?

(d) If an investigator does not intend to generalize over days, what are $\hat{\sigma}(\Delta)$ and $\widehat{\Phi}$ for a single replication, for both systolic and diastolic blood pressure?

5
Advanced Topics in Univariate Generalizability Theory

Chapter 3 discussed G studies for random models primarily, although Section 3.5 provided general procedures for estimating G study variance components for any model. By contrast, the D study procedures in Chapter 4 are based solely on using random effects G study estimated variance components. The obvious gap in these discussions is a consideration of general procedures for D studies. That is the subject of Section 5.1. Subsequent sections treat stratified objects of measurement, conditional standard errors of measurement, and several other issues.

5.1 General Procedures for D Studies

The procedures discussed in Chapter 4 are sufficient for estimating D study results for most univariate balanced designs in generalizability theory. However, these procedures have several limitations. First, they presume that G study variance components are estimated for a random model. Sometimes, however, one or more facets are clearly fixed for the universe of admissible observations. Second, for mixed model D studies, the simplified procedures in Section 4.3 do not provide estimates of the variance components *for the effects* in the linear model associated with the D study design and restricted universe of generalization, that is, a universe of generalization in which one or more facets is fixed. In this sense, these procedures are susceptible to misunderstanding. Third, the procedures discussed in Chapter 4 do not apply to D study designs that involve sampling from a finite universe.

Such designs are not common in generalizability theory, but they do occur occasionally.

Some additional notational conventions are required to accommodate distinctions drawn in this section. First, we use N' to designate the size of a facet in the universe of generalization. Second, we use $\sigma^2(\bar{\alpha}|M')$ to designate D study variance components for mean scores for a model M' that relates the D study design and the universe of generalization. Recall that $\sigma^2(\alpha|M)$ was introduced in Section 3.5 to designate G study variance components for a model M. Throughout this section τ designates the objects of measurement, $N'_\tau \to \infty$, and it is assumed that τ is *not* nested within any other facet.

5.1.1   D Study Variance Components

Before considering a general equation for obtaining $\sigma^2(\bar{\alpha}|M')$ from $\sigma^2(\alpha|M)$, two special cases are considered: (a) for each facet $N' = N$, which may or may not approach infinity; and (b) for each facet $N' \leq N \to \infty$. Unless otherwise noted, it is assumed in this section that the G and D study have the same design structure.

$N' = N$ for Each Facet

In this case, the universe of generalization and the universe of admissible observations are the same size, and

$$\sigma^2(\bar{\alpha}|M') = \frac{C(\bar{\alpha}|\tau)}{d(\bar{\alpha}|\tau)}\sigma^2(\alpha|M), \tag{5.1}$$

where $\sigma^2(\alpha|M)$ is discussed in Section 3.5,

$$d(\bar{\alpha}|\tau) = \begin{cases} 1 \text{ if } \bar{\alpha} = \tau, \text{ and, otherwise, the product} \\ \text{of the D study sample sizes } (n') \text{ for all} \\ \text{indices in } \bar{\alpha} \text{ except } \tau, \end{cases} \tag{5.2}$$

and

$$C(\bar{\alpha}|\tau) = \begin{cases} \text{the product of the terms } (1 - n'/N') \\ \text{for all } primary \text{ indices in } \bar{\alpha} \text{ except } \tau \\ [C(\bar{\alpha}|\tau) = 1 \text{ if } \bar{\alpha} = \tau]. \end{cases} \tag{5.3}$$

Recall that a primary index is one that appears before any colon in the notational representation of a component.

The term $C(\bar{\alpha}|\tau)$ is a finite universe correction factor (see, e.g., Cochran, 1977, pp. 23–25). When $N' \to \infty$, $C(\bar{\alpha}|\tau) = 1$ and Equation 5.1 becomes the familiar equation for a random effects D study estimated variance component. If $n' = N'$ for any *primary* index in $\bar{\alpha}$, then $C(\bar{\alpha}|\tau) = 0$ and $\sigma^2(\bar{\alpha}|M') = 0$.

$N' \leq N \to \infty$ for Each Facet

For this case, the universe of generalization can be viewed loosely as a subset[1] of the infinite universe of admissible observations, and the G study estimated variance components are for a random model; that is, $\sigma^2(\alpha|M)$ is $\sigma^2(\alpha)$. Under these circumstances, the D study variance components are:

$$\sigma^2(\bar{\alpha}|M') = \frac{C(\bar{\alpha}|\tau)}{d(\bar{\alpha}|\tau)} \left[\sum \frac{\sigma^2(\beta)}{\Pi'(\beta|\alpha)} \right], \qquad (5.4)$$

where β is any component that contains all the indices in α, and

$$\Pi'(\beta|\alpha) = \begin{cases} \text{the product of the D study universe sizes} \\ (N') \text{ for all indices in } \beta \text{ except those in } \alpha \\ [\Pi'(\beta|\alpha) = 1 \text{ if } \alpha = \beta]. \end{cases} \qquad (5.5)$$

The term in brackets in Equation 5.4 is identical to the right side of Equation 3.34, which is used to express G study variance components for any model M in terms of random effects variance components.

Consider, for example, the $p \times (I{:}H)$ design with $\tau = p$, and assume that $n_i' < N_i' = N_i \to \infty$ and $n_h' = N_h' < N_h \to \infty$ (i.e., a mixed model with H fixed, given estimated variance components for a random effects G study). Using H to designate the model M', $\hat{\sigma}^2(H|H)$ and $\hat{\sigma}^2(pH|H)$ are both zero because $C(H|p) = C(pH|p) = 0$; and $\hat{\sigma}^2(I{:}H|H)$ and $\hat{\sigma}^2(pI{:}H|H)$ are the same as for the random model. Also, since $d(p|p) = 1$ and $C(p|p) = 1$,

$$\begin{aligned} \hat{\sigma}^2(p|H) &= \frac{\sigma^2(p)}{\Pi'(p|p)} + \frac{\sigma^2(ph)}{\Pi'(ph|p)} \\ &= \sigma^2(p) + \frac{\sigma^2(ph)}{N_h'} \\ &= \sigma^2(p) + \frac{\sigma^2(ph)}{n_h'}. \end{aligned}$$

General Equation

Equations 5.1 and 5.4 are both special cases of the equation:

$$\sigma^2(\bar{\alpha}|M') = \frac{C(\bar{\alpha}|\tau)}{d(\bar{\alpha}|\tau)} \left[\sum \frac{K'(\beta|\alpha)}{\Pi'(\beta|\alpha)} \sigma^2(\beta|M) \right], \qquad (5.6)$$

where β is any component that contains all of the indices in α, and

$$K'(\beta|\alpha) = \begin{cases} \text{the product of the terms } (1 - N'/N) \text{ for} \\ \text{all } primary \text{ indices in } \beta \text{ that are } not \text{ in } \alpha \\ [K'(\beta|\alpha) = 1 \text{ if } \alpha = \beta]. \end{cases} \qquad (5.7)$$

---

[1]The word "subset" includes the possibility that $N' = N$ for one or more facets.

Equation 5.6 applies under the very general condition that $n' \leq N' \leq N$ for each facet, where N may or may not approach infinity.

5.1.2 Universe Score Variance and Error Variances

Let

$$\mathcal{R} = \begin{cases} \text{the set of all random facets in the universe} \\ \text{of generalization such that } n' < N', \end{cases}$$

$$\mathcal{F} = \begin{cases} \text{the set of all fixed facets (i.e., } n' = N') \text{ in} \\ \text{the universe of generalization,} \end{cases}$$

$$\overline{X}_\tau = \begin{cases} \text{an observable mean score for an object} \\ \text{of measurement } \tau, \text{ which is also denoted} \\ X_{\tau \mathcal{R} \mathcal{F}}, \text{ and} \end{cases}$$

$$\overline{X} = \begin{cases} \text{the average value of } \overline{X}_\tau \text{ over all sampled} \\ \text{objects of measurement.} \end{cases}$$

Note that in this section, a facet occurs in $\mathcal{R}$ if conditions of the facet are sampled from a larger, but possibly finite, set of conditions.

By definition, the universe score associated with an object of measurement is

$$\mu_\tau \equiv \underset{\mathcal{R}}{E}\, X_{\tau \mathcal{R} \mathcal{F}} = \underset{\mathcal{R}}{E}\, \overline{X}_\tau. \tag{5.8}$$

Note that, since an observed mean score involves all levels of all fixed facets $\mathcal{F}$, universe score does not involve taking an expectation over any of them. Absolute error is

$$\Delta_\tau \equiv \overline{X}_\tau - \mu_\tau, \tag{5.9}$$

relative error is

$$\delta_\tau \equiv (\overline{X}_\tau - \underset{\tau}{E}\, \overline{X}_\tau) - (\mu_\tau - \mu), \tag{5.10}$$

and the error associated with using $\overline{X}$ as an estimate of μ is simply $\overline{X} - \mu$.

To identify which score effects enter μ_τ, Δ_τ, and δ_τ, the observable quantities $\overline{X}_\tau$ and $\overline{X}$ can be replaced by their expressions in terms of score effects. Then the variances of μ_τ, Δ_τ, and δ_τ can be obtained algebraically. These variances will involve the $\sigma^2(\bar{\alpha}|M')$, as provided by Equations 5.1, 5.4, or 5.6. This tedious process can be circumvented using the following rules.

Rule 5.1.1: $\sigma^2(\tau)$ is the particular $\sigma^2(\bar{\alpha}|M')$ with $\bar{\alpha} = \tau$,

Rule 5.1.2: $\sigma^2(\Delta)$ is the sum of all the $\sigma^2(\bar{\alpha}|M')$ except $\sigma^2(\tau|M')$, and

Rule 5.1.3: $\sigma^2(\delta)$ is the sum of all the $\sigma^2(\bar{\alpha}|M')$ such that $\bar{\alpha}$ includes τ and at least one of the indices in $\mathcal{R}$.

TABLE 5.1. D Studies for the APL $p \times (I{:}H)$ Design for the Southern Region

$\hat{\sigma}^2(\alpha\|H)$	D Studies		
	n'_h	5	5
	n'_i	8	20
$\hat{\sigma}^2(p\|H) = .0378$	$\hat{\sigma}^2(p\|H)$	.0378	.0378
$\hat{\sigma}^2(h\|H) = .0013^a$	$\hat{\sigma}^2(H\|H)$	—	—
$\hat{\sigma}^2(i{:}h\|H) = .0259$	$\hat{\sigma}^2(I{:}H\|H)$	.0006	.0003
$\hat{\sigma}^2(ph\|H) = .0051$	$\hat{\sigma}^2(pH\|H)$	---	—
$\hat{\sigma}^2(pi{:}h\|H) = .1589$	$\hat{\sigma}^2(pI{:}H\|H)$	.0040	.0016
	$\hat{\sigma}^2(\tau)$	.0378	.0378
	$\hat{\sigma}^2(\delta)$	.0040	.0016
	$\hat{\sigma}^2(\Delta)$	.0046	.0019
	$E\hat{\rho}^2$	.905	.959
	$\hat{\Phi}$	.892	.952

^{a}Strictly speaking, $\hat{\sigma}^2(h\|H)$ is a quadratic form.

These rules, which bear an obvious similarity tp those for the random model, rely heavily on the fact that $\sigma^2(\bar{\alpha}|M') = 0$ when $n' = N'$ for a primary index in $\bar{\alpha}$ (except τ). Given these rules, the usual formulas apply for $\sigma^2(\overline{X})$, $E\rho^2$, and Φ (see Table 4.3 on page 109), as well as for signal–noise and error–tolerance ratios.

5.1.3 Examples

The procedures discussed in Sections 5.1.1 and 5.1.2 are general enough to apply to virtually all D studies for balanced designs, although the simplified procedures in Chapter 4 are often sufficient. The examples considered here are illustrative of situations in which the simplified procedures do not suffice.

APL Survey with $N' = N$ for each Facet

Table 3.7 on page 91 reported variance components for the APL Survey discussed in Section 3.5. These variance components are for a $p \times (i{:}h)$ design under a mixed model in which $n_i = 8$ items are nested within each of $n_h = N_h = 5$ fixed content categories. Table 5.1 reports D study results for the same mixed model; that is, $n'_h = N'_h = N_h = 5$ and $n'_i < N'_i = N_i \to \infty$. Both the G and D study perspectives on this example assume that the measurement procedure would always involve the same set of fixed categories, which means that generalization is intended over samples of items (within content areas) only.

Clearly, for this example, $\tau = p$ because persons are the objects of measurement. Since H is fixed, we designate the model simply as H. The D study estimated variance components $\hat{\sigma}^2(\bar{\alpha}|H)$ are obtained using Equation 5.1 since $N' = N$ for both facets. For example, for $\hat{\sigma}^2(I{:}H|H)$ and $\hat{\sigma}^2(pI{:}H|H)$, $d(\bar{\alpha}|p) = n_i'n_h'$, and $C(\bar{\alpha}|p) = 1$ because $N_i' \to \infty$. Therefore,

$$\hat{\sigma}^2(I{:}H|H) = \frac{\hat{\sigma}^2(i{:}h|H)}{n_i'n_h'} \quad \text{and} \quad \hat{\sigma}^2(pI{:}H|H) = \frac{\hat{\sigma}^2(pi{:}h|H)}{n_i'n_h'}.$$

For both $\hat{\sigma}^2(H|H)$ and $\hat{\sigma}^2(pH|H)$, $C(\bar{\alpha}|p) = 0$ because $n_h' = N_h'$. Therefore, both $\hat{\sigma}^2(H|H)$ and $\hat{\sigma}^2(pH|H)$ are zero. To appreciate why these two variance components disappear, recall the notion of replications of a measurement procedure. For this example, each replication yields person mean scores over samples of n_i' items from *every one* of the five content areas. Therefore, the set of content area means is constant for all replications. Effectively, this implies that μ_H gets absorbed into μ and μ_{pH} gets absorbed into μ_p. It follows that $\nu_H = \mu_H - \mu = 0$, which necessarily leads to $\sigma^2(H|H)$ being zero. Also,

$$\nu_{pH} = (\mu_{pH} - \mu_p) - (\mu_H - \mu) = 0 - 0 = 0,$$

which necessarily leads to $\sigma^2(pH|H)$ being zero.

In short, because generalization is over only one facet (items), there are only three nonzero variance components for the D study $p \times (I{:}H)$ design with H fixed. This is directly analogous to the fact that there are three variance components for the D study $p \times I$ design and its single-faceted universe of generalization.

It is important to note that the procedures employed in this example rely heavily on the fact that an equal number of items are nested within each content category. If that were not the case, then the design would be unbalanced and best treated using the multivariate generalizability theory procedures discussed in Chapters 9 and 11.[2]

Mixed Model D Study Given G Study Random Effects Variance Components

Using G study random effects variance components, Section 4.3 considered simplified procedures for obtaining mixed-model D study results. One of the examples involved the $p \times (R{:}T)$ design for Synthetic Data Set No. 4 in Table 3.2, under the assumption that an investigator wanted to generalize to a restricted universe of generalization with tasks fixed at $n_t' = N_t' = 3$. In considering this example, it was pointed out that the random effects estimated variance components $\hat{\sigma}^2(\alpha)$ in Table 4.8 are *not* associated with the restricted universe of generalization, even though they can be used to obtain unbiased estimates of universe score variance, error variances, and coefficients for the restricted universe.

---

[2]The APL example considered here is a very simple version of the table of specifications model outlined later in Section 9.1.

TABLE 5.2. D Studies for Synthetic Data Set No. 4 Using $p \times (R:T)$ Design with Tasks Fixed

$\hat{\sigma}^2(\alpha)$		$\hat{\sigma}^2(\alpha|T)$		D Studies				
				n'_t	3	3	3	3
				n'_r	1	2	3	4
$\hat{\sigma}^2(p) =$	.4731	$\hat{\sigma}^2(p|T) =$	.6596	$\hat{\sigma}^2(p|T)$	.660	.660	.660	.660
$\hat{\sigma}^2(t) =$	.3252	$\hat{\sigma}^2(t|T) =$	$.3252^a$	$\hat{\sigma}^2(T|T)$	—	—	—	—
$\hat{\sigma}^2(r{:}t) =$	.6475	$\hat{\sigma}^2(r{:}t|T) =$	.6475	$\hat{\sigma}^2(R{:}T|T)$	.216	.108	.072	.054
$\hat{\sigma}^2(pt) =$	.5596	$\hat{\sigma}^2(pt|T) =$	.5596	$\hat{\sigma}^2(pT|T)$	—	—	—	—
$\hat{\sigma}^2(pr{:}t) =$	2.3802	$\hat{\sigma}^2(pr{:}t|T) =$	2.3802	$\hat{\sigma}^2(pR{:}T|T)$	.793	.397	.265	.198
				$\hat{\sigma}^2(\tau)$	.66	.66	.66	.66
				$\hat{\sigma}^2(\delta)$	.79	.40	.26	.20
				$\hat{\sigma}^2(\Delta)$	1.01	.50	.34	.25
				$E\hat{\rho}^2$	.45	.62	.71	.77
				$\hat{\Phi}$	.40	.57	.66	.72

^{a}Strictly speaking, $\hat{\sigma}^2(t|T)$ is a quadratic form.

Table 5.2 provides estimated variance components for the actual mixed model. The first column provides the random effects estimated G study variance components $\hat{\sigma}^2(\alpha)$, which we assume were provided to the investigator. The second column provides estimated G study variance components for the mixed model that is actually of interest. These estimates can be obtained using Equation 3.34. The remaining columns provide estimated D study variance components for the mixed model, which can be obtained using Equation 5.4. Note, in particular, that for this mixed model $C(T|p) = C(pT|p) = 0$, which means that $\sigma^2(T|T)$ and $\sigma^2(pT|T)$ disappear, and R is the only facet over which generalization occurs.

The bottom of Table 5.2 provides D study statistics for the mixed model based on using Rules 5.1.1 to 5.1.3. These statistics are necessarily the same as those reported in Table 4.8.

Sampling from a Finite Universe

The equations, rules, and procedures at the beginning of this section apply to models involving sampling from a finite universe. Such models are rare in generalizability theory, but they do occur occasionally (see, e.g., Section 6.4 and Cronbach et al., 1997). Results for sampling from a finite universe will not differ greatly from results for an infinite universe if the correction factors

TABLE 5.3. D Studies for Synthetic Data Set No. 4 Using the $p \times (R:T)$ Design for Sampling $n_t' = 2$ Tasks from a Finite Universe of $N_t = N_t' = 3$ Tasks

$\hat{\sigma}^2(\alpha|N_t = 3)$	$d(\bar{\alpha}|p)$	$C(\bar{\alpha}|p)$	$\hat{\sigma}^2(\bar{\alpha}|N_t' = 3)$ for $n_t' = 2$ and $n_r' = 6$
		D Study	
$\hat{\sigma}^2(p|N_t = 3) = .6596$	1	1	$\hat{\sigma}^2(p|N_t' = 3) = .660$
$\hat{\sigma}^2(t|N_t = 3) = .3252$	2	1/3	$\hat{\sigma}^2(T|N_t' = 3) = .054$
$\hat{\sigma}^2(r:t|N_t = 3) = .6475$	12	1	$\hat{\sigma}^2(R:T|N_t' = 3) = .054$
$\hat{\sigma}^2(pt|N_t = 3) = .5596$	2	1/3	$\hat{\sigma}^2(pT|N_t' = 3) = .093$
$\hat{\sigma}^2(pr:t|N_t = 3) = 2.3802$	12	1	$\hat{\sigma}^2(pR:T|N_t' = 3) = .198$
			$\hat{\sigma}^2(\tau) = .66$
			$\hat{\sigma}^2(\delta) = .29$
			$\hat{\sigma}^2(\Delta) = .40$
			$E\hat{\rho}^2 = .69$
			$\hat{\Phi} = .62$

$C(\bar{\alpha}|\tau)$ are close to one. When correction factors are considerably smaller than one, however, differences can be important.

Consider, again, the $p \times (R:T)$ design and Synthetic Data Set No. 4 under the assumptions that:

1. each replication of the D study design involves sampling $n_t' = 2$ tasks from a *finite* facet of size $N_t' = 3$, and

2. each of the two tasks is evaluated by a random sample of $n_r' = 6$ raters from a facet of infinite size.

Presumably, the $N_t' = 3$ tasks in the universe of generalization are the actual $n_t = 3$ tasks in the G study.[3]

Under these assumptions, Table 5.3 provides both G and D study results.[4] The first column in Table 5.3 provides the G study estimated variance components for $N_t = 3$, which are identified as $\hat{\sigma}^2(\alpha|N_t = 3)$. They can be obtained using Equation 3.34. These estimates are necessarily identical to the $\hat{\sigma}^2(\alpha|H)$ in Table 5.2, because the facet sizes (N) for the universe (of admissible observation) are the same for both sets of estimates.

The last column in Table 5.3 provides estimates of the D study variance components. Note, in particular, that $\hat{\sigma}^2(T|N_t' = 3)$ and $\hat{\sigma}^2(pT|N_t' = 3)$ are both nonzero because their correction factors are nonzero. None of the estimated variance components disappear because sampling is involved for

---

[3] If this were not true, interpretations would be strained, at best.

[4] The G study random effects estimated variance components are, of course, the same as those in Table 5.2.

both facets in the universe of generalization, even though one of these facets (tasks) is of finite size.

It is instructive to compare D study statistics for sampling from a finite universe in Table 5.3 with the corresponding results for the random model and the mixed model. With $n'_r = 6$ randomly selected raters, and with $n'_t = 2$, these results are as follows.

Model	$\hat{\sigma}^2(\tau)$	$\hat{\sigma}^2(\delta)$	$\hat{\sigma}^2(\Delta)$	$Est[\boldsymbol{ES}^2(\tau)]$
Random $(N'_t \to \infty)$	.47	.48	.69	.95
Finite sampling $(N'_t = 3)$	.66	.29	.40	.95
Mixed $(N'_t = 2)$	.75	.20	.25	.95

The pattern of these results is predictable based solely on the size of N'_t:

- universe score variance increases as the size of the universe of generalization decreases,

- error variances decrease as the size of the universe of generalization decreases, and

- expected observed score variance is unaffected by the size of the universe of generalization.

5.1.4 Hidden Facets

Any reasonably complete discussion of generalizability theory distinguishes between universes of admissible observations and universes of generalization, as well as between G studies and D studies. Yet, it is evident that some applications of the theory do not require sharp distinctions among all four concepts. For example, when only one set of data is available and there is no intent to speculate about generalizability for different sample sizes, it can be awkward to talk about two studies. The role of "hidden" facets in generalizability theory, however, makes important use of all four concepts.

Essentially, a facet is hidden when there is only one sampled condition of the facet, which induces interpretational complexities and (usually) bias in statistics such as $\boldsymbol{E}\hat{\rho}^2$. It is important to understand that generalizability theory does not create these complexities; rather it helps an investigator understand these complexities, isolate sources of bias, and even estimate them. These matters are illustrated next using examples of a hidden fixed facet and a hidden random facet.

Hidden Fixed Facets and Internal Consistency Coefficients

A coefficient of internal consistency such as Cronbach's (1951) alpha or Kuder and Richardson's (1937) Formula 20 is typically obtained using data

TABLE 5.4. Two Perspectives on the $p \times I$ Design

$\hat{\sigma}^2(\alpha)$	$\hat{\sigma}^2(\alpha \mid N_o = 1)$	D Studies: $n'_o = 1$ and $n'_i = 4$	
		$\hat{\sigma}^2(\bar{\alpha} \mid N'_o = 1)$	$\hat{\sigma}^2(\bar{\alpha})$
$\hat{\sigma}^2(p) = .5528$	$\hat{\sigma}^2(p\mid o) = .6537$	$\hat{\sigma}^2(p\mid o) = .654$	$\hat{\sigma}^2(p) = .553$
$\hat{\sigma}^2(i) = .4417$	$\hat{\sigma}^2(i\mid o) = .5982$	$\hat{\sigma}^2(I\mid o) = .150$	$\hat{\sigma}^2(I) = .110$
$\hat{\sigma}^2(o) = .0074$			$\hat{\sigma}^2(o) = .007$
$\hat{\sigma}^2(pi) = .5750$	$\hat{\sigma}^2(pi\mid o) = 1.5102$	$\hat{\sigma}^2(pI\mid o) = .378$	$\hat{\sigma}^2(pI) = .144$
$\hat{\sigma}^2(po) = .1009$			$\hat{\sigma}^2(po) = .101$
$\hat{\sigma}^2(io) = .1565$			$\hat{\sigma}^2(Io) = .039$
$\hat{\sigma}^2(pio) = .1565$			$\hat{\sigma}^2(pIo) = .234$
		$\hat{\sigma}^2(\tau) = .65$	$\hat{\sigma}^2(\tau) = .55$
		$\hat{\sigma}^2(\delta) = .38$	$\hat{\sigma}^2(\delta) = .48$
		$\boldsymbol{E}\hat{\rho}^2 = .63$	$\boldsymbol{E}\hat{\rho}^2 = .54$

collected on a single occasion, although users often interpret such results as if they generalize to other occasions. Let us consider this situation from the perspective of generalizability theory.

Specifically, recall Synthetic Data Set No. 3 for the $p \times i \times o$ design in Table 3.1 on page 72, which has four items administered on two occasions. The first column in Table 5.4 reports the G study random effects estimated variance components. The last column reports the D study estimated variance components and typical statistics for the random model, under the assumptions that $n'_i = 4$ and a single occasion is randomly sampled from a universe of occasions. Let us contrast these results with those for a single fixed occasion.

The second column in Table 5.4 provides G study estimated variance components for $N_o = 1$, which can be obtained from the random effects variance components in the first column using Equation 3.34:

$$\begin{aligned}
\hat{\sigma}^2(p\mid o) &= \hat{\sigma}^2(p) + \hat{\sigma}^2(po) \\
\hat{\sigma}^2(i\mid o) &= \hat{\sigma}^2(i) + \hat{\sigma}^2(io) \quad\quad\quad (5.11) \\
\hat{\sigma}^2(pi\mid o) &= \hat{\sigma}^2(pi) + \hat{\sigma}^2(pio). \quad\quad\quad (5.12)
\end{aligned}$$

The third column provides D study estimated variance components for $N'_o = 1$, which can be obtained using Equation 5.4 or, more simply, by dividing Equations 5.11 and 5.12 by $n'_i = 4$. Note that, by comparison with the random model results, for a single fixed occasion,

- error variances are smaller, and

- universe score variance and coefficients are larger.

Often, of course, data are not collected on two occasions, and reliability is estimated simply by computing KR-20 or Cronbach's alpha. The results in Table 5.4 illustrate that if generalization is intended *over* occasions, then such coefficients will *over*estimate reliability. This conclusion directly contradicts the conventional wisdom that Cronbach's alpha is a lower limit to reliability![5]

The lower-limit argument is promulgated, for example, by Lord and Novick (1968) in their section on "Coefficient α and the Reliability of Composite Measurements." As discussed by Brennan (in press), the problem with the Lord and Novick proof and all others like it is *not* that there is a mathematical error; rather, the problem is that such proofs fail to differentiate between the universe of generalization that is of interest to an investigator and the characteristics of available data. In particular, if data are collected on a single occasion, an estimate of reliability based on such data will almost certainly overestimate reliability when interest is in generalizing *over* occasions.

When only one condition of a facet is sampled, that facet is *hidden* from the investigator in the sense that variance components associated with that facet are confounded with other variance components. If all observations are influenced by the *same* level of the facet, then analyses of the data will effectively treat the facet as fixed with $n = n' = N = N' = 1$. That is why we often refer to occasion as a hidden *fixed* facet when Cronbach's alpha is computed. However, a hidden facet is not always fixed.

Hidden Random Facets and Performance Assessments

Suppose, for example, that each student (p) at a piano recital played a musical selection (m) of his or her own choosing, with all students evaluated by the same two judges (j). Since each student played only one musical piece, the m facet is hidden. Importantly, however, students were not required to play the same musical selection and, therefore, m is *not* fixed. Rather, m is more reasonably considered random. To keep matters simple we assume that each student played a different musical selection, and the universe of generalization has musical selections crossed with judges; that is, in theory, each judge could evaluate performance for each musical selection.

For this universe of generalization with sample sizes of $n'_m = n_m = 1$ and $n'_j = n_j = 2$, the generalizability coefficient for the random effects D study $p \times m \times J$ design is

$$E\rho_1^2 = \frac{\sigma^2(p)}{\sigma^2(p) + \left[\dfrac{\sigma^2(pj)}{2} + \dfrac{\sigma^2(pm)}{1} + \dfrac{\sigma^2(pjm)}{2}\right]}. \tag{5.13}$$

Presumably, it is this coefficient that an investigator would like to estimate.

---

[5]There are certainly cases in which alpha is properly interpreted as a lower limit to reliability. Recall, for example, Exercise 4.10.

However, the recital has students (p) completely confounded with musical selections (m). Therefore, the $p \times j$ design that characterizes the available data can be denoted more explicitly $(p, m) \times j$, where the notation (p, m) signifies that p and m are completely confounded. This means that

$$\sigma^2(p, m) = \sigma^2(p) + \sigma^2(m) + \sigma^2(pm),$$

where variance components to the right are for the $p \times m \times j$ design. Similarly,

$$\sigma^2[(p, m)j] = \sigma^2(pj) + \sigma^2(jm) + \sigma^2(pjm).$$

It follows that $E\hat{\rho}^2$ for the $(p, m) \times J$ design that characterizes the available data is actually an estimate of

$$
\begin{aligned}
E\rho_2^2 &= \frac{\sigma^2(p, m)}{\sigma^2(p, m) + \left[\dfrac{\sigma^2[(p, m)j]}{2}\right]} \\
&= \frac{\sigma^2(p) + \sigma^2(m) + \sigma^2(pm)}{\sigma^2(p) + \sigma^2(m) + \sigma^2(pm) + \left[\dfrac{\sigma^2(pj)}{2} + \dfrac{\sigma^2(jm)}{2} + \dfrac{\sigma^2(pjm)}{2}\right]}.
\end{aligned}
$$
$$(5.14)$$

Equation 5.13, with the subscript 1, designates the conceptualization for the intended universe of generalization. Equation 5.14, with the subscript 2, designates the statistic that is estimated using available data. Clearly, Equation 5.14 is not associated with the intended universe of generalization. It is evident from these equations that $\sigma_1^2(\tau) \leq \sigma_2^2(\tau)$, but this does not guarantee that $E\rho_1^2 \leq E\rho_2^2$. However, if

$$\sigma^2(jm) \leq 2\,\sigma^2(pm),$$
$$(5.15)$$

then $\sigma_1^2(\delta) \geq \sigma_2^2(\delta)$ and $E\rho_1^2 \leq E\rho_2^2$. Conversely, if

$$\sigma^2(jm) \gg \sigma^2(m) + \sigma^2(pm),$$

then it can be true that $E\rho_1^2 > E\rho_2^2$.[6]

Clearly, when the m facet is hidden but random, the bias in $E\hat{\rho}_2^2$ using the $(p, m) \times J$ design can be either positive or negative, based largely on the magnitude of $\sigma^2(jm)$. This variance component reflects the extent to which judges rank order the difficulty of the musical selections differently. With well-trained judges, it seems unlikely that $\sigma^2(jm)$ would be very large, and the condition in Equation 5.15 is likely to be true. If so, $E\hat{\rho}_2^2$ using the $(p, m) \times J$ design will be an upper limit to reliability for the intended universe of generalization.

---

[6]The symbol $\gg$ means "very much bigger than."

This "musical selection" example reasonably well reflects some of the important considerations involved in assessing the reliability of some types of performance assessments, including portfolio assessments, that have a hidden random facet. It is particularly important to note that a hidden *random* facet induces bias different from that induced by a hidden *fixed* facet, and the magnitude (and sometimes the direction) of such bias cannot be ascertained from an analysis of data in which the facet is hidden. When decisions are to be made based on $n' = 1$ for some facet, a G study with $n > 1$ for that facet is extraordinarily useful, and often essential, because it permits disentangling sources of variance that are otherwise indistinguishable in analyses of the data with $n' = 1$.

The "musical selection" example can be extended further by hypothesizing a hidden occasion facet. That is, suppose generalization is intended over occasions, but the data used to estimate $E\rho^2$ come from only one recital. Then, there are two hidden facets that influence the magnitude $E\hat{\rho}^2$:

1. the hidden occasion facet that is fixed, which likely causes $E\hat{\rho}^2$ to be higher than it should be; and

2. the hidden musical-selection facet that is random, which can cause $E\hat{\rho}^2$ to be either higher or lower than it should be.

Clearly, for this example, a G study with $n_m > 1$ and $n_o > 1$ would be extraordinarily useful in that it would permit estimating the variance components required for an estimate of reliability for the intended universe of generalization. This thought experiment indicates how generalizability theory helps an investigator understand the consequences of hidden facets on characteristics of a measurement procedure.

Serious consideration of many, if not most, real-world studies reveals hidden facets. A desire for simplicity may lead investigators to disregard such facets, but doing so can lead to misleading interpretations.

5.2   Stratified Objects of Measurement

Often, objects of measurement are stratified with respect to some other variable, and an investigator may be interested in variability within levels of the stratification variable as well as variability across levels. To discuss and illustrate these matters, consider, again, the APL Survey. Previous results (see Tables 3.7 on page 91 and 5.1 on page 145) have focused on only one region of the country (south). Table 5.5 provides results for 608 persons in each of four regions (northeast, north central, south, and west) as well as a "global" analysis for the undifferentiated group of 2432 persons in all regions r_+. We assume here that regions (r) are fixed and, as before, content categories (h) are fixed.

TABLE 5.5. APL Survey Variance Components for All Four Regions

Effect	\multicolumn{5}{c}{$\hat{\sigma}^2(\alpha\|H)$}					\multicolumn{5}{c}{$\hat{\sigma}^2(\bar{\alpha}\|H)$}				
	r_1	r_2	r_3	r_4	r_+	r_1	r_2	r_3	r_4	r_+
p	.0446	.0418	.0378	.0403	.0425	.0446	.0418	.0378	.0403	.0425
h	.0024	.0024	.0013	.0037	.0024	—	—	—	—	—
$i{:}h$	.0272	.0244	.0259	.0203	.0242	.0007	.0006	.0006	.0005	.0006
ph	.0056	.0048	.0051	.0050	.0051	—	—	—	—	—
$pi{:}h$	.1597	.1431	.1589	.1383	.1503	.0040	.0036	.0040	.0035	.0038
$\overline{X}$	.6130	.6886	.6520	.7117	.6663					

$$\hat{\sigma}^2(\tau) \quad .0446 \quad .0418 \quad .0378 \quad .0403 \quad .0425$$
$$\hat{\sigma}^2(\delta) \quad .0040 \quad .0036 \quad .0040 \quad .0035 \quad .0038$$
$$\hat{\sigma}^2(\Delta) \quad .0047 \quad .0042 \quad .0046 \quad .0040 \quad .0044$$

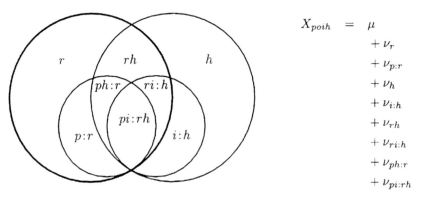

$$X_{poih} = \mu$$
$$+ \nu_r$$
$$+ \nu_{p:r}$$
$$+ \nu_h$$
$$+ \nu_{i:h}$$
$$+ \nu_{rh}$$
$$+ \nu_{ri:h}$$
$$+ \nu_{ph:r}$$
$$+ \nu_{pi:rh}$$

FIGURE 5.1. Representations of $(p{:}r) \times (i{:}h)$ design.

TABLE 5.6. APL Survey Variance Components for $(p{:}r) \times (i{:}h)$ Design

Effect	MS	$\hat{\sigma}^2(\alpha\|RH)$	\multicolumn{2}{l}{D Study $\sigma^2(\bar{\alpha}\|RH)$}
			Est.	Cornfield and Tukey Definitions
r	45.4125	.0018	.0018	$\sigma^2(r\|RH) = \sum_r \nu_r^2/(n_r - 1)$
$p{:}r$	1.7954	.0411	.0411	$\sigma^2(p{:}r\|RH) = \sum_r (E\nu_{p:r}^2)/n_r$
h	106.0222	.0024	—	
$i{:}h$	58.9846	.0242	.0006	$\sigma^2(I{:}H\|RH) = E\nu_{I:H}^2$
rh	.4078	.0000	—	
$ri{:}h$	.3609	.0003	.0000	$\sigma^2(rI{:}H\|RH) = \sum_r (E\nu_{rI:H}^2)/(n_r - 1)$
$ph{:}r$	.1910	.0051	—	
$pi{:}rh$	.1500	.1500	.0038	$\sigma^2(pI{:}rH\|RH) = \sum_r (E\nu_{pI:rH}^2)/n_r$

If regions were random, the Cornfield and Tukey definitions imply that the multiplicative factor f_r no longer applies or, stated differently, f_r should be set to unity. Obviously, this somewhat simplifies the relationships among variance components discussed in Section 5.2.1. Otherwise, however, the rules and relationships still apply when regions are random.

Sometimes variance components for a stratified analysis are available, but results for the individual strata are not. In this case, D study results for a randomly sampled stratum can be obtained using the variance components that do *not* contain the strata index in a primary position. For example, for the APL Survey, the $\hat{\sigma}^2(\alpha|RH)$ that do not contain r as a primary index are those for $p{:}r$, h, $i{:}h$, $ph{:}r$, and $pi{:}rh$ in Table 5.6. These variance components can be used with the rules and procedures in Section 5.1 (see Exercise 5.6).

Throughout this section it has been assumed that the design is balanced in two senses—an equal number of items within categories and an equal number of persons within regions. If either of these conditions is not fulfilled, then the design is unbalanced, the procedures discussed here do not apply, and appropriate procedures are much more complicated.

5.3   Conventional Wisdom About Group Means

Sometimes it is asserted that if a test is not reliable enough for making decisions about individuals, or if error variance for individuals is unacceptably large, then test scores can still be used for making decisions about groups. Implicit in such a statement is an assumption that reliability for groups is necessarily larger than reliability for persons, and/or error variance for groups is necessarily smaller than error variance for persons. Brennan (1995a), shows that this "conventional wisdom" about group means is not necessarily true. These results are summarized here.

5.3.1   Two Random Facets

Consider the random effects $(p{:}g) \times i$ design with persons (p) nested within groups (g) and crossed with items (i). Suppose that groups are the objects of measurement, and the universe of generalization has $P \times I$, with both facets being random. For this design and universe, as discussed in Section 4.4.4, the generalizability of group means is

$$E\rho_g^2 = \frac{\sigma^2(g)}{\sigma^2(g) + \sigma^2(p{:}g)/n_p + \sigma^2(gI) + \sigma^2(pI{:}g)/n_p}. \qquad (5.22)$$

By contrast, if persons within a single randomly selected group are the objects of measurement,

$$E\rho_{p:g}^2 = \frac{\sigma^2(p{:}g)}{\sigma^2(p{:}g) + \sigma^2(pI{:}g)}, \tag{5.23}$$

and if persons *over* groups are the objects of measurement,

$$E\rho_p^2 = \frac{\sigma^2(g) + \sigma^2(p{:}g)}{\sigma^2(g) + \sigma^2(p{:}g) + \sigma^2(gI) + \sigma^2(pI{:}g)}. \tag{5.24}$$

Usually, when comparative statements are made about reliability coefficients for groups and persons, the intended interpretation of reliability for persons is given by Equation 5.24. Therefore, the conventional wisdom is violated when $E\rho_g^2 < E\rho_p^2$. It is immediately obvious that this inequality is true under the trivial (although unlikely) conditions that $\sigma^2(g) = 0$ or $n_i' \to \infty$. Also, Brennan (1995a) shows that this inequality is satisfied when

$$\frac{\sigma^2(g)}{\sigma^2(g) + \sigma^2(p{:}g)} < \frac{\sigma^2(gi)}{\sigma^2(gi) + \sigma^2(pi{:}g)}. \tag{5.25}$$

That is, Inequality 5.25 is a sufficient condition for the conventional wisdom to be violated.

As an example, refer again to the Kane et al. (1976) course-evaluation-questionnaire study discussed in Section 4.4.4. Recall that for the attribute items, the G study estimated variance components were:[7]

$$\hat{\sigma}^2(g) = .03, \quad \hat{\sigma}^2(p{:}g) = .17, \quad \hat{\sigma}^2(i) = .13,$$

$$\hat{\sigma}^2(gi) = .05 \quad \text{and} \quad \hat{\sigma}^2(pi{:}g) = .28.$$

Using Inequality 5.25,

$$\frac{\hat{\sigma}^2(g)}{\hat{\sigma}^2(g) + \hat{\sigma}^2(p{:}g)} = \frac{.03}{.20} = .150 < \frac{\hat{\sigma}^2(gi)}{\hat{\sigma}^2(gi) + \hat{\sigma}^2(pi{:}g)} = \frac{.05}{.33} = .152.$$

Therefore, the sufficient condition is satisfied, and $E\rho_g^2 < E\rho_p^2$ for all pairs of values of n_p and n_i. For example, if $n_p' = 20$ and $n_i' = 8$,

$$E\hat{\rho}_g^2 = .65 < E\hat{\rho}_p^2 = .83.$$

Inequality 5.25 is a sufficient condition for the conventional wisdom about reliability coefficients to be violated, but it is not a *necessary* condition. That is, there are circumstances that lead to $E\rho_g^2 < E\rho_p^2$ even when Inequality 5.25 is not satisfied (see Exercise 5.7).

---

[7]Here group (g) is used instead of class (c).

The conventional wisdom also suggests that error variance for groups is less than error variance for persons. However, this conventional wisdom is not necessarily true, either. Relative error variance for group means is:

$$\sigma^2(\delta_g) = \sigma^2(p{:}g)/n_p + \sigma^2(gI) + \sigma^2(pI{:}g)/n_p, \qquad (5.26)$$

and relative error variance for person means over groups is:

$$\sigma^2(\delta_p) = \sigma^2(gI) + \sigma^2(pI{:}g). \qquad (5.27)$$

The conventional wisdom is violated when $\sigma^2(\delta_g) > \sigma^2(\delta_p)$. It can be shown (see Exercise 5.8) that a necessary condition for this inequality is

$$E\rho^2_{p{:}g} > \frac{n_p - 1}{n_p}. \qquad (5.28)$$

Clearly, when $n_p \to \infty$, Inequality 5.28 will not hold. So, for large values of n_p, it is reasonable to assume that relative error variance for persons is likely to be larger than relative error variance for groups, as the conventional wisdom suggests. However, for small values of n_p, this need not be so. Two sets of conditions that give $\sigma^2(\delta_g) > \sigma^2(\delta_p)$ are: (a) $n_p < 20$ and $E\rho^2_{p{:}g} = .95$, and (b) $n_p < 10$ and $E\rho^2_{p{:}g} = .90$. These examples are not so extreme as to be implausible, especially for long tests. Consequently, it is unwise to assume that error variance for person mean scores is always greater than error variance for group mean scores.

5.3.2   One Random Facet

When persons are fixed and items are random, as discussed in Section 4.4.4,

$$E\rho^2_g(I) = \frac{\sigma^2(g) + \sigma^2(p{:}g)/n_p}{\sigma^2(g) + \sigma^2(p{:}g)/n_p + [\sigma^2(gI) + \sigma^2(pI{:}g)/n_p]}. \qquad (5.29)$$

By comparing the relative error variance for group means in square brackets with the over-all-persons relative error variance in Equation 5.27, it is evident that the conventional wisdom holds for relative error variances. However, it does not always hold for reliability coefficients. Inequality 5.25 is a necessary condition for $E\rho^2_g(I) < E\rho^2_p$.

5.4   Conditional Standard Errors of Measurement

For many decades, it has been recognized that standard errors of measurement vary as a function of true scores. The importance of this issue is recognized by the current *Standards for Educational and Psychological Testing* (AERA/APA/NCME, 1999), as well as its predecessors. Perhaps the best known example of a *conditional* SEM formula was derived by Lord (1955, 1957) based on the binomial error model. Feldt and Brennan (1989) provide a review of many procedures for estimating conditional SEMs. Feldt and Qualls (1996) provide a more recent review. Brennan (1996a, 1998) provides an extensive discussion of conditional SEMs in generalizability theory. This section summarizes those parts of Brennan (1996a, 1998) that consider univariate balanced designs.[8] Later chapters treat conditional SEMs for unbalanced and multivariate designs.

5.4.1 Single-Facet Designs

Section 2.1 considered the assumptions for the single-facet G study $p \times i$ design:

$$X_{pi} = \mu + \nu_p + \nu_i + \nu_{pi},$$

where $\nu_{pi} = X_{pi} - \nu_p - \nu_i + \mu$ is the residual effect that includes effects attributable to the pi interaction as well as other unexplained sources of variability. In Section 2.1 it was noted that most effects are necessarily uncorrelated because score effects are defined in terms of mean scores in generalizability theory. For example,

$$\mathop{E}_{p} \nu_p\nu_i = \mathop{E}_{i} \nu_p\nu_i = \mathop{E}_{i}\mathop{E}_{p} \nu_p\nu_{pi} = \mathop{E}_{p}\mathop{E}_{i} \nu_i\nu_{pi} = 0.$$

Note, however, that $\mathbf{E}_p(\nu_p\nu_{pi})$ and $\mathbf{E}_i(\nu_i\nu_{pi})$ are not necessarily zero. As will be shown, the nonzero expectation of $\nu_i\nu_{pi}$ influences conditional relative SEMs, but it is irrelevant for conditional absolute SEMs.

Conditional Absolute SEM

Since absolute error for person p is $\Delta_p = X_{pI} - \mu_p$, the associated error variance is

$$\sigma^2(\Delta_p) \equiv \text{var}(X_{pI} - \mu_p|p), \tag{5.30}$$

which is the variance of the mean over n_i' items for person p. The average over persons of $\sigma^2(\Delta_p)$ is $\sigma^2(\Delta)$.

An unbiased estimator is

$$\hat{\sigma}^2(\Delta_p) = \frac{\text{var}(X_{pi}|p)}{n_i'} = \frac{\sum_i (X_{pi} - X_{pI})^2}{n_i'(n_i - 1)}, \tag{5.31}$$

where n_i' and n_i need not be the same. The average of the estimates using $\hat{\sigma}^2(\Delta_p)$ in Equation 5.31 will be the usual estimate of $\sigma^2(\Delta)$.

---

[8]Various parts of this section are taken from Brennan (1998) with the permission of the publisher, Sage.

Of course, the square root of Equation 5.31 provides an estimator of conditional absolute SEM:

$$\hat{\sigma}(\Delta_p) = \sqrt{\frac{\sum_i (X_{pi} - X_{pI})^2}{n'_i(n_i - 1)}}. \tag{5.32}$$

If items are scored dichotomously and $n'_i = n_i$, this estimator is Lord's (1955, 1957) conditional SEM:

$$\hat{\sigma}(\Delta_p) = \sqrt{\frac{X_{pI}(1 - X_{pI})}{n_i - 1}}. \tag{5.33}$$

Whether or not items are scored dichotomously, when $n'_i = n_i = 2$, as often occurs in performance assessments,

$$\hat{\sigma}(\Delta_p) = \sqrt{\frac{1}{2}\sum_{i=1}^{2}(X_{pi} - X_{pI})^2} = \frac{|X_{p1} - X_{p2}|}{2}.$$

These conditional absolute SEM results have been discussed for the $p \times I$ design, but results are identical for the nested $I:p$ design. That is, the *within*-person (i.e., conditional) absolute SEM is unaffected by the across-persons design.

Conditional Relative SEM

Relative error for person p is

$$\delta_p = (X_{pI} - \mu_I) - (\mu_p - \mu) = (X_{pI} - \mu_p) - (\mu_I - \mu).$$

It follows that

$$
\begin{aligned}
\sigma^2(\delta_p) &= \text{var}(X_{pI} - \mu_p | p) + \text{var}(\mu_I - \mu) - 2\,\text{cov}(X_{pI} - \mu_p, \mu_I - \mu | p) \\
&= \sigma^2(\Delta_p) + \sigma^2(I) - 2\,\text{cov}(\nu_I + \nu_{pI}, \nu_I | p) \\
&= \sigma^2(\Delta_p) + \sigma^2(I) - 2\left[\text{var}(\nu_I) + \text{cov}(\nu_I, \nu_{pI} | p)\right] \\
&= \sigma^2(\Delta_p) + \frac{\sigma^2(i)}{n'_i} - 2\left[\frac{\sigma^2(i)}{n'_i} + \frac{\sigma(i, pi|p)}{n'_i}\right],
\end{aligned}
\tag{5.34}
$$

where

$$\sigma(i, pi|p) = \text{cov}(\nu_i, \nu_{pi}|p) = \mathop{E}_i \nu_i \nu_{pi} \tag{5.35}$$

is the covariance between item and residual effects for person p. This covariance is not necessarily zero, as noted at the beginning of this section. The form of Equation 5.34 indicates that $\sigma^2(\delta_p)$ can be viewed as an adjustment to $\sigma^2(\Delta_p)$, an adjustment that can be positive or negative. The square root of Equation 5.34 is the conditional relative SEM for person p.

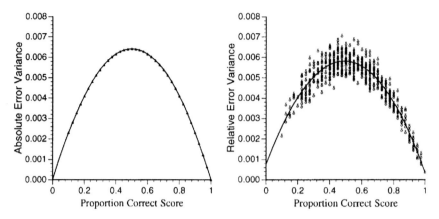

FIGURE 5.2. Conditional error variances for ITED Vocabulary Test.

Using results from Jarjoura (1986), Brennan (1998) shows that when n_p is large an approximate estimator of the conditional relative SEM is

$$\hat{\sigma}(\delta_p) \doteq \sqrt{\hat{\sigma}^2(\Delta_p) + \frac{\hat{\sigma}^2(i)}{n_i'} - \frac{2 \operatorname{cov}(X_{pi}, X_{Pi}|p)}{n_i'}}, \qquad (5.36)$$

where

$$\operatorname{cov}(X_{pi}, X_{Pi}|p) = \frac{\sum_i (X_{pi} - X_{pI})(X_{Pi} - X_{PI})}{n_i - 1} \qquad (5.37)$$

is the observed covariance over items between examinee p's item scores and the item mean scores. For dichotomously scored items, this is the covariance between examinee p's item scores and item difficulty levels. Whether or not items are scored dichotomously, when $n_i' = n_i = 2$,

$$\hat{\sigma}(\delta_p) \doteq \frac{|(X_{p1} - X_{P1}) - (X_{p2} - X_{P2})|}{2}.$$

Obviously, $\sigma(\delta_p)$ is complicated by the covariance term $\sigma(i, pi|p)$. Considerable simplification occurs if it is assumed that $\sigma(i, pi|p) = 0$. Then

$$\hat{\sigma}(\delta_p) = \sqrt{\hat{\sigma}^2(\Delta_p) - \hat{\sigma}^2(I)}. \qquad (5.38)$$

The principal theoretical problem with Equation 5.38 is that the adjustment to $\hat{\sigma}^2(\Delta_p)$ is always negative—namely, $\hat{\sigma}^2(I)$—although it is known from Equation 5.34 that the true adjustment may be positive or negative.

Examples

Figure 5.2 provides estimates of $\sigma^2(\Delta_p)$ and $\sigma^2(\delta_p)$ for a spaced sample of 493 eleventh graders in Iowa who took Level 17/18 (Form K) of the 40-item ITED Vocabulary test (Feldt et al., 1994) in Fall 1995. Because items

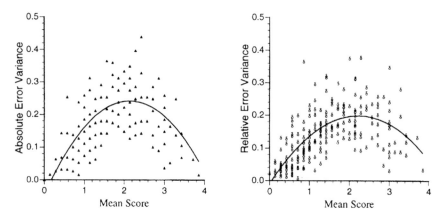

FIGURE 5.3. Conditional error variances for QUASAR Project.

were dichotomously scored, the functional form of absolute error variance conditional on universe score is a concave-down quadratic, $\mu_p(1-\mu_p)/n_i'$. A similar statement holds for the estimates (see Equation 5.33), as illustrated by the perfect quadratic fit in the left panel of Figure 5.2.

For conditional relative error variance, the estimates (conditional on $\overline{X}_p$) have considerable dispersion, as illustrated in the right panel of Figure 5.2. Since there is error in the $\hat{\sigma}^2(\delta_p)$, and since practical circumstances often require a single value of $\hat{\sigma}^2(\delta_p)$ for all examinees with the same $\overline{X}_p$, it is natural to consider using the best fitting quadratic, which is

$$\hat{\sigma}^2(\delta_p) = .00075 + .02063\,\overline{X}_p - .02101\,\overline{X}_p^2.$$

Figure 5.2 suggests that this quadratic provides a good fit, and, therefore, the square roots might be used as conditional relative SEMs. Alternatively, for $\hat{\sigma}(\delta_p)$, the simpler estimates based on Equation 5.38 might be used. For this example, both sets of estimates are quite similar (see Exercise 5.10). One advantage of quadratic fits (or any polynomial fit, for that matter) is that the average of the fitted values is identical to the average of the unfitted values.[9] The obvious advantage of Equation 5.38 is that it is simple.

Figure 5.3 provides estimates of conditional error variances, along with quadratic fits, for 229 seventh graders who took the mathematics performance tasks that are part of the QUASAR project (Lane et al., 1996). The estimates are for $n_t' = 9$ polytomously scored tasks. The most notable difference between Figures 5.3 and 5.2 is that, for specific values of $\overline{X}_p$, the $\hat{\sigma}^2(\Delta_p)$ for the QUASAR example are quite variable, whereas the $\hat{\sigma}^2(\Delta_p)$ for the ITED example are not. The explanation for this difference is that the QUASAR tasks are scored using a five-point holistic scale, whereas the ITED items are dichotomously scored.

---

[9]Experience suggests that quadratic fits are usually quite adequate for practical use.

Strictly speaking, conditional SEMs are conditional on examinee universe scores. Since universe scores are unknown, however, estimates of conditional SEMs are usually expressed with respect to $\overline{X}_p$, as in Figures 5.2 and 5.3.

5.4.2 Multifacet Random Designs

It is relatively straightforward to extend the notion of conditional absolute SEMs for single-facet designs to multifacet designs. The conceptual leap is facilitated by recalling that, for single-facet designs,

$$\sigma(\Delta_p) = \sqrt{\frac{\text{var(item scores for person } p)}{\text{number of items}}} \qquad (5.39)$$

$$= \text{SE(mean for person } p). \qquad (5.40)$$

That is, $\sigma(\Delta_p)$ is the standard error (SE) of the *within*-person mean.

Recall that, for random models, the standard error of the mean over all sampled conditions of all facets is

$$\sigma(\overline{X}) = \sqrt{\sum \frac{\sigma^2(\alpha)}{\pi'(\alpha)}}, \qquad (5.41)$$

where the summation is taken over all random effects variance components in the design, and $\pi'(\alpha)$ is the product of the D study sample sizes for all the indices in α. This standard error formula, when applied to the within-person design, gives $\sigma(\Delta_p)$. For example, if the across-persons D study design is $p \times I \times H$, then the within-person design is $I \times H$. It follows that

$$\sigma(\Delta_p) = \sigma(\overline{X}|p) = \sqrt{\frac{\sigma^2(i)_p}{n'_i} + \frac{\sigma^2(h)_p}{n'_h} + \frac{\sigma^2(ih)_p}{n'_i n'_h}},$$

where $\sigma^2(i)_p$, $\sigma^2(h)_p$, and $\sigma^2(ih)_p$ are estimated using data for person p only. In general, then, for any within-person random effects design, the absolute SEM for person p is

$$\sigma(\Delta_p) = \sigma(\overline{X}|p). \qquad (5.42)$$

Brennan (1996a, 1998) provides formulas for conditional relative SEMs for several multifacet random designs, but they are rather complicated. For practical use, a generalization of Equation 5.38 often provides an adequate estimator:

$$\hat{\sigma}(\delta_p) = \sqrt{\hat{\sigma}^2(\Delta_p) + [\hat{\sigma}^2(\Delta) - \hat{\sigma}^2(\delta)]}. \qquad (5.43)$$

Figure 5.4 provides two perspectives on $\hat{\sigma}^2(\Delta_p)$ for the QUASAR project introduced previously. In the left panel, the estimates are for 228 examinees whose responses to tasks were rated twice. Specifically, the within-person

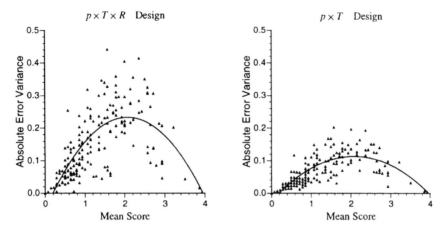

FIGURE 5.4. Conditional absolute error variances for QUASAR Project for two different designs.

design is $T \times R$ with $n'_t = 9$ tasks and $n'_r = 2$ raters. In the right panel, the estimates are for the undifferentiated set of $9 \times 2 = 18$ responses for each person. That is, the right panel provides the $\hat{\sigma}^2(\Delta_p)$ for the simplistic single-facet within-person design that does not differentiate between the contributions of the two facets. Clearly, these values of $\hat{\sigma}^2(\Delta_p)$ are much too small relative to those for the correctly specified $T \times R$ design. This is just one more example of the underestimation of error variance that tends to occur when the multifaceted nature of a universe of generalization is disregarded.

5.5   Other Issues

Generalizability theory is not without its critics, some of whom have objected rather strenuously to the theory or, at least, to certain elements of it. Cronbach et al. (1972, Chap. 11) provide responses to many of the objections that were raised by critics of the time. More recently, Brennan (in press) has discussed some misconceptions about the theory.

Cronbach (1976, p. 199) once made the provocative statement that generalizability theory

> ... has a protean quality. The procedures and even the issues take a new form in every context. [Generalizability] theory enables you to ask your questions better; what is most significant for you cannot be supplied from the outside.

In a sense, therefore, any treatment of generalizability theory—no matter how extensive—provides only a scaffolding. It cannot possibly address all

TABLE 5.7. Variance–Covariance Matrix for Person Scores for Synthetic Data Set No. 3

o_1				o_2			
i_1	i_2	i_3	i_4	i_1	i_2	i_3	i_4
3.1667	*1.0566*	*.8333*	*.8889*	**1.2778**	.6667	.9444	1.5000
	3.3889	*2.2778*	*-.1111*	-.1667	**1.6667**	.9444	.6111
		2.2778	*.0000*	-.7222	1.4444	**.6111**	.6111
			.6222	.7111	-.0444	.1333	**.9556**
				2.4556	*.1556*	*.2556*	*.7667*
					1.9556	*.8000*	*.2889*
	Symmetric					1.2111	*.6333*
							2.2333

issues that might be of concern to an investigator. In this section, a few potentially relevant issues are considered.

The formulas of generalizability theory are almost always presented in terms of raw scores, rather than transformed raw scores (i.e., scale scores). For a linear transformation $E\hat{\rho}^2$ and $\widehat{\Phi}$ are unchanged; and $\hat{\sigma}^2(\tau)$, $\hat{\sigma}^2(\delta)$, and $\hat{\sigma}^2(\Delta)$ are easily obtained by multiplying the raw score values of these statistics by the squared slope. For nonlinear transformations, however, complexities arise, especially when replications are not available, and inferences must be drawn based on a single instance of the measurement procedure. Most current research in this area has focused on conditional SEMs for scale scores (e.g., Kolen et al., 1992; Lee et al., 2000), but there are still complexities and unsolved problems in applying generalizability theory when average (or total) raw scores are subjected to a nonlinear transformation.

The estimation of variance components is clearly central to applications of generalizability theory, although the theory itself is essentially silent about which estimation procedures are preferable. To this point, for practical and historical reasons, the so-called ANOVA procedure for complete balanced designs has been emphasized. Using this procedure, the computer program GENOVA (see Appendix F) can be used for computations. In Chapter 7 other procedures for both balanced and unbalanced random effects designs are considered. The topic of estimating variance components is extraordinarily broad (see, e.g., Searle et al., 1992), and sometimes useful approaches for specific problems are embedded in other literature (see, e.g., Longford, 1995).

5.5.1   Covariances as Estimators of Variance Components

Relationships exist between random effects variance components, on the one hand, and variances and covariances of observable scores. In addition to their conceptual value, these relationships can be useful for conducting some generalizability analyses when only observed variances and covariances are available. Consider, for example, Table 5.7 which reports the observed variances and covariances *over persons* for the Synthetic Data No 3 in Table 3.1 on page 72, which was used to illustrate the $p \times i \times o$ design. Various combinations of the elements in this variance-covariance matrix can be used to estimate certain sums of variance components.

For the $p \times i \times o$ design,

$$\underset{i}{\boldsymbol{E}} \underset{o}{\boldsymbol{E}} \sigma^2(X_{pio}) = \sigma^2(p) + \sigma^2(pi) + \sigma^2(po) + \sigma^2(pio), \qquad (5.44)$$

where

$$\sigma^2(X_{pio}) = \underset{p}{\boldsymbol{E}}(X_{pio} - \mu_{io})^2.$$

Note that Equation 5.44 includes all variance components that contain the index p, the index over which variances and covariances are taken. The relationship in Equation 5.44 also holds for estimates; that is, the average of the variances on the diagonal of Table 5.7 equals the sum of the estimates of the variance components, as reported in Table 3.3:

$$
\begin{aligned}
Est[\underset{i}{\boldsymbol{E}} \underset{o}{\boldsymbol{E}} \sigma^2(X_{pio})] &= \hat{\sigma}^2(p) + \hat{\sigma}^2(pi) + \hat{\sigma}^2(po) + \hat{\sigma}^2(pio) \\
&= 0.5528 + 0.5750 + 0.1009 + 0.9352 \\
&= 2.164,
\end{aligned}
$$

which is identical to the average of the variances in the diagonal of Table 5.7:

$$(3.1667 + 3.3889 + \ldots + 2.2333)/8 = 2.164.$$

Three types of covariances are identifiable in Table 5.7:

1. covariances with levels of i in common, but different levels of o— these are in boldface;

2. covariances with levels of o in common, but different levels of i— these are italicized; and

3. covariances that have neither levels of i nor levels of o in common— these are in normal type above the main diagonal.

In terms of random effects variance components for the $p \times i \times o$ design, the expected value of the i-common covariances is:

$$\underset{i}{\boldsymbol{E}} \underset{o \neq o'}{\boldsymbol{E}} \sigma(X_{pio}, X_{pio'}) = \sigma^2(p) + \sigma^2(pi), \qquad (5.45)$$

where o' is used to designate a different condition of the o facet. This relationship also holds for estimates. Specifically,

$$Est[\underset{i}{\boldsymbol{E}} \underset{o \neq o'}{\boldsymbol{E}} \sigma(X_{pio}, X_{pio'})] = \hat{\sigma}^2(p) + \hat{\sigma}^2(pi) = 0.5528 + 0.5750 = 1.128,$$

which is identical to the average of the i-common covariances (boldface values) in Table 5.7:

$$[1.2778 + 1.6667 + 0.6111 + 0.9556]/4 = 1.128.$$

Similarly, for the o-common covariances,

$$\underset{i\neq i'}{E}\underset{o}{E}\,\sigma(X_{pio}, X_{pi'o}) = \sigma^2(p) + \sigma^2(po); \qquad (5.46)$$

and for the covariances that have neither i nor o in common,

$$\underset{i\neq i'}{E}\underset{o\neq o'}{E}\,\sigma(X_{pio}, X_{pi'o'}) = \sigma^2(p). \qquad (5.47)$$

From Equations 5.44 to 5.47, it is evident that a variance-covariance matrix over persons for the $p \times i \times o$ design provides sufficient information for an investigator to estimate all random effects variance components that contain the person index p. It is equally evident, however, that such a variance-covariance matrix does not provide a basis for estimating variance components that do not contain p, namely, $\sigma^2(i)$, $\sigma^2(o)$, and $\sigma^2(io)$. When persons are the objects of measurement, variance components that do not contain p enter $\sigma^2(\Delta)$ for the $p \times i \times o$ design. It follows that an analysis of the properties of a measurement procedure based solely on a variance-covariance matrix over persons is incapable of providing a basis for estimating absolute error variance $\sigma^2(\Delta)$.

Although the above discussion has been specifically related to a variance-covariance matrix over persons for the $p \times i \times o$ design, the basic procedures are applicable to a variance-covariance matrix for any design. That is, the covariances over a particular facet provide a sufficient basis for estimating the random effects variance components that involve that facet.

5.5.2 Estimators of Universe Scores

Observed scores are easy to compute and they are unbiased estimators of universe scores. Consequently, they are frequently used. However, there are other estimators that have characteristics that are sometimes judged desirable.

Regressed Score Estimates

For any design, the regression equation discussed by Cronbach et al. (1972) for estimating universe scores has the general form:

$$\hat{\mu}_p = \mu + \rho^2(\overline{X}_p, \mu_p)(\overline{X}_p - \underset{p}{E}\,\overline{X}_p), \qquad (5.48)$$

where the parameter μ is the grand mean for the population and universe, $\overline{X}_p$ is a person's observed mean score for the particular sample of conditions used in a D study, and the parameter $\rho^2(\overline{X}_p, \mu_p)$ is the squared correlation between observed and universe scores for the population. The error variance associated with using $\hat{\mu}_p$ as an estimate of μ_p is

$$\sigma^2(\mathcal{E}) = \boldsymbol{E}(\hat{\mu}_p - \mu_p)^2 = \sigma^2(p)(1 - \rho^2) \leq \sigma^2(\Delta), \qquad (5.49)$$

where ρ^2 is used here as an abbreviation for $\rho^2(\overline{X}_p, \mu_p)$. Since $\sigma^2(\mathcal{E})$ is generally smaller than $\sigma^2(\Delta)$, it is frequently stated that it is "better" to estimate μ_p using $\hat{\mu}_p$, rather than $\overline{X}_p$. Such a statement, however, applies to a randomly selected person, not necessarily to a particular person. Also, for a given person, a regressed score estimate is a biased estimate of universe score. The standard error of estimate $\sigma(\mathcal{E})$ can be used to establish *tolerance* intervals for $\hat{\mu}_p$, as discussed later in Section 12.2.4.

For the $I{:}p$ design, $\boldsymbol{E}_p \overline{X}_p = \mu$, and the regression equation is

$$\hat{\mu}_p = \mu + \rho^2(\overline{X}_p - \mu) = \rho^2 \overline{X}_p + (1 - \rho^2)\mu. \qquad (5.50)$$

To use this last equation for estimating μ_p we replace the parameters ρ^2 and μ with the estimates $\boldsymbol{E}\hat{\rho}^2$ and $\overline{X}$, respectively. This gives

$$\hat{\mu}_p = \boldsymbol{E}\hat{\rho}^2 \overline{X}_p + (1 - \boldsymbol{E}\hat{\rho}^2)\overline{X}, \qquad (5.51)$$

which is sometimes called an "estimation" equation to distinguish it from the regression Equation 5.50. Such point estimates of universe scores were originally proposed by Kelley (1947). Equation 5.51 has a Bayesian interpretation since any such estimate is based upon information about an individual (his or her observed score $\overline{X}_p$) as well as information about the group to which the individual belongs (e.g., a sample mean $\overline{X}$). Clearly, for a highly dependable measurement procedure, $\overline{X}_p$ is weighted very heavily, and $\overline{X}$ receives relatively little weight. By contrast, when $\boldsymbol{E}\hat{\rho}^2$ is small, $\hat{\mu}_p$ is heavily influenced by the group mean.

For the $p \times I$ design, $\boldsymbol{E}_p \overline{X}_p = \mu_I$, the regression equation is

$$\hat{\mu}_p = \mu + \rho^2(\overline{X}_p - \mu_I), \qquad (5.52)$$

and an associated estimation equation requires estimates of μ, μ_I, and ρ^2. Cronbach et al. (1972) suggest that μ can be estimated from the G study, the D study, or both; indeed, they consider myriad possibilities for estimating μ. Ideally they want both μ_I and ρ^2 to be estimated for the *fixed* set of items in a particular D study. In practice, however, their estimation equations usually necessitate one or more of the classical test theory assumptions of equal means, equal variances, and equal correlations of observed and universe scores. When all of these assumptions are made, the estimation equation for the $p \times I$ design is identical to Equation 5.51 for the $I{:}p$ design. When the classical assumptions are relaxed, the resulting estimation equations are probably best viewed as ad hoc approximations with properties that are difficult to ascertain (see Cronbach et al., 1972, pp. 138ff).

Although we have focused here on single-facet designs, Equations 5.48 and 5.49 apply to any design. In practice, estimates are required for $E\rho^2$ and μ, as well as $\mu_{\mathcal{R}}$ if the design is not fully nested.[10]

Best Linear Prediction

Jarjoura (1983) suggested that theoretical developments in the area of "best" linear prediction functions (see Searle, 1974; Searle et al., 1992) can be used to derive *predictors* of universe scores that have certain optimal properties. These prediction functions have some similarities with regressed score estimates, but neither the derivation nor the estimation of such prediction functions necessitates any of the classical test theory assumptions or any assumptions about distributional form. For a $p \times I$ design, assuming G and D study sample sizes are the same, the best linear prediction function for universe scores is

$$\dot{\mu}_p = \mu + A(\overline{X} - \mu) + B(\overline{X}_p - \overline{X}), \tag{5.53}$$

where $\overline{X}_p$ is a person's observed mean score for a particular sample of items (i.e., a particular test form), $\overline{X}$ is the mean of $\overline{X}_p$ over a sample of persons,

$$A = \frac{\sigma^2(p)/n_p}{\sigma^2(p)/n_p + \sigma^2(i)/n_i + \sigma^2(pi)/n_p n_i},$$

and

$$B = \frac{\sigma^2(p)}{\sigma^2(p) + \sigma^2(pi)/n_i}.$$

The error variance associated with using $\dot{\mu}_p$ as an estimate of μ_p is

$$E(\dot{\mu}_p - \mu_p)^2 = \sigma^2(p)[(1 - B) + (B - A)/n_p] \leq \sigma^2(\Delta). \tag{5.54}$$

The last term in the prediction function in Equation 5.53 adjusts $\overline{X}_p$ for the relative difficulty of a particular test form, and the second term can be viewed as a correction to this adjustment. Note that B has the form of a generalizability coefficient. For the nested design, the prediction function and regression equation are identical, even though the assumptions underlying the two approaches are different. However, for the crossed design, the regression equation has no term corresponding to the second term in the prediction function. Furthermore, for the crossed design, Cronbach et al. (1972) want ρ^2 in the regression equation to be dependent upon the particular form of a test used in a D study. By contrast, the parameters A and B in the best linear prediction function are for all randomly parallel forms of a test with a constant number of items.

---

[10]$\mu_{\mathcal{R}}$ is the mean for the random facets.

In practice, to use the prediction function in Equation 5.53, an investigator must estimate μ and the variance components $\sigma^2(p)$, $\sigma^2(i)$, and $\sigma^2(pi)$. It is highly desirable that estimates of these parameters be based on multiple test forms, perhaps by averaging the sample means and estimated variance components for each of the forms. This approach to estimating the parameters of the prediction function is conceptually simple, but it does require the actual existence of multiple test forms, not just an assumption about the possibility of multiple forms. There are other approaches to estimation, but any reasonably defensible approach would seem to necessitate a larger sample of items than those in a particular form of a test.

Subgroup Regressions

Nothing in the theory of regressed score estimates (or best linear predictors) requires categorizing a population of persons into different predefined subgroups and regressing the observed score for a person to the mean of the subgroup into which the person is categorized. However, this practice is common enough to merit some consideration here. Such practice forces an investigator to specify a variable (e.g., gender, race, or ethnicity) to be used as a basis for defining group membership, and the choice of a grouping variable can substantially influence the mean to which an individual examinee's observed score is regressed. Consequently, an individual examinee can have as many possibly different estimates of μ_p as there are subgroups into which such an examinee might be classified.

Furthermore, the use of regressed score estimates with subgroups can have unintended negative consequences. For example, suppose that members of a minority group score lower, on the average, than members of a majority group on some test. Also, suppose that ρ^2 is the same for both groups, and separate regression equations are used that regress minority group members to their mean and majority group members to their mean. Now, consider a minority and majority examinee who get the *same* observed score, which is between the mean for the minority and the majority groups. In this case, the majority examinee's observed score will be regressed upward, the minority examinee's observed score will be regressed downward, and the difference between their estimated universe scores could be substantial if the group means were quite different.

These comments do not necessarily mean that regressed score estimates are to be avoided when subgroups are involved. However, investigators are advised to give careful consideration to the consequences of using regressed score estimates (or best linear predictors) with subgroups.

5.5.3 Random Sampling Assumptions

Any theory necessarily involves certain assumptions that are idealized to some extent, and in inferential statistical theories, assumptions are always made about random sampling, independence, and/or uncorrelated effects. Some investigators find random sampling assumptions less appealing, but random sampling explanations are often used to justify assumptions of independence or uncorrelated effects.

Also, generalizability theory does *not* dictate that, say, all of the items in a test must be considered to be a random sample from the same undifferentiated universe of items. This point was illustrated in examples of tables of specifications. Almost always, one can structure a universe in such a way that random sampling assumptions become more acceptable. Doing so may render a universe more complex, but serious consideration of universes frequently reveals that they are indeed complex, without even considering sampling assumptions.

Furthermore, random sampling assumptions are not unique to generalizability theory. For example, in introducing their treatment of item sampling, Lord and Novick (1968, p. 235) offer the following rationale in support of random sampling assumptions in test theory.

> A possible objection to the item-sampling model (for example, see Loevinger, 1965) is that one does not ordinarily build tests by drawing items at random from a pool. There is, however, a similar and equally strong objection to classical test theory: Classical theory requires test forms that are strictly parallel, and yet no one has ever produced two strictly parallel forms for any ordinary paper-and-pencil test. Classical test theory is to be considered a useful idealization of situations encountered with actual mental tests. The assumption of random sampling of items may be considered in the same way. Further, even if the items of a particular test have not actually been drawn at random, we can still make certain interesting projections: We can conceive an item population from which the items of the test might have been randomly drawn and then consider the score the examinee would be expected to achieve over this population. The abundant information available on such expected scores enhances their natural interest to the examiner.

In a related vein, but in the broader context of inferential statistics, Cornfield and Tukey (1956, pp. 912–913) discuss a "bridge analogy" that has become a famous defense of random sampling assumptions.

> In almost any practical situation where analytical statistics is applied, the inference from the observations to the real conclusion has two parts, only the first of which is statistical. A genetic experiment on *Drosophila* will usually involve flies of a certain race of a certain species. The statistically based conclusions cannot extend beyond this race, yet the geneticist will usually, and often wisely, extend the conclusion to (a) the whole

species, (b) all *Drosophila*, or (c) a larger group of insects. This wider extension may be implicit or explicit, but it is almost always present. If we take the simile of the bridge crossing a river by way of an island, there is a statistical span from the near bank to the island, and a subject-matter span from the island to the far bank. Both are important.

By modifying the observation program and the corresponding analysis of the data, the island may be moved nearer to or farther from the distant bank, and the statistical span may be made stronger or weaker. In doing this it is easy to forget the second span, which usually can only be strengthened by improving the science or art on which it depends. Yet a balanced understanding of, and choice among, the statistical possibilities requires constant attention to the second span. It may often be worth while to move the island nearer to the distant bank, at the cost of weakening the statistical span—particularly when the subject-matter span is weak.

In an experiment where a population of C columns was specified, and a sample of c columns was randomly selected, it is clearly possible to make analyses where

    1) the c columns are regarded as a sample of c out of C, or

    2) the c columns are regarded as fixed.

The questions about these analyses is not their validity but their wisdom. ... The analyses will differ in the length of their inferences; both will be equally strong statistically. Usually it will be best to make analysis (1) where the inference is more general. Only if this analysis is entirely unrevealing on one or more points of interest are we likely to be wise in making analysis (2), whose limited inferences may be somewhat revealing.

But what if it is unreasonable to regard c columns as any sort of a fair sample from a population of C columns with $C > c$. We can (at least formally and numerically) carry out an analysis with, say, $C = \infty$. What is the logical position of such an analysis? It would seem to be much as follows: We cannot point to a specific population from which the c columns were a random sample, yet the final conclusion is certainly not to just these c columns. We are likely to be better off to move the island to the far side by introducing an unspecified population of columns "like those observed" and making the inference to the mean of this population. This will lengthen the statistical span at the price of leaving the location of the far end vague. Unless there is a known fixed number of reasonably possible columns, this lengthening and blurring is likely to be worthwhile.

In short, there is considerable precedent for random sampling assumptions in many statistical and psychometric theories. To say that such assumptions are idealized to some extent is hardly a compelling reason to reject them categorically, particularly in generalizability theory where the central focus is on generalization from a sample of conditions of measurement to a universe of conditions of measurement.

5.5.4   Generalizability and Other Theories

At various points in this book, relationships between generalizability theory and classical test theory have been discussed. Provided below are brief considerations of some similarities and dissimilarities between generalizability theory and three other psychometric theories.

Covariance Structure Analysis

The discussion of variance-covariance matrices in Section 5.5.1 has ties to covariance structure analysis and structural equation modeling (see, e.g., Jöreskog & Sörbom, 1979, and Bollen, 1989), although these methodologies have a much broader scope of inquiry than merely estimating random effects variance components (Linn & Werts, 1979) using computer programs such as LISREL (Jöreskog & Sörbom, 1993). Many applications of these methodologies are more directly related to multivariate generalizability theory (see Chapters 9 to 12), which involves specific variance and covariance components for different fixed strata. Even in such cases, however, the assumptions in covariance structure analysis and structural equation modeling are usually stronger than those in generalizability analyses. Also, covariance structure analyses and structural equation modeling do not emphasize generalizations from a sample of conditions of measurement to an explicitly defined multifaceted universe of conditions of measurement, at least not to the same extent that generalizability theory does. Even with these discontinuities, however, covariance structure analyses and structural equation modeling have potential applicability in generalizability theory.

In a sense, covariance structure analysis and structural equation modeling are extensions of procedures for examining the effects of measurement error on independent variables in regression analyses. The basic idea behind such analyses is summarized by Snedecor and Cochran (1980, pp. 171–172), and Fuller (1987) devotes much of his book on *Measurement Error Models* to this topic. The title of Fuller's book suggests a strong relationship between these models and generalizability theory. There are important differences, however. Foremost among them is that generalizability theory requires that the investigator carefully define the types of errors of interest, and generalizability theory focuses on quantifying and understanding the various sources of measurement error for all variables of interest. Still, there are similarities between some aspects of generalizability theory and regression analyses with fallible independent variables.

Multiple Matrix Sampling

There are some similarities between generalizability theory and multiple matrix sampling theory (see Lord & Novick, 1968, Chap. 11; and Sirotnik & Wellington, 1977). Both are sampling theories and both consider the estimation of variance components and their standard errors. Generalizability theory, however, involves a conceptual framework that is considerably broader than multiple matrix sampling. Also, the two theories emphasize different issues and frequently define and estimate parameters in a somewhat different manner. The simplest multiple matrix sampling design involves randomly partitioning the finite universe of items to k subuniverses, randomly partitioning the finite population of persons into k subpopulations, and then randomly pairing the subuniverses and subpopulations. These subuniverses and subpopulations are *not* fixed strata in the sense discussed in generalizability theory, however.

Also, for sampling from a finite population and/or universe, variance components are sometimes defined differently in multiple matrix sampling and generalizability theory. For example, Sirotnik and Wellington's (1977, p. 354) variance components (called θ^2-terms) are not identical to those based on the Cornfield and Tukey (1956) definitions, which are the definitions usually employed in generalizability theory.

In multiple matrix sampling, using generalized symmetric sums or means, variance components can be estimated for complicated designs with or without missing data, and higher-order moments of distributions can be estimated. Doing so is computationally intensive, but these procedures merit further consideration in the context of generalizability analyses.

Item Response Theory

Some believe that item response theory will replace most other measurement theories in the future. This author is unconvinced, however. It seems unarguable that, if certain (rather strong) assumptions are fulfilled, then item response theory provides a powerful model for addressing a number of important psychometric issues that are not easily treated with other theories, including generalizability theory. However, even if these assumptions are fulfilled, item response theory does not easily accommodate generalizations to multifaceted universes.

Item response theory pays attention to individual items as *fixed* entities without specific consideration of other conditions of measurement. By contrast, in generalizability theory, items are generally viewed as random and emphasis is placed on viewing them as *sampled* conditions of measurement. There are, of course, other differences between the two theories, but the fixed/random difference is fundamental. A few efforts have been made to link the theories (e.g., Kolen & Harris, 1987; Bock et al., 2000) but in this author's opinion, one theory is not properly viewed as a replacement for the other. Generalizability theory is primarily a sampling model, whereas item response theory is principally a scaling model.

5.6 Exercises

5.1* For the APL Survey D study design in Table 5.1, each person responds to the same set of items. Suppose, however, that each person took a different set of eight items for each content area. What is the notational representation of this design? What are the estimated G study and D study variance components? What are $\hat{\sigma}^2(\tau)$, $\hat{\sigma}^2(\delta)$, $\hat{\sigma}^2(\Delta)$, $E\hat{\rho}^2$, and $\widehat{\Phi}$?

5.2* Using the APL Survey Instrument discussed in Sections 3.5 and 5.1.3 a test–retest study was conducted in which 206 examinees took the same form of the Survey on two occasions. The resulting mean squares were as follows.

Effect	MS	Effect	MS	Effect	MS
p	2.9304	po	.2085	$oi{:}h$	.1742
o	.6068	ph	.2767	poh	.1089
h	18.4148	$pi{:}h$	.2406	$poi{:}h$	.0825
$i{:}h$	10.6986	oh	.5290		

Provide the notational representation, Venn diagram, and linear model for this G study design. What are the estimated G study variance components for the model in which content areas are fixed? If generalization is intended for a single randomly selected occasion and $n'_i = 8$, what are the D study variance components, $\hat{\sigma}^2(\tau)$, $\hat{\sigma}^2(\delta)$, and $E\hat{\rho}^2$?

5.3* Why is the value of $E\hat{\rho}^2$ from Exercise 5.2 considerably smaller than the value of $E\hat{\rho}^2 = .905$ in Table 5.1? If the results in Table 5.1 were unknown, but the results from Exercise 5.2 were available, what would be reasonable estimates of $\sigma^2(\tau)$, $\sigma^2(\delta)$, and $E\rho^2$ under the assumptions implicit in Table 5.1?

5.4 Verify that Equation Sets 5.16 and 5.17 apply to the estimates reported in Tables 5.5 and 5.6 for the APL Survey.

5.5 Given the Cornfield and Tukey definitions of variance components in Table 5.6 and the definition of $\sigma^2(\tau)$ in Equation 5.18, prove that

$$\sigma^2(\tau) = \frac{n_r - 1}{n_r}\sigma^2(r|RH) + \sigma^2(p{:}r|RH).$$

Hint: The proof involves careful consideration of model restrictions, as discussed in Section 3.5.

5.6 Using only the variance components in Table 5.6, what is an estimate of $\boldsymbol{E}\rho^2$ for a randomly selected region?

5.7* Shavelson et al. (1993) provide an example of a voluntary statewide science assessment program in which students (p) in approximately 600 schools (g) took five performance tasks (t). For a subset of their results:

$$\hat{\sigma}^2(g) = .07, \quad \hat{\sigma}^2(p{:}g) = .23, \quad \hat{\sigma}^2(gi) = .07, \quad \text{and} \quad \hat{\sigma}^2(pi{:}g) = .43.$$

For these data:

  (a) show that Inequality 5.25 is not satisfied; and

  (b) provide a few values of n_p and n_i that lead to a contradiction of the conventional wisdom that $\boldsymbol{E}\rho_g^2 > \boldsymbol{E}\rho_p^2$.

5.8 When both persons and items are random, prove that $\sigma^2(\delta_g) > \sigma^2(\delta_p)$ implies that Inequality 5.28 is satisfied.

5.9* Consider, again, Exercise 2.1. For each of the six persons determine $\hat{\sigma}(\Delta_p)$ for $n'_r = 3$ and verify that the average of the squares of these estimated conditional SEMs is $\hat{\sigma}^2(\Delta)$.

5.10 For the ITED Vocabulary example in Section 5.4.1, $\hat{\sigma}^2(\Delta) = .00514$ and $\hat{\sigma}^2(\delta) = .00475$. For examinees with $\overline{X}_p = .6$, show that the quadratic fit and Equation 5.38 give very similar values for $\hat{\sigma}(\delta_p)$.

5.11 In Section 5.5.1, the expressions for covariances in terms of random effects variance components are expressed for single conditions of facets (lowercase letters). These relationships also hold for means over sets of conditions for facets. For Synthetic Data No. 3 in Table 3.1 on page 72, verify that

$$Est[\underset{I}{\boldsymbol{E}}\ \underset{o \neq o'}{\boldsymbol{E}}\ \sigma(X_{pIo}, X_{pIo'})] = \hat{\sigma}^2(p) + \hat{\sigma}^2(pI).$$

6
Variability of Statistics in Generalizability Theory

Estimates of variance components, error variances, generalizability coefficients, and so on, like all statistics, are subject to sampling variability. Traditionally, such variability is quantified through estimated standard errors and/or confidence intervals. Cronbach et al. (1972) recognized the importance of this topic for generalizability analyses and gave it more than passing attention, although at that time statistical methodologies for addressing the topic were limited. Subsequently, in the generalizability theory literature, Smith (1978, 1982) considered standard errors of estimated variance components, Brennan (1992a) summarized some procedures for establishing standard errors and confidence intervals, Brennan et al. (1987) considered bootstrap and jackknife procedures, Betebenner (1998) examined a relatively new procedure for establishing confidence intervals, Wiley (2000) studied the bootstrap, and Gao and Brennan (2001) provided examples from the performance assessment literature.

In the statistical literature, assuming score effects are normally distributed, Searle et al. (1992) treat in detail standard errors for estimated variance components,[1] and Burdick and Graybill (1992) provide a very comprehensive and readable treatment of confidence intervals for variance components and various ratios of them.

This chapter summarizes most methodologies that have been developed to estimate standard errors and confidence intervals for (estimated) vari-

---

[1] In many respects, Searle et al. (1992) build on the seminal work reported two decades earlier by Searle (1971).

ance components. These are complicated matters with many unsolved problems, especially when score effects are not normally distributed. Most of the procedures discussed apply, at least in theory, to random effects variance components for any balanced design. However, examples, simulations, and exercises are largely specific to the $p \times i$ and $p \times I$ designs. Extensions to other designs are evident for some procedures, but largely unstudied for others. The goal of this chapter is simply to introduce students and practitioners of generalizability theory to the methodologies, as they apply in generalizability theory. Much research remains to be done.

Most notational conventions in this chapter are the same as those used in previous chapters. The most noteworthy exception is that a variation on the notation in Burdick and Graybill (1992) is employed for some procedures. This leads to some notational inconsistencies between parts of this chapter and the rest of this book. Such inconsistencies are identified when they arise.[2]

6.1   Standard Errors of Estimated Variance Components

In principal, standard errors of estimated variance components can be estimated by replicating a study and computing the standard deviation of the resulting sets of estimates. In theory, this procedure has much to recommend it (see Brennan, in press). It is directly related to the basic notion of generalizing over randomly parallel instances of a measurement procedure, it is computationally straightforward, and it is empirically based without distributional-form assumptions. For example, Table 6.1 provides estimated variance components for nine forms on one mathematics subtest of the ACT Assessment (ACT, 1997). The data were collected for an equating study in which all forms were administered to randomly equivalent groups. Consequently, the standard deviations in the next-to-the-last row are direct empirical estimates of the standard errors of the estimated variance components.

Estimated standard errors such as those in Table 6.1 are close to ideal in the sense that they require minimal assumptions. In many contexts, however, replications are simply not available. In the absence of replications, there are two general classes of procedures for estimating standard errors. One class makes assumptions about the distributional form of the effects, usually multivariate normality. The other class involves resampling methods such as the jackknife and bootstrap.

---

[2]It is possible to resolve these notational inconsistencies, but doing so would complicate matters substantially for readers who study the literature cited in this chapter.

TABLE 6.1. G Study $p \times i$ Results for Nine Forms of ACT Math Pre-Algebra/Elementary Algebra Subtest with 24 Items

Form	n_p	$MS(p)$	$MS(i)$	$MS(pi)$	$\hat{\sigma}^2(p)$	$\hat{\sigma}^2(i)$	$\hat{\sigma}^2(pi)$
A	3388	.9472	85.9556	.1768	.0321	.0253	.1768
B	3363	1.0621	91.5600	.1517	.0379	.0272	.1517
C	3458	.8872	127.2861	.1421	.0310	.0368	.1421
D	3114	.9031	96.5736	.1671	.0307	.0310	.1671
E	3428	1.1280	63.5186	.1712	.0399	.0185	.1712
F	3318	.9601	136.5013	.1599	.0333	.0411	.1599
G	3257	.8828	95.1346	.1717	.0296	.0292	.1717
H	3221	.8448	161.2862	.1579	.0286	.0500	.1579
I	3178	.8646	94.8867	.1724	.0288	.0298	.1724
Mean					.0324	.0321	.1634
SD					.0040	.0093	.0114
Ave SE[a]					.0010	.0094	.0008

[a]Square root of the average of the squared estimated standard errors using Equation 6.1.

6.1.1 Normal Procedure

The traditional approach to estimating standard errors of estimated variance components assumes that score effects have a multivariate normal distribution. Under this assumption, it can be shown (see Searle 1971, pp. 415–417; Searle et al. 1992, pp. 137–138) that an estimator of the standard error of an estimated variance component is

$$\hat{\sigma}[\hat{\sigma}^2(\alpha|M)] = \sqrt{\sum_{\beta} \frac{2[f(\beta|\alpha)MS(\beta)]^2}{df(\beta) + 2}}, \tag{6.1}$$

where M designates the model, β indexes the mean squares that enter $\hat{\sigma}^2(\alpha|M)$, and $f(\beta|\alpha)$ is the coefficient of $MS(\beta)$ in the linear combination of mean squares that gives $\hat{\sigma}^2(\alpha|M)$. The square of the right-hand side of Equation 6.1 is an unbiased estimator of the variance of $\hat{\sigma}^2(\alpha|M)$. For a random effects model, a slightly simpler version of Equation 6.1 is

$$\hat{\sigma}[\hat{\sigma}^2(\alpha)] = \frac{1}{\pi(\dot{\alpha})} \sqrt{\sum_{\beta} \frac{2[MS(\beta)]^2}{df(\beta) + 2}}, \tag{6.2}$$

where, as in previous chapters, $\pi(\dot{\alpha})$ is the product of the sample sizes for indices not in α.[3]

---

[3]Equation 6.1 applies to both G study and D study variance components (including error variances) that can be expressed as linear combinations of mean squares. Equation 6.2, however, should be used only for G study random effects variance components.

TABLE 6.2. G Study Results for Synthetic Data Set No. 4 Recast as a $p \times i$ Design

	df	MS	$\hat{\sigma}^2$	Estimated Standard Errors		
				Normal	Jackknife	Ratio[a]
p	9	10.2963	.6258	.3673	.4254	1.5420
i	11	11.6273	.8840	.4577	.5551	1.7774
pi	99	2.7872	2.7872	.3922	.5569	7.0356

[a]Ratio of estimated variance component to estimated standard error, with the latter computed using df rather than $df + 2$ in Equation 6.1.

It is evident from the form of Equation 6.1 that estimated standard errors decrease as degrees of freedom (i.e., sample sizes) increase. Also, other things being equal, estimated standard errors are likely to be smaller for estimated variance components that involve fewer mean squares. For this reason, all other things being equal, the estimated variance component for a nested effect involving k indices (e.g., $i{:}h$) will have a smaller standard error than for an interaction effect involving the same k indices (e.g., ih).

Consider again the Synthetic Data Set No. 4 in Table 3.2 on page 73 recasting it as a $p \times i$ design with $n_p = 10$ persons and $n_i = 12$ items. Table 6.2 provides the typical ANOVA results, the estimated variance components, and the estimates of the standard errors of the estimated variance components using Equation 6.2. As might be expected with such small sample sizes, the estimated standard errors are substantial. Of course, we do not know whether the normality assumptions are sensible for these data.

When normality assumptions are violated, Equations 6.1 and 6.2 may give misleading results. Consider, again, the results in Table 6.1 for nine forms of an ACT Assessment Mathematics subtest. Each form is based on 24 dichotomously scored items. The last row in Table 6.1 reports the averages of the standard errors estimated using Equation 6.2, although there is good reason to believe that normality assumptions are not fulfilled for these data. Clearly, for $\hat{\sigma}^2(p)$ and particularly $\hat{\sigma}^2(pi)$, the normality-based average estimates in the last row are substantially smaller than the preferable estimates in the next-to-the-last row.

6.1.2 Jackknife Procedure

Quenouille (1949) suggested a nonparametric estimator of bias. Tukey (1958) extended Quenouille's idea to a nonparametric estimator of the standard error of a statistic. The theory underlying the jackknife is discussed extensively by Shao and Tu (1995). Here, we briefly outline the basics of the theory and then discuss its application to estimated variance components for the $p \times i$ design.

Overview

Suppose a set of s data points is used to estimate some parameter θ. The general steps in using the jackknife to estimate the standard error of the jackknife estimate of θ are:

1. obtain $\hat{\theta}$ for all s data points;

2. obtain the s estimates of θ that result from deleting each one of the data points, and let each such estimate be designated $\hat{\theta}_{-j}$;

3. for each of the s data points, obtain $\hat{\theta}_{*j} = \hat{\theta} + (s-1)(\hat{\theta} - \hat{\theta}_{-j})$, which are called "pseudovalues";

4. obtain the mean of the pseudovalues $\hat{\theta}_J$, which is the jackknife estimator of θ; and

5. obtain the jackknife estimate of the standard error of $\hat{\theta}$:

$$\hat{\sigma}(\hat{\theta}_J) = \sqrt{\frac{1}{s(s-1)} \sum_{j=1}^{s} (\hat{\theta}_{*j} - \hat{\theta}_J)^2}, \tag{6.3}$$

which is the standard error of the mean for the pseudovalues.

The essence of the jackknife procedure is a conjecture by Tukey (1958) that the pseudovalues $\hat{\theta}_{*j}$ have approximately the same variance as $\sqrt{s}\hat{\theta}$.

Application to $p \times i$ Design

It is not so simple, however, to extend these general steps to obtain estimated standard errors for ANOVA estimated variance components. In a sense, the basic difficulty is that the sampling process involves more than one dimension (i.e., facet). Based on advice from Tukey, Cronbach et al. (1972, pp. 54–57, 66, 70–72) provide a jackknife procedure for the $p \times i$ design. That procedure is outlined here for the general case of sampling from a finite population and/or universe, as originally provided by Brennan et al. (1987).

Using the Cornfield and Tukey (1956) definitions, the parameters are defined as

$$\sigma^2(p) = \frac{1}{N_p - 1} \sum_{p=1}^{N_p} (\mu_p - \mu)^2 \tag{6.4}$$

$$\sigma^2(i) = \frac{1}{N_i - 1} \sum_{i=1}^{N_i} (\mu_i - \mu)^2 \tag{6.5}$$

$$\sigma^2(pi) = \sum_{p=1}^{N_p} \sum_{i=1}^{N_i} \frac{(X_{pi} - \mu_p - \mu_i + \mu)^2}{(N_p - 1)(N_i - 1)}. \tag{6.6}$$

Letting $c_p = 1 - n_p/N_p$ and $c_i = 1 - n_i/N_i$ be finite population/universe correction factors, the estimators of the variance components are:

$$\hat{\sigma}^2(p) = [MS(p) - c_i MS(pi)]/n_i \tag{6.7}$$
$$\hat{\sigma}^2(i) = [MS(i) - c_p MS(pi)]/n_p \tag{6.8}$$
$$\hat{\sigma}^2(pi) = MS(pi). \tag{6.9}$$

Obviously, these equations give the usual random effects estimates when $c_p = c_i = 1$.

Consider the following notational conventions.

$\hat{\theta}$ = any estimated variance component for the $p \times i$ design based on analyzing the full $n_p \times n_i$ matrix [i.e., $\hat{\theta}$ could be $\hat{\sigma}^2(p)$, $\hat{\sigma}^2(i)$, or $\hat{\sigma}^2(pi)$];

$\hat{\theta}_{-pi}$ = value of $\hat{\theta}$ for the $(n_p - 1) \times (n_i - 1)$ matrix that results from deleting person p and item i;

$\hat{\theta}_{-p0}$ = value of $\hat{\theta}$ for the $(n_p - 1) \times n_i$ matrix that results from deleting person p;

$\hat{\theta}_{-0i}$ = value of $\hat{\theta}$ for the $n_p \times (n_i - 1)$ matrix that results from deleting item i; and

$\hat{\theta}_{-00}$ = value of $\hat{\theta}$ for the original $n_p \times n_i$ matrix (i.e., $\hat{\theta} = \hat{\theta}_{-00}$).

Given these notational conventions, the pseudovalue for person p and item i is

$$\hat{\theta}_{*pi} = n_p n_i \hat{\theta}_{-00} - (n_p - 1)n_i \hat{\theta}_{-p0} - n_p(n_i - 1)\hat{\theta}_{-0i} + (n_p - 1)(n_i - 1)\hat{\theta}_{-pi}, \tag{6.10}$$

and the mean of the pseudovalues is the jackknife estimator of θ:

$$\hat{\theta}_J = \frac{1}{n_p n_i} \sum_{p=1}^{n_p} \sum_{i=1}^{n_i} \hat{\theta}_{*pi}. \tag{6.11}$$

For the ANOVA estimators of variance components, $\hat{\theta}_J = \hat{\theta}$.

The standard error of $\hat{\theta}_J = \hat{\theta}$ is estimated using the matrix of pseudovalues, which has n_p rows and n_i columns. Note that there is one such matrix for each of the three estimated variance components. For any one of these estimated variance components, let $\hat{\sigma}^2(rows)$, $\hat{\sigma}^2(cols)$, and $\hat{\sigma}^2(res)$ be the estimated variance components for the matrix of pseudovalues. Then, the estimated standard error of $\hat{\theta}_J$ is[4]

$$\hat{\sigma}(\hat{\theta}_J) = \sqrt{\frac{c_p\hat{\sigma}^2(rows)}{n_p} + \frac{c_i\hat{\sigma}^2(cols)}{n_i} + \frac{\hat{\sigma}^2(res)}{n_pn_i}}. \tag{6.12}$$

Table 6.2 provides the jackknife estimates of the three variance components for the synthetic data example. The principal advantage of these estimates is that they are not based on any distributional-form assumptions. The principal disadvantage is that extensive computations are required. Even for this very small set of synthetic data, each of the three jackknife estimates requires computing $11 \times 13 = 143$ sets of estimated variance components that are the basis for the final computations using Equation 6.3.

6.1.3 Bootstrap Procedure

The bootstrap is similar to the jackknife in that both are resampling procedures and both are primarily nonparametric methods for assessing the accuracy of a particular $\hat{\theta}$ as an estimate of θ. A principal difference between the two procedures is that the bootstrap employs sampling with replacement, whereas the jackknife employs sampling without replacement. Efron (1982) provides an early theoretical treatment of the bootstrap; Efron and Tibshirani (1986) provide a simpler and more applied treatment; and Shao and Tu (1995) provide a recent extensive treatment.

Overview

For a statistic based on s observations, the bootstrap algorithm is based on multiple bootstrap samples, with each such sample consisting of a random sample of size s *with replacement* from the original sample. Using the bootstrap, estimation of the standard error of a statistic $\hat{\theta}$ involves these steps:

1. using a random number generator, independently draw a large number of bootstrap samples, say B of them;

2. for each sample, evaluate the statistic of interest, say $\hat{\theta}_b$ ($b = 1, \ldots, B$); and

3. calculate the sample standard deviation of the $\hat{\theta}_b$:

$$\hat{\sigma}(\hat{\theta}_b) = \sqrt{\frac{1}{B-1}\sum_{b=1}^{B}(\hat{\theta}_b - \hat{\theta}_B)^2}, \tag{6.13}$$

---

[4]In their discussion of the jackknife, Cronbach et al. (1972, pp. 56 and 71) incorrectly suggest that this result be divided by $\sqrt{n_pn_i}$. Also, they suggest jackknifing the logarithms of the estimated variance components rather than the estimates themselves; this author's research does not support doing so.

where

$$\hat{\theta}_B = \frac{1}{B} \sum_{b=1}^{B} \hat{\theta}_b \qquad (6.14)$$

is the bootstrap estimate of θ.

Note that the bootstrap standard error in Equation 6.13 has the form of a standard deviation, whereas the formula for the jackknife standard error in Equation 6.3 has the form of a standard error of a mean.

Application to $p \times i$ Design

Although the bootstrap is conceptually simple, it is not unambiguously clear how to extend it to the random effects $p \times i$ design, which involves two dimensions/facets. Difficulties were originally discussed by Brennan et al. (1987) and subsequently by Wiley (2000).[5] The crux of the matter is specifying how to draw bootstrap samples from the $n_p \times n_i$ matrix of observed scores. It might seem that the obvious approach is to:

1. draw a sample of n_p persons with replacement from the sampled persons;

2. draw an independent random sample of n_i items with replacement from the sampled items; and

3. let the bootstrap sample consist of the responses of the sampled persons to the sampled items.

This double sampling procedure is designated "boot-p, i."

It is important to note that the boot-p, i procedure involves sampling with replacement, which means that any bootstrap sample likely will contain some repeated persons and some repeated items. This characteristic of the boot-p, i procedure introduces bias in estimates of some variance components (i.e., $\hat{\theta}_B$ in Equation 6.14 is sometimes biased), which certainly casts doubt on the accuracy of any associated confidence intervals. For example, in a random effects design, the ANOVA estimate of $\hat{\sigma}^2(p)$ is usually expressed as $[MS(p) - MS(pi)]/n_i$, but an equivalent expression is

$$\hat{\sigma}^2(p) = \frac{1}{n_i(n_i - 1)} \sum_{i \neq i'} \sum \left[\sum_p \frac{(X_{pi} - \overline{X}_i)(X_{pi'} - \overline{X}_{i'})}{n_i - 1} \right], \qquad (6.15)$$

which is the average of the estimates of the item covariances. When items are repeated, Equation 6.15 suggests that $\hat{\sigma}^2(p)$ is likely to be an inflated estimate of $\sigma^2(p)$, especially when n_i is relatively small. A similar statement holds for $\hat{\sigma}^2(i)$.

---

[5]See also Leucht and Smith (1989) and Othman (1995).

Brennan et al. (1987) suggested three other procedures for obtaining bootstrap samples.[6] The boot-p procedure involves sampling n_p persons with replacement, but not items. The boot-i procedure involves sampling n_i items with replacement, but not persons. The boot-p and boot-i procedures keep items and persons fixed, respectively, in obtaining bootstrap samples. Since results are desired for a design in which both persons and items are random, it is to be expected that neither of these procedures will be completely satisfactory for estimating standard errors for all estimated variance components.

The final procedure suggested by Brennan et al. (1987) is the boot-p, i, r procedure in which persons, items, and residuals are each sampled with replacement. Recall from Section 3.3.2 that, in terms of observed-score effects,

$$\begin{aligned} X_{pi} &= \overline{X} + x_p + x_i + r_{pi} \\ &= \overline{X} + (\overline{X}_p - \overline{X}) + (\overline{X}_i - \overline{X}) + (X_{pi} - \overline{X}_p - \overline{X}_i + \overline{X}), \end{aligned}$$

where, for our purposes here, r_{pi} is used to designate the residual effect. The boot-p, i, r procedure involves using a random sample with replacement of size n_p from the set of x_p, an independent random sample with replacement of size n_i from the set of x_i, and an independent random sample with replacement of size $n_p n_i$ from the set of r_{pi}. It is important to note that the sampling of the three effects is independent. So, for example, if the first person sampled for x_p were 12, and the first item sampled for x_i were 15, only by chance would the sampled residual be the residual associated with person 12 and item 15 in the original data.

Intuition and logic such as that discussed in conjunction with Equation 6.15 suggest that none of these bootstrap procedures is likely to lead to unbiased estimates of variance components. Wiley (2000) formalized these intuitive notions and derived the results in Figure 6.1 that adjust for bias induced by the various bootstrap sampling procedures. For example, if boot-p is used to estimate $\sigma^2(p)$ or the standard error of $\hat{\sigma}^2(p)$, then the bootstrap estimates should be multiplied by $n_p/(n_p - 1)$.

For the synthetic data example introduced in Section 6.1.1, Table 6.3 provides both unadjusted and adjusted bootstrap means and estimated standard errors for $\hat{\sigma}^2(p)$, $\hat{\sigma}^2(i)$, and $\hat{\sigma}^2(pi)$ based on the four bootstrap sampling procedures introduced above. The results in Table 6.3 are based on 1000 replications. For example, the adjusted boot-p value of .6185 for the mean of $\hat{\sigma}^2(p)$ was obtained as follows.

1. Obtain a bootstrap sample using the boot-p procedure,

2. for this bootstrap sample, compute the usual ANOVA estimate of $\sigma^2(p)$ and designate it $\hat{\sigma}^2(p|\text{boot-}p)$,

---

[6]Wiley (2000) has suggested a few other bootstrap procedures not considered here.

Adjusted Estimates Based on boot-p

$$\hat{\sigma}^2(p) \;=\; \frac{n_p}{n_p - 1}\, \hat{\sigma}^2(p|\text{boot-}p)$$

$$\hat{\sigma}^2(i) \;=\; \hat{\sigma}^2(i|\text{boot-}p) - \frac{1}{n_p - 1}\, \hat{\sigma}^2(pi|\text{boot-}p)$$

$$\hat{\sigma}^2(pi) \;=\; \frac{n_p}{n_p - 1}\, \hat{\sigma}^2(pi|\text{boot-}p)$$

Adjusted Estimates Based on boot-i

$$\hat{\sigma}^2(i) \;=\; \frac{n_i}{n_i - 1}\, \hat{\sigma}^2(i|\text{boot-}i)$$

$$\hat{\sigma}^2(p) \;=\; \hat{\sigma}^2(p|\text{boot-}i) - \frac{1}{n_i - 1}\, \hat{\sigma}^2(pi|\text{boot-}i)$$

$$\hat{\sigma}^2(pi) \;=\; \frac{n_i}{n_i - 1}\, \hat{\sigma}^2(pi|\text{boot-}i)$$

Adjusted Estimates Based on boot-p, i

$$\hat{\sigma}^2(p) \;=\; \frac{n_p}{n_p - 1}\, \hat{\sigma}^2(p|\text{boot-}p, i) - \frac{n_p}{(n_p - 1)(n_i - 1)}\, \hat{\sigma}^2(pi|\text{boot-}p, i)$$

$$\hat{\sigma}^2(i) \;=\; \frac{n_i}{n_i - 1}\, \hat{\sigma}^2(i|\text{boot-}p, i) - \frac{n_i}{(n_p - 1)(n_i - 1)}\, \hat{\sigma}^2(pi|\text{boot-}p, i)$$

$$\hat{\sigma}^2(pi) \;=\; \frac{n_p n_i}{(n_p - 1)(n_i - 1)}\, \hat{\sigma}^2(pi|\text{boot-}pi)$$

Adjusted Estimates Based on boot-p, i, r

$$\hat{\sigma}^2(p) \;=\; \frac{n_p}{n_p - 1}\, \hat{\sigma}^2(p|\text{boot-}p, i, r) - \frac{n_p}{(n_p - 1)(n_i - 1)}\, \hat{\sigma}^2(pi|\text{boot-}p, i, r)$$

$$\hat{\sigma}^2(i) \;=\; \frac{n_i}{n_i - 1}\, \hat{\sigma}^2(i|\text{boot-}p, i, r) - \frac{n_i}{(n_p - 1)(n_i - 1)}\, \hat{\sigma}^2(pi|\text{boot-}p, i, r)$$

$$\hat{\sigma}^2(pi) \;=\; \frac{n_p n_i}{(n_p - 1)(n_i - 1)}\, \hat{\sigma}^2(pi|\text{boot-}p, i, r)$$

FIGURE 6.1. Wiley's adjustments for bias in bootstrap procedures for estimating variance components and their standard errors.

TABLE 6.3. Bootstrap Means and Estimated Standard Errors for G Study Variance Components Based on 1000 Replications for Synthetic Data Set No. 4 Recast as a $p \times i$ Design

	Means			Est. Standard Errors		
	$\hat{\sigma}^2(p)$	$\hat{\sigma}^2(i)$	$\hat{\sigma}^2(pi)$	$\hat{\sigma}^2(p)$	$\hat{\sigma}^2(i)$	$\hat{\sigma}^2(pi)$
Original Data[a]	.6258	.8840	2.7872			
Unadjusted						
boot-p	.5567	1.1409	2.5170	.1886	.4508	.4035
boot-i	.8789	.8242	2.5353	.4535	.4445	.4288
boot-p, i	.7754	1.0509	2.3042	.5147	.6751	.5977
boot-p, i, r	.7577	1.0818	2.3156	.3225	.5649	.3178
Adjusted						
boot-p	.6185	.8613	2.7967	.2096	.4534	.4483
boot-i	.6484	.8991	2.7658	.4809	.4850	.4678
boot-p, i	.6288	.8671	2.7929	.5976	.7387	.7245
boot-p, i, r	.6080	.8994	2.8068	.3607	.6193	.3853

[a]Usual ANOVA estimates (see Table 6.3).

3. multiply $\hat{\sigma}^2(p|\text{boot-}p)$ by the adjustment factor $n_p/(n_p - 1)$,

4. repeat the first three steps $B = 1000$ times,

5. compute the mean of the 1000 estimates.

The boot-p value of .2096 for the standard error of $\hat{\sigma}^2(p)$ was obtained using the same steps, except that Step 5 involved computing the standard deviation rather than the mean of the 1000 estimates.

It is often judged that B in the range of 50 to 200 is adequate for estimating standard errors (see, e.g., Efron & Tibshirani (1986, p. 56). $B = 1000$ is used here for several reasons. First, estimated variance components, using $B = 1000$ generally give results that are closer to the parameters. Second, it is usually suggested that $B = 1000$ is required for confidence intervals (discussed later), and using $B = 1000$ for both seems preferable in terms of consistency. Third, with today's computers, using $B = 1000$ is not prohibitively time consuming.

Since we do not know the values of the parameters for this synthetic data example, we cannot draw firm conclusions about which bootstrap procedure is preferable for which estimate. However, for each of the three estimated variance components, it is clear that the unadjusted mean estimates vary considerably, while all the adjusted estimates are close to the estimate for the original data. This surely suggests that Wiley's bias adjustments are working properly. Knowing this, however, does not help in deciding which bootstrap procedure is preferable for estimating standard errors. Some guidance is provided by simulations treated in Section 6.4 (see also Wiley, 2000) and the discussion in Section 6.5.

6.2   Confidence Intervals for Estimated Variance Components

An estimate of a standard error is sometimes sufficient for an investigator's judgments about the variability of estimated variance components. More often than not, however, an investigator wants to establish a confidence interval. The first three procedures discussed next for establishing confidence intervals assume that score effects are normally distributed; the last two (jackknife and bootstrap) do not. With minor exceptions (see Searle, 1971, p. 414), all procedures give results that are only approximate, even when normality assumptions are fulfilled.[7]

6.2.1   Normal Procedure

When score effects have a multivariate normal distribution and degrees of freedom are very large, a simple normal approximation to a $100(\gamma)\%$ confidence interval for $\sigma^2(\alpha|M)$ is

$$\hat{\sigma}^2(\alpha|M) \pm z \, \hat{\sigma}[\hat{\sigma}^2(\alpha|M)], \qquad (6.16)$$

where z is the normal deviate corresponding to the upper $(1+\gamma)/2$ percentile point. For moderate degrees of freedom, z can be replaced by Student's t. In any case, however, the computed upper limit is likely to be too low, which gives an interval that is too short.

6.2.2   Satterthwaite Procedure

Under the assumption of multivariate normality, a usually better procedure is one developed by Satterthwaite (1941, 1946) and summarized by Graybill (1976, pp. 642–643). Under the Satterthwaite procedure, an approximate $100(\gamma)\%$ confidence interval on a variance component is

$$\text{Prob}\left[\frac{\hat{\sigma}^2(\alpha|M)\nu}{\chi_U^2(\nu)} \leq \sigma^2(\alpha|M) \leq \frac{\hat{\sigma}^2(\alpha|M)\nu}{\chi_L^2(\nu)}\right] \doteq \gamma, \qquad (6.17)$$

---

[7]Although the notation in this section occasionally recognizes that a random effects variance component may come from a mixed model, little explicit attention is given to this issue. See Burdick and Graybill (1992, Chap. 7) for a statistical discussion of confidence intervals for variance components in mixed models.

where $\chi_U^2(\nu)$ and $\chi_L^2(\nu)$ are the *lower* $U = (1 + \gamma)/2$ and $L = (1 - \gamma)/2$ percentile points of the chi-squared distribution with ν degrees of freedom, called the "effective" degrees of freedom; and

$$\nu = \frac{[\sum_\beta f(\beta|\alpha)MS(\beta)]^2}{\sum_\beta [f(\beta|\alpha)MS(\beta)]^2/df(\beta)}, \tag{6.18}$$

using the notational conventions in Section 6.1.1.

As discussed by Brennan (1992a), the numerator of Equation 6.18 is simply $[\hat{\sigma}^2(\alpha|M)]^2$, and the denominator is $\hat{\sigma}^2[\hat{\sigma}^2(\alpha|M)]/2$ provided $df(\beta)+2$ in Equation 6.1 is replaced by $df(\beta)$ (see Equation 6.1). Letting the ratio of the estimated variance component to its estimated standard error be

$$r = \frac{\hat{\sigma}^2(\alpha|M)}{\hat{\sigma}[\hat{\sigma}^2(\alpha|M)]}, \tag{6.19}$$

it follows that $\nu = 2r^2$ and $r = \sqrt{\nu/2}$. Consequently, for a $100(\gamma)\%$ confidence interval, the limits are

$$\text{lower limit} = \hat{\sigma}^2(\alpha|M)\left[\frac{2r^2}{\chi_U^2(2r^2)}\right] \tag{6.20}$$

and

$$\text{upper limit} = \hat{\sigma}^2(\alpha|M)\left[\frac{2r^2}{\chi_L^2(2r^2)}\right]. \tag{6.21}$$

The multiplicative terms in square brackets in Equations 6.20 and 6.21 are tabulated in Appendix D for $r = 2$ to 200. Since r in Equation 6.19 is easily computed, obtaining a Satterthwaite confidence interval is relatively straightforward.

Table 6.2 provides r for each of the variance components for the synthetic data. For example, $r = 7.0356$ for $\hat{\sigma}^2(pi)$, which means that the estimated variance component is about seven times larger than its estimated standard error. For this value of r and an 80% confidence interval, the tabled multiplicative factors in Appendix D are approximately .843 and 1.216. It follows that an 80% confidence interval is

$$(2.7872 \times .843, 2.7872 \times 1.216) = (2.350, 3.388),$$

or about $(2.4, 3.4)$.

6.2.3   Ting et al. Procedure

Suppose that expressing a variance component in terms of expected mean squares requires subtracting one or more of them. For example, in the $p \times i$ design, $\hat{\sigma}^2(p) = [EMS(p) - EMS(pi)]/n_i$. Under these circumstances, when normality assumptions are fulfilled, Burdick and Graybill (1992) claim that the Satterthwaite intervals tend to be too liberal (i.e., too wide) leading to actual coverage being larger than nominal coverage. For a random model and a balanced design, they recommend a procedure developed by Ting et al. (1990). The Ting et al. procedure is succinctly summarized by Burdick and Graybill (1992, pp. 36–42). (See also Betebenner, 1998.)

The discussion here employs most of the notational conventions in Burdick and Graybill, which are slightly different (in obvious ways) from the notation used in other parts of this book. Note, especially, that in this section, $1 - \alpha$, rather than γ, designates a confidence coefficient.[8]

The general problem Ting et al. address is the construction of a confidence interval on the parameter

$$\psi = \sum_{q=1}^{P} k_q \boldsymbol{EM}_q - \sum_{r=P+1}^{Q} k_r \boldsymbol{EM}_r, \qquad (6.22)$$

where $k_q, k_r \geq 0$, the $\boldsymbol{EM}$ are Q expected mean squares that contribute to ψ, P is the number of expected mean squares with a positive contribution to ψ, and $Q - P$ is the number of expected mean squares with a negative contribution to ψ. Although the theory permits ψ to be a linear combination of random effects variance components, in generalizability theory often ψ is a single variance component.

To clarify the notational conventions in Equation 6.22, consider the $p \times i$ design. Using the ANOVA procedure,

$$\sigma^2(p) = \frac{\boldsymbol{EMS}(p) - \boldsymbol{EMS}(pi)}{n_i},$$

which, in terms of the notation in Equation 6.22, means that $Q = 2$, $P = 1$, $\boldsymbol{EM}_1 = \boldsymbol{EMS}(p)$, $\boldsymbol{EM}_2 = \boldsymbol{EMS}(pi)$, and $k_1 = k_2 = 1/n_i$. When $Q = 2$, constructing a confidence interval using the Ting et al. procedure is considerably simpler than when $Q > 2$.

The distribution of estimated variance components generally is not symmetric. For that reason, Ting et al. discuss two one-sided intervals based on obtaining

1. the lower bound L on an upper $1 - \alpha$ confidence interval on ψ, and

2. the upper bound U on a lower $1 - \alpha$ confidence interval on ψ.

The first interval is (L, ∞) with probability α, and the second interval is $(-\infty, U)$ with probability α.[9] This general procedure permits obtaining a two-sided interval (L, U) with different probabilities in the two tails. More

---

[8] In almost all other parts of this book α is a generic designator for an effect.

[9] Although variance components cannot be negative, combinations of variance components can be. That is why the second interval is identified as $(-\infty, U)$ rather than $(0, U)$.

often than not, investigators want intervals with equal probabilities in the two tails. If so, then α should be halved in the following formulas.

Intervals for Q = 2

For $Q = 2$, the lower bound on an upper $1 - \alpha$ confidence interval on ψ is

$$L = \hat{\psi} - \sqrt{V_L}, \tag{6.23}$$

where

$$
\begin{aligned}
\hat{\psi} &= k_1 M_1 - k_2 M_2, \\
V_L &= G_1^2 k_1^2 M_1^2 + H_2^2 k_2^2 M_2^2 + G_{12} k_1 k_2 M_1 M_2, \\
G_1 &= 1 - \frac{1}{F_{\alpha:\eta_1,\infty}}, \\
H_2 &= \frac{1}{F_{1-\alpha:\eta_2,\infty}} - 1, \quad \text{and} \\
G_{12} &= \frac{(F_{\alpha:\eta_1,\eta_2} - 1)^2 - G_1^2 F_{\alpha:\eta_1,\eta_2}^2 - H_2^2}{F_{\alpha:\eta_1,\eta_2}},
\end{aligned}
$$

with η designating degrees of freedom. Note also that in specifying the F distribution, the Burdick and Graybill (1992) notation uses the convention that α is the area to the *right* of the percentile point. So, for example, for an upper $1 - \alpha = .9$ confidence interval, $F_{\alpha:\eta_1,\infty}$ in G_1 is the percentile point such that the area to the right is .1

For $Q = 2$, the upper bound on a lower $1 - \alpha$ confidence interval on ψ is

$$U = \hat{\psi} + \sqrt{V_U}, \tag{6.24}$$

where

$$
\begin{aligned}
V_U &= H_1^2 k_1^2 M_1^2 + G_2^2 k_2^2 M_2^2 + H_{12} k_1 k_2 M_1 M_2, \\
H_1 &= \frac{1}{F_{1-\alpha:\eta_1,\infty}} - 1, \\
G_2 &= 1 - \frac{1}{F_{\alpha:\eta_2,\infty}}, \quad \text{and} \\
H_{12} &= \frac{(1 - F_{1-\alpha:\eta_1,\eta_2})^2 - H_1^2 F_{1-\alpha:\eta_1,\eta_2}^2 - G_2^2}{F_{1-\alpha:\eta_1,\eta_2}}.
\end{aligned}
$$

Intervals for Q > 2

Formulas for constructing confidence intervals for $Q > 2$ are succinctly summarized by Burdick and Graybill (1992, pp. 39–42), but they are formidable. For any value of Q, computations can be performed with a hand calculator, but doing so is both tedious and error prone. (Computer programs are discussed in Section 6.5.)

The lower bound for an upper $1 - \alpha$ interval on ψ is

$$L = \hat{\psi} - \sqrt{V_L}, \qquad (6.25)$$

where

$$\hat{\psi} = \sum_{q=1}^{P} k_q M_q - \sum_{r=P+1}^{Q} k_r M_r,$$

$$V_L = \sum_{q=1}^{P} G_q^2 k_q^2 M_q^2 - \sum_{r=P+1}^{Q} H_r^2 k_r^2 M_r^2$$

$$+ \sum_{q=1}^{P} \sum_{r=P+1}^{Q} G_{qr} k_q k_r M_q M_r + \sum_{q=1}^{P-1} \sum_{t>q}^{P} G_{qt}^* k_q k_t M_q M_t,$$

$$G_q = 1 - \frac{1}{F_{\alpha:\eta_q,\infty}} \quad (q = 1, \ldots, P),$$

$$H_r = \frac{1}{F_{1-\alpha:\eta_r,\infty}} - 1 \quad (r = P+1, \ldots, Q),$$

$$G_{qr} = \frac{(F_{\alpha:\eta_q,\eta_r} - 1)^2 - G_q^2 F_{\alpha:\eta_q,\eta_r}^2 - H_r^2}{F_{\alpha:\eta_q,\eta_r}}, \quad \text{and}$$

$$G_{qt}^* = \frac{1}{P-1} \left[\left(1 - \frac{1}{F_{\alpha:\eta_q+\eta_t,\infty}} \right)^2 \frac{(\eta_q + \eta_t)^2}{\eta_q \eta_t} - \frac{G_q^2 \eta_q}{\eta_t} - \frac{G_t^2 \eta_t}{\eta_q} \right]$$
$$(t = q+1, \ldots, P).$$

If $P = 1$ then G_{qt}^* is defined to be zero.

The upper bound for a lower $1 - \alpha$ interval on ψ is

$$U = \hat{\psi} + \sqrt{V_U}, \qquad (6.26)$$

where

$$V_U = \sum_{q=1}^{P} H_q^2 k_q^2 M_q^2 - \sum_{r=P+1}^{Q} G_r^2 k_r^2 M_r^2$$

$$+ \sum_{q=1}^{P} \sum_{r=P+1}^{Q} H_{qr} k_q k_r M_q M_r + \sum_{r=P+1}^{Q} \sum_{u>r}^{Q-1} H_{ru}^* k_r k_u M_r M_u,$$

$$H_q = \frac{1}{F_{1-\alpha:\eta_q,\infty}} - 1 \quad (q = 1, \ldots, P),$$

$$G_r = 1 - \frac{1}{F_{\alpha:\eta_r,\infty}} \quad (r = P+1, \ldots, Q),$$

$$H_{qr} = \frac{(F_{1-\alpha:\eta_q,\eta_r} - 1)^2 - H_q^2 F_{1-\alpha:\eta_q,\eta_r}^2 - G_r^2}{F_{1-\alpha:\eta_q,\eta_r}}, \quad \text{and}$$

TABLE 6.4. Estimated 80% Confidence Intervals for G Study Variance Components for Synthetic Data Set No. 4 Recast as a $p \times i$ Design

	$\sigma^2(p)$	$\sigma^2(i)$	$\sigma^2(pi)$
Normal	(.152, 1.099)	(.294, 1.474)	(2.282, 3.293)
Satterthwaite	(.335, 2.022)	(.504, 2.326)	(2.350, 3.388)
Ting et al.	(.289, 1.619)	(.456, 2.013)	(2.350, 3.388)
Jackknife	(.078, 1.174)	(.169, 1.599)	(2.070, 3.505)
boot-p^a	(.348, .888)	(.378, 1.536)	(2.196, 3.366)
boot-i^a	(.066, 1.323)	(.160, 1.506)	(2.169, 3.348)
boot-p, i^a	(−.068, 1.435)	(.012, 1.925)	(1.878, 3.737)
boot-p, i, r^a	(.166, 1.127)	(.131, 1.677)	(2.310, 3.266)

[a]Based on adjusted estimates with 1000 replications.

$$H_{ru}^* = \frac{1}{Q - P - 1} \left[\left(1 - \frac{1}{F_{\alpha : \eta_r + \eta_u, \infty}} \right)^2 \frac{(\eta_r + \eta_u)^2}{\eta_r \eta_u} - \frac{G_r^2 \eta_r}{\eta_u} - \frac{G_u^2 \eta_u}{\eta_r} \right]$$

$$(u = r + 1, \dots, Q).$$

If $Q = P + 1$ then H_{ru}^* is defined to be zero.

When $Q = P > 2$ (which means there are no negative expected mean squares in Equation 6.22) and a two-sided $1 - 2\alpha$ interval on ψ is desired, Burdick and Graybill suggest using the Graybill and Wang (1980) interval

$$\left(\hat{\psi} - \sqrt{\sum_{q=1}^{Q} G_q^2 k_q^2 M_q^2}, \quad \hat{\psi} + \sqrt{\sum_{q=1}^{Q} H_q^2 k_q^2 M_q^2} \right), \qquad (6.27)$$

where G_q and H_q are defined as in the Ting et al. procedure. The method in Equation 6.27 is sometimes called the Modified Large Sample Method.

Example

Table 6.4 provides the Ting et al. 80% confidence intervals for the synthetic data example, along with 80% confidence intervals for all procedures discussed in this chapter. Note that the Ting et al. intervals for $\sigma^2(p)$ and $\sigma^2(i)$ are both narrower than the corresponding Satterthwaite intervals. For $\sigma^2(pi)$ both intervals are the same, because they involve only one mean square, namely, $MS(pi)$.

6.2.4 Jackknife Procedure

The jackknife procedure discussed in Section 6.1.2 for estimating standard errors of estimated variance components is nonparametric. However, to establish a confidence interval using the jackknife, typically a distributional-form assumption is required.[10] Usually, normality is assumed, and Student's t distribution is employed. Thus, a $100(\gamma)\%$ confidence interval for θ is

$$\hat{\theta}_J - t\,\hat{\sigma}(\hat{\theta}_J) \leq \theta \leq \hat{\theta}_J + t\,\hat{\sigma}(\hat{\theta}_J),$$

where θ can be any one of the variance components and t is the $(1+\gamma)/2$ percentile point of the t distribution with $n_p n_i - 1$ degrees of freedom (see Collins, 1970, p. 29).

For the synthetic data example, Table 6.4 provides the jackknife 80% confidence intervals. It is noteworthy that they are narrower than Satterthwaite and Ting et al. intervals.

6.2.5 Bootstrap Procedure

An appealing characteristic of the bootstrap algorithm is that it can be used almost automatically to obtain an approximate confidence interval, provided the number of bootstrap samples is $B \geq 1000$ (see Efron & Tibshirani, 1986, p. 67). For example, a simple approach to obtaining an 80% approximate confidence interval for θ is to use the 10th and 90th percentile points of the distribution of the $\hat{\theta}_b$ discussed in Section 6.1.3. More generally, an approximate $100(\gamma)\%$ confidence interval can be defined as the $[100(1-\gamma)/2]\%$th and $[100(1+\gamma)/2]\%$th percentile points for $\hat{\theta}_b$. (See Shao & Tu, 1995, and Wiley, 2000, for possibly better, but more complicated, approaches.) For the synthetic data example, Table 6.4 provides 80% confidence intervals for each of the four procedures for obtaining bootstrap samples.

6.3 Variability of D Study Statistics

The methods and procedures discussed in Sections 6.1 and 6.2 can be applied easily to D study variance components that involve nothing more than division of G study components by sample sizes. For the $p \times I$ design, this means that the standard errors for $\hat{\sigma}^2(I)$ and $\hat{\sigma}^2(pI) = \hat{\sigma}^2(\delta)$ are obtained by dividing the standard errors for $\hat{\sigma}^2(i)$ and $\hat{\sigma}^2(pi)$, respectively, by n_i'. A similar statement applies to the limits of the confidence intervals.

---

[10]The version of the jackknife discussed here is more specifically a delete-1 jackknife procedure. Theory suggests that there may be a delete-d version that could be used to obtain nonparametric estimates of confidence intervals for estimated variance components (see the Shao & Tu, 1995, discussion of jackknife histograms on pp. 50, 55–60).

6.3.1   Absolute Error Variance

Estimated absolute error variance for the $p \times I$ design is

$$
\hat{\sigma}^2(\Delta) \;=\; \frac{\hat{\sigma}^2(i) + \hat{\sigma}^2(pi)}{n_i'}
$$

$$
= \frac{MS(i) - MS(pi)}{n_p n_i'} + \frac{MS(pi)}{n_i'}
$$

$$
= \frac{MS(i) + (n_p - 1)MS(pi)}{n_p n_i'}. \tag{6.28}
$$

Therefore, using Equation 6.1 with $\alpha = \Delta$, the estimated standard error of $\hat{\sigma}^2(\Delta)$ is

$$
\hat{\sigma}[\hat{\sigma}^2(\Delta)] = \frac{1}{n_p n_i'} \sqrt{\frac{2[MS(i)]^2}{(n_i - 1) + 2} + \frac{2[(n_p - 1)MS(pi)]^2}{(n_p - 1)(n_i - 1) + 2}}. \tag{6.29}
$$

Satterthwaite's procedure and the Ting et al. procedure can be used to obtain confidence intervals for $\sigma^2(\Delta)$. Also, a normal approximation might be employed, using Equation 6.16 with $\hat{\sigma}^2(\Delta)$ replacing $\hat{\sigma}^2(\alpha|M)$.

The jackknife and bootstrap procedures can both be used to obtain estimated standard errors and confidence intervals for D study quantities. For example, for the jackknife, the $\hat{\theta}_{*j}$ discussed in Section 6.1.2 can be pseudovalues for $\sigma^2(\Delta)$, $E\rho^2$, or Φ. Similarly, for the bootstrap, $\hat{\theta}_b$ can be estimates of $\sigma^2(\Delta)$, $E\rho^2$, or Φ.

When these procedures are applied to the synthetic data (see Table 6.2), Table 6.5 reports the resulting estimated standard errors for $\hat{\sigma}^2(\Delta)$, $E\hat{\rho}^2$, or $\widehat{\Phi}$, and estimated 80% confidence intervals for $\sigma^2(\Delta)$, $E\rho^2$, or Φ. Clearly, the four different bootstrap procedures give substantially different estimated standard errors and confidence intervals.

Wiley's bias adjustments in Figure 6.1 on page 188 were employed to obtain the results in Table 6.5. However, there is no reason to believe that these adjustments remove all bias in estimates of *ratios* of variance components, such as $E\rho^2$ and Φ. Note also that, for boot-i and boot-p, i there were estimates of $E\rho^2$ and Φ that were smaller than 0 or larger than 1; these were set to 0 or 1, respectively.

Note that the upper limit of the normal interval for $\sigma^2(\Delta)$ is smaller than the upper limit of both the Satterthwaite and Ting et al. intervals. Also, the Ting et al. interval for $\sigma^2(\Delta)$ is *wider* than the Satterthwaite interval. Ting et al. intervals tend to be narrower than Satterthwaite intervals when ψ in Equation 6.22 has one or more $k_r > 0$; that is, $\hat{\psi}$ involves subtracting one or more mean squares. However, both mean squares incorporated in $\hat{\sigma}^2(\Delta)$ in Equation 6.28 have a positive contribution.

Table 6.6 provides D study results for the same nine forms of an ACT Assessment Mathematics subtest that were considered in Table 6.1. It is

TABLE 6.5. D Study Estimated Standard Errors and Confidence Intervals for Synthetic Data Set No. 4 Recast as a $p \times I$ Design with $n_i' = 12$, $\hat{\sigma}^2(\Delta) = .3059$, $E\hat{\rho}^2 = .729$, and $\widehat{\Phi} = .672$

	Means			Est. Standard Errors		
	$\hat{\sigma}^2(\Delta)$	$E\hat{\rho}^2$	$\widehat{\Phi}$	$\hat{\sigma}^2(\Delta)$	$E\hat{\rho}^2$	$\widehat{\Phi}$
Normal				.0481		
Jackknife				.0763	.0902	.1280
boot-p^a	.3048	.7074	.6524	.0502	.0984	.1000
boot-i^a	.3054	.6336	.5848	.0597	.2565	.2512
boot-p, i^a	.3050	.6103	.5622	.0831	.3006	.2939
boot-p, i, r^a	.3089	.6577	.6037	.0578	.1930	.1921

	$\sigma^2(\Delta)$	$E\rho^2$	Φ
Normal	(.244, .368)		
Feldt		(.541, .876)	
Arteaga et al.[b]			(457, .845)
Satterthwaite	(.251, .385)		
Ting et al.	(.255, .410)		
Jackknife	(.208, .404)	(.665, .898)	(.559, .889)
boot-p^a	(.242, .372)	(.595, .805)	(.533, .755)
boot-i^a	(.226, .384)	(.186, .876)	(.145, .838)
boot-p, i^a	(.207, .411)	(.000, .895)	(.000, .855)
boot-p, i, r^a	(.233, .384)	(.403, .829)	(.343, .788)

[a]Based on 1000 replications; $E\hat{\rho}^2$ and $\widehat{\Phi}$ constrained to be between 0 and 1.

[b]Using the transformation in Equation 6.36.

evident that the empirically based estimate of $\sigma[\hat{\sigma}^2(\delta)]$ in the next-to-the-last row is 10 times larger than the average of the normality-based estimates in the last row. By contrast, the empirically based estimate of $\sigma[\hat{\sigma}^2(\Delta)]$ is quite close to the average of the normality-based estimates. This last result is encouraging since there is good reason to believe that the score effects for these real data do not have a multivariate normal distribution.

6.3.2   Feldt Confidence Interval for $E\rho^2$

Under normality assumptions, Burdick and Graybill (1992) provide a detailed discussion of procedures for establishing confidence intervals on ratios of variance components. One special case they consider (see pp. 128–129) is the two-sided $100(1 - 2\alpha)\%$ exact confidence interval for $\zeta = \sigma^2(p)/\sigma^2(pi)$ in a $p \times i$ design:

$$\left(\frac{L^* - 1}{n_i} , \frac{U^* - 1}{n_i} \right), \tag{6.30}$$

TABLE 6.6. D Study $p \times I$ Results for Nine Forms of ACT Math Pre-Algebra/Elementary Algebra Subtest with $n'_i = 24$ Items

Form	n_p	$MS(i)$	$MS(pi)$	$\hat{\sigma}^2(\delta)$	$\hat{\sigma}^2(\Delta)$	$E\hat{\rho}^2$	$\hat{\Phi}$
A	3388	85.95564	.17681	.00737	.00842	.81335	.79217
B	3363	91.55995	.15167	.00632	.00745	.85719	.83580
C	3458	127.28606	.14213	.00592	.00745	.83980	.80638
D	3114	96.57363	.16712	.00696	.00825	.81496	.78794
E	3428	63.51859	.17122	.00713	.00790	.84821	.83454
F	3318	136.50134	.15989	.00666	.00837	.83346	.79925
G	3257	95.13462	.17172	.00716	.00837	.80547	.77971
H	3221	161.28615	.15793	.00658	.00866	.81306	.76761
I	3178	94.88673	.17237	.00718	.00842	.80063	.77395
Mean				.00681	.00814	.82513	.79748
SD				.00048	.00044	.02006	.02451
Ave SE[a]				.00004	.00039		

[a]Square root of the average of the squared estimated standard errors using Equation 6.1.

where

$$L^* = \frac{M_p}{M_{pi} F_{\alpha:\eta_p, \eta_{pi}}} \quad \text{and} \quad U^* = \frac{M_p}{M_{pi} F_{1-\alpha:\eta_p, \eta_{pi}}}, \qquad (6.31)$$

where $M_p = MS(p)$, $M_{pi} = MS(pi)$, and η designates degrees of freedom. Obviously, as in Section 6.2.3, α is *not* being used here to designate an effect; rather $1 - 2\alpha$ is the confidence coefficient expressed as a proportion.

Note that ζ is a signal–noise ratio for $n'_i = 1$. A generalizability coefficient for a test of length n'_i in terms of ζ is

$$E\rho^2 = \frac{n'_i \zeta}{1 + n'_i \zeta}. \qquad (6.32)$$

This transformation can be applied to the endpoints in Equation 6.30 to obtain a confidence interval for $E\rho^2$. The resulting interval for $E\rho^2$ is identical to that derived by Feldt (1965) and is identified as such in subsequent tables and discussion.

For the synthetic data, $M_p = MS(p) = 10.2963$ with 9 degrees of freedom, and $M_{pi} = MS(pi) = 2.7872$ with 99 degrees of freedom. For a two-sided 80% confidence interval, $\alpha = .1$. Since $F_{.9:9,99} = 1.6956$ and $F_{.1:9,99} = .4567$, it follows that $L^* = 2.1787$ and $U^* = 8.0888$. Consequently, an 80% confidence interval for ζ is (.0982, .5907). Transforming these limits using Equation 6.32 gives (.541, .876) as an 80% interval for $E\rho^2$.

When n_p is very large, this author's experience suggests that a Feldt confidence interval for $E\rho^2$ is likely to be too narrow for real test forms based on dichotomously scored items. Consider, again, the ACT Assessment Mathematics results in Table 6.6. The range of the nine estimates of $E\rho^2$ is from about .80 to .86.[11] However, since $n_p > 3000$ and $n_i = 24$ for all forms, the degrees of freedom for both p and pi are very large, both $F_{\alpha:\eta_p,\eta_{pi}}$ and $F_{1-\alpha:\eta_p,\eta_{pi}}$ are close to unity for moderately sized confidence coefficients (e.g., 80%), and the resulting intervals tend to be narrower than the empirical data in Table 6.6 would suggest.

6.3.3   Arteaga et al. Confidence Interval for Φ

As discussed by Burdick and Graybill (1992, p. 129), assuming score effects are normally distributed, Arteaga et al. (1982) developed the following approximate two-sided $100(1 - 2\alpha)\%$ confidence interval for the ratio $\Lambda = \sigma^2(p)/[\sigma^2(p) + \sigma^2(i) + \sigma^2(pi)]$ in a $p \times i$ design.

$$\left(\frac{n_p L_p}{n_p L_p + n_i} \; , \; \frac{n_p U_p}{n_p U_p + n_i} \right), \tag{6.33}$$

where

$$L_p = \frac{M_p^2 - F_{\alpha:\eta_p,\infty} M_p M_{pi} + (F_{\alpha:\eta_p,\infty} - F_{\alpha:\eta_p,\eta_{pi}}) F_{\alpha:\eta_p,\eta_{pi}} M_{pi}^2}{(n_p - 1) F_{\alpha:\eta_p,\infty} M_p M_{pi} + F_{\alpha:\eta_p,\eta_i} M_p M_i} \tag{6.34}$$

and

$$U_p = \frac{M_p^2 - F_{1-\alpha:\eta_p,\infty} M_p M_{pi} + (F_{1-\alpha:\eta_p,\infty} - F_{1-\alpha:\eta_p,\eta_{pi}}) F_{1-\alpha:\eta_p,\eta_{pi}} M_{pi}^2}{(n_p - 1) F_{1-\alpha:\eta_p,\infty} M_p M_{pi} + F_{1-\alpha:\eta_p,\eta_i} M_p M_i}, \tag{6.35}$$

with $M_p = MS(p)$, $M_i = MS(i)$, $M_{pi} = MS(pi)$, and η designating degrees of freedom.[12]

It is evident that Λ is Φ for the $p \times I$ design with $n_i' = 1$. For a test of length n_i',

$$\Phi = \frac{n_i' \Lambda}{1 + (n_i' - 1)\Lambda}, \tag{6.36}$$

which is simply an application of the Spearman–Brown Formula to Λ. Therefore, applying the transformation in Equation 6.36 to the limits in Equation 6.33 gives a confidence interval for Φ. For the synthetic data, an 80% confidence interval is (.457, 845).

---

[11] Since the various forms were administered to randomly equivalent groups of examinees, it is unlikely that this variability in estimates is attributable to systematic differences in the samples of persons.

[12] As in Sections 6.2.3 and 6.3.2, α does *not* designate an effect here.

6.4   Some Simulation Studies

The real-data example in Tables 6.1 and 6.6 illustrates that normality-based procedures for estimating standard errors and/or confidence intervals may not always work well with dichotomous data. The synthetic data example illustrates that different procedures for estimating standard errors and confidence intervals can give quite different results, and there is no obvious basis for deciding which estimates are best. An investigator faced with results such as those in Tables 6.2 through 6.4 may be tempted to pick modal results or discard results that appear atypical. There is no guarantee, however, that either strategy is defensible, let along optimal.

In this section four simulation studies are discussed that may assist investigators in choosing among methods and procedures. The first two studies involve estimated G study variance components for the $p \times i$ design. The next two studies involve D study statistics for the $p \times I$ design. All bootstrap results employ Wiley's adjustments. Wiley (2000) provides some additional simulations of G study results.

6.4.1   G Study Variance Components

Two simulation studies are discussed here. The first is based on normally distributed score effects; the other is based on dichotomous data from an actual testing program.

Normal Data

Table 6.7 provides estimated standard error and confidence interval results for a simulation study using sample sizes of $n_p = 100$ and $n_i = 20$, based on the assumption that score effects are normally distributed with variance components of $\sigma^2(p) = 4$, $\sigma^2(i) = 16$, and $\sigma^2(pi) = 64$. Specifically, each observable person-item score was generated using the formula:

$$X_{pi} = \mu + \sigma(p)z_p + \sigma(i)z_i + \sigma(pi)z_{pi},$$

where z_p, z_i, and z_{pi} are randomly and independently sampled values from a unit normal distribution. (The parameter μ is irrelevant for our purposes here.) This data generation procedure assumes that $N_p \rightarrow \infty$ and $N_i \rightarrow \infty$. Under these assumptions, the actual standard errors of the estimated variance components are given by Equation 6.1 with mean squares replaced by expected mean squares.[13] These parameters are reported at the top of Table 6.7.

The simulation involved 1000 trials. That is, 1000 matrices of size $n_p \times n_i$ were generated. For each trial, the variance components were estimated.

---

[13] Degrees of freedom are *not* increased by two.

TABLE 6.7. Simulation of G Study Estimated Standard Errors and Confidence Interval Coverage with Normal Data for $n_p = 100$, $n_i = 20$, and 1000 Trials

	Means			Standard Errors		
	$\hat{\sigma}^2(p)$	$\hat{\sigma}^2(i)$	$\hat{\sigma}^2(pi)$	$\hat{\sigma}^2(p)$	$\hat{\sigma}^2(i)$	$\hat{\sigma}^2(pi)$
Parameters	4.0000	16.0000	64.0000	1.0287	5.3988	2.0869
Empirical	4.0402	15.7535	64.0082	1.0000	5.4046	2.0326
Normal				1.0338	5.3267	2.0871
Jackknife	4.0402	15.7535	64.0082	1.0327	5.4893	2.0842
boot-p^a	4.0395	15.7550	64.0094	1.0337	*1.4966*	2.0764
boot-i^a	4.0413	15.7536	64.0064	*1.5038*	5.2319	2.2956
boot-p, i^a	4.0407	15.7572	64.0055	*2.1075*	5.6380	*3.6983*
boot-p, i, r^a	4.0365	15.7532	64.0117	*1.4751*	5.4412	2.0869

	Nominal 66.7 Percent			Nominal 90 Percent		
	$\sigma^2(p)$	$\sigma^2(i)$	$\sigma^2(pi)$	$\sigma^2(p)$	$\sigma^2(i)$	$\sigma^2(pi)$
Normal	66.5	62.9	67.8	90.9	83.9	91.1
Satterthwaite	69.6	66.6	67.4	92.1	90.1	90.8
Ting et al.	67.7	66.4	67.4	91.0	89.5	90.8
Jackknife	65.9	61.8	66.9	90.0	83.0	88.4
boot-p^a	65.4	*20.9*	66.8	90.4	*35.2*	89.7
boot-i^a	*84.0*	60.0	71.4	*98.3*	82.1	92.3
boot-p, i^a	*94.9*	63.3	*91.4*	*100.0*	84.7	*99.4*
boot-p, i, r^a	*84.8*	62.1	67.8	*98.2*	83.6	90.3

[a]Based on 1000 replications.

Their average values are reported in the "Empirical" row in Table 6.7. The empirical standard errors in the same row are simply the standard deviations of the 1000 estimates. The standard errors in the "Normal" row were computed using Equation 6.1, which assumes normality.[14] For an infinitely long simulation, the parameters, empirical estimates, and normal estimates should all be the same. The differences provide a type of yardstick for judging the credibility of the simulation results.

For the jackknife results, *each* of the 1000 trials involved computing $101 \times 21 = 2121$ sets of estimated variance components. For each bootstrap procedure, 1000 replications were employed for *each* of the 1000 trials. This means that the results for each of the bootstrap procedures were based on one million $p \times i$ analyses.

---

[14]The reported values for the normal standard errors were obtained by taking the square root of the average of the 1000 squared standard errors.

Results that appear particularly extreme to this author are in italics in Table 6.7. Since the simulation conditions are in accord with major assumptions of the normal, Satterthwaite, and Ting et al. procedures, it is encouraging that these procedures give results that are close to the parameters and nominal coverages. As predicted by theory, the normal intervals are a bit too conservative and the Satterthwaite intervals are a bit too liberal, but not by much. Although the jackknife procedure makes no normality assumptions, its results appear quite good, at least for practical purposes.

By contrast, the bootstrap results are mixed; it appears that:

- boot-p works well for $\sigma^2(p)$ and $\sigma^2(pi)$;

- boot-i works reasonably well for $\sigma^2(i)$ and better for $\sigma^2(pi)$;

- boot-p, i works reasonably well for $\sigma^2(i)$; and

- boot-p, i, r works reasonably well for $\sigma^2(i)$ and quite well for $\sigma^2(pi)$.

Note, in particular, that for $\sigma^2(i)$, each of the bootstrap procedures involving sampling items (i.e., boot-i, boot-p, i, and boot-p, i, r) works reasonably well; however, it is not true that for $\sigma^2(p)$, each of bootstrap procedures involving sampling persons works well. Since the $p \times i$ design is symmetric, it seems sensible to conclude that at least some of the asymmetry in these results is attributable to the five-to-one ratio of the sample sizes. Apparently, the applicability of the bootstrap depends to some extent on both the bootstrap sampling procedure employed and the pattern of sample sizes.

Based on the results in Table 6.7, one rule that might be considered, at least tentatively, is to use boot-p for $\sigma^2(p)$, boot-i for $\sigma^2(i)$, and boot-p, i, r for $\sigma^2(pi)$ when it can be assumed that the score effects are approximately normal.

Nonnormal Dichotomous Data

In generalizability theory, there is often good reason to believe that score effects are not normally distributed, particularly when data are dichotomous or fall into an otherwise small number of discrete categories. For such data, it is risky to use a simulation such as that in Table 6.7 as a basis for judgments about the applicability of procedures. By way of comparison, Table 6.8 provides the results of a simulation using dichotomous data with the same sample sizes as those in Table 6.7.

Specifically, in Table 6.8 each of the 1000 trials involved sampling $n_p = 100$ persons without replacement from a population of size $N_p = 2000$ and $n_i = 20$ items without replacement from a universe of size $N_i = 200$. The 2000×200 data set is from an actual testing program.[15] Obviously, this simulation involves sampling from a finite population and universe,

---

[15]Several tests were concatenated to give the entire universe of 200 items.

TABLE 6.8. Simulation of G Study Estimated Standard Errors and Confidence Interval Coverage with Dichotomous Data for $n_p = 100$, $n_i = 20$, $N_p = 2000$, $N_i = 200$, and 1000 Trials

	Means			Standard Errors		
	$\hat{\sigma}^2(p)$	$\hat{\sigma}^2(i)$	$\hat{\sigma}^2(pi)$	$\hat{\sigma}^2(p)$	$\hat{\sigma}^2(i)$	$\hat{\sigma}^2(pi)$
Parameters[a]	.0264	.0245	.1856	.0053	.0068	.0099
Empirical	.0264	.0244	.1854	.0053	.0068	.0103
Normal				.0050	.0083	*.0061*
Jackknife	.0264	.0244	.1854	.0056	.0075	.0107
boot-p[b]	.0264	.0244	.1854	*.0044*	*.0033*	*.0056*
boot-i[b]	.0264	.0244	.1854	.0057	.0071	.0098
boot-p, i[b]	.0264	.0244	.1854	*.0080*	.0084	.0121
boot-p, i, r[b]	.0264	.0244	.1854	.0057	.0078	*.0045*

	Nominal 66.7 Percent			Nominal 90 Percent		
	$\sigma^2(p)$	$\sigma^2(i)$	$\sigma^2(pi)$	$\sigma^2(p)$	$\sigma^2(i)$	$\sigma^2(pi)$
Normal	63.6	74.5	*43.0*	85.8	90.4	*68.0*
Satterthwaite	65.2	76.7	*43.2*	87.2	96.0	*67.3*
Ting et al.	63.2	76.4	*43.2*	86.1	95.4	*67.3*
Jackknife	66.5	69.1	67.9	88.0	87.8	90.3
boot-p[b]	*58.9*	*36.7*	*39.5*	*81.4*	*57.8*	*63.1*
boot-i[b]	68.9	65.4	63.7	90.3	86.3	86.7
boot-p, i[b]	*84.3*	74.3	74.8	*98.9*	93.7	94.4
boot-p, i, r[b]	70.8	71.0	*33.0*	90.1	90.5	*53.1*

[a] Based on 5000 trials.

[b] Based on 1000 replications.

and computational procedures reflected this design characteristic (e.g., the parameters were defined by Equations 6.4 to 6.6; and Equations 6.7 to 6.9 used $c_p = .95$ and $c_i = .90$, as did Equation 6.12). This approach to conducting the simulation was taken because there is no simple method for generating dichotomous data for a model with prespecified values for the variance components.

The standard error "parameters" were not determined by theory, because no such theory exists for dichotomous data and a $p \times i$ design. Rather, the standard deviations of the estimated variance components for 5000 trials were used as parameters. Otherwise, however, the simulation procedures and formulas used to give the results in Table 6.8 mirror those that led to the results in Table 6.7.

Perhaps the most obvious and encouraging similarity between the results in Tables 6.7 and 6.8 is that the jackknife procedure works quite well for both normal and dichotomous data, for all three variance components. Perhaps the most obvious difference is that for the normality-based procedures (i.e., normal, Satterthwaite, and Ting et al.) the confidence interval coverage for $\sigma^2(pi)$ is much too low for dichotomous data, and the coverage for $\sigma^2(i)$ is usually too high.

The bootstrap results are mixed; it appears that:

- boot-p understates the standard errors and nominal confidence coefficients for all variance components, although results for $\sigma^2(p)$ are much more accurate than for $\sigma^2(i)$ and $\sigma^2(pi)$;

- boot-i results are reasonably accurate for all variance components;

- boot-p, i results are somewhat accurate for $\sigma^2(i)$ and $\sigma^2(pi)$; and

- boot-p, i, r works quite well for $\sigma^2(p)$ and $\sigma^2(i)$, but very poorly for $\sigma^2(pi)$.

Clearly, conclusions based on the normal-data simulation do not all generalize to the dichotomous-data simulation. In particular, boot-p, i, r works well for $\sigma^2(pi)$ with normal data, but very poorly with dichotomous data.

6.4.2 D Study Statistics

As noted at the beginning of Section 6.3, for the $p \times I$ design, standard errors for $\hat{\sigma}^2(\delta)$ are obtained by dividing standard errors for $\hat{\sigma}^2(pi)$ by n_i', and a similar statement applies to confidence interval limits for $\sigma^2(\delta)$. Therefore, the simulation results for $\sigma^2(pi)$ in Tables 6.7 and 6.8 can be used directly to obtain results for $\sigma^2(\delta)$. Note, in particular, that the confidence interval coverage results for $\sigma^2(pi)$ are necessarily identical to those for $\sigma^2(\delta)$.

Normal Data

Under the assumption that score effects are normally distributed, the simulation results provided in Table 6.9 are for D study statistics for the $p \times I$ design with $n_i' = 20$. The simulation conditions are exactly the same as those in Table 6.7 but, of course, the statistics under consideration are different, and there are two additional procedures: Feldt's confidence interval for $E\rho^2$, and the Arteage et al. confidence interval for Φ.

The bootstrap estimates, standard errors, and confidence intervals for these D study statistics employed the Wiley-adjusted bootstrap estimates of variance components discussed in Section 6.4.1. For example, for each bootstrap sample, the boot-p estimate of $E\rho^2$ was computed as the boot-p estimate of $\sigma^2(p)$ divided by itself plus the boot-p estimate of $\sigma^2(pi)/20$.[16]

---

[16]For $E\hat{\rho}^2$ and $\widehat{\Phi}$, any bootstrap estimate smaller than 0 or larger than 1 was set to 0 or 1, respectively.

TABLE 6.9. Simulation of D Study Estimated Standard Errors and Confidence Interval Coverage with Normal Data for $n_p = 100$, $n_i = 20$, and 1000 Trials

	Means			Standard Errors		
	$\hat{\sigma}^2(\Delta)$	$E\hat{\rho}^2$	$\widehat{\Phi}$	$\hat{\sigma}^2(\Delta)$	$E\hat{\rho}^2$	$\widehat{\Phi}$
Parameters[a]	4.0000	.5556	.5000	.2922	.0679	.0698
Empirical	3.9881	.5495	.4961	.2864	.0643	.0659
Normal				.2857		
Jackknife	3.9881	.5495	.4961	.2928	.0608	.0648
boot-p[b]	3.9882	.5403	.4879	*.1263*	.0685	.0676
boot-i[b]	3.9880	.5295	.4790	.2873	*.1052*	*.1037*
boot-p, i[b]	3.9881	.5110	.4629	.3373	*.1536*	*.1480*
boot-p, i, r[b]	3.9882	.5293	.4787	.2906	*.1062*	*.1039*

	Nominal 66.7 Percent			Nominal 90 Percent		
	$\sigma^2(\Delta)$	$E\rho^2$	Φ	$\sigma^2(\Delta)$	$E\rho^2$	Φ
Normal	64.0			86.2		
Feldt		67.9			90.4	
Arteaga et al.[c]			74.3			93.7
Satterthwaite	66.0			87.8		
Ting et al.	66.8			90.2		
Jackknife	63.9	62.4	65.3	84.2	85.1	85.9
boot-p[b]	*33.3*	65.9	66.0	*53.2*	89.9	88.7
boot-i[b]	64.1	*84.6*	*83.2*	83.7	*98.5*	*98.1*
boot-p, i[b]	72.3	*95.8*	*94.8*	91.8	*100.0*	*99.9*
boot-p, i, r[b]	65.2	*83.9*	*84.3*	85.8	*98.6*	*98.2*

[a]5000 trials   [b]1000 replications   [c]Using Equation 6.36

All procedures based on normality assumptions appear to perform reasonably well for this normal-data simulation. Also, the nonparametric jackknife procedure works quite well, although actual confidence interval coverage is a bit low. Again, the bootstrap procedure provides mixed results, with boot-p working very well for $E\rho^2$ and Φ; and boot-i, boot-p, i, and boot-p, i, r working reasonably well for $\sigma^2(\Delta)$. Note that boot-p gives very inaccurate results for $\sigma^2(\Delta)$.

Nonnormal Dichotomous Data

Table 6.10 extends the dichotomous data simulation in Table 6.8 to D study statistics with a sample size of $n_i' = 20$. Again, the jackknife works very well for all statistics. The normality-based procedures work reasonably well for estimating the standard error of $\hat{\sigma}^2(\Delta)$ and for obtaining a confidence

TABLE 6.10. Simulation of D Study Estimated Standard Errors and Confidence Interval Coverage with Dichotomous Data for $n_p = 100$, $n_i = 20$, $N_p = 2000$, $N_i = 200$, and 1000 Trials

	Means			Standard Errors		
	$\hat{\sigma}^2(\Delta)$	$E\hat{\rho}^2$	$\hat{\Phi}$	$\hat{\sigma}^2(\Delta)$	$E\hat{\rho}^2$	$\hat{\Phi}$
Parameters[a]	.00946	.7596	.7363	.00047	.0411	.0452
Empirical	.00947	.7544	.7308	.00048	.0418	.0459
Normal				.00046		
Jackknife	.00947	.7544	.7308	.00050	.0438	.0483
boot-p[b]	.00944	.7501	.7264	*.00026*	.0360	.0385
boot-i[b]	.00944	.7468	.7229	.00046	.0478	.0517
boot-p, i[b]	.00944	.7387	.7147	.00056	*.0739*	*.0781*
boot-p, i, r[b]	.00944	.7467	.7230	.00039	.0468	.0495

	Nominal 66.7 Percent			Nominal 90 Percent		
	$\sigma^2(\Delta)$	$E\rho^2$	Φ	$\sigma^2(\Delta)$	$E\rho^2$	Φ
Normal	68.8			89.4		
Feldt		60.2			84.0	
Arteaga et al.[c]			64.1			87.2
Satterthwaite	69.8			90.7		
Ting et al.	69.4			91.5		
Jackknife	69.4	68.4	68.1	90.1	90.1	89.5
boot-p[b]	*43.0*	58.0	57.0	*65.7*	81.9	81.0
boot-i[b]	65.9	67.9	67.9	88.8	90.8	90.4
boot-p, i[b]	75.7	*84.0*	*83.9*	95.0	*98.8*	*98.7*
boot-p, i, r[b]	61.1	66.7	66.1	85.0	89.3	88.7

[a]5000 trials   [b]1000 replications   [c]Using Equation 6.36

interval for $\sigma^2(\Delta)$. Confidence intervals using Feldt's procedure for $E\rho^2$ and the Arteaga et al. procedure for Φ are less accurate, but still probably usable for most practical purposes.

For all variance components, the bootstrap results using boot-p and boot-p, i are discouraging, but the boot-i results are very good. Also, the boot-p, i, r results are quite good for the two coefficients. Clearly, the bootstrap normal-data simulation results in Table 6.9 are often not confirmed by the bootstrap dichotomous-data simulations in Table 6.10. Apparently, the nature of the underlying data influences the choice of an optimum bootstrap procedure.

6.5   Discussion and Other Issues

As noted at the beginning of this chapter, the subject of variability of statistics in generalizability theory is complicated, and there are many unanswered questions. Also, it is evident that many procedures require extensive computations. For balanced designs, both GENOVA and urGENOVA (see Appendices F and G) provide G study estimated standard errors of the type discussed in Section 6.1.1, and GENOVA provides D study estimated standard errors such as those in Section 6.3.1. In addition, for balanced designs urGENOVA provides both the Satterthwaite and Ting et al. confidence intervals for G study variance components.

Tentative Conclusions

The simulations discussed in this chapter are the only ones yet available that systematically compare so many procedures for estimating the variability of statistics in both G and D studies in generalizability theory. These simulations have obvious limitations, however. They apply to single facet crossed designs only, they involve only one pattern of sample sizes, and they are based on only normal and dichotomous data.

   Still, on balance, these tentative conclusions seem warranted:

- the jackknife procedure works quite well for all statistics and for different types of data; this claim cannot be made for any other procedure;

- normality-based procedures do not work well for estimating standard errors or confidence intervals for $\sigma^2(pi)$ or $\sigma^2(\delta)$ with dichotomous data;

- the Ting et al. procedure works somewhat better than the Satterthwaite procedure but not so much better that the simpler Satterthwaite procedure should be abandoned entirely; and

- the Feldt and Arteaga et al. procedures work reasonably well for $E\rho^2$ and Φ, respectively, even when score effects are not normally distributed.

Admittedly, some of these conclusions involve a rather liberal interpretation of "works well." In most cases, when standard errors and confidence intervals are used in generalizability theory, great precision is not required. It is usually sufficient that investigators not be misled by gross inaccuracies.

   An important additional conclusion is that the bootstrap provides mixed results. Sometimes boot-p, boot-i, boot-p, i, and/or boot-p, i, r works quite well, but there is a notable lack of consistency across different statistics, types of data, and sample sizes. It is easy to be misled by the bootstrap, at least as it has been used here, and it is difficult to make statements about its general applicability with any great degree of confidence. This is unfortunate since the bootstrap is appealing not only because of its freedom from distributional-form assumptions, but also because it appears relatively easy to generalize the bootstrap to more complicated designs (see, for example, Wiley, 2000).

Multifacet Designs

For the most part, the normality-based procedures in this chapter have been provided at a level of detail that makes them applicable to any balanced design. The confidence interval procedures for $E\rho^2$ and Φ are obvious exceptions.

The Ting et al. and Satterthwaite procedures apply to any balanced design. Betebenner (1998) provides simulation results for $\sigma^2(\delta)$ and $\sigma^2(\Delta)$ for various two-facet designs. The Satterthwaite procedure is particularly easy to use, as discussed in Section 6.2.2. Under the assumption of normally distributed score effects, there are theoretical reasons for concluding that sometimes the Satterthwaite procedure will give intervals that are too liberal (i.e., too wide), as confirmed by the normal-data simulations in this chapter. However, with large numbers of degrees of freedom, any bias in the Satterthwaite procedure is not likely to be very big. For the single-facet simulations in this chapter involving nonnormal data, the Satterthwaite and Ting et al. coverages are largely indistinguishable (see, e.g., Table 6.8), and sometimes the Satterthwaite coverage appears slightly better.

Extending the jackknife to multifacet crossed designs appears straightforward, although the computations may be formidable. By contrast, it is not unambiguously clear how to extend the jackknife to nested designs such as $p \times (i:h)$. The essence of the problem centers on $\hat{\theta}_{-0ih}$ and $\hat{\theta}_{-pih}$. If one level of i is to be eliminated from one level of h then the resulting design is unbalanced. If one level of i is be eliminated from each level of h, the algorithm does not specify which level of i should be eliminated for each h. Note also that, when samples sizes are even moderately large, computational demands for the jackknife can be excessive. One or more variations on the delete-d jackknife (see Shao & Tu, 1995) may be advisable.

Other Procedures

Sirotnik and Wellington (1977, p. 346) presented a framework that can be used to estimate variance components and their standard errors using generalized symmetric means with what they called "incidence samples." An incidence sample refers "to the configuration of data points or entries sampled from a matrix population." Boodoo (1982) considered these procedures in the specific context of generalizability theory. Standard errors can be estimated using this theoretical framework without reverting to normality assumptions, but computations are very complicated. The Sirotnik and Wellington procedures have not been widely used in generalizability theory, but they may have considerable promise if their computational burden can be overcome.

The confidence intervals discussed in this chapter are one-at-a-time intervals. See Burdick and Graybill (1992, pp. 18–20, 51–56) for a discussion of simultaneous confidence intervals. Under normality assumptions, Khuri (1981) developed an elegant but complicated procedure for simultaneous confidence intervals for balanced random models, and Bell (1986) discussed the Khuri procedure in the specific context of generalizability theory. The Bonferroni inequality applied to one-at-a-time intervals is a simpler approach that often leads to shorter intervals. If Q intervals with a common coverage probability are to be established such that

$$\Pr(L_q \leq \psi_q \leq U_q) = 1 - 2\alpha \quad (q = 1, \ldots, Q),$$

then the Bonferroni inequality states that

$$\Pr[(L_1 \leq \psi_1 \leq U_1) \quad \text{and} \quad (L_2 \leq \psi_2 \leq U_2) \quad \text{and} \quad \ldots \quad (L_Q \leq \psi_Q \leq U_Q)]$$

$$\geq 1 - Q(2\alpha). \tag{6.37}$$

So, if an investigator wants a simultaneous coverage probability of at least SPr for Q intervals, then each of the individual intervals should use

$$\alpha = \frac{1 - \text{SPr}}{2Q},$$

which gives an individual coverage probability of

$$\text{PR} = 1 - \frac{1 - \text{SPr}}{Q}.$$

For example, if it is desired that the overall coverage be SPr $= .80$ for $Q = 3$ intervals, then individual intervals should use $\alpha = .0333$, which gives an individual coverage probability of Pr $= .9333$.

"Achilles Heel" Argument

It has almost become conventional wisdom to believe that the variability of estimated variance components is the "Achilles heel" of generalizability theory. Commenting on this, Brennan (2000a, p. 8) states:

> Because a generalizability analysis focuses on estimating variance components, the credibility of any conclusions rests on the extent to which estimated variance components are reasonably accurate reflections of the parameters. However, it is well known that estimated variance components can be quite unstable when the number of conditions of measurement is small, which is often the case.

To the extent that the "Achilles heel" argument is valid, it applies, of course, not only to (generalizability) theory but also to any variance components model. Such models are very prevalent in statistics, and investigators do not discard them simply because some of the estimated parameters may contain sizable random error. In this sense, it is easy to exaggerate the importance of the "Achilles heel" argument, at least relative to what is accepted practice in other areas of statistical inference.

More importantly, however, the "Achilles heel" argument, as it is usually stated, can be misleading. The argument typically focuses on estimated variance components for single conditions of measurement, whereas decisions about examinees are made with respect to mean or total scores over all sampled conditions of measurement. ... These can be, and often are, quite small even when the standard errors for single conditions of measurement are relatively large.

Even with (these) caveats ..., the "Achilles heel" argument is probably more helpful than harmful. It properly encourages investigators to pay attention to sampling error, and it provides a strong although indirect challenge to investigators to gather as much data as possible. After all, the ultimate solution to large standard errors is the collection of more data, not simply better estimation procedures.

It is easy to be discouraged by the complexities or inaccuracies of various methods discussed in this chapter, and the occasional inconsistent results across methods. Still, in generalizability theory it is almost always advisable to employ one or more methods to estimate the variability of statistics used to make important decisions, even if the methods chosen are less than optimal. The message of this chapter is that results should be interpreted with caution, not that the topic of variability in estimates should be disregarded.

6.6   Exercises

6.1* Verify the result for $\hat{\sigma}[\hat{\sigma}^2(\Delta)]$ reported in Table 6.2.

6.2* Verify the 80% Satterthwaite confidence interval for $\sigma^2(\Delta)$ reported in Table 6.2.

6.3 Verify the lower limit of the Ting et al. confidence interval for $\sigma^2(p)$ in Table 6.4.

6.4 Given the results for the $p \times i$ design in Table 6.2, what is an 80% confidence interval for $\sigma(\Delta)$ if the D study uses the $I{:}p$ design with

$n_i' = 6$? Provide answers for both the mean-score metric and the total-score metric.

6.5* The Wiley adjustments discussed in Section 6.1.3 involve multiplying the boot-p means and estimated standard errors for $\hat{\sigma}^2(p)$ by the correction factor $n_p/(n_p - 1)$. Consider the following data for a $p \times i$ design.

	i_1	i_2	i_3	i_4
p_1	5	8	5	7
p_2	7	5	5	4
p_3	5	2	1	0

Verify that use of the adjustment factor with these data gives an unbiased estimate of $\sigma^2(p)$. Why is it sensible to use the same correction factor with the bootstrap estimate of the standard error of $\hat{\sigma}^2(p)$?

6.6* Suppose an individual's responses to three performance assessment tasks are each evaluated by the same four raters, leading to the following matrix of scores for the individual.

	r_1	r_2	r_3	r_4
t_1	5	8	5	7
t_2	7	5	5	4
t_3	5	2	1	0

(a) What is the absolute error variance for the individual?

(b) Under normality assumptions, what is the estimated standard error of this absolute error variance?

(c) What are the answers to (a) and (b) for $n_t' = 6$ and $n_r' = 2$?

6.7 Feldt's confidence interval limits for $E\rho^2$ discussed in Section 6.3.2 assume $N_i \to \infty$, whereas the dichotomous data simulation in Table 6.10 requires limits for sampling from a finite universe. Show that for $n_i' = n_i$ the limits are obtained by using

$$\frac{n_i}{c_i}\left(\zeta + \frac{1}{N_i}\right)$$

in place of $n_i'\zeta$ in Equation 6.32, where $c_i = 1 - n_i/N_i$.

6.8* Verify the confidence interval for Φ reported in Table 6.5.

6.9 Winer (1971, p. 288) provides these data for a $p \times i$ design:

	i_1	i_2	i_3	i_4
p_1	2	4	3	3
p_2	5	7	5	6
p_3	1	3	1	2
p_4	7	9	9	8
p_5	2	4	6	1
p_6	6	8	8	4

For these data, the usual ANOVA results are:

Effect	df	MS	$\hat{\sigma}^2$
p	5	24.5000	5.8167
i	3	5.8333	.7667
pi	15	1.2333	1.2333

What is the minimum number of items needed to be 90% confident that $E\rho^2$ is at least .85?

7
Unbalanced Random Effects Designs

To this point, our discussion of generalizability theory has been restricted to balanced designs. Doing so substantially decreases a number of statistical complexities, and even some conceptual ones. In practice, however, generalizability analyses with real data are often characterized by unbalanced designs. Unbalanced random effects designs are the subject of this chapter. Unbalanced mixed effects designs are treated using multivariate generalizability theory, which is discussed in Chapters 9 to 12.[1] The discussion here of unbalanced random effects designs is relevant for both univariate and multivariate generalizability theory.

There is a vast statistical literature on estimating variance components with unbalanced random effects designs. Most of this literature is extensively reviewed by Searle et al. (1992), with Searle (1971, Chap. 10) providing an earlier treatment. Jarjoura and Brennan (1981) and Brennan (1994) provide brief summaries. One reason the literature is so extensive is that there are numerous, and sometimes conflicting, perspectives on the subject. As vast as the literature is, however, for our purposes it is limited in one respect—it deals almost exclusively with G study issues, although, of course, discussions in statistical literature do not use such terminology.

In this chapter the estimation of G study variance components is treated primarily in terms of one particular method called the *analogous-ANOVA*

---

[1] In multivariate generalizability theory, the fixed effects in a mixed model play the role of multiple dependent variables, and there is a random effects design (balanced or unbalanced) associated with each variable. Therefore, in generalizability theory we seldom need to treat unbalanced mixed models in the conventional statistical sense.

procedure. General equations are provided, along with results and illustrations for several G study designs. Matters are even more complicated for D studies, and general procedures applicable to any unbalanced D study design are unknown. Consequently, in this chapter estimators of error variances and coefficients are derived and illustrated for several, frequently encountered unbalanced D study designs.

7.1   G Study Issues

It is important to recall that the estimation problems induced by unbalanced designs do not affect the definition of G study variance components. That is, the variance components that characterize the universe of admissible observations are blind to the complexities of estimating random effects variance components, and the definitions discussed in Section 3.4 still apply.

In this book, to estimate random effects variance components for balanced designs we have employed primarily the ANOVA procedure, which involves solving the simultaneous linear equations that result from equating mean squares with their expected values. Since sums of squares are linear combinations of mean squares, the same estimators of variance components would be obtained by equating sums of squares with their expected values. Similarly, since the T terms discussed in Section 3.3 are linear combinations of sums of squares (and mean squares), the same estimators of variance components would be obtained by equating T terms with their expected values. Whichever of these sets of statistics is used, the resulting estimators of variance components are unbiased; indeed, they are called best quadratic unbiased estimators (BQUE).

Mean squares, sums of squares, and T terms are examples of *quadratic forms*, that is, statistics that involve *squared* values of the observations. A number of other quadratic forms might be employed to estimate variance components with balanced designs, but a great many of them lead to the same estimators as those obtained from the ANOVA procedure. For example, with the $p \times i$ and $n_i = 2$, the covariance of person mean scores is identical to the estimator of $\sigma^2(p)$ obtained from the ANOVA procedure.

The fact that many quadratic forms give the same estimators of variance components with balanced designs is not a statistical virtue per se, but it does provide a sense of security that evaporates when unbalanced designs are encountered. For unbalanced designs, it is rare for different quadratic forms to lead to the same estimators of variance components, even when the various estimators are unbiased. Furthermore, many properties of the estimators are usually unknown. For unbalanced designs, then, the principal problem is not obtaining estimators of variance components; rather, the principal problem is that there are many estimators and no obvious statistical basis for choosing among them.

From one perspective, the various procedures for estimating random effects variance components can be split into two types: those that assume normality of score effects and those that do not. For example, maximum likelihood procedures almost always assume normality. The normality-based procedures are usually complicated, but more important, for many generalizability analyses the normality assumptions seem highly suspect. For these reasons, the normality-based procedures are not emphasized here.

Although procedures that do not make normality assumptions seem more appropriate for most generalizability analyses, many of them require operations with matrices that have dimensions that can be as large as the number of cells in the design. Almost always these dimensions are huge in generalizability analyses, because the designs employed usually have single observations per cell, with large numbers of cells (i.e., one or more facets have a large number of conditions). Even with powerful computers, such procedures often are not viable from a practical point of view.

Here we focus on the so-called *analogous*-ANOVA procedure for estimating variance components with unbalanced designs, using the quadratic forms associated with Henderson's (1953) Method 1. (A number of other procedures are briefly discussed later in Section 7.3.1.) The resulting estimates are unbiased, but other properties are generally unknown, except in a few special cases. In the analogous-ANOVA procedure the total sums of squares for an unbalanced design is decomposed in a manner analogous to the decomposition for a balanced design. For example, for the balanced $i\!:\!p$ design the total sums of squares is decomposed as follows.

$$\sum_{p=1}^{n_p}\sum_{i=1}^{n_i}(X_{pi}-\overline{X})^2 = n_i\sum_{p=1}^{n_p}(\overline{X}_p-\overline{X})^2 + \sum_{p=1}^{n_p}\sum_{i=1}^{n_i}(X_{pi}-\overline{X}_p)^2.$$

For the unbalanced $i\!:\!p$ design the decomposition in terms of analogous sums of squares is

$$\sum_{p=1}^{n_p}\sum_{i=1}^{n_{i:p}}(X_{pi}-\overline{X})^2 = \sum_{p=1}^{n_p}n_{i:p}(\overline{X}_p-\overline{X})^2 + \sum_{p=1}^{n_p}\sum_{i=1}^{n_{i:p}}(X_{pi}-\overline{X}_p)^2.$$

Although the analogous-ANOVA procedure is often described in terms of sums of squares (more specifically, *corrected* sums of squares), it is often easier to use analogous T terms (i.e., analogous *uncorrected* sums of squares), as discussed in the next section.

7.1.1 Analogous-ANOVA Procedure

In terms of the notation introduced in Section 3.3, the T term for an effect α in a balanced design is

$$T(\alpha) = \pi(\dot{\alpha})\sum_{\alpha}\overline{X}_{\alpha}^2,$$

where $\pi(\dot{\alpha})$ is the product of the sample sizes for the indices not in α, and the summation is over all of the indices in α. This means that each of the $\overline{X}^2_\alpha$ terms is multiplied by the same constant $\pi(\dot{\alpha})$, which is the number of observations involved in computing each mean $\overline{X}_\alpha$. For an unbalanced design, however, the multiplier is not necessarily a constant. Consider, for example, the unbalanced $p \times (i{:}h)$ design in which the number of items within each level of h ($n_{i:h}$) need not be the same. For $T(h)$ in this design, $\overline{X}_h$ is based on $n_p n_{i:h}$ observations, which is not a constant for all levels of h.

So-called *analogous* T terms are defined as

$$T(\alpha) = \sum_\alpha \tilde{n}_\alpha \overline{X}^2_\alpha = \sum_\alpha \left[\frac{\left(\sum_\varepsilon X_{\alpha\varepsilon} \right)^2}{\tilde{n}_\alpha} \right], \tag{7.1}$$

where

$\tilde{n}_\alpha$ is the total number of observations for a given level of α,

ε is the set of all indices that are not in α,

$\alpha\varepsilon$ means all of the indices in the design, and

the last term is a frequently used computational formula.

So, for example, for the unbalanced $p \times (i{:}h)$ design,

$$T(h) = \sum_h n_p n_{i:h} \overline{X}^2_h = \sum_h \left[\frac{\left(\sum_p \sum_i X_{pih} \right)^2}{n_p n_{i:h}} \right].$$

It is important to note that $\overline{X}_\alpha$ in Equation 7.1 is defined as a sum divided by the total number of observations that are summed. So, for example, for the unbalanced $p \times (i{:}h)$ design,

$$\overline{X}_p = \frac{\sum\limits_{h=1}^{n_h} \sum\limits_{i=1}^{n_{i:h}} X_{pih}}{\sum\limits_{h=1}^{n_h} n_{i:h}}.$$

That is, $\overline{X}_p$ is the sum of all the observations for a level of p divided by the total number of such observations. By contrast, $\overline{X}_p$ is *not* defined as

$$\frac{1}{n_h} \sum_h \left(\frac{1}{n_{i:h}} \sum_i X_{pih} \right),$$

which is the average over levels of h of the average over levels of i. Of course, these two definitions are equivalent for balanced designs.

Estimating random effects variance components using analogous T terms involves these steps:

1. obtain the expected value of each T term with respect to μ^2, the variance components, and their multipliers (i.e., coefficients), and

2. use matrix procedures, or traditional algebraic procedures, to estimate the variance components.

Both steps are usually tedious, but obtaining the coefficients of μ^2 and the variance components is considerably simplified by general results discussed by Searle (1971, p. 431) and Searle et al. (1992, p. 186).

In our notation, the coefficient of μ^2 in the expected value of every T term is simply

$$k[\mu^2, \boldsymbol{E}T(\alpha)] = n_+, \tag{7.2}$$

where n_+ is the total number of observations in the design.[2] The coefficients of the variance components are often more complicated, however. In general, the coefficient of $\sigma^2(\alpha)$ in the expected value of the T term for β is

$$k[\sigma^2(\alpha), \boldsymbol{E}T(\beta)] = \sum_\beta \left(\sum_\gamma \frac{\tilde{n}_{\beta\gamma}^2}{\tilde{n}_\beta} \right), \tag{7.3}$$

where

γ is the set of all indices in α that are not in β (if $\beta = \mu$, then $\gamma = \alpha$);

$\tilde{n}_{\beta\gamma}$ is the total number of observations for a given combination of levels of β and γ; and

$\tilde{n}_\beta$ is the total number of observations for a given level of β (note that $\tilde{n}_\beta = \sum_\gamma \tilde{n}_{\beta\gamma}$).

One useful special case of Equation 7.3 is

$$k[\sigma^2(\alpha), \boldsymbol{E}T(\mu)] = \sum_\alpha \frac{\tilde{n}_\alpha^2}{n_+}, \tag{7.4}$$

where $\tilde{n}_\alpha$ is the total number of observations for a given level of α. Note also that

$$k[\sigma^2(\alpha), \boldsymbol{E}T(\alpha)] = n_+, \tag{7.5}$$

and

$$k[\sigma^2(\alpha), \boldsymbol{E}T(\omega)] = n_+, \tag{7.6}$$

where ω is the effect associated with all the indices in the design.

It is important to note that derivation of the k terms given by Equations 7.2 to 7.6 assumes that the sample sizes are uncorrelated with the

---

[2]The coefficient k in this section should not be confused with the coefficients k_q and k_r in Section 6.2.3.

TABLE 7.1. Analogous T Terms for Unbalanced $i:p$ Design

Effect	df	T	SS
p	$n_p - 1$	$\sum_p n_{i:p} \overline{X}_p^2$	$T(p) - T(\mu)$
$i:p$	$n_+ - n_p$	$\sum_p \sum_i X_{pi}^2$	$T(i:p) - T(p)$
Mean(μ)	1	$n_+ \overline{X}^2$	

Note. $n_+ = \sum_p n_{i:p}$.

effects in the linear model for the design. Also, it is assumed that the sample sizes, and sample size patterns for unbalanced facets, are the same over replications.

In the following sections, the analogous-ANOVA procedure is illustrated for the unbalanced $i:p$ design, the unbalanced $p \times (i:h)$ design, and the $p \times i$ design with missing data. Appendix E provides results for the unbalanced $i:h:p$ and $(p:c) \times i$ designs.

7.1.2   Unbalanced $i:p$ Design

Table 7.1 provides the T terms for the unbalanced $i:p$ design. Consider the coefficient of $\sigma^2(p)$ in the expected value of $T(\mu)$. Using Equation 7.4 with $\alpha = p$,

$$k[\sigma^2(p), ET(\mu)] = \sum_p \frac{\tilde{n}_p^2}{n_+} = \sum_p \frac{n_{i:p}^2}{n_+}.$$

As another example, consider the coefficient of $\sigma^2(i:p)$ in the expected value of $T(p)$. To obtain this coefficient, use $\alpha = i:p$ and $\beta = p$ in Equation 7.3:

$$k[\sigma^2(i:p), ET(p)] = \sum_p \left(\sum_i \frac{\tilde{n}_{pi}^2}{\tilde{n}_p} \right) = \sum_p \left(\sum_i \frac{1}{n_{i:p}} \right) = \sum_p \frac{n_{i:p}}{n_{i:p}} = n_p.$$

Note that $\tilde{n}_{pi}$ is the number of observations in each pi combination, which is necessarily one, because this design has only a single observation for each cell. Derivation of the other coefficients is straightforward.

The full set of expected T terms is

$$
\begin{aligned}
ET(p) &= n_+ \mu^2 + n_p \sigma^2(i:p) + n_+ \sigma^2(p) \\
ET(i:p) &= n_+ \mu^2 + n_+ \sigma^2(i:p) + n_+ \sigma^2(p) \\
ET(\mu) &= n_+ \mu^2 + \sigma^2(i:p) + r_i \sigma^2(p),
\end{aligned}
\tag{7.7}
$$

where

$$r_i = \sum_p \frac{n_{i:p}^2}{n_+}. \tag{7.8}$$

TABLE 7.2. Synthetic Data Example for Unbalanced $i:p$ Design

p	Scores	$n_{i:p}$	$\overline{X}_p$	$\hat{\sigma}^2(\Delta_p)$
1	2 6 7	3	5.0	2.3333
2	4 5 6 7 6	5	5.6	.2600
3	5 5 4 6 5	5	5.0	.1000
4	5 9 8 6 5	5	6.6	.6600
5	4 3 5 6	4	4.5	.4167
6	4 4 4	3	4.0	.0000
7	3 6 6 5	4	5.0	.5000
8	3 5 4	3	4.0	.3333
9	6 8 7 6 6	5	6.6	.1600

Effect	df	T	MS
p	8	1069.40	3.8912
$i:p$	28	1120.00	1.8071
μ	1	1038.27	

$\hat{\sigma}^2(p) = .5098 \quad \hat{\sigma}^2(i:p) = 1.8071$

$$n_p = 9$$
$$n_+ = 37$$
$$r_i = 4.2973$$
$$\dot{n}_i = 3.9130$$

Note that $r_i = n_i$ for a balanced design. That is why the subscript i is attached to r.

Since there are an equal number of equations (the three ET terms) and unknowns (μ^2 and the two variance components), matrix procedures can be used to obtain estimates of the variance components; the ET terms are simply replaced with computed numerical values. Alternatively, conventional algebraic procedures can be employed to express estimators of the variance components in terms of analogous T terms, which gives

$$\hat{\sigma}^2(i:p) = \frac{T(i:p) - T(p)}{n_+ - n_p} \tag{7.9}$$

$$\hat{\sigma}^2(p) = \frac{T(p) - T(\mu) - (n_p - 1)\hat{\sigma}^2(i:p)}{n_+ - r_i}. \tag{7.10}$$

Estimators can also be obtained in terms of analogous sums of squares or analogous mean squares. Analogous sums of squares are obtained by using analogous T terms in the usual formulas for sums of squares. Analogous mean squares are analogous sums of squares divided by their degrees of freedom.[3]

Using the expected T terms in Equation Set 7.7 it is easy to derive the expected values of the analogous mean squares. For example,

$$EMS(p) = \frac{ET(p) - ET(\mu)}{n_p - 1}$$

---

[3]For a nested main effect, the degrees of freedom equals the sample size associated with all indices in the effect *minus* the sample size associated with the nesting indices in the effect. The degrees of freedom for an interaction effect is the product of the degrees of freedom for the main effects that make up the interaction effect.

$$= \frac{(n_+ - n_+)\mu^2 + (n_p - 1)\sigma^2(i{:}p) + (n_+ - r_i)\sigma^2(p)}{n_p - 1}$$

$$= \sigma^2(i{:}p) + t_i\sigma^2(p),$$

where

$$t_i = \frac{n_+ - r_i}{n_p - 1}.$$

Similarly,

$$EMS(i{:}p) = \frac{ET(i{:}p) - ET(p)}{n_+ - n_p}$$

$$= \frac{(n_+ - n_+)\mu^2 + (n_+ - n_p)\sigma^2(i{:}p) + (n_+ - n_+)\sigma^2(p)}{n_+ - n_p}$$

$$= \sigma^2(i{:}p).$$

It follows immediately that the estimators of the variance components in terms of analogous mean squares are

$$\hat{\sigma}^2(p) = [MS(p) - MS(i{:}p)]/t_i \qquad (7.11)$$
$$\hat{\sigma}^2(i{:}p) = MS(i{:}p). \qquad (7.12)$$

The estimators are unbiased, as are all analogous-ANOVA estimators. In addition, for this design, Henderson's (1953) Methods 1 and 3 give the same decomposition of the sums of squares. Also, assuming effects are normally distributed, Searle et al. (1992, Chap. 4) discuss other properties of these estimators.[4]

Table 7.2 provides a small synthetic data example that illustrates the computation of estimated variance components for the unbalanced $i{:}p$ design. Additional aspects of this example are discussed later in Section 7.2.

7.1.3   Unbalanced $p \times (i{:}h)$ Design

Table 7.3 provides degrees of freedom and T-terms for each effect in the unbalanced $p \times (i{:}h)$ design, and Table 7.4 provides all of the coefficients for the expected values of the T terms for this design. Consider, for example, the coefficient of $\sigma^2(h)$ in the expected value of $T(p)$ for the unbalanced $p \times (i{:}h)$ design. To obtain this coefficient, use $\alpha = h$ and $\beta = p$ in Equation 7.3. Since α contains only h, and h is not one of the indices in β, it follows that $\gamma = h$. Therefore,

$$k[\sigma^2(h), ET(p)] = \sum_p \left(\sum_h \frac{\widetilde{n}_{ph}^2}{\widetilde{n}_p} \right) = \sum_p \left(\sum_h \frac{n_{i{:}h}^2}{n_{i+}} \right) = n_p \left(\sum_h \frac{n_{i{:}h}^2}{n_{i+}} \right),$$

---

[4]At the beginning of this chapter it was noted that the properties of analogous-ANOVA estimators of variance components are generally unknown for unbalanced designs, except for a few special cases. The unbalanced $i{:}p$ design is one such exception.

TABLE 7.3. Analogous T Terms for Unbalanced $p \times (i{:}h)$ Design

Effect	df	T
p	$n_p - 1$	$n_{i+} \sum_p \overline{X}_p^2$
h	$n_h - 1$	$n_p \sum_h (n_{i:h} \overline{X}_h^2)$
$i{:}h$	$n_{i+} - n_h$	$n_p \sum_h \sum_i \overline{X}_{i:h}^2$
ph	$(n_p - 1)(n_h - 1)$	$\sum_h (n_{i:h} \sum_p \overline{X}_{ph}^2)$
$pi{:}h$	$(n_p - 1)(n_{i+} - n_h)$	$\sum_p \sum_h \sum_i X_{pih}^2$
Mean(μ)	1	$n_+ \overline{X}^2$

Note. $n_{i+} = \sum_h n_{i:h}$.

TABLE 7.4. Coefficients of μ^2 and Variance Components in Expected Values of T Terms for Unbalanced $p \times (i{:}h)$ Design

				Coefficients		
ET term	μ^2	$\sigma^2(pi{:}h)$	$\sigma^2(ph)$	$\sigma^2(i{:}h)$	$\sigma^2(h)$	$\sigma^2(p)$
$ET(p)$	n_+	n_p	$n_p r_i$	n_p	$n_p r_i$	n_+
$ET(h)$	n_+	n_h	n_{i+}	$n_p n_h$	n_+	n_{i+}
$ET(i{:}h)$	n_+	n_{i+}	n_{i+}	n_+	n_+	n_{i+}
$ET(ph)$	n_+	$n_p n_h$	n_+	$n_p n_h$	n_+	n_+
$ET(pi{:}h)$	n_+	n_+	n_+	n_+	n_+	n_+
$ET(\mu)$	n_+	1	r_i	n_p	$n_p r_i$	n_{i+}

Note. $n_{i+} = \sum_h n_{i:h}$ and $r_i = \sum_h n_{i:h}^2 / n_{i+}$.

where n_{i+} is the total number of levels of i over all levels of h; that is, $n_{i+} = \sum_h n_{i:h}$. Note that $\tilde{n}_{ph}$ is the number of observations in each ph combination, which is $n_{i:h}$, and $\tilde{n}_p$ is the number of observations for a level of p, which is n_{i+}. As another example, the coefficient of $\sigma^2(i{:}h)$ in the expected value of $T(p)$ is

$$k[\sigma^2(i{:}h), ET(p)] = \sum_p \left(\sum_h \sum_i \frac{\tilde{n}_{pih}^2}{\tilde{n}_p} \right) = \sum_p \left(\sum_h \sum_i \frac{1}{n_{i+}} \right) = n_p.$$

In effect, Table 7.4 provides six equations (the ET terms) in six unknowns (μ^2 and the variance components). Replacing the ET terms with computed numerical values, matrix procedures can be used to obtain estimates of the variance components.

Alternatively, using Table 7.4, the expected mean square equations can be obtained, and they can be used in turn to get estimators of the variance components in terms of mean squares. Specifically, the expected mean

TABLE 7.5. Synthetic Data Example for Unbalanced $p \times (i{:}h)$ Design

p	h_1 i_1	i_2	h_2 i_1	i_2	i_3	i_4	h_3 i_1	i_2	$\overline{X}_{p1}$	$\overline{X}_{p2}$	$\overline{X}_{p3}$	$\overline{X}_p$	$\hat{\sigma}^2(\Delta_p)$
1	4	5	3	3	5	4	5	7	4.50	3.75	6.00	4.500	.4800
2	2	1	2	3	1	4	4	6	1.50	2.50	5.00	2.875	.9656
3	2	4	4	7	6	5	8	7	3.00	5.50	7.50	5.375	1.4906
4	1	3	5	4	5	5	4	5	2.00	4.75	4.50	4.000	.7900
5	3	3	6	7	5	7	8	9	3.00	6.25	8.50	6.000	2.2900
6	1	2	5	6	4	4	5	6	1.50	4.75	5.50	4.125	1.4156
7	3	5	6	8	6	7	7	8	4.00	6.75	7.50	6.250	1.0425
8	0	1	1	2	0	4	7	8	0.50	1.75	7.50	2.875	4.3856
Mean	2	3	4	5	4	5	6	7	2.50	4.50	6.50	4.500	1.6075

$$n_p = 8 \qquad n_{i+} = 8 \qquad r_i = 3.0000 \qquad \breve{n}_h = 2.6667$$

Effect	df	T	MS	$\hat{\sigma}^2$
p	7	1390.0	13.4286	1.2014
h	2	1424.0	64.0000	2.9161
$i{:}h$	5	1440.0	3.2000	.2946
ph	14	1564.5	3.3214	.9913
$pi{:}h$	35	1610.0	.8429	.8429
μ	1	1296.0		

square equations are

$$\begin{aligned}
\mathbf{E}MS(p) &= \sigma^2(pi{:}h) + r_i\,\sigma^2(ph) + n_{i+}\,\sigma^2(p) \\
\mathbf{E}MS(h) &= \sigma^2(pi{:}h) + t_i\,\sigma^2(ph) + n_p\,\sigma^2(i{:}h) + n_p t_i\,\sigma^2(h) \\
\mathbf{E}MS(i{:}h) &= \sigma^2(pi{:}h) + n_p\,\sigma^2(i{:}h) \\
\mathbf{E}MS(ph) &= \sigma^2(pi{:}h) + t_i\,\sigma^2(ph) \\
\mathbf{E}MS(pi{:}h) &= \sigma^2(pi{:}h),
\end{aligned} \qquad (7.13)$$

where

$$r_i = \sum_h \frac{n_{i{:}h}^2}{n_{i+}} \quad \text{and} \quad t_i = \frac{n_{i+} - r_i}{n_h - 1}.$$

Note that these values of r_i and t_i are different from those for the $i{:}p$ design.

Using Equation Set 7.13, estimators of the variance components in terms of mean squares are

$$\begin{aligned}
\hat{\sigma}^2(p) &= [MS(p) - r_i MS(ph)/t_i + (r_i - t_i)MS(pi{:}h)/t_i]/n_{i+} \\
\hat{\sigma}^2(h) &= [MS(h) - MS(i{:}h) - MS(ph) + MS(pi{:}h)]/n_p t_i
\end{aligned}$$

$$\hat{\sigma}^2(i{:}h) = [MS(i{:}h) - MS(pi{:}h)]/n_p \qquad (7.14)$$
$$\hat{\sigma}^2(ph) = [MS(ph) - MS(pi{:}h)]/t_i$$
$$\hat{\sigma}^2(pi{:}h) = MS(pi{:}h).$$

The results in Equation Set 7.14 were first reported by Jarjoura and Brennan (1981).

Table 7.5 provides a small synthetic data example of the unbalanced $p \times (i{:}h)$ design.[5] Certain aspects of this example are discussed later when we consider D study issues for unbalanced designs.

7.1.4 Missing Data in the $p \times i$ Design

The unbalanced $i{:}p$ and $p \times (i{:}h)$ designs described in Sections 7.1.2 and 7.1.3, respectively, are unbalanced with respect to nesting. By contrast, suppose some data are missing at random from the $p \times i$ design. Let $\tilde{n}_p$ be the number of nonmissing observations for person p and $\tilde{n}_i$ be the number of nonmissing observations for item i. For this design, Table 7.6 provides formulas for degrees of freedom, analogous T terms, and sums of squares. Note that the degrees of freedom for the pi effect is not the product of the degrees of freedom for the p and i effects. Rather, the degrees of freedom for pi is the total number of degrees of freedom $(n_+ - 1)$ minus the degrees of freedom for p and i. In this sense, the pi effect is to be interpreted as a residual effect.

The expected T-term equations are

$$\begin{aligned}
\boldsymbol{ET}(p) &= n_+\mu^2 + n_p\sigma^2(pi) + n_p\sigma^2(i) + n_+\sigma^2(p) \\
\boldsymbol{ET}(i) &= n_+\mu^2 + n_i\sigma^2(pi) + n_+\sigma^2(i) + n_i\sigma^2(p) \\
\boldsymbol{ET}(pi) &= n_+\mu^2 + n_+\sigma^2(pi) + n_+\sigma^2(i) + n_+\sigma^2(p) \\
\boldsymbol{ET}(\mu) &= n_+\mu^2 + \sigma^2(pi) + r_p\sigma^2(i) + r_i\sigma^2(p),
\end{aligned} \qquad (7.15)$$

where

$$r_p = \sum_i \frac{\tilde{n}_i^2}{n_+} \qquad \text{and} \qquad r_i = \sum_p \frac{\tilde{n}_p^2}{n_+}.$$

Given these expected T-term equations, it is straightforward to show that the expected mean squares are

---

[5]The data are from Rajaratnam et al. (1965), although the analysis here is different from theirs.

TABLE 7.6. Analogous T Terms for $p \times i$ Design with Missing Data

Effect	df	T	SS
p	$n_p - 1$	$\sum_p \tilde{n}_p \overline{X}_p^2$	$T(p) - T(\mu)$
i	$n_i - 1$	$\sum_i \tilde{n}_i \overline{X}_i^2$	$T(i) - T(\mu)$
pi	$n_+ - n_p - n_i + 1$	$\sum_p \sum_i X_{pi}^2$	$T(pi) - T(p) - T(i) + T(\mu)$
μ	1	$n_+ \overline{X}^2$	

$$
\begin{aligned}
EMS(p) &= \sigma^2(pi) + \left(\frac{n_p - r_p}{n_p - 1}\right) \sigma^2(i) + \left(\frac{n_+ - r_i}{n_p - 1}\right) \sigma^2(p) \\[2mm]
EMS(i) &= \sigma^2(pi) + \left(\frac{n_+ - r_p}{n_i - 1}\right) \sigma^2(i) + \left(\frac{n_i - r_i}{n_i - 1}\right) \sigma^2(p) \\[2mm]
EMS(pi) &= \sigma^2(pi) + \left(\frac{r_p - n_p}{n_+ - n_p - n_i + 1}\right) \sigma^2(i) \\[2mm]
&\quad + \left(\frac{r_i - n_i}{n_+ - n_p - n_i + 1}\right) \sigma^2(p).
\end{aligned}
\tag{7.16}
$$

Note that there is a nonzero coefficient for each of the three variance components in each of the expected mean squares. By contrast, for designs that are unbalanced with respect to nesting, only, $EMS(\beta)$ involves solely those variance components that include all of the indices in β.

Because the expected mean square equations in Equation Set 7.16 are quite complicated, expressions for estimators of the variance components in terms of mean squares are quite complicated, too. Consequently, it is simpler to express estimators of the variance components with respect to T terms. Letting

$$
\lambda_p = \frac{n_+ - r_p}{n_+ - n_p} \quad \text{and} \quad \lambda_i = \frac{n_+ - r_i}{n_+ - n_i},
$$

the estimators of the variance components are

$$
\begin{aligned}
\hat{\sigma}^2(pi) &= \frac{\lambda_p[T(pi) - T(p)] + \lambda_i[T(pi) - T(i)] - [T(pi) - T(\mu)]}{n_+ - r_p - r_i + 1} \\[2mm]
\hat{\sigma}^2(p) &= \frac{T(pi) - T(i)}{n_+ - n_i} - \hat{\sigma}^2(pi) \\[2mm]
\hat{\sigma}^2(i) &= \frac{T(pi) - T(p)}{n_+ - n_p} - \hat{\sigma}^2(pi).
\end{aligned}
\tag{7.17}
$$

These estimators were reported by Huynh (1977) based on somewhat more general equations in Searle (1971, p. 487). Table 7.7 provides a simple synthetic data example from Huynh (1977) that illustrates the computation of estimates of variance components for a $p \times i$ design with missing data.

TABLE 7.7. Huynh (1977) Example of Missing-Data $p \times i$ Design

p	\multicolumn{6}{c}{Items}	$\tilde{n}_p$	$\overline{X}_p$
	1	2	3	4	5	6		
1	1	1	1	1	0	1	6	.83
2	1	0	1	1	1	1	6	.83
3		1	1	1	0	0	5	.60
4	0			1	0	1	4	.50
5	1		0	1	0		4	.50
6	1	1	1	1	0	1	6	.83
7	0		1		0	0	4	.25
8	0		1	1	1	1	5	.80
9	1	1	0	1		0	5	.60
10		0	0	0		0	4	.00
11	1	1	1	1	1	1	6	1.00
12		0		0	0	1	4	.25
$\tilde{n}_i$	9	8	10	11	10	11	59	

Effect	df	T	$\hat{\sigma}^2$
p	11	27.8000	.0473
i	5	24.7432	.0117
pi	42	37.0000	.1840
μ	1	23.2034	

$$n_p = 12$$
$$n_i = 6$$
$$n_+ = 59$$
$$r_p = 9.9492$$
$$r_i = 5.0678$$
$$\ddot{n}_i = 4.7682$$

The analogous-ANOVA procedure can also be used with more complicated missing-data designs. The computations are more complex, but the basic procedure is the same. It is almost always easier to work with T terms and their expected values rather than with SS or MS terms.

7.2   D Study Issues

The analogous-ANOVA procedure for estimating variance components takes into account the missing data pattern in obtaining the estimates, but the estimates themselves are estimates of population/universe parameters that are defined in such a way that they do not depend on whether data are missing in obtaining the estimates. So, for example, for the $p \times (i:h)$ design, whether the G study is balanced or unbalanced, for the universe of admissible observations the definition of $\sigma^2(p)$ is $\boldsymbol{E}\nu_p^2$, the definition of $\sigma^2(h)$ is $\boldsymbol{E}\nu_h^2$, and the definition of $\sigma^2(i:h)$ is $\boldsymbol{E}\nu_{i:h}^2$. Similarly, for the universe of generalization, the definitions of the variance components do not depend on whether the D study design is balanced or unbalanced. It follows that the occurrence of missing data in the D study has no effect on universe score variance. Error variances, however, are affected by whether or not the D study design is balanced.

Suppose the G study design is unbalanced, and consider several possibilities. First, the D study might use a balanced design. In this case, the estimates of the variance components from the unbalanced G study are used in exactly the same way as discussed in Chapters 4 and 5. Second, the D study might use the same sample sizes and sample size patterns as in the G study; that is, decisions might be based on the same data as in the G study. In this case, error variances are affected by the unbalanced characteristics of the G study. Third, the D study might use an unbalanced design with sample sizes and/or sample size patterns different from those in the G study. In this case, error variances will be affected by the unbalanced characteristics of the D study.

In generalizability theory, then, unbalanced D study designs often involve conceptual and statistical complexities over and beyond those addressed in statistical treatments of variance components (e.g., Searle, 1971, and Searle et al., 1992). In effect, such treatments consider G study issues, only. By way of illustration, a few of the complexities arising from the unbalanced characteristics of several D study designs are considered in the following sections. The only designs treated explicitly are the unbalanced $I:p$, $p \times (I:H)$, $(P:c) \times I$, and $p \times I$ designs.

For balanced designs, as discussed in Chapters 4 and 5, it is possible to provide relatively simple, general formulas for obtaining D study results for any design. For unbalanced designs, formulas are generally much more complicated. Often it is necessary to derive results for the specific design and circumstances under consideration, as illustrated in the following sections and in the exercises at the end of this chapter.

7.2.1   Unbalanced $I:p$ Design

By definition, absolute error is the difference between the observed and universe score for an object of measurement. For the unbalanced $I:p$ design, assuming the G and D study sample sizes are the same,

$$
\begin{aligned}
\Delta_p &= \overline{X}_p - \mu_p \\
&= \left(\frac{1}{n_{i:p}} \sum_{i=1}^{n_{i:p}} X_{pi} \right) - \mu_p \\
&= \left(\mu + \nu_p + \frac{\sum_i \nu_{i:p}}{n_{i:p}} \right) - (\mu + \nu_p) \\
&= \frac{\sum_i \nu_{i:p}}{n_{i:p}} .
\end{aligned}
$$

The expected value of Δ_p^2 is absolute error variance:

$$
\begin{aligned}
\sigma^2(\Delta) &= E \left(\frac{\sum_i \nu_{i:p}}{n_{i:p}} \right)^2 \\
&= E \left(\frac{\sum_i \nu_{i:p}^2}{n_{i:p}^2} + \frac{\sum_i \sum_{i'} \nu_{i:p}\nu_{i':p}}{n_{i:p}^2} \right) ,
\end{aligned}
$$

where $i \neq i'$. The expected values of the cross-product terms are all zero, and the expected value of each squared term is the same. It follows that

$$\sigma^2(\Delta) = E\left(\frac{\nu_{i:p}^2}{n_{i:p}}\right),$$

which clearly depends upon the $n_{i:p}$. Recall that $\sigma^2(i:p) = E\nu_{i:p}^2$, and assume that the distribution of the $n_{i:p}$ in the population mirrors that in the sample. It follows that

$$\begin{aligned}
\sigma^2(\Delta) &= \frac{1}{n_p}\sum_p\left[\frac{\sigma^2(i:p)}{n_{i:p}}\right] \\
&= \frac{\sigma^2(i:p)}{\ddot{n}_i}, \quad\quad (7.18)
\end{aligned}$$

where $\ddot{n}_i$ is the harmonic mean of the $n_{i:p}$; namely,

$$\ddot{n}_i = \left[\frac{1}{n_p}\sum_p\left(\frac{1}{n_{i:p}}\right)\right]^{-1}. \quad\quad (7.19)$$

The error variance $\sigma^2(\Delta)$ in Equation 7.18 was derived without distinguishing between G and D study sample sizes. However, they need not be equal. The variance component $\sigma^2(i:p)$ can be estimated using the G study sample sizes, and it can be used in Equation 7.18 along with the harmonic mean for the D study sample sizes $\ddot{n}_i'$.

An alternative perspective on $\sigma^2(\Delta)$ is to view it as the average of the conditional absolute error variances, in the manner discussed in Section 5.4. For the $I:p$ design, conditional absolute error variance for a person is simply the variance of the mean of the item scores for the person. Consequently, assuming the G and D study sample sizes are the same,

$$\sigma^2(\Delta) = \frac{1}{n_p}\sum_p\frac{\text{var}(X_{pi}|p)}{n_{i:p}}. \quad\quad (7.20)$$

This equation cannot be simplified using an harmonic mean because both $\text{var}(X_{pi}|p)$ and $n_{i:p}$ could vary for each person.

For balanced $I:p$ designs, using ANOVA estimates of variance components in Equations 7.18 and 7.20 gives the same estimate of absolute error variance. This equality does not necessarily hold for unbalanced designs, however. Consider again the synthetic data example in Table 7.2 on page 221. Using Equation 7.18, $\hat{\sigma}^2(\Delta) = 1.8071/3.9130 = .4618$, but using Equation 7.20, the reader can verify that the average of the conditional absolute error variances is $\hat{\sigma}^2(\Delta) = .5293$.

For the unbalanced $I:p$ design, $\delta_p = \Delta_p$, as it does for the balanced design. Viewing a generalizability coefficient as the ratio of universe score variance to itself plus relative error variance, we obtain different estimates depending on whether we use Equation 7.18 or 7.20. For the synthetic data in Table 7.2, using Equation 7.18 gives $E\hat{\rho}^2 = .525$, and using Equation 7.20 gives $E\hat{\rho}^2 = .491$.

A generalizability coefficient can be viewed also as the ratio of universe score variance to expected observed score variance. From this perspective, the investigator might estimate a generalizability coefficient as the ratio of $\hat{\sigma}^2(p)$ to the variance of the observed mean scores for persons:

$$S^2(p) = \frac{\sum_p (\overline{X}_p - \overline{X})^2}{n_p - 1} = \frac{\sum_p \overline{X}_p^2 - n_p \overline{X}^2}{n_p - 1}, \tag{7.21}$$

where

$$\overline{X}_p = \frac{1}{n_{i:p}} \sum_{i=1}^{n_{i:p}} X_{pi} \quad \text{and} \quad \overline{X} = \frac{1}{n_p} \sum_{p=1}^{n_p} \overline{X}_p. \tag{7.22}$$

It is important to recognize that $S^2(p)$ in Equation 7.21 is not as simple a statistic as it is for balanced designs. Complexities arise for two related reasons: the mean scores for different persons can be based on different numbers of items; and the grand mean $\overline{X}$ is an unweighted average of the person mean scores, not $\sum_p \sum_i X_{pi}/n_+$.

It can be shown that

$$E\left(\sum_p \overline{X}_p^2\right) = n_p \mu^2 + n_p \sigma^2(p) + \frac{n_p \sigma^2(i:p)}{\ddot{n}_i}, \tag{7.23}$$

and

$$E(\overline{X}^2) = \mu^2 + \frac{\sigma^2(p)}{n_p} + \frac{\sigma^2(i:p)}{n_p \ddot{n}_i}. \tag{7.24}$$

It follows that the expected value of the observed score variance in Equation 7.21 is[6]

$$ES^2(p) = \sigma^2(p) + \frac{\sigma^2(i:p)}{\ddot{n}_i}. \tag{7.25}$$

The form of the expected value of the observed score variance in Equation 7.25 clearly indicates that it is $\sigma^2(p)$ plus the absolute error variance in Equation 7.18. Consequently, with respect to *parameters*

$$E\rho^2 = \frac{\sigma^2(p)}{ES^2(p)} = \frac{\sigma^2(p)}{\sigma^2(p) + \sigma^2(\Delta)},$$

where $\sigma^2(\Delta)$ is given by Equation 7.18. This equality in terms of parameters does not necessarily hold for the estimators that have been discussed here,

---

[6] As noted previously, Cronbach et al. (1972) and Brennan (1992a) designate expected observed score variance using $E\sigma^2(X)$. In this book, however, $ES^2(p)$ is used for expected observed score variance, which simplifies some notational conventions, especially in Chapters 9 to 12.

however. Specifically, replacing the mean-square estimators of $\sigma^2(p)$ and $\sigma^2(i{:}p)$ (see Equations 7.11 and 7.12, respectively) in Equation 7.25 does not lead to the observed score variance in Equation 7.21. For example, for the synthetic data in Table 7.2 on page 221, the variance of the observed person mean scores is .9428, which leads to $\boldsymbol{E}\hat{\rho}^2 = .541$ using Equation 7.25, whereas $\boldsymbol{E}\hat{\rho}^2 = .525$ using Equation 7.18 for $\hat{\sigma}^2(\Delta)$.

In short, for the unbalanced $I{:}p$ design, different perspectives on $\sigma^2(\Delta)$ and $\boldsymbol{E}\rho^2$ lead to different estimates. These differences in estimates are associated with the fact that, for unbalanced designs, different quadratic forms of the observations often lead to different estimates of variance components and statistics formed from them.

7.2.2   Unbalanced $p \times (I{:}H)$ Design

Assuming the G and D study sample sizes are the same, the usual estimator of universe score for the D study $p \times (I{:}H)$ design is

$$\overline{X}_p = \frac{1}{n_{i+}} \sum_{h=1}^{n_h} \sum_{i=1}^{n_{i:h}} X_{pih}.$$

For this estimator, absolute error is

$$\begin{aligned}
\Delta_p &= \overline{X}_p - \mu_p \\
&= \frac{1}{n_{i+}} \sum_h \sum_i (\nu_h + \nu_{i:h} + \nu_{ph} + \nu_{pi:h}),
\end{aligned}$$

and absolute error variance is

$$\begin{aligned}
\sigma^2(\Delta) &= \boldsymbol{E}(\overline{X}_p - \mu_p)^2 \\
&= \boldsymbol{E}\left[\frac{1}{n_{i+}} \sum_h \sum_i (\nu_h + \nu_{i:h} + \nu_{ph} + \nu_{pi:h}) \right]^2.
\end{aligned}$$

Since the expected value of the product of different effects (e.g., $\boldsymbol{E}\nu_h\nu_{i:h}$) is zero,

$$\sigma^2(\Delta) = \frac{1}{n_{i+}^2}\boldsymbol{E}\left[\left(\sum_h n_{i:h}\nu_h\right)^2 + \left(\sum_h \sum_i \nu_{i:h}\right)^2 + \left(\sum_h n_{i:h}\nu_{ph}\right)^2 + \left(\sum_h \sum_i \nu_{pi:h}\right)^2 \right]. \quad (7.26)$$

Consider the expected value of the first term in large parentheses:

$$\begin{aligned}
\boldsymbol{E}\left(\sum_h n_{i:h}\nu_h\right)^2 &= \boldsymbol{E}\left(\sum_h n_{i:h}^2\nu_h^2 + \sum_h \sum_{h \neq h'} n_{i:h}n_{i:h'}\nu_h\nu_{h'}\right) \\
&= \textstyle\sum_h n_{i:h}^2\sigma^2(h),
\end{aligned}$$

because $E\nu_h\nu_{h'} = 0$ for $h \neq h'$. By contrast, the expected value of the second term in large parentheses in Equation 7.26 is simply $n_{i+}\sigma^2(i{:}h)$.

It follows that

$$\sigma^2(\Delta) = \frac{\sigma^2(h)}{\breve{n}_h} + \frac{\sigma^2(i{:}h)}{n_{i+}} + \frac{\sigma^2(ph)}{\breve{n}_h} + \frac{\sigma^2(pi{:}h)}{n_{i+}}, \tag{7.27}$$

where

$$\breve{n}_h = \frac{n_{i+}^2}{\sum_h n_{i{:}h}^2}, \tag{7.28}$$

which equals n_h for balanced designs. A similar derivation for relative error variance leads to

$$\sigma^2(\delta) = \frac{\sigma^2(ph)}{\breve{n}_h} + \frac{\sigma^2(pi{:}h)}{n_{i+}}. \tag{7.29}$$

If the G and D study sample sizes differ, then the D study sample sizes should be used for n_{i+} and $n_{i{:}h}$ in Equations 7.27–7.29.

Overall absolute error variance $\sigma^2(\Delta)$ can be viewed also as the average of the conditional absolute error variances $\sigma^2(\Delta_p)$. For the unbalanced $p \times (I{:}H)$ design, $\sigma^2(\Delta_p)$ for a particular person p is the variance of the mean for the unbalanced $I{:}H$ design for that person. Letting $\sigma^2(h|p)$ and $\sigma^2(i{:}h|p)$ represent the variance components for the within-person G study $I{:}H$ design for person p, conditional absolute error variance for the person is

$$\sigma^2(\Delta_p) = \frac{\sigma^2(h|p)}{\breve{n}_h} + \frac{\sigma^2(i{:}h|p)}{n_{i+}}. \tag{7.30}$$

Consider, for example, the synthetic data for the first person in Table 7.5 on page 224. For these data, $MS(h) = 3.375$, $MS(i{:}h) = 1.050$, and $t_i = 2.5$. Estimates of $\sigma^2(h|p)$ and $\sigma^2(i{:}h|p)$ can be obtained using Equations 7.11 and 7.12, giving $\hat{\sigma}^2(h|p) = .93$ and $\hat{\sigma}^2(i{:}h|p) = 1.05$. Using these estimated variance components with $n_{i+} = 8$ and $\breve{n}_h = 64/24 = 2.6667$ in Equation 7.30 we obtain

$$\hat{\sigma}^2(\Delta_p) = \frac{.93}{2.6667} + \frac{1.05}{8} = .48.$$

It is straightforward to verify that the average of the eight estimates of conditional error variance in Table 7.5 is 1.6075, which is identical to the result obtained using Equation 7.27:

$$\hat{\sigma}^2(\Delta) = \frac{2.9161}{2.6667} + \frac{.2946}{8} + \frac{.9913}{2.6667} + \frac{.8429}{8} = 1.6075.$$

Recall that this equivalence of estimates does not occur for the $I{:}p$ design. The equivalence occurs for the $p \times (I{:}H)$ design partly because the within-person design is the same for each person (including the same sample sizes), even though the design is unbalanced.

A generalizability coefficient can be viewed as the ratio of universe score variance to itself plus relative error variance. For the synthetic data in Table 7.5, replacing parameters with estimates in Equation 7.29 gives

$$\hat{\sigma}^2(\delta) = \frac{.9913}{2.6667} + \frac{.8429}{8} = .4771. \tag{7.31}$$

It follows that an estimated generalizability coefficient is

$$E\hat{\rho}^2 = \frac{1.2014}{1.2014 + .4771} = .716. \tag{7.32}$$

A generalizability coefficient can be viewed also as the ratio of universe score variance to expected observed score variance. For the data in Table 7.5, the reader can verify that the variance of the observed mean scores is $S^2(p) = 1.6786$, which leads to $E\hat{\rho}^2 = 1.2014/1.6786 = .716$, the same value obtained previously. A partial explanation for this equivalence is that, for both relative error variance and observed score variance, the variance is taken over persons, and for the $p \times (I:H)$ design the sample sizes are the *same* for each person.

7.2.3   Unbalanced $(P:c) \times I$ Design

From the perspective of the universe of admissible observations and a G study, the $(p:c) \times i$ design is formally identical to the $p \times (i:h)$ design; that is, results for the $(p:c) \times i$ design can be obtained by using c, p, and i in place of h, i, and p, respectively, in the $p \times (i:h)$ design. Therefore, with these notational changes, the G study results in Section 7.1.3 also apply to the $(p:c) \times i$ design. However, this formal identity is not always easy to perceive, and for this reason, Appendix E provides these G study results using the $(p:c) \times i$ notation.

This formal identity in terms of G studies does not extend to the D study issues discussed in Section 7.2.2, however. In Section 7.2.2 p plays the role of objects of measurement, whereas in this section c plays that role. The important distinction is not that the letters are different, but rather that all facets are crossed with p in the D study $p \times (I:H)$ design, whereas c has a facet nested within it in the $(P:c) \times I$ design.

Assuming the G and D study sample sizes are the same, the usual estimator of universe score for class c is the class mean

$$\overline{X}_c = \frac{1}{n_{p:c} n_i} \sum_{p=1}^{n_{p:c}} \sum_{i=1}^{n_i} X_{cpi}.$$

For this estimator, absolute error is

$$\begin{aligned}
\Delta_c &= \overline{X}_c - \mu_c \\
&= \frac{\sum_p \nu_{p:c}}{n_{p:c}} + \frac{\sum_i \nu_i}{n_i} + \frac{\sum_i \nu_{ci}}{n_i} + \frac{\sum_p \sum_i \nu_{pi:c}}{n_{p:c} n_i}.
\end{aligned}$$

Using a derivation similar to that in Sections 7.2.1 and 7.2.2, absolute error variance is

$$\sigma^2(\Delta) = \frac{\sigma^2(p{:}c)}{\ddot{n}_p} + \frac{\sigma^2(i)}{n_i} + \frac{\sigma^2(ci)}{n_i} + \frac{\sigma^2(pi{:}c)}{\ddot{n}_p n_i}, \qquad (7.33)$$

where $\ddot{n}_p$ is the harmonic mean of the $n_{p{:}c}$; namely,

$$\ddot{n}_p = \left[\frac{1}{n_c} \sum_c \left(\frac{1}{n_{p{:}c}} \right) \right]^{-1}. \qquad (7.34)$$

Alternatively, $\sigma^2(\Delta)$ can be viewed as the average of the conditional absolute error variance for each class $\sigma^2(\Delta_c)$. When the G and D study sample sizes are the same, $\sigma^2(\Delta_c)$ is simply the variance of the mean for the $P \times I$ design associated with each class. Therefore, letting $\sigma^2(\alpha|c)$ denote the variance components for the $p \times i$ design for class c, conditional absolute error variance for class c is

$$\sigma^2(\Delta_c) = \frac{\sigma^2(p|c)}{n_{p{:}c}} + \frac{\sigma^2(i|c)}{n_i} + \frac{\sigma^2(pi|c)}{n_{p{:}c} n_i}, \qquad (7.35)$$

and the unweighted mean is

$$\sigma^2(\Delta) = \frac{1}{n_c} \sum_c \sigma^2(\Delta_c). \qquad (7.36)$$

For balanced $(P{:}c) \times I$ designs, using ANOVA estimates of variance components in Equations 7.33 and 7.36 gives the same estimate of absolute error variance. This equality does not necessarily hold for unbalanced designs, however.

A generalizability coefficient for the $(P{:}c) \times I$ design can be viewed as the ratio of universe score variance $\sigma^2(c)$ to itself plus relative error variance:

$$\sigma^2(\delta) = \frac{\sigma^2(p{:}c)}{\ddot{n}_p} + \frac{\sigma^2(ci)}{n_i} + \frac{\sigma^2(pi{:}c)}{\ddot{n}_p n_i}. \qquad (7.37)$$

Alternatively, a generalizability coefficient can be viewed as the ratio of $\sigma^2(c)$ to expected observed score variance:

$$\boldsymbol{E}S^2(c) = \sigma^2(c) + \frac{\sigma^2(p{:}c)}{\ddot{n}_p} + \frac{\sigma^2(ci)}{n_i} + \frac{\sigma^2(pi{:}c)}{\ddot{n}_p n_i}. \qquad (7.38)$$

Note that observed score variance is defined here as the variance of the unweighted distribution of class means:

$$S^2(c) = \frac{\sum_c (\overline{X}_c - \overline{X})^2}{n_c - 1} = \frac{\sum_c \overline{X}_c^2 - n_c \overline{X}^2}{n_c - 1}, \qquad (7.39)$$

where

$$\overline{X}_c = \frac{1}{n_{p:c}n_i} \sum_{p=1}^{n_{p:c}} \sum_{i=1}^{n_i} X_{pci} \quad \text{and} \quad \overline{X} = \frac{1}{n_c} \sum_{c=1}^{n_c} \overline{X}_c. \tag{7.40}$$

Obviously, these two perspectives on a generalizability coefficient are equivalent in terms of parameters, and for balanced designs they lead to the same estimates. This equality does not necessarily hold for unbalanced designs, however. That is, the analogous-ANOVA estimate of $\sigma^2(c)$ divided by the observed score variance in Equation 7.39 gives an estimated generalizability coefficient that does not necessarily equal $\hat{\sigma}^2(c)/[\hat{\sigma}^2(c) + \hat{\sigma}^2(\delta)]$, when variance components are estimated using the analogous-ANOVA procedure.[7]

Estimating a generalizability coefficient using observed score variance as the denominator is possible, of course, only if the observed score variance is known. Usually it is known only when the G and D study sample sizes are the same.

7.2.4 Missing Data in the $p \times I$ Design

For the $p \times I$ design with missing data, a formula for absolute error variance is obtained using a derivation like that used with the $I:p$ design. The result is

$$\sigma^2(\Delta) = \frac{\sigma^2(i)}{\ddot{n}_i} + \frac{\sigma^2(pi)}{\ddot{n}_i}, \tag{7.41}$$

where

$$\ddot{n}_i = \left[\frac{1}{n_p} \sum_p \left(\frac{1}{\tilde{n}_p} \right) \right]^{-1} \tag{7.42}$$

is the harmonic mean of the $\tilde{n}_p$, which are the numbers of items responded to by the various persons (recall the notational conventions in Section 7.1.1). For example, using the synthetic data in Table 7.7 on page 227, Equation 7.41 gives

$$\hat{\sigma}^2(\Delta) = (.0117 + .1840)/4.7682 = .0410.$$

The derivation of Equation 7.41 assumes that the G and D study sample sizes are the same. If they differ, the harmonic mean in Equation 7.41 should be based on the D study sample sizes.

An alternative perspective on absolute error variance is to view it as the average of the conditional absolute error variances. When the G and D

---

[7]This lack of equivalence is not caused by the analogous-ANOVA procedure; it is a consequence of the fact that, with unbalanced designs, different estimation procedures give different results.

studies are the same and the data are dichotomous, an estimator of conditional absolute error variance for a given person is $\overline{X}_p(1 - \overline{X}_p)/(\tilde{n}_p - 1)$, as discussed in Section 5.4.1. Consequently, for dichotomous data an estimator of overall absolute error variance, in the sense of an equally weighted average of the conditional error variances, is

$$\hat{\sigma}^2(\Delta) = \frac{1}{n_p} \sum_p \frac{\overline{X}_p(1 - \overline{X}_p)}{\tilde{n}_p - 1}. \tag{7.43}$$

This estimator of $\sigma^2(\Delta)$ is not a simple function of $\hat{\sigma}^2(i)$ and $\hat{\sigma}^2(pi)$, as is the case for the estimator based on Equation 7.41. For example, using the synthetic data in Table 7.7, Equation 7.43 gives $\hat{\sigma}^2(\Delta) = .0446$, which is clearly different from $\hat{\sigma}^2(\Delta) = .0410$ obtained using Equation 7.41.

A generalizability coefficient for the $p \times I$ design often is viewed as the ratio of $\sigma^2(p)$ to itself plus $\sigma^2(pi)/n_{i'}$. If all persons take all $n_{i'}$ items in the D study, this expression is still appropriate, even if the G study variance components are estimated based on a $p \times i$ design with missing data (e.g., using Equation Set 7.17).

Often, of course, the available data, with their missing patterns, are the only data available, and an investigator is interested in generalizability with respect to the sample sizes and sample size patterns in these data. Presumably, such an investigator believes that a replication of the design (in the D study sense) would involve much the same pattern of missing data. Under these circumstances, the investigator might estimate a generalizability coefficient as the ratio of $\hat{\sigma}^2(p)$ to the variance of the observed mean scores for persons. This observed variance is given by $S^2(p)$ in Equation 7.21, but for the $p \times I$ design with missing data

$$\overline{X}_p = \frac{1}{\tilde{n}_p} \sum_{i=1}^{\tilde{n}_p} X_{pi} \quad \text{and} \quad \overline{X} = \frac{1}{n_p} \sum_{p=1}^{n_p} \overline{X}_p. \tag{7.44}$$

The expected value of the observed variance for the unbalanced $I{:}p$ design is relatively simple—the only complexity being that an harmonic mean is used in place of a constant sample size (see Equation 7.25). For the unbalanced $p \times I$ design, however, the expected value of the observed variance is a complicated expression involving all three estimated variance components:

$$\boldsymbol{E}S^2(p) = \sigma^2(p) + \left[\frac{1}{\ddot{n}_i} - \frac{\sum_p \sum_{p'} \tilde{n}_{pp'}/\tilde{n}_p\tilde{n}_{p'}}{n_p(n_p - 1)} \right] \sigma^2(i) + \frac{\sigma^2(pi)}{\ddot{n}_i}, \tag{7.45}$$

where $\tilde{n}_{pp'}$ is the number of items responded to by both persons p and p' $(p \neq p')$, and the term in square brackets multiplying $\sigma^2(i)$ is positive.[8]

---

[8]For balanced designs, the harmonic mean of the $\tilde{n}_p$ is $\ddot{n}_i = n_i$, which is why i is used here as the subscript of $\ddot{n}$. Also, for balanced designs, the term in square brackets in Equation 7.45 [the multiplier of $\sigma^2(i)$] is zero.

Using Equation 7.45 with the synthetic data in Table 7.5 gives an estimated generalizability coefficient of $E\hat{\rho}^2 = .0473/.0896 = .528$.

The last two terms in Equation 7.45 are an expression for δ-type error variance for the unbalanced $p \times I$ design, where

$$\delta_p = (\overline{X}_p - \overline{X}) - (\mu_p - \mu), \tag{7.46}$$

with $\overline{X}_p$ and $\overline{X}$ defined as in Equation 7.44.[9] Note, in particular, that a fraction of $\sigma^2(i)$ is included in this version of $\sigma^2(\delta)$.[10]

For the synthetic data in Table 7.5, this perspective on $\sigma^2(\delta)$ leads to an estimate of

$$\hat{\sigma}^2(\delta) = S^2(p) - \hat{\sigma}^2(p) = .0896 - .0473 = .0423.$$

Note that this estimate of $\sigma^2(\delta)$ is actually *larger* than the estimate of $\sigma^2(\Delta)$ based on Equation 7.41, which is .0410! This apparent contradiction is attributable to the fact that different sets of quadratic forms are used to obtain the two estimates. The T terms discussed in Section 7.1.4 are used in obtaining the value of .0410 based on Equation 7.41. By contrast, the quadratic forms $\sum \overline{X}_p^2$ and $\overline{X}^2$ (based on Equation Set 7.44), as well as the T terms, are used in obtaining the value of .0423. Again, with unbalanced designs, different sets of quadratic forms often lead to different estimates of variance components, and statistics formed from them.

These are only a few of the complexities that arise when D study designs are unbalanced in the sense of missing data. Such designs usually involve more statistical difficulties than designs that are unbalanced with respect to nesting. And, of course, complexities escalate as missing-data designs get more involved than the single-facet $p \times I$ design.

7.2.5 Metric Matters

Since G study variance components are for single conditions of facets, there is no meaningful distinction between the mean-score metric and the total-score metric. Metric issues arise only when D study issues are considered.

By the conventions of generalizability theory, D study variance components and error variances are usually expressed in the mean-score metric. For balanced designs it is easy to convert them into corresponding results for the total-score metric. For example, for the balanced $I : p$ design, assuming G and D study sample sizes are the same, universe score variance

---

[9] Strictly speaking, the last two terms in Equation 7.45 are only an approximation of the variance of δ in Equation 7.46, because these two terms do not take into account $E(\overline{X} - \mu_I)^2$.

[10] Other versions of $\sigma^2(\delta)$ are possible. For example, observed deviation scores could be weighted proportional to the $\tilde{n}_p$, which would lead to a different expression for $\sigma^2(\delta)$.

and absolute error variance in the total-score metric are

$$\sigma^2(\tau^+) = n_i^2 \sigma^2(p),$$

and

$$\sigma^2(\Delta^+) = n_i^2 \sigma^2(\Delta) = n_i \sigma^2(i{:}p),$$

respectively, where the superscript "+" designates total-score metric. It follows, of course, that $E\rho^2$ is the same for both the mean-score and total-score metrics. For unbalanced designs, however, the two metrics can lead to different values for $E\rho^2$ (and their signal–noise counterparts) whenever different numbers of conditions are nested within the objects of measurement. A similar statement applies to the Φ coefficient.

As becomes evident soon, when there are different numbers of conditions nested within the objects of measurement, the total-score metric may not be very sensible. Still, D study quantities can be defined and estimated using this metric. Consider, again, the unbalanced $I{:}p$ design, assuming G and D study sample sizes are the same. In the total-score metric, the observed score for person p is

$$X_p = \sum_{i=1}^{n_{i:p}} X_{pi} = n_{i:p}\mu + n_{i:p}\nu_p + \sum_i \nu_{i:p}.$$

One perspective on the universe score for a person is that it is the expected value of the person's observed scores over replications of the measurement procedure. Under the assumptions of the unbalanced $I{:}p$ design, the distribution of the $n_{i:p}$ is independent of the effects in the model, which means, among other things, that there is no "linkage" between any particular $n_{i:p}$ and any particular person. It follows that the universe score for person p in the total score metric is

$$\tau_p^+ = EX_p = \bar{n}_i \mu_p,$$

where

$$\bar{n}_i = \frac{1}{n_p} \sum_p n_{i:p}$$

is the arithmetic mean of the $n_{i:p}$. Consequently, universe score variance is simply

$$\sigma^2(\tau^+) = \bar{n}_i^2 \sigma^2(p). \tag{7.47}$$

Absolute error for person p in the total-score metric is

$$\Delta_p^+ = \sum_i \nu_{i:p},$$

and absolute error variance is

$$\sigma^2(\Delta^+) = E(\Delta_p^+)^2 = E\left(\sum_i \nu_{i:p}\right)^2 = E\left(\sum_i \nu_{i:p}^2\right). \tag{7.48}$$

TABLE 7.8. D Study Statistics in Total-Score Metric for Data in Table 7.2

p	Scores	$n_{i:p}$	X_p	$\hat{\sigma}^2(\Delta_p^+)$		
1	2 6 7	3	15	21.0000		
2	4 5 6 7 6	5	28	6.5000	n_p =	9
3	5 5 4 6 5	5	25	2.5000	n_+ =	37
4	5 9 8 6 5	5	33	16.5000		
5	4 3 5 6	4	18	6.6667	$\bar{n}_i$ =	4.1111
6	4 4 4	3	12	0.0000	r_i =	4.2973
7	3 6 6 5	4	20	8.0000	$\overline{X}^+$ =	21.7778
8	3 5 4	3	12	3.0000	$S^2(p)$ =	69.4444
9	6 8 7 6 6	5	33	4.0000		
Mean		4.1111	21.7778	7.5741		

For any particular value of $n_{i:p}$, Equation 7.48 gives $n_{i:p}\sigma^2(i:p)$, and since the distribution of the $n_{i:p}$ is independent of the effects in the model,

$$\sigma^2(\Delta^+) = \bar{n}_i\sigma^2(i:p). \qquad (7.49)$$

Using Equations 7.47 and 7.49, a generalizability coefficient is

$$E\rho^2 \;=\; \frac{\sigma^2(\tau^+)}{\sigma^2(\tau^+) + \sigma^2(\Delta^+)} \qquad (7.50)$$

$$\;=\; \frac{\bar{n}_i^2\sigma^2(p)}{\bar{n}_i^2\sigma^2(p) + \bar{n}_i\sigma^2(i:p)}. \qquad (7.51)$$

Consider, again, the synthetic data in Table 7.2 on page 221, and assume that the G and D study sample sizes are the same. It is easy to verify that for these data, $\bar{n}_i = 4.1111$, $\hat{\sigma}^2(\tau^+) = 8.6163$, and $\hat{\sigma}^2(\Delta^+) = 7.4292$. Using these values with Equation 7.51 gives $E\hat{\rho}^2 = .537$. Recall that the corresponding mean-score metric value reported in Section 7.2.1 was $E\hat{\rho}^2 = .525$. Since both results use the analogous-ANOVA G study estimated variance components, this difference is not attributable to estimation procedures; it is a consequence of the difference in metrics.

An alternative expression for $E\rho^2$ can be obtained using the average of the conditional absolute error variances in place of $\sigma^2(\Delta^+)$ in Equation 7.50. For the data in Table 7.2, Table 7.8 provides these conditional absolute error variances and their average (7.5741), which gives $E\hat{\rho}^2 = .532$. Recall that the corresponding mean-score metric value reported in Section 7.2.1 was $E\hat{\rho}^2 = .491$.

Another expression for $E\rho^2$ can be obtained by replacing the denominator of Equation 7.50 with the expected observed score variance based on persons' total scores. As shown in Table 7.8, the observed score variance, in the total-score metric, for these synthetic data is 69.4444, which gives

$E\hat{\rho}^2 = .124$. This is dramatically different from the corresponding mean-score metric value of $E\hat{\rho}^2 = .541$. Why? A principal reason is that, for the total-score metric, expected observed score variance is a function of not only the variance components, but also μ^2, which tends to be quite large. In particular, when the G and D study sample sizes are the same, it can be shown that expected observed score variance is

$$ES^2(p) = \frac{n_+}{n_p} \left[\frac{n_p r_i - n_+}{n_p - 1}(\mu^2) + r_i \sigma^2(p) + \sigma^2(i{:}p) \right], \qquad (7.52)$$

where $r_i = \sum_p n_{i:p}^2 / n_+$ (as in Equation 7.8).

The fact that $ES^2(X_p)$ involves μ^2 is indicative of the questionable status of the total-score metric with an unbalanced $I{:}p$ design. Furthermore, since the $n_{i:p}$ are independent of the effects in the model, a high total score for a person is likely to be associated with a large number of observations, and a low total score is likely to be associated with a small number of observations. In this sense, "high" and "low" do not have the same meaning for the total-score and mean-score metrics.

The problematic status of the total-score metric applies to designs in which different numbers of conditions are associated with the objects of measurement. This includes designs in which the lack of balance is with respect to a facet that is nested within the objects of measurement [e.g., the unbalanced $I{:}p$ and $(P{:}c) \times I$ designs], as well as designs that are unbalanced in the missing-data sense (e.g., unbalanced $p \times I$ design). For other unbalanced designs, however, no particular problems are encountered using the total-score metric. Consider, for example, the unbalanced $p \times (I{:}H)$ design in Section 7.2.2. Note that there are n_{i+} observations for each object of measurement (i.e., person). It follows that the mean-score metric D study variance components and error variances can be converted to their total-score metric counterparts simply by multiplying by n_{i+}^2.

7.3 Other Topics

Under normality assumptions, Searle et al. (1992) provide estimators of the variances of the estimated variance components for a number of unbalanced designs including the $i{:}p$ design (pp. 427–428), the $p \times i$ design with missing data (pp. 439–440), and the $i{:}h{:}p$ design (pp. 429–430). For the vast majority of designs that occur in generalizability theory, however, estimators have not been derived, even under normality assumptions.

It was stated in the introduction to this chapter that the complexities of unbalanced designs do not affect the definition of G study variance components. This statement does not extend unambiguously to D study variance components, however. For example, for the unbalanced $p \times (I{:}H)$ design, every instance of the measurement procedure involves a specific and *unequal* pattern of $n'_{i:h}$ sample sizes. Consequently, error variances are conditional on the imbalance of the D study design, and error variances involve D study variance components that are functions of the $n'_{i:h}$.

Analogous-ANOVA estimates of G study variance components are un-biased provided sample sizes are fixed. That is, the expected value of analogous-ANOVA estimates of a variance component, over replications involving the *same* sample sizes, equals the variance component. Conse-quently, D study variance components are not necessarily unbiased when they employ the G study estimates, and the D study sample sizes are *different* from those used in the G study.

This chapter has emphasized the analogous-ANOVA procedure for es-timating random effects variance components with unbalanced G study designs. There are other procedures that have been suggested in the litera-ture, however. The next section provides a brief review of these alternative procedures, followed by a discussion of computer programs for obtaining estimates. Some other issues involving unbalanced designs are considered in the examples in Chapter 8.

7.3.1 Estimation Procedures

Searle et al. (1992) provide an in-depth discussion of most procedures that have been proposed in the statistical literature for estimating variance com-ponents with unbalanced designs. Brennan (1994) briefly reviews some of these procedures, and mentions a few others, from the perspective of their applicability in generalizability analyses. Terse descriptions of most of these methods are provided next. Although the principal intent here is to focus on unbalanced designs and random models, there is occasional reference to balanced designs and/or mixed models. To avoid such considerations entirely would create needless discontinuities in discussions. Consequently, some matters covered in Chapter 3 are repeated here.

ANOVA-Like Procedures

There are a number of procedures for estimating variance components that involve the following steps: (i) calculate certain quadratic forms of the ob-servations; (ii) determine the expected values of the quadratic forms in terms of the variance components in the model;[11] and (iii) solve the set of simultaneous linear equations that result from equating the numerical values of the quadratic forms to the estimates of their expected values. We call such procedures ANOVA-like because, when used with balanced de-signs, they reduce to the basic ANOVA procedure. For unbalanced designs, however, different procedures (i.e., different quadratic forms) can give quite

---

[11] Hartley's (1967) method of synthesis is often helpful for this step.

different estimates. For both balanced and unbalanced designs, ANOVA-like procedures do not require assumptions about distributional form.

The best-known ANOVA-like procedures are three methods proposed by Henderson (1953). Henderson's Method 1 is called the analogous-ANOVA procedure in this chapter. As such, it has been discussed extensively already. Although it leads to unbiased estimates of variance components for random models, it usually produces biased estimates with models involving fixed effects. Henderson's Methods 2 and 3 were developed for use with mixed models. Method 2 is not applicable to models that include interactions between fixed and random effects. This restriction is removed in Method 3 which provides unbiased estimates of variance components for any mixed model.

Henderson's Method 3 is often referred to as the fitting constants method because it employs reductions in sums of squares due to fitting specified submodels. The investigator must specify the order of fitting the submodels, however, and a change in the order can lead to a change in the variance component estimates. For single-observations-per-cell designs that are unbalanced with respect to nesting, only, Henderson's Methods 1 and 3 often give the same estimates for all orderings,[12] which lends further credence to the use of the analogous-ANOVA procedure in generalizability theory.

The analysis of means is another ANOVA-like procedure. This procedure hides design imbalance by using subclass means; that is, a subclass mean is treated as a single observation. For generalizability analyses, however, the procedure is often not defensible. In a G study, using average scores over conditions of a facet effectively confounds that facet with others. Section 8.4 provides an example of the problems encountered in generalizability theory with the analysis-of-means procedure.

Koch (1968) discusses an ANOVA-like procedure based on certain "symmetric sums" of products of the observations. This procedure makes use of the fact that expected values of products of observations are linear functions of the variance components. Incidence sampling procedures (Sirotnik & Wellington, 1977) are similar to the symmetric sums procedure, but incidence sampling procedures tend to be more complex. Another procedure based on so-called "C terms" is discussed later in Section 11.1.3.

Maximum Likelihood Procedures

The maximum likelihood (ML) procedures discussed in the literature assume normality of the score effects. There are two basic types: general ML and restricted maximum likelihood (REML). Using general ML, fixed effects and variance components are estimated simultaneously, while they are estimated separately in REML. With balanced designs, REML estimates and analogous-ANOVA estimates are the same, except when neg-

---

[12]This is known to be true for the $i{:}p$, $p \times (i{:}h)$, and $(p{:}c) \times i$ designs.

ative variance component estimates are obtained. A principal advantage of maximum likelihood procedures is that the estimated variance components have certain desirable statistical properties, and asymptotic variances and covariances of the estimates are immediately available. However, these properties are purchased at the price of the normality assumptions.

MINQUE

Minimum norm quadratic unbiased estimation (MINQUE) requires no assumptions about distributional form, but estimating variance components does require specifying a priori weights that correspond to the relative sizes of the variance components. Hartley et al. (1978) developed a special case of MINQUE, sometimes referred to as MINQUE(0), that assigns a priori weights of zero to the variance components, except for the residual which is assigned unit weight. MINQUE(0) was developed to provide a computationally efficient method for estimating variance components when the number of observations is large. It is a special case of the first iterate solution of REML. Estimates obtained using MINQUE(0) are asymptotically consistent and locally best quadratic unbiased estimates. However, these characteristics are conditional on the a priori weights, and different weights generally give different results. An alternative to MINQUE(0) is MINQUE(1) in which the a priori weights are all unity.

In the last two decades, MINQUE(0) has achieved considerable popularity as a method for estimating random effects variance components with unbalanced designs. In part, this is probable because it is the default procedure in SAS. Also, Bell (1985) suggests that, all things considered, MINQUE(0) is the best general-purpose alternative for unbalanced designs in generalizability theory. Searle et al. (1992, p. 398), however, are less enthusiastic because of the dependence of MINQUE(0) (or any special case of MINQUE) on the choice of a particular set of a priori weights. This author shares that concern, unless there is some substantive basis for defending the choice of the particular a priori weights.

Other Procedures and Comparisons

For unbalanced designs, Cronbach et al. (1972) suggested grouping conditions into sets, such as half-tests, and performing analyses in terms of these sets of conditions. This procedure is similar to the analysis-of-means procedure, but it does not involve the type of confounding induced by the analysis-of-means procedure. Of course, estimates obtained using this grouped-conditions procedure can be influenced by which conditions happen to be assigned to which group.

No discussion of estimating variance components in unbalanced situations would be complete without referring to the frequently employed procedure of randomly discarding data to make an inherently unbalanced situation into an apparently balanced one. This procedure, which is exam-

TABLE 7.9. Estimated Variance Components Using Various Methods with a $p \times (i{:}h)$ Design Based on 2600 Observations

Method	$\hat{\sigma}^2(p)$	$\hat{\sigma}^2(h)$	$\hat{\sigma}^2(i{:}h)$	$\hat{\sigma}^2(ph)$	$\hat{\sigma}^2(pi{:}h)$
Henderson's Method 1[a]	.02472	-.00092	.02490	.00931	.18454
ML	.02448	0	.02341	.00945	.18446
REML	.02457	0	.02420	.00945	.18446
MINQUE(0)	.02456	-.00075	.02476	.01001	.18399
MINQUE(1)	.02430	-.00104	.02490	.00934	.185[b]

[a]Henderson's Method 1 is the analogous-ANOVA procedure. For this design, it is also Henderson's Method 3.

[b]MINQUE(1) results are from SPSS which provided only three decimal digits of accuracy for $\hat{\sigma}^2(pi{:}h)$.

TABLE 7.10. Estimated Variance Components Using Various Methods with a $(r{:}p) \times i$ Design Based on 5092 Observations

Method	$\hat{\sigma}^2(p)$	$\hat{\sigma}^2(r{:}p)$	$\hat{\sigma}^2(i)$	$\hat{\sigma}^2(pi)$	$\hat{\sigma}^2(ri{:}p)$
Henderson's Method 1[a]	.07163	.23108	.02235	.00229	.19072
ML	.06850	.23075	.02192	.00348	.18955
REML	.07004	.23077	.02235	.00348	.18955
MINQUE(0)	.07810	.22472	.02233	.00355	.18949
MINQUE(1)	.06897	.231[b]	.02780	.00134	.191[b]

[a]Henderson's Method 1 is the analogous-ANOVA procedure. For this design, it is also Henderson's Method 3.

[b]MINQUE(1) results are from SPSS which provided only three decimal digits of accuracy.

ined in some detail in Sections 8.2 and 8.3, is questionable if it requires eliminating large amounts of data, or leads to a reduced data set that is not representative of the full data set.

For many of the procedures discussed in this section, Table 7.9 provides a comparison of estimated variance components for a $p \times (i{:}h)$ design using data for 100 examinees who took a test consisting of 26 dichotomously scored items, each of which was nested within one of four passages (6, 6, 7, and 7 items, respectively). For this example, the estimates from the various procedures are not very different, due in part to the fact that the design is only slightly unbalanced.

Table 7.10 provides another comparison of estimated variance components. These estimates are for an unbalanced $(r{:}p) \times i$ design, based on a random sample of 50 persons from a larger data set of polytomously scored items that is discussed extensively in Section 8.3. This particular design is considerably more unbalanced than that in Table 7.9, and the number of observations is nearly twice as large as that in Table 7.9.

For both examples, the various estimation procedures give similar results, due in part to the relatively large numbers of observations (2600 and 5092, respectively). It should not be assumed, however, that the different estimation procedures will always yield similar results.

In general, with unbalanced designs, there is no unambiguously preferable procedure for estimating random effects variance components with unbalanced designs. In particular circumstances, of course, one or more procedures may be preferable. This book emphasizes Henderson's Method 1 not because it is always preferable, but rather because it makes no distributional-form assumptions, it requires no a priori judgments about weights, and it is a practical procedure no matter how large the data set may be.

7.3.2   Computer Programs

In evaluating computer packages or programs for estimating variance components in generalizability theory, it is important to know what assumptions are made in the estimation procedures. Sometimes this is difficult to ascertain, in part because not all terminology is standardized. Also, it is often useful to know how procedures are implemented by the particular package or program. Some procedures necessarily make heavy demands on computer resources (e.g., ML and REML); and some procedures are often implemented in a manner that is computationally burdensome, if not impractical, in many generalizability analyses. Designs in generalizability theory usually have (at most) one observation per cell, with large numbers of observations (10,000–50,000 is not that unusual). This means that design matrices are often huge. When the implementation of a procedure uses matrix operators on such design matrices, the procedure may be impractical for many generalizability analyses. With large amounts of computer memory and/or virtual memory, large design matrices are less problematic, but computational efficiency is still a relevant concern.

Appendix G describes the computer program urGENOVA (Brennan, 2001a), which uses the analogous-ANOVA procedure (Henderson's Method 1) to estimate random effects variance components. It was designed primarily for use with designs that are unbalanced with respect to nesting. However, it can be used with balanced designs and, to a limited extent, with designs that are unbalanced with respect to missing data. For designs that are unbalanced with respect to nesting, urGENOVA can process data sets of almost unlimited size. Designs with thousands of observations typically require only a second or two of computer time.

Table 7.11 lists procedures in SAS (SAS Institute, 1996), SPSS (SPSS, 1997), and S-Plus (Mathsoft, 1997) that provide estimates of random effects

TABLE 7.11. Comparison of Computerized Procedures for Estimating Variance Components for Unbalanced Designs

Statistical Package	Procedure[a]	Estimation Method[b]
urGENOVA	analogous ANOVA	Henderson's Method 1
SAS	MIVQUE0	MINQUE(0)
	TYPE1	Henderson's Method 3
	ML	ML
	REML	REML
SPSS	MINQUE(0)	MINQUE(0)
	MINQUE(1)	MINQUE(1)
	TYPE I	Henderson's Method 3
	ML	ML
	REML	REML
S-Plus	MINQUE0	MINQUE(0)
	ML	ML
	REML	REML

[a]Procedure name in the statistical package.
[b]Estimation method in the terminology of Section 7.3.1.

variance components.[13] The third column provides the estimation method in the terminology of this section. Note that SAS TYPE1 and SPSS TYPE I are *not* Henderson's Method 1. Also, the SAS MIVQUE0 procedure is MINQUE(0).

Processing times depend on many considerations, including the amount of memory available, the complexity of the design, the numbers of observations, the algorithms employed, and so on. Furthermore, processing times for the different procedures often vary substantially. For example, for the results reported in Table 7.10, Henderson's Method 1 estimates (using urGENOVA) took less than a second, MINQUE(0) estimates (using SAS) also took less than one second, Henderson's Method 3 estimates (using SAS) took nearly a minute, and ML and REML (using SAS) took nearly 12 hours.

Processing time is a relevant concern, but not the only issue of importance. With some commercially available statistical packages, "insufficient memory" error messages are common with even moderately large generalizability analyses, especially with SPSS. Also, memory requirements and processing time are likely to be prohibitive for ML and REML when the number of observations exceeds 10,000, which is not that large for a gener-

---

[13]SAS ANOVA provides analogous-ANOVA sums of squares, but not estimated variance components.

alizability analysis. For example, the results reported in Table 7.10 are for a subset of the data discussed in Section 8.3. For the full data set of 175 persons and 17,195 observations, SAS did not produce results for the ML and REML procedures even after 20 hours of processing time.

The discussion of computer programs in this section relates to G study issues only. None of these packages or programs provides D study results. However, the computer program mGENOVA, which is described in Appendix H, provides some D study results for unbalanced $I\!:\!p$, $p \times (I\!:\!H)$, and $(P\!:\!c) \times I$ designs.

7.4   Exercises

7.1* Derive "from scratch" the results for $ET(p)$ in Table 7.4. That is, express $T(p)$ in terms of score effects, and then take the expected value.

7.2* Using the expected values of the T terms in Table 7.4, derive the expected mean square equations for the unbalanced $p \times (i\!:\!h)$ design.

7.3* Using the expected mean square equations derived in Exercise 7.2, verify Equation Set 7.14 for the estimators of the variance components in the unbalanced $p \times (i\!:\!h)$ design.

7.4* Show that the expected mean square equations given by Equation Set 7.14 are identical to those for the balanced $p \times (i\!:\!h)$ design when the $n_{i:h}$ are the same for all levels of h.

7.5* Using the expected T terms for the unbalanced $p \times i$ design in Equation Set 7.15, verify the estimators of the variance components given in Equation Set 7.17.

7.6* Verify the T terms and estimates of the variance components in Table 7.7.

7.7* For the $I\!:\!p$ design, derive $\boldsymbol{E}\left(\sum_p \overline{X}_p^2\right)$ in Equation 7.23 and $\boldsymbol{E}(\overline{X}^2)$ in Equation 7.24.

7.8 Derive Equation 7.45. (Hint: The derivation proceeds much like that used in Exercise 7.7.)

7.9 Prove Equation 7.52.

8

Unbalanced Random Effects Designs—Examples

This chapter provides several examples of generalizability analyses that involve estimating random effects variance components for unbalanced designs. Largely, these examples are applications of the theoretical results discussed in Chapter 7. Some additional topics related to unbalanced designs are introduced as well.

8.1 ACT *Science Reasoning*

The ACT Assessment *Science Reasoning* Test is a 40-item, 35-minute test that "measures the student's interpretation, analysis, evaluation, reasoning, and problem-solving skills required in the natural sciences (ACT, 1997, p.16)." Each form consists of seven test units that provide scientific information (the stimuli) along with a set of multiple-choice test questions. The stimuli can be from various content areas (biology, earth/space sciences, chemistry, or physics), and they may be presented in various formats (data representation, research summaries, and conflicting viewpoints). In this sense, the stimuli are quite heterogenous.

Table 8.1 provides estimated G study variance components and various D study statistics for five forms of the *Science Reasoning* Test. For each of these forms, three stimuli (h) had five items, three had six items, and one had seven items. The D study statistics are based on the same numbers of items, which means that $\breve{n}_h$ in Equation 7.28 is 6.8966. Each form

TABLE 8.1. ACT *Science Reasoning* Example

Statistic	Form 1	Form 2	Form 3	Form 4	Form 5	Mean	S.D.
$\hat{\sigma}^2(p)$	.0245	.0202	.0246	.0207	.0196	.0219	.0024
$\hat{\sigma}^2(h)$	.0183	.0195	.0106	.0212	.0112	.0161	.0049
$\hat{\sigma}^2(i{:}h)$	.0151	.0178	.0211	.0153	.0191	.0177	.0025
$\hat{\sigma}^2(ph)$	.0113	.0084	.0080	.0101	.0110	.0098	.0015
$\hat{\sigma}^2(pi{:}h)$	.1624	.1724	.1740	.1715	.1749	.1710	.0050
$\hat{\sigma}^2(\delta)$	.0057	.0055	.0055	.0058	.0060	.0057	.0002
$\hat{\sigma}^2(\Delta)$	.0087	.0088	.0076	.0092	.0081	.0085	.0006
$E\hat{\rho}^2$	.811	.785	.817	.782	.767	.793	.021
$\widehat{S/N}(\Delta)$	2.807	2.230	3.255	2.246	2.426	2.607	.423

Note. D study statistics are based on $n'_h = 7$ passages with 5, 5, 5, 6, 6, 6, and 7 items associated with the various passages.

was administered to over 3000 examinees in a randomly equivalent groups design.

Also provided in Table 8.1 are the means and standard deviations (S.D.s) of the various statistics. The standard deviations are particularly informative, because they are direct empirical estimates of the standard errors of the estimated variance components and the other statistics. Note, in particular, that these estimated standard errors are not based on any normality assumptions. It is evident that the standard errors are quite small relative to the magnitudes of the estimates, which affords a measure of confidence about the stability of the estimates for the various forms.

Although the standard errors of the estimated G study variance components are rather small, $\hat{\sigma}[\hat{\sigma}^2(h)]$ is about twice as large as the next largest estimated standard error, and the values of $\hat{\sigma}^2(h)$ themselves are relatively large. This means that there is considerable variability, both within and across forms, in the overall difficulty of the stimuli. By contrast, the person-stimuli interactions $\hat{\sigma}^2(ph)$ tend to be substantially smaller and less variable across forms.

The relatively large values for $\hat{\sigma}^2(h)$, contrasted with the relatively small values for $\hat{\sigma}^2(ph)$, are the principal reason that $\hat{\sigma}^2(\Delta)$ tends to be about 50% larger than $\hat{\sigma}^2(\delta)$. As noted above, the relatively large values for the $\hat{\sigma}^2(h)$ mean that stimuli differ in average difficulty. The substantially smaller values for $\hat{\sigma}^2(ph)$ mean that persons would not be rank ordered too differently on the various stimuli or, equivalently, the relative difficulties of the various stimuli tend to be similar for most persons. The forms of the ACT Assessment undergo an elaborate equating process. This process is unlikely to adjust much for $\hat{\sigma}^2(ph)$, but it eliminates (or, at least mitigates) constant effects such as those captured by $\hat{\sigma}^2(h)$. Consequently, it is likely that the differences between $\hat{\sigma}^2(\Delta)$ and $\hat{\sigma}^2(\delta)$ have no particularly important consequences for the reported scores.

Contrary to some tenets of conventional wisdom, reliabilities lower than .80 are not necessarily indicative of poor tests. Such results are to be expected for relatively short tests that sample a heterogeneous content domain, as does the ACT *Science Reasoning* Test. (One of the exercises at the end of this chapter explores the sample sizes that might be used to obtain higher reliability with an increased time limit.) Note, also, that the absolute-error SEM is $\sqrt{.0085} \doteq .09$, which is only about a tenth of a raw-score point, even with reliabilities slightly less than .80.

In some testing programs, the user of examinee data knows which form of the test was taken by each examinee. Generally, this is not the case for testing programs such as the ACT Assessment, and usually the examinees in any particular data set have taken different forms. Consequently, "mean" statistics such as those of Table 8.1 provide relevant answers to questions about error variances and coefficients. These mean results are actually more stable than the standard deviations in the last column suggest. The standard deviations are estimated standard errors for a single form. The estimated standard errors for the "mean" statistics are the standard deviations divided by the square root of five.

It is evident that the means of the G study estimated variance components are quite stable, which makes them particularly useful in predicting D study results for different sample sizes and sample size patterns. Table 8.2 provides estimated D study results for 40 and 35 items, for four different numbers of stimuli (4, 5, 6, and 7), and for two patterns of sample sizes. The first pattern consists of sample sizes that are about as equal as possible, which gives a value of $\breve{n}'_h$ that is quite close to n'_h. The second pattern of sample sizes is more diverse, resulting in a smaller value of $\breve{n}'_h$. The first line in Table 8.2 is a baseline; it provides results using the mean of the G study estimated variance components in Table 8.1 and the sample sizes for the current ACT Science Test. It is evident from Table 8.2 that for fixed values of n'_i and n'_h, modest variation in sample size patterns does not make much difference. However, decreases in n'_{i+}, and especially decreases in n'_h, cause notable increases in error variances.

8.2   District Means for ITED *Vocabulary*

As an illustration of an unbalanced design in which a facet is nested within the objects of measurement, consider a G study $(p{:}d) \times i$ design for pupils (p) within school districts (d) in Iowa who took the 40-item (i) ITED

TABLE 8.2. Different Sample Size Patterns for ACT *Science Reasoning* Example

n'_{i+}	n'_h	$n'_{i:h}$	$\breve{n}'_h$	$\hat{\sigma}^2(\delta)$	$\hat{\sigma}^2(\Delta)$	$E\hat{\rho}^2$	$\widehat{S/N}(\Delta)$
40	7	5,5,5,6,6,6,7	6.8966	.0057	.0085	.793	2.607
40	6	6,6,7,7,7,7	5.9702	.0059	.0091	.788	2.421
40	6	5,6,7,7,7,8	5.8824	.0059	.0091	.787	2.403
40	5	8,8,8,8,8	5.0000	.0062	.0099	.779	2.215
40	5	6,7,8,9,10	4.8485	.0063	.0101	.777	2.179
40	4	10,10,10,10	4.0000	.0067	.0112	.765	1.958
40	4	8,10,10,12	3.9216	.0068	.0113	.764	1.936
35	7	5,5,5,5,5,5,5	7.0000	.0063	.0091	.777	2.411
35	7	3,4,4,5,6,6,7	6.5508	.0064	.0093	.775	2.346
35	6	5,6,6,6,6,6	5.9756	.0065	.0097	.771	2.254
35	6	4,5,5,6,7,8	5.6977	.0066	.0099	.769	2.206
35	5	7,7,7,7,7	5.0000	.0068	.0106	.762	2.073
35	5	5,6,7,8,9	4.8039	.0069	.0108	.760	2.033
35	4	8,9,9,9	3.9902	.0073	.0119	.749	1.845
35	4	6,8,9,12	3.7692	.0075	.0123	.746	1.787

Note. The first line of results is based on the means of the G study estimated variance components in Table 8.1 using the sample sizes for the current ACT *Science Reasoning* Test.

Vocabulary Test (Form K, Level 17/18) (Feldt et al., 1994) administered in the Fall of 1997. Table 8.3 provides the results of such a G study based on a sample of about one-third of the pupils in approximately one-third of the districts that took the ITED that year. For illustrative purposes we assume here that these data can be viewed as a random effects design, without taking into account the complicating possibility of sampling from a finite population of pupils and/or districts.[1]

The degree to which the design is unbalanced is partly indicated by the grouped frequency distribution of the $n_{p:d}$ sample sizes at the bottom of Table 8.3. At a finer level of detail, the range of the 108 sample sizes was 4 to 309, with a mean of 28.6 and a median of 17. The difference between the mean and the median is largely attributable to several districts with quite large values for $n_{p:d}$.

Probably the most dramatic aspect of the G study results is the relatively large variance component for pupils, $\hat{\sigma}^2(p:d) \doteq .034$ compared to the variance component for districts, $\hat{\sigma}^2(d) \doteq .002$. Apparently, pupil variability (within districts) is about 17 times greater than district variability. This

---

[1] Although the data are real, there are several aspects of this example that are artificial. For example, the analysis here is restricted to raw scores. Also, it is not too likely that an actual investigation would involve sampling one-third of the districts and one-third of the pupils within each district.

TABLE 8.3. G Study Analysis of a Sample of District Means for ITED Vocabulary

Effect	df	T	SS	MS	$\hat{\sigma}^2$
d	107	37013.57628	429.66310	4.01554	.00213
$p{:}d$	2983	41666.42500	4652.84872	1.55979	.03412
i	39	38417.57198	1833.65881	47.01689	.01509
di	4173	39928.01358	1080.77849	.25899	.00228
$pi{:}d$	116337	67255.00000	22674.13770	.19490	.19490
μ	1	36583.91318			

$n_{p{:}d}$	freq	Number-Correct Metric	
		Mean	Variance
4–10	20	19.6459	48.4318
11–20	43	21.6574	55.7458
21–40	37	20.9605	59.3581
> 40	8	22.4115	66.4076

$$n_{p+} = 3091 \quad \text{and} \quad \sum_d n^2_{p{:}d}/n_{p+} = 81.62245$$

fact suggests that district-level coefficients may be smaller than pupil-level coefficients. This is not a certainty, however, because we do not yet know how large district error variances are relative to pupil error variances. (See Exercise 8.4.)

Using the same sample sizes as those in the G study, let us consider a D study $(P{:}d) \times I$ design, with districts as the objects of measurement. General results for this design are treated in Section 7.2.3, with d here playing the role of c in Section 7.2.3. The harmonic mean of the $n_{p{:}d}$ is 15.213. Using Equation 7.33, absolute error variance is

$$\hat{\sigma}^2(\Delta) = \frac{.03412}{15.213} + \frac{.01509}{40} + \frac{.00228}{40} + \frac{.19490}{608.520} = .00300.$$

Recall from Section 7.2.3 that this result is not necessarily equal to the average of the 108 conditional absolute error variances (see Equation 7.35). For these data, this average is .00266. In short, in terms of SEMs, $\hat{\sigma}(\Delta)$ is .0548, while the average $\hat{\sigma}(\Delta_d)$ is .0516. These estimates are quite similar, given the substantial lack of balance in the data.

The variance of the unweighted distribution of district means is .00476, where each mean is expressed in the proportion-correct metric. It follows that an estimated generalizability coefficient for the G study sample sizes is

$$E\hat{\rho}^2 = \frac{\hat{\sigma}^2(d)}{S^2(d)} = \frac{.00213}{.00476} = .4471. \tag{8.1}$$

Alternatively, using Equation 7.37 relative error variance is estimated to be

$$\hat{\sigma}^2(\delta) = \frac{.03412}{15.213} + \frac{.00228}{40} + \frac{.19490}{608.520} = .00262,$$

and an estimated generalizability coefficient is

$$E\hat{\rho}^2 = \frac{\hat{\sigma}^2(d)}{\hat{\sigma}^2(d) + \hat{\sigma}^2(\delta)} = \frac{.00213}{.00213 + .00262} = .4479. \tag{8.2}$$

The estimates in Equations 8.1 and 8.2 are remarkably similar, given the substantial lack of balance in the data.

Recall that these D study results are based on one-third of the tested pupils in about one-third of the districts. An investigator might well be interested in the generalizability of the 108 district means when each mean is based on all tested examinees. The harmonic mean of these sample sizes is 45.640, which can be used in Equation 7.33 to obtain $\hat{\sigma}^2(\Delta) = .00129$ and in Equation 7.37 to obtain $\hat{\sigma}^2(\delta) = .00091$. The latter leads to an estimated generalizability coefficient of $E\hat{\rho}^2 = .7004$, which is a "stepped-up" version of the result in Equation 8.2 for the smaller sample sizes. We are not able to estimate a generalizability coefficient in the manner of Equation 8.1 because we have no direct estimate of observed score variance for the larger sample sizes.

Strictly speaking, the results discussed in the previous paragraph are based on only a sample of the districts. It is unlikely, however, that much different results would be obtained by using a larger sample of districts, or even all the districts. Furthermore, as long as the number of districts is large enough to lead to an acceptably stable estimate of universe score variance, the magnitude of an estimated coefficient is not likely to be much affected by increasing the number of districts.

If the G study were based on all the districts in Iowa, and these districts were viewed as the entire population of interest, there is an additional statistical issue that merits consideration. In this case, it can be argued rather convincingly that districts are fixed, which suggests that the G study should use a mixed-model analysis, not a random one. In most cases, with unbalanced designs, the estimation of variance components for mixed models is an even more thorny subject than their estimation for random models. One approach that does not make distributional-form assumptions (e.g., normality) is to use Henderson's (1953) Method 3. For the $(p{:}d) \times i$ design, however, Henderson's Method 3 leads to the same decomposition of the total sums of squares as does Method 1 (the analogous-ANOVA procedure). Consequently, equating sums of squares to their expected values leads to the same estimates of $\sigma^2(p{:}d)$, $\sigma^2(i)$, $\sigma^2(di)$, and $\sigma^2(pi{:}d)$ for both Methods 1 and 3.

Under the mixed model, conventional statistical terminology would not refer to a variance component for districts; rather, it would be called a quadratic form. Setting aside this terminological distinction, under the mixed model, given the conventions adopted in Section 3.5, $\hat{\sigma}^2(d)$ that results from equating sums of squares to their expected values, is an estimate of

$$\sigma^2(d) = \sum_{d=1}^{n_d} \frac{\nu_d^2}{n_d - 1}.$$

Clearly, under the mixed model assumptions, $\hat{\sigma}^2(d)$ can be multiplied by $(n_d - 1)/n_d$ to obtain a statistic with the conventional form of a variance.

This sidetrack into mixed models is something of a statistical technicality that has very little influence on estimates of error variances and coefficients. It is especially important to note that the fixed part of the mixed model discussed in this section is associated with the objects of measurement, not the universe of generalization. A fixed facet in the universe of generalization is an entirely different matter, as already discussed in Chapter 4, and as discussed more fully when we treat multivariate generalizability theory in subsequent chapters.

Given the complexities and ambiguities involved in estimating variance components for unbalanced G study designs, it seems natural to consider the possibility of randomly discarding data until a balanced design is obtained. Such an ad hoc procedure hardly deserves to be called a "design," but it is a frequently employed strategy. Let us consider what might happen if a balanced design were employed for this ITED *Vocabulary* example.

The total number of pupils in the data set for the 108 districts is 3091. Obviously, if we want to obtain a balanced design, we would like to eliminate relatively few districts and pupils. For a particular number of pupils, say n_p^*, to obtain a balanced design all districts with fewer than n_p^* pupils are eliminated; the data are retained for all districts with exactly n_p^* pupils; and for districts with more than n_p^* pupils, a random sample of n_p^* pupils is drawn. So, for example, we could use all districts, but this would require using $n_p^* = 4$, a very small value, that results in including only 432 pupils (14%). If $n_p^* = 10$, then the design will include 93 districts (86%) with a total of 930 pupils (30%). If $n_p^* = 15$, then the design will include a much smaller number of districts, 66 (61%), with a slightly larger total number of pupils, 990 (32%). Let us consider using $n_p^* = 10$.

Obviously, different estimated variance components will be obtained depending on which 10 pupils are sampled for districts having more than 10 pupils in the data set. This is illustrated by the box-and-whiskers plots in Figure 8.1 based on 250 replications of a balanced design with 10 pupils per district. For each box, the lowest, middle, and highest lines are at the 25th, 50th, and 75th percentile points, respectively. The bottom whisker is at the 10th percentile point, and the top one at the 90th percentile point. The box-and-whisker plots are ordered from left to right in terms of vertical axis lengths. For d, di, i the length is .0025; for $p{:}d$ and $pi{:}d$ the length is doubled to .005. Therefore, if the first three plots were displayed using the

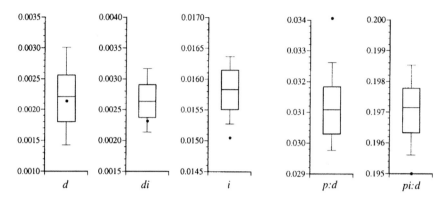

FIGURE 8.1. Box-and-whisker plots of estimated variance components for ITED Vocabulary based on a balanced design with 10 students per district and 250 replications.

scale of the last two, the first three plots would be half as tall as they are depicted in Figure 8.1.

It is important to note that the variability of the estimated variance components in Figure 8.1 is not the result of true replications of a balanced design with 10 pupils per district. First, the districts and the items in each pseudoreplication are the same. Second, for each pseudoreplication, the data are identical for those districts that have exactly 10 pupils. Third, the potentially variable data result from a stratified random sampling process that is conditional on the specific, *fixed* sizes for pupils in the original data set. Although these three characteristics tend to make the pseudoreplications much more similar than they otherwise would be, there is still evidence of variability for the estimates in Figure 8.1.

The dots in Figure 8.1 approximately locate the analogous-ANOVA estimates of the variance components using the original data. Except for $\hat{\sigma}^2(d)$, these estimates are rather distant from the median of the pseudoreplications estimates. Indeed, $\hat{\sigma}^2(p{:}d)$ and $\hat{\sigma}^2(pi{:}d)$ are very different from the pseudoreplications estimates.[2] Why? There are two classes of reasons: the characteristics of the eliminated data and differences in estimation procedures.

The pseudoreplications in Figure 8.1 are based on eliminating all districts with fewer than 10 pupils. Clearly this may distort estimates if the eliminated districts are unlike the retained ones. The bottom part of Table 8.3 provides relevant information. For example, there were 20 districts with sample sizes of 4 to 10 pupils per district. The unweighted average, over these 20 districts, of the mean number-correct score for each district

---

[2]The absolute magnitudes of the differences might not be judged very large, but the relative magnitudes are considerable.

was 19.6459 items correct. Similarly, the average of the observed variances was 48.4318. It is apparent from Table 8.3 that the observed means and variances tend to increase with increases in $n_{p:d}$. This clearly suggests that districts with 10 students are not interchangeable with districts with larger numbers of students, which will cause discrepancies between the estimates for the unbalanced and balanced designs.

The positive relationship between $n_{p:d}$ and the sample statistics will cause other differences, too. For example, the few large districts, which have the largest means and variances, will have less of an influence on estimated variance components for the balanced data set.

The above arguments should not be construed as meaning that the full-data estimates in Table 8.3 are necessarily "better" than estimates based on a balanced design with, say, 10 students per district. We may argue that the full-data estimates are more stable because they are based on larger sample sizes, but that does not necessarily mean that the full-data estimates are closer to the parameters. The essence of the problem is that we have differences not only in sample sizes but also in estimation procedures. We know a great deal about the characteristics of the estimates for the balanced design with the smaller sample sizes, but very little about the analogous-ANOVA estimates for the full data set.

Still, "squaring off" an unbalanced data set by randomly eliminating data has two problematic characteristics: it frequently results in discarding a large amount of data, and the process used to eliminate data is conditional on sample sizes in the data set and can lead easily to eliminating data that are not representative of the full data set. For these reasons, this author is not sanguine about the procedure, except when only a small amount of data needs to be eliminated and there is strong reason to believe that the eliminated data are representative.

8.3   Clinical Clerkship Performance

Kreiter et al. (1998) discuss certain psychometric characteristics of a standardized clinical evaluation form (CEF) used to evaluate the performance of medical students in various clinical clerkships at a large midwestern medical school. The CEF consists of 19 items measuring behaviorally specific skills judged to be important in clinical performance (e.g. "Accepts responsibility for actions"). Ratings are based on a five-point scale ranging from "unacceptable" to "outstanding." Standardized CEFs are often the only formal feedback that students receive about their clinical performance in the clerkships. As such, the information provided by the CEFs serves an important role in student assessment, including grading.

Typically, each student is observed by a small number of raters demonstrating his or her clinical skills in real-time and real-world settings. So,

TABLE 8.4. G Study for Surgery from Kreiter et al. (1998)

Effect	df	T	SS	MS	$\hat{\sigma}^2$
p	174	303099.82461	1993.57134	11.45731	.07066
$r\!:\!p$	730	306388.57895	3288.75434	4.50514	.22713
i	18	301484.60221	378.34894	21.01939	.02300
pi	3132	304111.86118	633.68763	.20233	.00245
$ri\!:\!p$	13140	309893.00000	2492.38448	.18968	.18968
μ	1	301106.25327			

$n_{r:p}$	freq	mean	$n_{r:p}$	freq	mean
2	6	4.2675	7	16	4.1612
3	24	4.3187	8	4	4.3914
4	19	4.1025	9	4	4.4240
5	56	4.1752	10	1	3.4632
6	44	4.1583	11	1	1.0096

$$n_{r+} = 905 \quad \text{and} \quad \sum n_{r:p}^2/n_{r+} = 5.64972$$

while the instrument is standardized, many other conditions of measurement are not. The Kreiter et al. (1998) study was designed to examine: (i) the magnitudes of the variance components associated with the CEF and the population; (ii) the effect on measurement precision of varying the numbers of raters and items; and (iii) the minimum number of CEF observations required to obtain a reasonably precise measure of performance for a student. Here, these issues are summarized for surgery only, which is one of six clerkships that students typically experience in a clinical year.

Medical faculty, residents, and adjunct faculty who worked with students during the surgery clerkship filled out a CEF for each student. As is often the case in real-world settings, there was no fully formalized design that assigned specific raters, or a specific number of raters, to each student. A total of 175 students was evaluated by 76 raters, with a mean of 7.1 ratings per student. No rater evaluated all students, but most raters evaluated a number of students. Clearly, raters are not crossed with students, but neither are they completely nested within students. Kreiter et al. (1998) treat the data as a G study for the $(r\!:\!p) \times i$ design. This choice is partly one of convenience, but it does not seem too unreasonable under the circumstances. (Some consequences of this choice are considered in Exercise 8.7.) The $(r\!:\!p) \times i$ design is formally identical to both the $p \times (i\!:\!h)$ and the $(p\!:\!c) \times i$ designs; so the results in Section 7.1.3 and Appendix E, respectively, apply with obvious changes in notation.

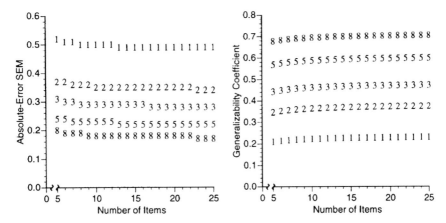

FIGURE 8.2. Estimated absolute error SEMs and generalizability coefficients for surgery from Kreiter et al. (1998) with $n'_r = 1, 2, 3, 5$, and 8 raters and $5 \leq n'_i \leq 25$ items.

Table 8.4 on the preceding page provides random-model G study results for the Kreiter et al. (1998) surgery data. The distribution of number of ratings is provided at the bottom of the table, which is one way of characterizing the extent to which the data are unbalanced. The number of ratings ranged from 2 to 11, but over 80% of the students were evaluated by 3 to 6 raters. The largest estimated variance component is for raters, which is not unexpected given the variabilities of their backgrounds, their sometimes limited familiarity with the students, and the nonstandardized circumstances of their ratings.

Since $\hat{\sigma}^2(r{:}p)$ is so large, it seems unlikely that a single rating would lead to acceptably reliable scores for students. This intuitive judgment and other results are formalized in Figure 8.2, which provides random-model absolute error SEMs, $\hat{\sigma}(\Delta)$, and estimated generalizability coefficients for 5 to 25 items and 1, 2, 3, 5, and 8 raters.[3] Kreiter et al. (1998) concluded from a slightly different presentation of these results that three or more raters provide an acceptably reliable measure of student performance using the full-length CEF. They also suggested that, from the perspective of measurement precision, the CEF could be shortened. (Other substantive validity concerns might not support doing so).

It is important to note that the estimation of the G study variance components was based on an unbalanced design, not because that is the optimal way to collect data, but rather because that is the manner in which the

---

[3] It is assumed here that items are random. If the only items of interest were those 19 items in the CEF, as specified, then it might be argued that items should be treated as fixed. This author seldom treats items as fixed, however, because almost always specific items are representative of some larger set.

data arose. In this sense, the lack of balance did not arise by "design." On the other hand, the D study considerations in the previous paragraph were addressed using a balanced design, because the principal questions involve recommended use of the CEF, and the psychometric consequences of such use.

There is no logic in the results presented here to suggest that the D study design should be unbalanced. Of course, in practice, operational use of the instrument might result in an unbalanced design. If so, an estimated conditional SEM, $\hat{\sigma}(\Delta_p)$, could be obtained for each student based on the actual number of ratings for that student. Also, the equations in Section 7.2.3, with obvious changes in notation, could be used to estimate overall absolute error variance, generalizability coefficients, and other parameters. For example, suppose the operational use of the CEF generally required a minimum of three ratings per student, but more could be used, and two ratings would be accepted in unusual circumstances. What would be the estimated absolute-error SEM for 200 students if 15% had two ratings, 75% had three ratings, and 10% had four ratings? Under these circumstances, the harmonic mean is $\ddot{n}_r = 2.85714$, and using Equation 7.33 we obtain $\hat{\sigma}(\Delta) \doteq .29$.

As we did in the previous section, let us now consider what estimated variance components might have resulted from a balanced design created by randomly discarding data. The lower part of Table 8.4 provides frequencies and means for the numbers of ratings (2 to 11) made for the 175 students. The total number of ratings is $n_{r+} = 905$. If we want to eliminate ratings to create a balanced design, a little computation leads to the conclusion that the fewest ratings are eliminated using a balanced design with five raters. Under these circumstances, all students with fewer than five ratings are eliminated; the ratings are retained for students with exactly five ratings; and for students having more than five ratings, a random sample of five ratings is used. This results in retaining 126 students (72%) and 630 ratings (about 70%).

Obviously, different estimated variance components will be obtained depending on which five scores are sampled for students having more than five ratings. This is illustrated by the box-and whiskers plots in Figure 8.3 based on 250 replications of a balanced design with five ratings per student. The plots are ordered from left to right in terms of vertical axis lengths. For p, $r{:}p$, and $ri{:}p$ the length is .02; for i and pi the length is five times smaller. Therefore, if the last two plots were displayed using the scale of the first three, the last two plots would be five times shorter than they are depicted in Figure 8.3.

These are pseudoreplications, not true ones. The students and the items in each pseudoreplication are the same, and the sampling process is conditional on the specific *fixed* sample sizes (6, 7, ..., 11) in the original data set. Although the pseudoreplications are much more similar than true ones,

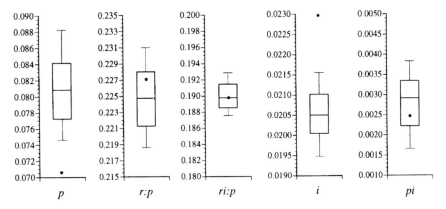

FIGURE 8.3. Box-and-whisker plots of estimated variance components for surgery from Kreiter et al. (1998) based on a balanced design with five raters per person and 250 replications.

there is still evidence of variability in some of the estimates in Figure 8.3, particularly $\hat{\sigma}^2(p)$.

The dots in Figure 8.3 approximately locate the analogous-ANOVA estimates of the variance components based on the original data. Except for $\hat{\sigma}^2(ri:p)$, these estimates are rather distant from the median of the pseudoreplications estimates. Indeed, $\hat{\sigma}^2(p)$ and $\hat{\sigma}^2(i)$ are quite different from the pseudoreplications estimates.[4] Why? As discussed in Section 8.2 for the ITED *Vocabulary* example, there are two classes of reasons: the characteristics of the eliminated data and differences in estimation procedures.

The pseudoreplications in Figure 8.3 are based on eliminating all students with fewer than five ratings. Clearly this may distort estimates if the eliminated students are unlike the retained ones. Given the means for the $n_{r:p}$ in Table 8.4, it is easily determined that the 49 eliminated students have a mean rating of 4.2286, while the mean for the 126 retained students is 4.1515—a rather large difference in the context of the means in Table 8.4. This explanation may appear to imply that it would be better to use all students by simply sampling two ratings for each of them. However, doing so would lead to using only about 39% of the original data to estimate variance components![5]

These arguments do not guarantee that the estimates in Table 8.4 are "better" than estimates based on a balanced design. Still, all things considered, this author prefers those in Table 8.4.

---

[4]The absolute magnitudes of the differences might not be judged very large, but the relative magnitudes are considerable.

[5]Estimates for 1000 replications were obtained for two ratings per student. The estimates were more variable than those in Figure 8.3 for five ratings, as is to be expected given the reduced sample size. Furthermore, the two-ratings estimates tended to be even more different from the full-data estimates than did the five-ratings estimates.

8.4   Testlets

Testlets, as the name implies, have been defined as small part-tests in a larger test (Wainer & Kiely, 1987; Wainer & Lewis, 1990; Sireci et al., 1991). Lee et al. (2001) have defined a testlet as a subset of one or more items in a test form, which is treated as a measurement unit in test construction, administration, and/or scoring. This definition differs from others that have been proposed in two important ways: it refers explicitly to test forms, and the definition permits single-item testlets. In effect, the ACT *Science Reasoning* example in Section 8.1 is an analysis of a test composed of testlets.

The discussion of testlets here involves consideration of both balanced and unbalanced random effects designs, as well as a mixed model that foreshadows a multivariate design. As such, this section can be viewed as the beginning of a transition to multivariate generalizability theory.

There are a number of analytic procedures that have been proposed and used to analyze testlets, most of them based on an IRT model of one kind or another. Here, we consider testlets from the perspective of generalizability theory. In particular, we concentrate on sample size issues, aggregation issues, and fixed-versus-random issues.

In considering testlets, it is important to distinguish between a test and a form of a test. A test is a generic characterization of the multiple forms that might be created using the test specifications. So, for example, "ACT *Science Reasoning*" is an identifier of a test, but there are many extant forms of this test, and many other forms that could be created. (Even if only one form of a test is available, the concept of a form is still relevant, because other forms could be created from the same specifications.)

Four Designs

Let us reconsider the synthetic data in Table 7.5 for a test form composed of eight items that are distributed across three testlets. We consider four analyses of these data, with the discussion framed in terms of reliability considerations. For the moment, we are not arguing which of these analyses is preferable; we are simply examining the assumptions and consequences of these analyses.

1. *D Study p × H Design.* Suppose this test is conceptualized as being composed of three random testlets, with the scores for the three testlets used as the sole basis for analysis. In this case, the D study design is $p \times H$, with $n'_h = 3$, but there are still two possible analyses. It is easily verified that, if total scores for the three testlets are employed, then $E\hat{\rho}^2 = .598$; if unweighted mean scores are used, $E\hat{\rho}^2 = .684$.

2. *D Study $p \times (I:H)$ Random Effects Design.* Section 7.2.2 provides results for the unbalanced $p \times (I:H)$ random effects design. In particular, Equation 7.32 gives $E\hat{\rho}^2 = .716$.

3. *D Study $p \times I$ Design.* If the testlets are ignored, with the individual items treated as an undifferentiated set, then $E\hat{\rho}^2 = .885$.

4. *D Study $p \times (I:H)$ Design with H Fixed.* The testlets could be viewed as fixed. Since the design is unbalanced, this possibility is not easily accommodated in univariate generalizability theory. A full discussion awaits our consideration of multivariate generalizability theory in Chapters 9 and 10. Here, we simply note that stratified alpha (Cronbach et al., 1965) is an appropriate generalizability coefficient, and it is .941 for these data.

Since the estimated generalizability coefficients for these four designs are based on a very small set of highly synthetic data, no particular conclusions should be drawn based on the absolute magnitude of the estimates. However, the relative magnitudes of the coefficients (1 = lowest; 4 = highest) are largely predictable from by understanding what is fixed and what is random in the various designs, what effects are confounded in certain designs, and how sample sizes affect the coefficients.

The analyses using the $p \times H$ design are the simplest in the sense that they require the least amount of computation, but they involve conceptual complexities and ambiguities. One potential ambiguity is that $E\hat{\rho}^2$ is different for the total and mean score metrics, because different numbers of items are summed or averaged for the three testlets. Other complexities arise because the role of items is hidden in the design, which leads to confounded effects. For example, $\hat{\sigma}^2(pH)$ from the $p \times H$ design represents the confounding $\hat{\sigma}^2(pH)$ and $\hat{\sigma}^2(pI:H)$ from the $p \times (I:H)$ random effects design.

The $p \times I$ analysis is computationally simple but, for the data in Table 7.5, the analysis is also conceptually simplistic, because it assumes that the eight items represent a single random facet. This design effectively assumes that a different form of the test would include a different set of eight undifferentiated items, without any consideration of their role in testlets. The higher value of the estimated generalizability coefficient for the $p \times I$ design, compared to the $p \times H$ design, is attributable largely to the fact that the average covariance for the eight item scores (1.30) is larger than the average covariance for the three testlet mean scores (.98).

For the $p \times (I:H)$ random effects design, both items and testlets are treated explicitly as random, with a recognition that there are different numbers of items associated with testlets. This means that a different form of the test would involve different items and different testlets, but the sample size patterns would be the same. The analysis may be somewhat complex, but there are no particular conceptual ambiguities, and the analysis faithfully reflects the structure of the test form. Since items and testlets are random, variability in both facets contributes to error, and the analysis reflects this explicitly.

Cases 2 and 4 have the same design structure, $p \times (I : H)$; that is, they both bundle the items in the same manner. However, in Case 2, the testlets are a random facet, whereas they are fixed in Case 4. Case 4 would be applicable, for example, if items were associated with three content categories, all of which would be involved in every form of a test. Since H is fixed, variability in scores attributable to content categories does not affect error variance. Consequently, a generalizability coefficient for this design with H fixed will be larger than when H is random. The univariate $p \times (I : H)$ design with H fixed can be viewed from the perspective of multivariate generalizability theory in which the content categories are the multiple dependent variables, and there is a $p \times I$ design associated with each category, as discussed more fully later in Section 9.1. This multivariate perspective overcomes the unbalanced-design problems of a univariate mixed model.

It may not be clear why the $p \times (I : H)$ design with H fixed leads to a larger estimated generalizability coefficient than for the $p \times I$ design. After all, both involve the same number of items, and both view items as random. The reason for the difference is that the variance attributable to the interaction of persons and testlets is effectively part of error variance for the $p \times I$ design, whereas it is part of universe score variance for the $p \times (I : H)$ design with H fixed.

In considering these four testlet analyses of the same data, it is natural to ask which one is best. There is no unequivocal answer to this question without an unambiguous specification of the universe of generalization including, in particular, the identification of which facets are fixed and which are random. However, both the $p \times H$ and $p \times I$ analyses have the distinct disadvantage of hiding a facet.

Single-Item Testlets

Suppose the test specifications require that each form of a geometry test contain an item that tests the Pythagorean theorem. From the perspective of the content categories in a table of specifications, the Pythagorean Theorem item in a particular form represents a fixed condition of measurement. However, the particular Pythagorean Theorem items in the various forms of the test will be different. From this perspective, the Pythagorean Theorem items constitute a random facet. In a sense, then, when we focus on a single form of the test, the single Pythagorean Theorem item can be viewed as having both fixed and random features. Any particular analysis of the single-form data, however, will necessarily treat the item as either fixed or random.

For example, if such an item is included in a test form that is analyzed using the $p \times I$ design, then the item will be treated as random. By contrast, in IRT, items are effectively fixed, or to be more specific, the item parameters are fixed, and a replication would involve identically the same item parameters, resulting in strictly parallel forms. This difference in underlying assumptions has a number of consequences that are rarely recognized. For example, estimated SEMs under the two models have different statistical characteristics (see, e.g., Lee et al., 2000).

It is important to understand that the fixed and random features of our single Pythagorean item are totally confounded in examinee responses. To disconfound these features, at a minimum we need data for at least two Pythagorean items taken by the same examinees. Then we can determine how much variability in examinee responses to all of the items in a test form is attributable to the Pythagorean Theorem fixed category, and how much variability is attributable to different items that test the Pythagorean Theorem. Knowing this, we may then estimate D study results when a form includes a single-item Pythagorean Theorem testlet.

As a simple example in a somewhat different context, consider a writing assessment in which each form contains one narrative prompt and one informative prompt, and suppose we are willing to assume that variance attributable to raters is negligible. Presumably, the universe of generalization has $I\!:\!H$, with H (genre) fixed and I (items-within-genre) random. To disconfound the fixed/random aspect of the genre/prompt distinction, suppose we conduct a G study using the $p \times (i\!:\!h)$ design in which there are two prompts for each genre. Then, using the balanced-design mixed model D study results in Chapter 4, the genre effects are distinguishable from the prompt effects, and an estimated generalizability coefficient for a test form with one prompt for each genre is

$$\boldsymbol{E}\rho^2 = \frac{\sigma^2(p) + \sigma^2(ph)/2}{\sigma^2(p) + \sigma^2(ph)/2 + \sigma^2(pi\!:\!h)/2}.$$

Of course, in many realistic contexts that involve single-item testlets, the designs are unbalanced. For example, a 20-item test form based on specifications for five fixed content categories might consist of 5 items from one category, 6 from another, 7 from a third, and 2 single-item testlets. For such complex cases, procedures in multivariate generalizability theory could be employed.

8.5 Exercises

8.1* Form 1 of the ACT Assessment *Science Reasoning* Test in Table 8.1 was administered to 3458 examinees. The resulting analogous T terms were:

$$T(p) = 61977.95000 \qquad T(h) = 60279.46591$$
$$T(i{:}h) = 62005.88433 \qquad T(ph) = 69160.59524$$
$$T(pi{:}h) = 89416.00000 \qquad T(\mu) = 57802.35003.$$

Verify the results reported in Table 8.1 for Form 1.

8.2* Using only the "mean" results in Table 8.1 for the ACT *Science Reasoning* Test, what is an estimated value of $E\rho^2$ for the $p \times I$ design? Why is this a questionable estimate of reliability for this test?

8.3* For the ACT *Science Reasoning* Test discussed in Section 8.1, assume that each passage takes two minutes to read, and each item requires an additional half-minute to answer. Under these assumptions, what sample sizes (n'_h and n'_{i+}) and sample size patterns (the $n'_{i:h}$) would give $E\hat{\rho}^2 \geq .85$ with testing time no longer than 50 minutes? Use the "mean" results in Table 8.1 to answer this question. Of these alternatives, which is likely the most acceptable from a practical perspective? Why?

8.4 For the ITED *Vocabulary* example in Section 8.2, assuming G and D study sample sizes are the same, what is the signal–noise ratio for district means, using absolute error variance as noise? What is this signal–noise ratio for pupils in a randomly selected district? What is this signal–noise ratio for pupils across all districts?

8.5 For the medical clerkship example in Section 8.3, suppose that the grade for a student is the rounded average of three ratings based on using the current 19-item CEF. Under the assumption that errors are normally distributed, what is the probability that students with a universe score of 4 will receive a grade of 3 or lower?

8.6 For the circumstances and assumptions in Exercise 8.5, what is the probability that two examinees with the same universe score will be assigned different grades? For two examinees, p and q, assume this question is asking for $\Pr\left(|\overline{X}_p - \overline{X}_q| > .5 \mid \mu_p = \mu_q\right)$, where $\overline{X}_p$ and $\overline{X}_q$ are the unrounded average ratings.

8.7* In discussing the medical clerkship example in Section 8.3, it was noted that, in the actual data set, raters were neither fully nested nor fully crossed. The analysis, however, was conducted as if the ratings were fully nested. Discuss the G study and D study consequences of this decision, under the assumption that there is no systematic assignment of raters to students. In particular, consider what $\hat{\sigma}(\Delta)$ and $E\rho^2$ might be if the D study design had raters crossed with students.

8.8* Verify the numerical results reported in Section 8.4 for testlets.

9
Multivariate G Studies

It took Cronbach and his colleagues nearly 10 years to write their 1972 monograph. This rather long time period is directly related to their efforts in developing multivariate generalizability theory, which they regard as the principal novel contribution of their work. In multivariate generalizability theory, each object of measurement has multiple universe scores, each of which is associated with a condition of one or more fixed facets, and there is a random-effects variance components design associated with each fixed condition. These random-effects designs are statistically "linked" through covariance components to yield a multivariate design.

Obviously, a univariate random model is a special case of a multivariate model. Also, a univariate mixed model is merely a simplification of a multivariate model. It follows that generalizability theory is essentially a random effects theory, and univariate generalizability theory is best viewed as a special case of multivariate generalizability theory.

Cronbach et al. (1972, Chaps. 9 and 10) provide an extensive treatment of multivariate generalizability theory. Introductions are provided by Brennan (1992a), Shavelson and Webb (1981), Shavelson et al. (1989), and Webb et al. (1983). The treatment of multivariate generalizability theory in this book is heavily influenced by Cronbach et al., but there are notable differences in emphasis, notation, and scope.

Multivariate generalizability theory is extraordinarily powerful and flexible. However, this power and flexibility are purchased at the price of challenging conceptual and statistical issues. Many of the conceptual challenges are associated largely with characterizing multivariate designs and universes. Most of the statistical challenges focus on the estimation of covariance components. For these reasons, this chapter is dominated by a detailed discussion of multivariate G study designs and procedures for estimating G study covariance components.

This chapter's treatment of G study design structures includes both balanced and unbalanced situations, but procedures for estimating covariance components are restricted to balanced designs. Estimation procedures for unbalanced designs are considered in detail in Chapter 11.

We begin with a brief overview of multivariate generalizability theory primarily using the so-called "table of specifications" model as an example. In this example, we consider G study issues and a subset of the D study issues that might be addressed. The remainder of the chapter is devoted to G study issues only.

9.1   Introduction

In the mid 1960s Cronbach, Gleser, and Rajaratnam published a series of three papers in which they outlined the basics of generalizability theory. Each of them was the first author for one paper. Two papers (Cronbach et al., 1963; and Gleser et al., 1965) focus on univariate generalizability, while the third (Rajaratnam et al., 1965) provides a simple but elegant snapshot of their early thinking about multivariate generalizability theory. In a sense, this third paper is a generalizability theory perspective on what is now often called stratified alpha, which was also considered from a classical perspective by Cronbach et al. (1965) at about the same time. In the early 1980s, Jarjoura and Brennan (1982, 1983) extended the basic model in the Rajaratnam et al. (1965) paper. They referred to it as the "table of specifications" model.

In this model, there is a different set of items nested within each of the levels of a fixed facet, such as the content categories in a table of specifications. To say the content categories are fixed is to say that every form of the test would involve the same categories of items. The items themselves are random in the sense that each form of the test would involve a different set of items for each category. For different categories of items, v and v', the model equations would be

$$X_{piv} = \mu_v + \nu_p + \nu_i + \nu_{pi} \tag{9.1}$$

and

$$X_{piv'} = \mu_{v'} + \xi_p + \xi_i + \xi_{pi}, \tag{9.2}$$

where ν and ξ designate effects for v and v', respectively.[1] That is, there is a random effects $p \times i$ design associated with each level of the fixed facet. The fixed levels are linked in the sense that the same persons respond to all items in both levels. Note that the items in v and v' are not the same; nor is it required that the G study number of items be the same for v and v'. We designate this multivariate design as $p^{\bullet} \times i^{\circ}$, with the number of levels of the fixed facet being n_v.

The parameters under consideration are variance and covariance components for the population and universe of admissible observations, which can be grouped into the following three symmetric matrices.

$$\Sigma_p = \begin{bmatrix} \sigma_v^2(p) & \sigma_{vv'}(p) \\ \sigma_{vv'}(p) & \sigma_{v'}^2(p) \end{bmatrix} \qquad (9.3)$$

$$\Sigma_i = \begin{bmatrix} \sigma_v^2(i) & \\ & \sigma_{v'}^2(i) \end{bmatrix} \qquad (9.4)$$

$$\Sigma_{pi} = \begin{bmatrix} \sigma_v^2(pi) & \\ & \sigma_{v'}^2(pi) \end{bmatrix} \qquad (9.5)$$

The diagonal elements are the variance components for v and v'. The covariance between universe scores in the off-diagonal positions of Σ_p is

$$\sigma_{vv'}(p) \equiv E(\mu_{pv} - \mu_v)(\mu_{pv'} - \mu_{v'}) = E\nu_p \xi_p. \qquad (9.6)$$

The Σ_i and Σ_{pi} matrices are diagonal because each level of v has a different set of items associated with it. The fact that these matrices are represented using only two rows and two columns does not mean that there are necessarily only two levels of the fixed facet. This compact form simply indicates the notation used to represent the elements of the $n_v \times n_v$ matrices.

As an example, consider the synthetic data in Table 9.1 where $n_v = 3$. This is the same data set used by Rajaratnam et al. (1965) to illustrate what they called the "generalizability of stratified-parallel tests." Also, these data were used in Section 8.4 to illustrate considerations about testlets. The bottom part of Table 9.1 provides the mean squares and estimated variance components for each of the three levels of v.[2] For any pair of levels of v, the observed covariance is

$$S_{vv'}(p) = \frac{n_p}{n_p - 1} \left(\frac{\sum_p \overline{X}_{pv} \overline{X}_{pv'}}{n_p} - \overline{X}_v \overline{X}_{v'} \right), \qquad (9.7)$$

---

[1] Readers should be careful not to confuse the lowercase Roman letter v with the lowercase Greek letter ν. Note, also, that in the Cronbach et al. (1972) notational system, v and v' would be used as prescripts.

[2] Sometimes v refers to the entire set of fixed conditions, and sometimes it refers to a specific condition. This notational liberty simplifies notational conventions. The context makes it clear how v is to be interpreted.

TABLE 9.1. Synthetic Data Example for $p^\bullet \times i^\circ$ Design

p	v_1 i_1 i_2	v_2 i_1 i_2 i_3 i_4	v_3 i_1 i_2	$\overline{X}_{p1}$ $\overline{X}_{p2}$ $\overline{X}_{p3}$	$\overline{X}_{p1}\overline{X}_{p2}$	$\overline{X}_{p1}\overline{X}_{p3}$	$\overline{X}_{p2}\overline{X}_{p3}$
1	4 5	3 3 5 4	5 7	4.50 3.75 6.00	16.8750	27.0000	22.5000
2	2 1	2 3 1 4	4 6	1.50 2.50 5.00	3.7500	7.5000	12.5000
3	2 4	4 7 6 5	8 7	3.00 5.50 7.50	16.5000	22.5000	41.2500
4	1 3	5 4 5 5	4 5	2.00 4.75 4.50	9.5000	9.0000	21.3750
5	3 3	6 7 5 7	8 9	3.00 6.25 8.50	18.7500	25.5000	53.1250
6	1 2	5 6 4 4	5 6	1.50 4.75 5.50	7.1250	8.2500	26.1250
7	3 5	6 8 6 7	7 8	4.00 6.75 7.50	27.0000	30.0000	50.6250
8	0 1	1 2 0 4	7 8	0.50 1.75 7.50	.8750	3.7500	13.1250
Mn	2 3	4 5 4 5	6 7	2.50 4.50 6.50	12.5469	16.6875	30.0781

	v_1	v_2	v_3	
$MS(p)$	3.7143	12.2143	4.1423	
$MS(i)$	4.0000	2.6667	4.0000	$S_{12}(p) = 1.4821$
$MS(pi)$	.5714	1.0714	.4286	$S_{13}(p) = .5000$
				$S_{23}(p) = .9464$
$\hat{\sigma}^2(p)$	1.5714	2.7857	1.8571	
$\hat{\sigma}^2(i)$	.4286	.1994	.4464	
$\hat{\sigma}^2(pi)$	.5714	1.0714	.4286	

which is an unbiased estimator of the covariance between universe scores. That is, for the $p^\bullet \times i^\circ$ design,

$$\hat{\sigma}_{vv'}(p) = S_{vv'}(p). \tag{9.8}$$

These estimated variance and covariance components can be displayed in the following three matrices.

$$\widehat{\Sigma}_p = \begin{bmatrix} 1.5714 & 1.4821 & .5000 \\ 1.4821 & 2.7857 & .9464 \\ .5000 & .9464 & 1.8571 \end{bmatrix}$$

$$\widehat{\Sigma}_i = \begin{bmatrix} .4286 & & \\ & .1994 & \\ & & .4464 \end{bmatrix}$$

$$\widehat{\Sigma}_{pi} = \begin{bmatrix} .5714 & & \\ & 1.0714 & \\ & & .4286 \end{bmatrix}.$$

Suppose the universe of generalization consists of randomly parallel tests with two, four, and two items for v_1, v_2, and v_3, respectively. This means that the G study and D study sample sizes are the same, and the estimated

D study variance-and-covariance matrices are:

$$\widehat{\Sigma}_p = \begin{bmatrix} 1.5714 & 1.4821 & .5000 \\ 1.4821 & 2.7857 & .9464 \\ .5000 & .9464 & 1.8571 \end{bmatrix}$$

$$\widehat{\Sigma}_I = \begin{bmatrix} .2143 & & \\ & .0499 & \\ & & .2232 \end{bmatrix}$$

$$\widehat{\Sigma}_{pI} = \begin{bmatrix} .2857 & & \\ & .2679 & \\ & & .2143 \end{bmatrix}.$$

The vth diagonal element of $\widehat{\Sigma}_I$ is obtained by dividing the vth diagonal element in $\widehat{\Sigma}_i$ by n_{iv}.[3] The $\widehat{\Sigma}_{pI}$ matrix is obtained similarly.

Using the D study matrices, it is easy to obtain the universe score, relative error, and absolute error matrices:

$$\Sigma_\tau = \Sigma_p, \qquad \Sigma_\delta = \Sigma_{pI}, \qquad \text{and} \qquad \Sigma_\Delta = \Sigma_I + \Sigma_{pI}.$$

For the synthetic data,

$$\widehat{\Sigma}_\tau = \begin{bmatrix} 1.5714 & 1.4821 & .5000 \\ 1.4821 & 2.7857 & .9464 \\ .5000 & .9464 & 1.8571 \end{bmatrix}$$

$$\widehat{\Sigma}_\delta = \begin{bmatrix} .2857 & & \\ & .2679 & \\ & & .2143 \end{bmatrix}$$

$$\widehat{\Sigma}_\Delta = \begin{bmatrix} .5000 & & \\ & .3177 & \\ & & .4375 \end{bmatrix}.$$

Occasionally, the above matrices are the primary (or even the sole) statistics of interest. This might occur, for example, when profiles of universe scores are of principal concern. More frequently, however, some composite of universe scores is of interest. Let us suppose that the composite of interest is a weighted average of the universe scores for the levels of v, with the weights proportional to the numbers of items in each level. That is, suppose the composite of interest is $\mu_{pC} = \sum_v w_v \mu_{pv}$, where $w_v = n_{iv}/n_{i+}$, with n_{i+} designating the total number of items over all levels of v. Under these circumstances, composite universe score variance is the following weighted

---

[3]The notation $n_{i:v}$ could be used rather than n_{iv}, but for multivariate designs we usually do not use a colon to designate the number of levels of a random facet nested within a level of a fixed facet.

sum of all of the elements in Σ_p:

$$\sigma_C^2(p) = \sum_v w_v^2 \sigma_v^2(p) + \sum_v \sum_{v \neq v'} w_v w_{v'} \sigma_{vv'}(p). \qquad (9.9)$$

For the synthetic data, the w_v weights are .25, .50, and .25, and

$$
\begin{aligned}
\hat{\sigma}_C^2(p) &= (.25)^2(1.5714) + (.50)^2(2.7857) + (.25)^2(1.8571) \\
&\quad + 2(.25)(.50)(1.4821) + 2(.25)(.25)(.5000) + 2(.50)(.25)(.9464) \\
&= 1.5804.
\end{aligned}
$$

Assuming the estimator of composite universe score is $\overline{X}_{pC} = \sum_v w_v \overline{X}_{pv}$, relative error variance for the composite is the following weighted sum of the diagonal elements of Σ_δ,

$$\sigma_C^2(\delta) = \sum_v w_v^2 \, \sigma_v^2(\delta) = \sum_v \frac{w_v^2}{n_{iv}} \sigma_v^2(pi); \qquad (9.10)$$

and absolute error variance for the composite is the following weighted sum of the diagonal elements of Σ_Δ,

$$\sigma_C^2(\Delta) = \sum_v w_v^2 \, \sigma_v^2(\Delta) = \sum_v \frac{w_v^2}{n_{iv}} \left[\sigma_v^2(i) + \sigma_v^2(pi) \right]. \qquad (9.11)$$

For the synthetic data in Table 9.1,

$$\hat{\sigma}_C^2(\delta) = (.25)^2(.2857) + (.50)^2(.2679) + (.25)^2(.2143) = .0982$$

and

$$\hat{\sigma}_C^2(\Delta) = (.25)^2(.5000) + (.50)^2(.3177) + (.25)^2(.4375) = .1380.$$

A multivariate generalizability coefficient can be defined as the ratio of composite universe score variance to itself plus composite relative error variance. Similarly, a multivariate phi coefficient can be defined as the ratio of composite universe score variance to itself plus composite absolute error variance. For the synthetic data example considered here, the estimates of these coefficients are $E\hat{\rho}^2 = .941$ and $\hat{\Phi} = .920$. With the w weights being proportional to the sample sizes, $E\hat{\rho}^2$ is stratified α.

The example that has been considered in this section involves unequal numbers of conditions nested within each of the levels of a fixed facet. From a univariate perspective, this would require a mixed-model analysis of an unbalanced $p \times (i{:}v)$ design. Using such an analysis to estimate variance components would be quite complicated, whereas the multivariate analysis is rather straightforward.

When there are an equal number of conditions within each level of the fixed facet, the univariate analysis is rather simple (see, e.g., Sections 4.4.2

and 5.1.3), and the ANOVA variance component estimates for the univariate analysis $\sigma^2(\alpha|V)$ are easily expressed in terms of ANOVA variance and covariance component estimates for a multivariate analysis. In particular, $\hat{\sigma}^2(p|V)$ is the average of the elements in $\widehat{\boldsymbol{\Sigma}}_p$; that is,

$$\hat{\sigma}^2(p|V) = \sum_v \sum_{v'} \frac{\hat{\sigma}_{vv'}(p)}{n_v^2}, \tag{9.12}$$

where $\hat{\sigma}_{vv'}(p) = \hat{\sigma}_v^2(p)$ when $v = v'$. Also, $\hat{\sigma}^2(pv|V)$ is the difference between the average of the variance components in $\widehat{\boldsymbol{\Sigma}}_p$ and the average of the covariance components; that is,

$$\hat{\sigma}^2(pv|V) = \sum_v \frac{\hat{\sigma}_v^2(p)}{n_v} - \sum_{v \neq v'} \sum \frac{\hat{\sigma}_{vv'}(p)}{n_v(n_v - 1)}. \tag{9.13}$$

Finally, $\hat{\sigma}^2(i{:}v|V)$ and $\hat{\sigma}^2(pi{:}v|V)$ are the averages of the variance components in $\widehat{\boldsymbol{\Sigma}}_i$ and $\widehat{\boldsymbol{\Sigma}}_{pi}$, respectively:

$$\hat{\sigma}^2(i{:}v|V) \;\;=\;\; \sum_v \frac{\hat{\sigma}_v^2(i)}{n_v} \tag{9.14}$$

$$\hat{\sigma}^2(pi{:}v|V) \;\;=\;\; \sum_v \frac{\hat{\sigma}_v^2(pi)}{n_v}. \tag{9.15}$$

These equalities illustrate that variance components from a univariate balanced design can be viewed as simple averages over the conditions of a fixed facet. This simplicity disappears for unbalanced designs, however. More importantly, the univariate analysis does not permit differentiated consideration of each of the levels of the fixed facet.

The relationships in Equations 9.12 to 9.15 are reminiscent of Scheffé's (1959) treatment of mixed models. Indeed, Scheffé's perspective on mixed models is not unlike that of a multivariate G study.

9.2   G Study Designs

Table 9.2 provides a listing of single-facet and some two-facet multivariate designs. A superscript filled circle • designates that the facet is crossed with the fixed multivariate variables. Such a facet is sometimes referred to as being *linked*. A superscript empty circle ∘ designates that the facet is nested within the fixed multivariate variables. When a multivariate design is represented in the manner indicated in the first column of Table 9.2, there is a variance-covariance matrix associated with each letter and with each combination of letters. For example, for the $p^{\bullet} \times i^{\bullet}$ design the matrices are $\boldsymbol{\Sigma}_p$, $\boldsymbol{\Sigma}_i$, and $\boldsymbol{\Sigma}_{pi}$.

TABLE 9.2. Some Multivariate Designs and Their Univariate Counterparts

Multivariate Design	Variance Components Design	Covariance Components Design	Univariate Counterpart
$p^\bullet \times i^\bullet$	$p \times i$	$p \times i$	$p \times i \times v$
$p^\bullet \times i^\circ$	$p \times i$	p	$p \times (i{:}v)$
$i^\bullet {:} p^\bullet$	$i{:}p$	$i{:}p$	$(i{:}p) \times v$
$i^\circ {:} p^\bullet$	$i{:}p$	p	$i{:}(p \times v)$
$p^\bullet \times i^\bullet \times h^\bullet$	$p \times i \times h$	$p \times i \times h$	$p \times i \times h \times v$
$p^\bullet \times i^\bullet \times h^\circ$	$p \times i \times h$	$p \times i$	$p \times i \times (h{:}v)$
$p^\bullet \times i^\circ \times h^\bullet$	$p \times i \times h$	$p \times h$	$p \times (i{:}v) \times h$
$p^\bullet \times i^\circ \times h^\circ$	$p \times i \times h$	p	$p \times [(i \times h){:}v]$
$p^\bullet \times (i^\bullet {:} h^\bullet)$	$p \times (i{:}h)$	$p \times (i{:}h)$	$p \times (i{:}h) \times v$
$p^\bullet \times (i^\circ {:} h^\bullet)$	$p \times (i{:}h)$	$p \times h$	$p \times [i{:}(h \times v)]$
$p^\bullet \times (i^\circ {:} h^\circ)$	$p \times (i{:}h)$	p	$p \times (i{:}h{:}v)$
$(p^\bullet {:} c^\bullet) \times i^\bullet$	$(p{:}c) \times i$	$(p{:}c) \times i$	$(p{:}c) \times i \times v$
$(p^\bullet {:} c^\bullet) \times i^\circ$	$(p{:}c) \times i$	$p{:}c$	$(p{:}c) \times (i{:}v)$
$(p^\circ {:} c^\bullet) \times i^\circ$	$(p{:}c) \times i$	c	$(p \times i){:}(c \times v)$

Note. The univariate counterparts of the multivariate designs use v to designate levels of a fixed facet.

For each of the multivariate designs in Table 9.2 the *variance components design* is the univariate design associated with each one of the fixed variables. The *covariance components design* is the univariate design associated with pairs of fixed variables. The *univariate counterpart* provides a univariate perspective on the multivariate design—a perspective that is closely associated with a Venn diagram representation of the multivariate design.[4]

It is somewhat inaccurate to say that a multivariate design involves a variance components design and a covariance components design. Strictly speaking, letting n_v be the number of levels of the fixed facet, a multivariate design involves n_v variance components designs and $n_v(n_v - 1)/2$ covariance components designs. We refer to the multiple variance and covariance designs only if doing so is required by the particular context.

---

[4]These notational and terminological conventions are somewhat different, at least in emphasis, from those used by Cronbach et al. (1972). For example, they routinely refer to *linked conditions* or *joint sampling* when we use •, and they routinely refer to *independent conditions* or *independent sampling* when we use ○. Such terminology is used much less frequently in this book.

The designs in Table 9.2 are illustrated in the subsections that follow. Some of these illustrations involve modeling specifications that characterize many large-scale educational testing programs. These particular illustrations are extensions of the simple table of specifications model discussed in Section 9.1. Several other illustrations are variations on a design discussed by Brennan et al. (1995). Of course, the designs themselves are not restricted to these particular contexts. Although the designs in Table 9.2 are only a subset of the possible multivariate designs, they are a rich subset. Only rarely is multivariate generalizability theory employed with a more complicated design.

Unless otherwise noted, p stands for persons, c stands for an aggregation of persons (e.g., classes), i and h are random facets in the universe of admissible observations, and v (or occasionally v') is used to designate fixed conditions of measurement. Sometimes we refer to a fixed condition as a *level of a fixed multivariate variable*, or simply a *variable* or *dimension*. Note that the set of fixed conditions in a single multivariate analysis may represent more than one facet, as the term *facet* is typically used in a univariate analysis. For example, suppose a and b are fixed facets in the usual sense. Then, assuming the facets are crossed, there are $n_a \times n_b$ fixed conditions of measurement. A multivariate analysis would simply involve $n_a \times n_b$ variables or fixed levels, and, in the notational conventions of this book, each of these fixed levels would be indexed with the letter v. In this sense, multiple fixed facets are easily accommodated in a multivariate analysis.

Formulating and understanding multivariate designs involves many challenging conceptual issues. To circumvent at least one complexity, we assume in this section that the structure of the universe of admissible observations[5] and the G study are the same. Later in Section 9.5 we briefly consider a broader perspective on these issues.

9.2.1 Single-Facet Designs

For multivariate designs, the phrase *single-facet designs* means, more specifically, designs with a single random facet. There are four single-facet multivariate designs. For two of them, the conditions of the single facet are crossed with persons. For the other two, conditions are nested within persons.

$p^\bullet \times i^\circ$ Design

The simple table of specifications model considered in Section 9.1 is an example of the $p^\bullet \times i^\circ$ design. In this design, a different set of items is

---

[5]Cronbach et al. (1972) call this the universe of admissible vectors when the design is multivariate.

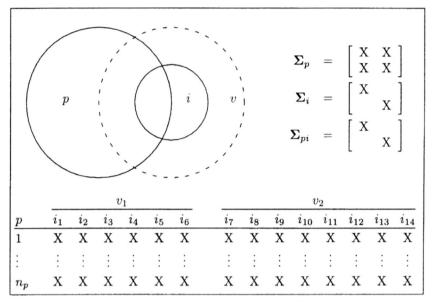

FIGURE 9.1. Representations of $p^\bullet \times i^\circ$ design.

associated with each of the fixed content categories in a table of specifications. Consequently, within each category there is a $p \times i$ design, and there are covariance components associated with p, only. From a univariate perspective, the design is $p \times (i{:}v)$. The multivariate $p^\bullet \times i^\circ$ design is more useful and powerful than the $p \times (i{:}v)$ design for two reasons: often there are unequal numbers of items within content categories, which creates complexities and ambiguities for a univariate analysis; and a univariate analysis does not provide separate estimates of variance and covariance components for each of the levels of the fixed facet.

Figure 9.1 provides three representations of the $p^\bullet \times i^\circ$ design. The upper left-hand corner is a Venn diagram that is identical to a $p \times (i{:}v)$ Venn diagram, except that the diagram in Figure 9.1 has a dashed circle for the fixed facet v, and interactions are not explicitly identified.

The solid p and i circles, and their pi interaction, are visually associated with the three variance-covariance matrices in the upper right-hand corner of the figure. That is, there is a variance-covariance matrix for each of the Venn-diagram areas bounded entirely by one or more solid lines. Here, for purposes of simplicity, these matrices employ only two levels of v.

The Σ_i and Σ_{pi} matrices have empty cells in their off-diagonal positions, indicating that the associated covariance components are zero. These zero covariances are associated with the fact that the i and pi areas of the Venn diagram do not intersect the v circle. By contrast, the p circle does intersect the v circle, signifying potentially non zero covariance components for Σ_p.

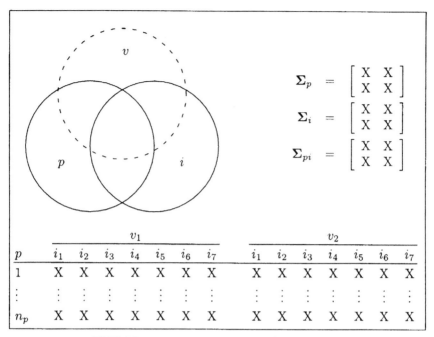

FIGURE 9.2. Representations of $p^{\bullet} \times i^{\bullet}$ design.

The bottom of Figure 9.1 provides a typical record layout for data for the $p^{\bullet} \times i^{\circ}$ design. To highlight the fact that different items are associated with each of the levels of v, the item numbers for v_1 are different from those for v_2. Again, the choice of two levels of v with six and eight items is for illustrative purposes, only.

$p^{\bullet} \times i^{\bullet}$ Design

The principal difference between the $p^{\bullet} \times i^{\circ}$ and $p^{\bullet} \times i^{\bullet}$ designs is that for the $p^{\bullet} \times i^{\bullet}$ design, each level of i is associated with each and every level of v. Consider the following examples: (i) examinees are administered the same set of items on two occasions, with the occasions considered fixed (perhaps a pretest and a posttest experiment); (ii) participants in a job analysis evaluate each of a set of tasks with respect to frequency and criticality; and (iii) examinee responses to each of several writing exercises are scored with respect to the same fixed dimensions (e.g., ideas, organization, voice, and conventions).[6]

---

[6]The third example begs many questions about how the ratings are obtained. In particular, does the same rater rate all examinees on both dimensions, or are multiple raters involved in some way? In both theory and practice, these are important issues, but they are disregarded for purposes of simplicity here.

In all of these examples, the random facet i is crossed with the fixed facet v, and we assume here that p is crossed with both facets. Under these circumstances, the multivariate design is $p^\bullet \times i^\bullet$, its univariate counterpart is $p \times i \times v$, there is a variance components $p \times i$ design for each level of v, and there is a covariance components $p \times i$ design for each pair of levels of v.

Figure 9.2 provides three representations on the $p^\bullet \times i^\bullet$ design. As in Figure 9.1, the fixed multivariate variable v is represented by a dashed circle in the Venn diagram. Because both p and i are crossed with v, the three matrices are full. This is represented in the Venn diagram by the v circle intersecting the p and i circles (and their interaction). The record layout representation at the bottom of Figure 9.2 indicates that the same number of items (seven) is associated with each level of v. To highlight that these seven items are the same for v_1 and v_2, the same item numbers are used. (Note the difference between this record layout and the one in Figure 9.1 for the $p^\bullet \times i^\circ$ design.) The use of seven items and two levels of v is purely illustrative.

$i^\bullet : p^\bullet$ and $i^\circ : p^\bullet$ Designs

For both the $i^\bullet : p^\bullet$ and $i^\circ : p^\bullet$ designs, each person takes different items and possibly different numbers of items. The two designs differ in terms of whether each item is associated with all levels of v.

For the $i^\bullet : p^\bullet$ design, each item is associated with the same fixed dimensions. That is, the random i facet is crossed with the fixed v facet. This means that $\mathbf{\Sigma}_p$ and $\mathbf{\Sigma}_{i:p}$ are both full matrices, and the univariate $(i : p) \times v$ design is the counterpart of the multivariate $i^\bullet : p^\bullet$ design.

By contrast, for the $i^\circ : p^\bullet$ design, each item is associated with only one level of v. That is, the random i facet is nested within the fixed v facet, $\mathbf{\Sigma}_{i:p}$ is a diagonal matrix, and the univariate $i : (p \times v)$ design is the counterpart of the multivariate $i^\circ : p^\bullet$ design. This design is analogous to the simple table of specifications model in Section 9.1, the only difference being that each person takes a different set of items for the $i^\circ : p^\bullet$ design, whereas each person takes the same set of items for the $p^\bullet \times i^\circ$ design.

9.2.2   Two-Facet Crossed Designs

There are four two-facet crossed designs, each of which involves seven variance-covariance matrices. From a statistical perspective, these designs differ in terms of which matrices are full and which are diagonal.

$p^\bullet \times i^\bullet \times h^\bullet$ Design

For the $p^\bullet \times i^\bullet \times h^\bullet$ design, all matrices are full, as indicated in Figure 9.3. That is, for each level of v there is a $p \times i \times h$ variance components design, and for each pair of levels of v there is a $p \times i \times h$ covariance components

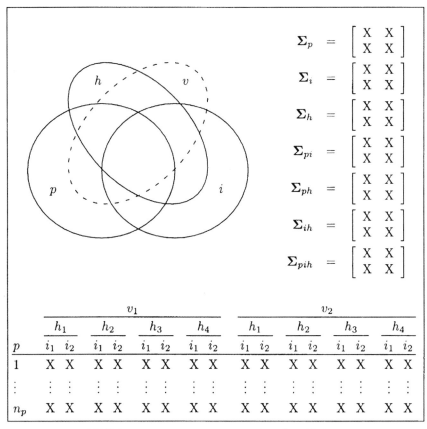

$$\Sigma_p = \begin{bmatrix} X & X \\ X & X \end{bmatrix}$$

$$\Sigma_i = \begin{bmatrix} X & X \\ X & X \end{bmatrix}$$

$$\Sigma_h = \begin{bmatrix} X & X \\ X & X \end{bmatrix}$$

$$\Sigma_{pi} = \begin{bmatrix} X & X \\ X & X \end{bmatrix}$$

$$\Sigma_{ph} = \begin{bmatrix} X & X \\ X & X \end{bmatrix}$$

$$\Sigma_{ih} = \begin{bmatrix} X & X \\ X & X \end{bmatrix}$$

$$\Sigma_{pih} = \begin{bmatrix} X & X \\ X & X \end{bmatrix}$$

	v_1								v_2							
	h_1		h_2		h_3		h_4		h_1		h_2		h_3		h_4	
p	i_1	i_2	i_1	i_2	i_1	i_2	i_1	i_2	i_1	i_2	i_1	i_2	i_1	i_2	i_1	i_2
1	X	X	X	X	X	X	X	X	X	X	X	X	X	X	X	X
$\vdots$	$\vdots$	$\vdots$	$\vdots$	$\vdots$	$\vdots$	$\vdots$	$\vdots$	$\vdots$	$\vdots$	$\vdots$	$\vdots$	$\vdots$	$\vdots$	$\vdots$	$\vdots$	$\vdots$
n_p	X	X	X	X	X	X	X	X	X	X	X	X	X	X	X	X

FIGURE 9.3. Representations of $p^{\bullet} \times i^{\bullet} \times h^{\bullet}$ design.

design. The univariate $p \times i \times h \times v$ design is the counterpart of the multi-variate $p^{\bullet} \times i^{\bullet} \times h^{\bullet}$ design, which is associated with the fact that the Venn diagram in Figure 9.3 has four intersecting ellipses.

For example, suppose examinees (p) are all administered the same 12 tasks (i), and the responses to each and every task are rated by the same three raters (h) on two fixed dimensions (v). This is a very powerful design in that it permits estimation of all variance components and all covariance components. However, this design is rarely employed because collecting such data is usually difficult, time-consuming, and/or expensive. Nuβbaum (1984) provides a real-data example in which 60 fourth-grade students (p) were asked to create watercolor paintings on four topics (i). Each painting was evaluated by 25 art students (h); that is, each art student evaluated all 240 paintings. Evaluations were made on three different 10-point scales (v):

1. Are persons and things represented in an objective way?

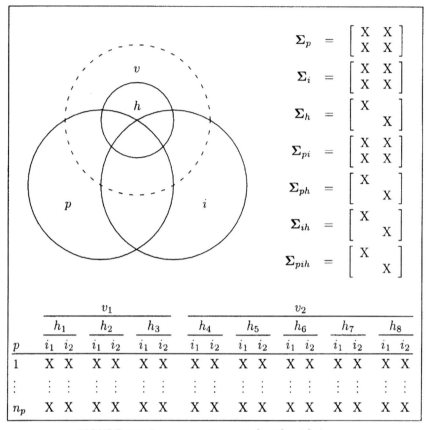

FIGURE 9.4. Representations of $p^\bullet \times i^\bullet \times h^\circ$ design.

2. Is the background appropriate?

3. Do relations between the objects become clear to the viewer?

For later reference, we refer to this as the *Painting* example.

$p^\bullet \times i^\bullet \times h^\circ$ Design

For the $p^\bullet \times i^\bullet \times h^\circ$ design different levels of h are associated with each of the fixed multivariate levels. Therefore, for each level of v there is a $p \times i \times h$ variance components design, and for each pair of levels of v there is a $p \times i$ covariance components design. Consequently, Σ_h, Σ_{ph}, Σ_{ih}, and Σ_{pih} are diagonal matrices; that is, each of the matrices involving h is diagonal. The univariate $p \times i \times (h{:}v)$ design is the counterpart of the multivariate $p^\bullet \times i^\bullet \times h^\circ$ design, and the univariate counterpart is clearly associated with the the Venn diagram in Figure 9.4.

Brennan et al. (1995) provide a real-data example of this design in which 50 examinees (p) each listened to 12 tape-recorded messages (i). Examinees were told to take notes while each message was played. At the conclusion of each message, examinees were told to use their notes to construct a written message. The written messages were scored by trained raters on a five-point holistic scale for both listening skills and writing skills. The listening score reflected the accuracy and completeness of the information provided in the written messages. The writing score reflected other aspects of the "quality" of the writing. There were two distinct sets of raters (h). Three raters evaluated the written messages for listening, and a different three raters evaluated the messages for writing. Subsequently, we refer to this as the *LW* (i.e., *Listening* and *Writing*) example.

$p^\bullet \times i^\circ \times h^\bullet$ and $p^\bullet \times i^\circ \times h^\circ$ Designs

In terms of the *LW* example, if *different* tasks were used for listening and writing, but the *same* raters were used for both dimensions, then the design would be $p^\bullet \times i^\circ \times h^\bullet$. This design has the same structure as the $p^\bullet \times i^\bullet \times h^\circ$ design. The only difference between the two is that the roles of i and h are interchanged.

Returning again to the *LW* example, suppose that different tasks *and* different raters were used for both listening and writing. Then the design would be $p^\bullet \times i^\circ \times h^\circ$, and all matrices except Σ_p would be diagonal.

9.2.3   Two-Facet Nested Designs

For discussion purposes, it is convenient to group multivariate two-facet nested designs into those that involve nesting within the universe of admissible observations $[p^\bullet \times (i^\bullet : h^\bullet), p^\bullet \times (i^\circ : h^\bullet),$ and $p^\bullet \times (i^\circ : h^\circ)]$, and those that involve nesting within the population of objects of measurement, which are designated here as c for these designs $[(p^\bullet : c^\bullet) \times i^\bullet, (p^\bullet : c^\bullet) \times i^\circ,$ and $(p^\circ : c^\bullet) \times i^\circ]$.[7]

$p^\bullet \times (i^\bullet : h^\bullet)$ Design

For the $p^\bullet \times (i^\bullet : h^\bullet)$ design, the variance components design is $p \times (i : h)$, the covariance components design is also $p \times (i : h)$, all five matrices are full, and the univariate counterpart is $p \times (i : h) \times v$. Returning to the *LW* example, suppose each of the three raters evaluated only four tasks, but each rater provided both listening and writing scores. In this case, tasks (i) would be nested within raters (h), both tasks and raters would be crossed with the fixed facet v, and the multivariate design would be $p^\bullet \times (i^\bullet : h^\bullet)$.

---

[7]Strictly speaking, of course, the objects of measurement are not identified until the D study and universe of generalization are defined, but for discussion purposes here it is convenient to think of c as the objects of measurement.

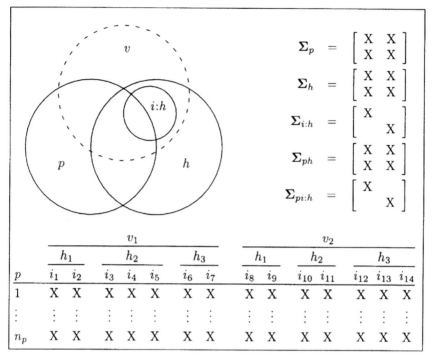

FIGURE 9.5. Representations of $p^\bullet \times (i^\circ : h^\bullet)$ design.

$p^\bullet \times (i^\circ : h^\bullet)$ Design

For the $p^\bullet \times (i^\circ : h^\bullet)$ design, the variance components design is $p \times (i:h)$, the covariance components design is $p \times h$, and the univariate counterpart is $p \times [i:(h \times v)]$. Using hypothetical sample sizes, Figure 9.5 provides three representations of this design. Note in particular that i is nested within both h and v.

Often this design reasonably well reflects reading tests that consist of several passages (h) in which items (i) are nested within passages, and each item contributes to only one content/process category. For example, consider the *Ability to Interpret Literary Materials* test of the *Iowa Tests of Educational Development* (ITED) (Feldt et al., 1994). Form L, Level 17/18, contains five passages with 9, 8, 9, 8, and 10 items, respectively. In addition to being nested within a passage, each item also can be viewed as being nested within one of two fixed process categories.[8] The numbers of items in each of the passages associated with the first category are 4, 4, 7, 2, and 6,

---

[8] Actually, there are five process categories: constructing factual/literal meaning, constructing nonliteral meaning, constructing inferential/interpretative meaning, generalizing themes and ideas, and recognizing literary techniques and tone. Here, to simplify matters, the first three are combined and the last two are combined.

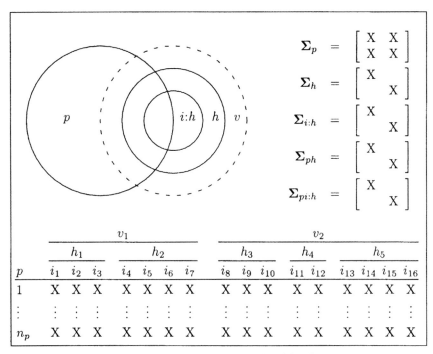

FIGURE 9.6. Representations of $p^\bullet \times (i^\circ : h^\circ)$ design.

respectively; the numbers of items in each of the passages associated with the second category are 5, 4, 2, 6, and 4, respectively. Since each passage contributes items to both categories, and each person responds to items in both categories, the covariance components design is $p \times h$; that is, both p and h are crossed with v, and p is crossed with h. We refer to this as the *LM* (...*Literary Materials* ...) example.

Note that, for the *LM* example, the design is unbalanced in two senses. First, for each process category, there are unequal numbers of items within passages (4, 4, 7, 2, 6 and 5, 4, 2, 6, 4), which means that there is an unbalanced *random* effects $p \times i \times h$ design associated with each process (i.e., each level of the fixed facet). This unbalanced aspect of the design affects the estimation of variance components, only, which can be accomplished using the procedures in Chapter 7. Second, although the same number of passages (5) is associated with each process, there are unequal numbers of items (23 and 21) nested within each of the processes, which affects the estimation of covariance components.

$p^\bullet \times (i^\circ : h^\circ)$ Design

For the $p^\bullet \times (i^\circ : h^\circ)$ design, the variance components design is $p \times (i : h)$, the only covariance components are those associated with p, and the univariate counterpart is $p \times (i : h : v)$. Using hypothetical sample sizes, Figure 9.6 provides three representations of this design. Note in particular that both i and h are nested within v.

Consider, for example, the *Maps and Diagrams* test of the *Iowa Tests of Basic Skills* (ITBS) (Hoover et al., 1993). As the name suggests, in this test there are two distinct types of stimulus materials: maps and diagrams. Specifically, for Form K, Level 10, there are two maps and two diagrams. For each of the two maps there are six and seven items, respectively; similarly, there are six and seven items, respectively, for each of the two diagrams. Subsequently, we refer to this as the *MD* (i.e., *Maps and Diagrams*) example.

Class-Means Designs: $(p^{\bullet}:c^{\bullet}) \times i^{\bullet}$, $(p^{\bullet}:c^{\bullet}) \times i^{\circ}$, and $(p^{\circ}:c^{\bullet}) \times i^{\circ}$

The $(p^{\bullet}:c^{\bullet}) \times i^{\bullet}$ and $(p^{\bullet}:c^{\bullet}) \times i^{\circ}$ designs can be viewed as "class-means" versions of the $p^{\bullet} \times i^{\bullet}$ and $p^{\bullet} \times i^{\circ}$ designs, respectively. For the $(p^{\bullet}:c^{\bullet}) \times i^{\bullet}$ design, as well as the $(p^{\bullet}:c^{\bullet}) \times i^{\circ}$ design, all persons (p) in all classes (c) take the same items (i). Furthermore, all persons and all classes contribute data to all levels of v, which means that Σ_c and $\Sigma_{p:c}$ are full matrices. The only difference between these two designs is that the same items are associated with each level of v for the $(p^{\bullet}:c^{\bullet}) \times i^{\bullet}$ design, whereas different items are associated with each level of v for the $(p^{\bullet}:c^{\bullet}) \times i^{\circ}$ design.

In most large-scale testing programs, the data collection procedure is one in which students are clearly nested within units (e.g., classes or schools), and all students take the same items. In this sense, the $(p^{\bullet}:c^{\bullet}) \times i^{\bullet}$ and $(p^{\bullet}:c^{\bullet}) \times i^{\circ}$ designs are often a more accurate reflection of the realities of data collection than are the $p^{\bullet} \times i^{\bullet}$ and $p^{\bullet} \times i^{\circ}$ designs. Usually, however, aggregations of persons are explicitly represented in designs only if D study interest will focus on the aggregated objects of measurement (e.g., classes).

For the $(p^{\circ}:c^{\bullet}) \times i^{\circ}$ design, different groups of persons within each class respond to different sets of items, and each group–set combination is associated with only one level of v. In describing the design this way, it is important to note that neither "group" nor "set" is a facet. Rather these words are used simply to characterize random samples of persons and items, respectively. As such, the only full matrix is Σ_c, and the design can be viewed as a matrix sampling version of the simple table of specifications model. Seldom, however, are different persons in the same class administered items in entirely different content categories; consequently, the $(p^{\circ}:c^{\bullet}) \times i^{\circ}$ design is rare.

9.3   Defining Covariance Components

A covariance component is the expected value of a product of two effects. Letting ν designate random effects for variable v, and ξ designate random effects for variable v' ($v \neq v'$), a covariance component for the effect α is defined as

$$\sigma_{vv'}(\alpha) \equiv \boldsymbol{E}\nu_\alpha\xi_\alpha. \tag{9.16}$$

Consider, for example, the model equations for two levels, v and v', of the $p^\bullet \times (i^\circ : h^\bullet)$ design:

$$X_{pihv} = \mu_v + \nu_p + \nu_h + \nu_{i:h} + \nu_{ph} + \nu_{pi:h} \tag{9.17}$$

and

$$X_{pihv'} = \mu_{v'} + \xi_p + \xi_h + \xi_{i:h} + \xi_{ph} + \xi_{pi:h}, \tag{9.18}$$

where the levels of i are different for v and v'. As discussed previously, the effects within a level of the fixed facet are uncorrelated,[9] which means that

$$\boldsymbol{E}\nu_\alpha\nu_\beta = \boldsymbol{E}\xi_\alpha\xi_\beta = 0, \tag{9.19}$$

for $\alpha \neq \beta$. Also,

$$\boldsymbol{E}\nu_\alpha\xi_\beta = 0, \tag{9.20}$$

for $\alpha \neq \beta$.

For the effects in Equations 9.17 and 9.18, there are three covariances that are not necessarily zero—one for p, one for h, and one for ph. These covariance components are denoted

$$\sigma_{vv'}(p) = \boldsymbol{E}\nu_p\xi_p,$$

$$\sigma_{vv'}(h) = \boldsymbol{E}\nu_h\xi_h,$$

and

$$\sigma_{vv'}(ph) = \boldsymbol{E}\nu_{ph}\xi_{ph}.$$

The covariance components for $i{:}h$ and $pi{:}h$ are both zero, because the levels of i are different for v and v' in the $p^\bullet \times (i^\circ : h^\bullet)$ design.[10]

The definitions of covariance components are unaffected by whether a G study design is balanced or unbalanced. However, estimating covariance components is much more complicated for unbalanced designs than for balanced ones. For that reason, estimation procedures for unbalanced designs are treated later in Chapter 11.

---

[9] Most effects are necessarily uncorrelated because of the manner in which score effects are defined in generalizability theory.

[10] Cronbach et al. (1972) would use different symbols to designate items for the two levels (e.g., i for v and j for v'), because the items are indeed different for the $p^\bullet \times (i^\circ : h^\bullet)$ design. We do not follow this convention in this book, because generalizing this convention to all designs that involve nesting leads to considerable notational complexity.

9.4 Estimating Covariance Components for Balanced Designs

It is important to understand the distinction between a balanced and an unbalanced covariance components design, as these terms are used in this book. Consider any two levels of a fixed facet. In the terminology of this book, to say that the covariance components design is balanced for these two levels means that, for *both* levels of the fixed facet, the variance components design is balanced. If the variance components design involves nesting, it is *not* required that an equal number of conditions be nested within each of the two levels of the fixed facet. So, for example, if the variance components design for each level of the fixed facet is $p \times (i : h)$ then n_i must be a constant for each level of h within a particular level of v. However, n_i need not be the same for *every* level of v. For this reason, we sometimes use n (with an appropriate subscript) to designate the G study sample size for one level of v, and m (with an appropriate subscript) to designate a possibly different sample size for a different level of v.

As an example of the distinction between balanced and unbalanced covariance components designs, consider the $p^{\bullet} \times (i^{\circ} : h^{\bullet})$ design, and suppose there are five levels of v, each of which has $n_h = 2$ levels of h (say, passages). Suppose further that the numbers of items associated with the two passages for each of the five levels of v are, respectively, 4/4, 4/4, 6/6, 3/5, and 2/6, where the slash simply separates sample sizes. Any pairing of the first three levels of v (i.e., 1 with 2, 1 with 3, or 2 with 3) constitutes a balanced covariance components design in the terminology of this book, because the two variance components designs are balanced. Any pairing involving either of the last two levels of v constitutes an *unbalanced* covariance components design. Clearly, then, for the multivariate $p^{\bullet} \times (i^{\circ} : h^{\bullet})$ design, estimation of some of the covariance components may involve a balanced design, while estimating others may involve an unbalanced design. As noted earlier, it is somewhat inaccurate to say that the multivariate $p^{\bullet} \times (i^{\circ} : h^{\bullet})$ design involves the $p \times h$ covariance components design. Strictly speaking, this multivariate design involves $n_v(n_v - 1)/2$ covariance components designs of the form $p \times h$.

For balanced multivariate designs with a single linked facet (e.g., the $p^{\bullet} \times i^{\circ}$ design), the observed covariance in Equation 9.7 is an unbiased estimate of the covariance component for the linked facet. For balanced designs with more than one linked facet, we employ a procedure that is very much like that used to estimate variance components for univariate balanced designs. For such designs, Sections 3.3 and 3.4 provide equations for obtaining sums of squares, mean squares, and expected-mean-square equations with respect to random effects variance components. Given the expected-mean-square equations, it is straightforward to obtain estimators of the variance components for univariate balanced designs. A corresponding procedure discussed by Cronbach et al. (1972) can be used to obtain estimators of covariance components when the covariance components design is balanced. For this procedure, sums of products (SP) replace sums of squares (SS), mean products (MP) replace mean squares (MS), expected mean products ($\boldsymbol{EMP}$) replace expected mean squares ($\boldsymbol{EMS}$), and covariance components are estimated using the $\boldsymbol{EMP}$ equations. This method for estimating covariance components is also discussed by Bock (1975, pp. 433ff.).

9.4.1 An Illustrative Derivation

Consider the $p^{\bullet} \times (i^{\circ} : h^{\bullet})$ multivariate design in which the covariance components design is $p \times h$. The covariance components for $\boldsymbol{\Sigma}_p$, $\boldsymbol{\Sigma}_h$, and $\boldsymbol{\Sigma}_{ph}$, respectively, are $\sigma_{vv'}(p)$, $\sigma_{vv'}(h)$, and $\sigma_{vv'}(ph)$. Let us focus on estimating the covariance component for persons for levels v and v' of the fixed facet. The sum of products for persons is

$$SP_{vv'}(p) = n_h \sum_p \left(\overline{X}_{pv} - \overline{X}_v \right) \left(\overline{X}_{pv'} - \overline{X}_{v'} \right) \tag{9.21}$$

$$= n_h \sum_p \overline{X}_{pv} \overline{X}_{pv'} - n_p n_h \overline{X}_v \overline{X}_{v'} \tag{9.22}$$

$$= TP_{vv'}(p) - TP_{vv'}(\mu), \tag{9.23}$$

where the TP terms have obvious similarities with the T terms discussed in Section 3.3. The mean product for persons is

$$MP_{vv'}(p) = \frac{SP_{vv'}(p)}{n_p - 1}. \tag{9.24}$$

From Equations 9.24 and 9.22 it is evident that the expected value of the mean product for persons is

$$\boldsymbol{EMP}_{vv'}(p) = \frac{1}{n_p - 1} \left[n_h \sum_p \boldsymbol{E} \left(\overline{X}_{pv} \overline{X}_{pv'} \right) - n_p n_h \boldsymbol{E} \left(\overline{X}_v \overline{X}_{v'} \right) \right], \tag{9.25}$$

which depends on the expected value of two cross-product terms.

To determine the first of these terms, $\boldsymbol{E} \left(\overline{X}_{pv} \overline{X}_{pv'} \right)$, note that $\overline{X}_{pv}$ is simply the average over all levels of i and h of the observed scores for person p on variable v. Similarly, $\overline{X}_{pv'}$ is the average over all levels of i and h of the observed scores for person p on variable v'. Formally,

$$\overline{X}_{pv} = \frac{1}{n_h n_i} \sum_{h=1}^{n_h} \sum_{i=1}^{n_i} X_{pihv} \quad \text{and} \quad \overline{X}_{pv'} = \frac{1}{n_h m_i} \sum_{h=1}^{n_h} \sum_{i=1}^{m_i} X_{pihv'}. \tag{9.26}$$

Using Equations 9.17 and 9.18 in these expressions, we obtain

$$\overline{X}_{pv} = \mu_v + \nu_p + \frac{\sum_h \nu_h}{n_h} + \frac{\sum_h \sum_i \nu_{i:h}}{n_h n_i} + \frac{\sum_h \nu_{ph}}{n_h} + \frac{\sum_h \sum_i \nu_{pi:h}}{n_h n_i} \tag{9.27}$$

and

$$\overline{X}_{pv'} = \mu_{v'} + \xi_p + \frac{\sum_h \xi_h}{n_h} + \frac{\sum_h \sum_i \xi_{i:h}}{n_h m_i} + \frac{\sum_h \xi_{ph}}{n_h} + \frac{\sum_h \sum_i \xi_{pi:h}}{n_h m_i}. \quad (9.28)$$

The expected value of the product of Equations 9.27 and 9.28, namely, $E\left(\overline{X}_{pv}\overline{X}_{pv'}\right)$, has $6 \times 6 = 36$ terms. Most of them are zero because, in general,

$$E\left(\mu_v \xi_\alpha\right) = E\left(\mu_{v'}\nu_\alpha\right) = 0, \quad (9.29)$$

$$E\left(\nu_\alpha \xi_\beta\right) = 0 \text{ when } \alpha \neq \beta, \text{ and} \quad (9.30)$$

$$E\left(\nu_\alpha \xi_\alpha\right) = 0 \left\{ \begin{array}{l} \text{when } \alpha \text{ is not an effect in the} \\ \text{covariance components design.} \end{array} \right. \quad (9.31)$$

Note that Equation 9.31 implies that $E\left(\nu_{i:h}\xi_{i:h}\right) = E\left(\nu_{pi:h}\xi_{pi:h}\right) = 0$. Since the possibly different values of n_i and m_i occur only in conjunction with the nested-effects terms in Equations 9.27 and 9.28, and since both $E\left(\nu_{i:h}\xi_{i:h}\right)$ and $E\left(\nu_{pi:h}\xi_{pi:h}\right)$ equal zero, any difference between the values of n_i and m_i has no effect on $E\left(\overline{X}_{pv}\overline{X}_{pv'}\right)$. This is a mathematical demonstration of the earlier statement that, if the multivariate design involves nesting, it is *not* required that an equal number of conditions be nested within each of the levels of the multivariate fixed facet.

It follows from Equations 9.29 to 9.31 that

$$E\left(\overline{X}_{pv}\overline{X}_{pv'}\right) = \mu_v\mu_{v'} + E\nu_p\xi_p + E\left(\frac{\sum_h \nu_h}{n_h}\right)\left(\frac{\sum_h \xi_h}{n_h}\right)$$
$$+ E\left(\frac{\sum_h \nu_{ph}}{n_h}\right)\left(\frac{\sum_h \xi_{ph}}{n_h}\right). \quad (9.32)$$

Each of the last two terms in Equation 9.32 expands into n_h^2 terms, most of which are zero. Specifically, $\left(\sum_h \nu_h\right)\left(\sum_h \xi_h\right)$ involves $n_h\left(n_h - 1\right)$ terms of the form $\nu_h\xi_{h'}$ $(h \neq h')$, each of which has an expected value of zero. Similarly, $\left(\sum_h \nu_{ph}\right)\left(\sum_h \xi_{ph}\right)$ involves $n_h\left(n_h - 1\right)$ terms of the form $\nu_{ph}\xi_{ph'}$ $(h \neq h')$, each of which has an expected value of zero. These are two instances of the fact that

$$E\left(\nu_\alpha \xi_{\alpha'}\right) = 0 \left\{ \begin{array}{l} \text{when } \alpha \text{ and } \alpha' \text{ involve differ-} \\ \text{ent levels of the same effect.} \end{array} \right. \quad (9.33)$$

It follows that Equation 9.32 simplifies to

$$E\left(\overline{X}_{pv}\overline{X}_{pv'}\right) = \mu_v\mu_{v'} + \sigma_{vv'}(p) + \frac{\sigma_{vv'}(h)}{n_h} + \frac{\sigma_{vv'}(ph)}{n_h}. \quad (9.34)$$

A similar derivation for $E\left(\overline{X}_v\overline{X}_{v'}\right)$ in Equation 9.25 gives

$$E\left(\overline{X}_v\overline{X}_{v'}\right) = \mu_v\mu_{v'} + \frac{\sigma_{vv'}(p)}{n_p} + \frac{\sigma_{vv'}(h)}{n_h} + \frac{\sigma_{vv'}(ph)}{n_p n_h}, \quad (9.35)$$

and replacing Equations 9.34 and 9.35 in Equation 9.25 leads to the final result for the expected mean product for persons:

$$EMP_{vv'}(p) = \sigma_{vv'}(ph) + n_h \sigma_{vv'}(p).$$

(9.36)

In a corresponding manner it can be shown that the expected mean product for h is

$$EMP_{vv'}(h) = \sigma_{vv'}(ph) + n_p \sigma_{vv'}(h),$$

(9.37)

and the expected mean product for ph is

$$EMP_{vv'}(ph) = \sigma_{vv'}(ph).$$

(9.38)

Replacing parameters with estimators in the expected mean product Equations 9.36 to 9.38, it is easy to obtain the estimators of the covariance components for the $p \times h$ covariance components design:

$$\hat{\sigma}_{vv'}(p) \quad = \quad \frac{MP_{vv'}(p) - MP_{vv'}(ph)}{n_h}$$

(9.39)

$$\hat{\sigma}_{vv'}(h) \quad = \quad \frac{MP_{vv'}(h) - MP_{vv'}(ph)}{n_p}$$

(9.40)

$$\hat{\sigma}_{vv'}(ph) \quad = \quad MP_{vv'}(ph).$$

(9.41)

9.4.2   General Equations

The form of Equations 9.36 to 9.38 is analogous to the expected-mean-square equations for the $p \times h$ design, and the form of Equations 9.39 to 9.41 is analogous to equations for the estimated random effects variance components for the $p \times h$ design. That is, covariance components in EMP equations play a role that parallels that of variance components in EMS equations. Indeed, general equations for estimating covariance components for balanced multivariate designs closely parallel the general equations in Section 3.4 for estimating random effects variance components for a univariate design.

For any component α in the covariance components design,[11] the sum of products is

$$SP_{vv'}(\alpha) = \pi(\dot{\alpha}) \sum_\alpha x_{\alpha v} \, x_{\alpha v'},$$

(9.42)

where $x_{\alpha v}$ and $x_{\alpha v'}$ are deviation scores of the type discussed in Section 3.3.2, the summation is taken over all indices in α, and the multiplier

---

[11]It is important to note that the α terms involved here are those for the covariance components design only.

is[12]

$$\pi(\dot{\alpha}) = \begin{cases} \text{the product of the sample sizes for all} \\ \text{indices in the covariance components} \\ \text{design } \textit{except} \text{ those indices in } \alpha. \end{cases} \quad (9.43)$$

For the $p^{\bullet} \times (i^{\circ} : h^{\bullet})$ multivariate design in which the covariance components design is $p \times h$, Equation 9.21 is an instance of Equation 9.42. For the same design, another instance is

$$SP_{vv'}(ph) = \sum_p \sum_h \left(\overline{X}_{phv} - \overline{X}_{pv} - \overline{X}_{hv} + \overline{X}_v \right)$$

$$\times \left(\overline{X}_{phv'} - \overline{X}_{pv'} - \overline{X}_{hv'} + \overline{X}_{v'} \right). \quad (9.44)$$

Since $SP_{vv'}(\alpha)$ in Equation 9.42 is a product of deviation scores, it can be called the *corrected* sum of products, as opposed to the *uncorrected* sum of products that is usually easier to compute:

$$TP_{vv'}(\alpha) = \begin{cases} \pi(\dot{\alpha}) \sum_{\alpha} \overline{X}_{\alpha v} \overline{X}_{\alpha v'} & \text{if } \alpha \neq \mu \\ \pi(all) \overline{X}_v \overline{X}_{v'} & \text{if } \alpha = \mu, \end{cases} \quad (9.45)$$

where $\pi(all)$ is the product of the sample sizes for all indices in the covariance components design.[13] To express any SP term with respect to a set of TP terms, the algorithm in Section 3.2.3 can be used, with SP terms replacing SS terms, and TP terms replacing T terms. Also, using the same replacements, for one- and two-facet designs, Appendix A provides SP terms with respect to a set of TP terms.

For the $p^{\bullet} \times (i^{\circ} : h^{\bullet})$ multivariate design in which the covariance components design is $p \times h$, Equation 9.23 expresses $SP_{vv'}(p)$ with respect to TP terms. For the same design, $SP_{vv'}(ph)$ in Equation 9.44 can be expressed as

$$SP_{vv'}(ph) = TP_{vv'}(ph) - TP_{vv'}(p) - TP_{vv'}(h) + TP_{vv'}(\mu).$$

The mean product for a component α is simply

$$MP_{vv'}(\alpha) = \frac{SP_{vv'}(\alpha)}{df(\alpha)}. \quad (9.46)$$

Making use of the zero expectations given by Equations 9.29, 9.30, 9.31, and 9.33, it can be shown that the expected value of the mean product for

---

[12]$\pi(\dot{\alpha})$ has been used previously in Equation 3.9 to designate the corresponding quantity for variance components designs.

[13]In this chapter occasionally α is permitted to represent μ as well as the effects associated with covariance components. This notational liberty is adopted to avoid introducing additional notational conventions.

a component β is

$$EMP_{vv'}(\beta) = \sum \pi(\dot{\alpha})\,\sigma_{vv'}(\alpha), \tag{9.47}$$

where α is any component in the covariance components design that contains at least *all of the indices* in β. That is, $EMP_{vv'}(\beta)$ is a weighted sum of each of the covariance components that contain all of the indices in β. As such, Equation 9.47 is directly analogous to Equation 3.24, which expresses expected mean squares in terms of random effects variance components.

To estimate the covariance components, the parameters in the EMP equations are replaced by estimates. Then, the procedures discussed in Section 3.4.3, the algorithm discussed in Section 3.4.4, or the matrix procedures in Appendix C can be used, through replacement of MS terms with MP terms, and the replacement of $\sigma^2(\alpha)$ with $\sigma_{vv'}(\alpha)$. With the same replacements, the equations in Appendix B provide estimators of covariance components for certain specific designs.

As an example, consider the $p^{\bullet} \times i^{\bullet}$ design. The linear model equations for v and v' are the same as those in Section 9.1 for the $p^{\bullet} \times i^{\circ}$ design (i.e., Equations 9.1 and 9.2). However, for the $p^{\bullet} \times i^{\bullet}$ design, there are covariance components for p, i, and pi—not just p, as is the case for the $p^{\bullet} \times i^{\circ}$ design. It follows that the variance-covariance matrices for the $p^{\bullet} \times i^{\bullet}$ design, in our compact notation, are

$$\Sigma_p = \begin{bmatrix} \sigma_v^2(p) & \sigma_{vv'}(p) \\ \sigma_{vv'}(p) & \sigma_{v'}^2(p) \end{bmatrix} \tag{9.48}$$

$$\Sigma_i = \begin{bmatrix} \sigma_v^2(i) & \sigma_{vv'}(i) \\ \sigma_{vv'}(i) & \sigma_{v'}^2(i) \end{bmatrix} \tag{9.49}$$

$$\Sigma_{pi} = \begin{bmatrix} \sigma_v^2(pi) & \sigma_{vv'}(pi) \\ \sigma_{vv'}(pi) & \sigma_{v'}^2(pi) \end{bmatrix}. \tag{9.50}$$

To estimate the covariance components, we use the following TP terms.

$$TP_{vv'}(p) = n_i \sum_p \overline{X}_{pv} \overline{X}_{pv'} \qquad TP_{vv'}(i) = n_p \sum_i \overline{X}_{iv} \overline{X}_{iv'}$$

$$TP_{vv'}(pi) = \sum_p \sum_i X_{piv} X_{piv'} \qquad TP_{vv'}(\mu) = n_p n_i \overline{X}_v \overline{X}_{v'}.$$

Then, for example, the estimator of the covariance component for p is

$$\hat{\sigma}_{vv'}(p) = \frac{MP_{vv'}(p) - MP_{vv'}(pi)}{n_i}, \tag{9.51}$$

where

$$MP_{vv'}(p) = \frac{TP_{vv'}(p) - TP_{vv'}(\mu)}{n_p - 1}$$

TABLE 9.3. Synthetic Data Example for Balanced $p^\bullet \times i^\bullet$ Design

p	v_1						v_2						$\overline{X}_{p1}$	$\overline{X}_{p2}$	$\overline{X}_{p1}\overline{X}_{p2}$
	i_1	i_2	i_3	i_4	i_5	i_6	i_1	i_2	i_3	i_4	i_5	i_6			
1	6	4	3	5	4	4	6	4	5	6	4	5	4.3333	5.0000	21.6667
2	3	2	2	4	5	5	6	4	5	6	2	5	3.5000	4.6667	16.3333
3	6	5	7	5	4	3	6	5	8	4	6	3	5.0000	4.8333	24.1667
4	4	2	2	3	3	5	5	4	2	4	3	5	3.1667	3.8333	12.1389
5	4	4	3	5	4	6	4	5	5	3	3	7	4.3333	4.5000	19.5000
6	8	5	4	7	5	4	9	6	6	5	7	7	5.5000	6.6667	36.6667
7	5	4	5	7	4	4	4	5	6	7	4	5	4.8333	5.1667	24.9722
8	4	5	3	4	5	6	6	7	6	6	5	6	4.5000	6.0000	27.0000
9	7	5	4	6	6	5	6	7	3	5	5	6	5.5000	5.3333	29.3333
10	5	3	3	7	4	5	6	5	4	6	2	6	4.5000	4.8333	21.7500
Sum	52	39	36	53	44	47	58	52	50	52	38	55	45.1667	50.8333	233.5278

$\overline{X}_1 = 4.5167$ $\qquad$ $TP_{12}(p) = 1401.1667$ $\qquad$ $TP_{12}(pi) = 1441.0000$
$\overline{X}_2 = 5.0833$ $\qquad$ $TP_{12}(i) = 1385.7000$ $\qquad$ $TP_{12}(\mu) = 1377.5833$

$$\mathbf{M}_p = \begin{bmatrix} 3.4611 & 2.6204 \\ 2.6204 & 3.7500 \end{bmatrix} \qquad \widehat{\boldsymbol{\Sigma}}_p = \begin{bmatrix} .3682 & .3193 \\ .3193 & .3689 \end{bmatrix}$$

$$\mathbf{M}_i = \begin{bmatrix} 4.6967 & 1.6233 \\ 1.6233 & 4.7367 \end{bmatrix} \qquad \widehat{\boldsymbol{\Sigma}}_i = \begin{bmatrix} .3444 & .0919 \\ .0919 & .3200 \end{bmatrix}$$

$$\mathbf{M}_{pi} = \begin{bmatrix} 1.2522 & .7048 \\ .7048 & 1.5367 \end{bmatrix} \qquad \widehat{\boldsymbol{\Sigma}}_{pi} = \begin{bmatrix} 1.2522 & .7048 \\ .7048 & 1.5367 \end{bmatrix}$$

and

$$MP_{vv'}(pi) = \frac{TP_{vv'}(pi) - TP_{vv'}(p) - TP_{vv'}(i) + TP_{vv'}(\mu)}{(n_p - 1)(n_i - 1)}.$$

Note, in particular, that for both the $p^\bullet \times i^\circ$ and the $p^\bullet \times i^\bullet$ designs, the definition of the covariance component for p is given by Equation 9.6, but the estimators are different for the two designs. For the $p^\bullet \times i^\circ$ design, the estimator is simply the observed covariance in Equation 9.7. For the $p^\bullet \times i^\bullet$ design, the estimator is $\hat{\sigma}_{vv'}(p)$ in Equation 9.51, which involves the difference between two mean products.

Table 9.3 provides a synthetic data example of the computations required to estimate variance and covariance components for the $p^\bullet \times i^\bullet$ design. This synthetic data set is for 10 persons and six items, with two levels of a fixed facet. The bottom of the table provides the mean-squares-and-mean-product matrices ($\mathbf{M}_p$, $\mathbf{M}_i$, $\mathbf{M}_{pi}$) in which the diagonal elements are mean squares and the off-diagonal elements are mean products. The estimated variance and covariance components are provided in the matrices $\widehat{\Sigma}_p$, $\widehat{\Sigma}_i$, and $\widehat{\Sigma}_{pi}$.

9.4.3 Standard Errors of Estimated Covariance Components

Estimated covariance components are subject to sampling variability, of course, but very little has been published about this topic. One result discussed by Kolen (1985, p. 215) and based on Kendall and Stuart (1977, p. 250) is that, if Y and Z have a bivariate normal distribution, then for samples of size n,

$$\text{var}[\hat{\sigma}(Y, Z)] = \frac{\sigma^2(Y)\sigma^2(Z) + \sigma^2(Y, Z)}{n}. \tag{9.52}$$

We cannot use this result directly for our purposes here because, among other things, it is expressed in terms of parameters rather than the estimators discussed in this chapter. However, the form of Equation 9.52 is evident in the expression discussed below for the estimator of the standard error of an estimated covariance component.

For v and v', let us assume that corresponding effects have a bivariate normal distribution, and different effects are uncorrelated. For the $p^{\bullet} \times i^{\bullet}$ design, this means that (ν_p, ξ_p), (ν_i, ξ_i), and (ν_{pi}, ξ_{pi}) have bivariate normal distributions; and

$$\boldsymbol{E}\nu_\alpha\nu_\beta = \boldsymbol{E}\xi_\alpha\xi_\beta = \boldsymbol{E}\nu_\alpha\xi_\beta = 0$$

for $\alpha \neq \beta$. Under these assumptions, using procedures discussed by Searle (1971, pp. 64–66), an approximate estimator of the standard error of an estimated covariance component is

$$\hat{\sigma}[\hat{\sigma}_{vv'}(\alpha)] = \sqrt{\sum_{\beta} \frac{MS_v(\beta)MS_{v'}(\beta) + [MP_{vv'}(\beta)]^2}{[\pi(\dot{\alpha})]^2[df(\beta) + 2]}}, \tag{9.53}$$

where β indexes the mean products that enter $\hat{\sigma}_{vv'}(\alpha)$, and $\pi(\dot{\alpha})$ is given by Equation 9.43. Note that when $v = v'$, $\hat{\sigma}[\hat{\sigma}_{vv'}(\alpha)]$ in Equation 9.53 is identical to Equation 6.2 for the estimated standard error of an estimated variance component for a random model.

For the $p^{\bullet} \times i^{\bullet}$ design, $\hat{\sigma}_{vv'}(p)$ is a linear combination of $MP_{vv'}(p)$ and $MP_{vv'}(pi)$ (see Equation 9.51), and $\hat{\sigma}[\hat{\sigma}_{vv'}(p)]$ is

$$\sqrt{\frac{MS_v(p)MS_{v'}(p) + [MP_{vv'}(p)]^2}{n_i^2[(n_p - 1) + 2]} + \frac{MS_v(pi)MS_{v'}(pi) + [MP_{vv'}(pi)]^2}{n_i^2[(n_p - 1)(n_i - 1) + 2]}}. \tag{9.54}$$

For the synthetic data in Table 9.3, it is easy to verify that $\hat{\sigma}[\hat{\sigma}_{vv'}(p)] = .227$.

The standard error in Equation 9.54 is interpretable as the standard deviation over replications of $\hat{\sigma}_{vv'}(p) = [MP_{vv'}(p) - MP_{vv'}(pi)]/n_i$, where each such estimate is obtained for a different random sample of n_p persons and n_i items from the population and universe. For each replication, if the square of the quantity in Equation 9.54 were obtained, the expected value over replications of these squared quantities would be the variance of $\hat{\sigma}_{vv'}(p)$.

A special case of Equation 9.53 arises for balanced multivariate designs with a single linked facet, say p. Under the assumptions of bivariate normality and uncorrelated effects, an estimate of the standard error of $\hat{\sigma}_{vv'}(p)$ is

$$\hat{\sigma}[\hat{\sigma}_{vv'}(p)] = \sqrt{\frac{S_v^2(p) \, S_{v'}^2(p) + [S_{vv'}(p)]^2}{(n_p - 1) + 2}}, \tag{9.55}$$

where $S_v^2(p)$ and $S_{v'}^2(p)$ are unbiased estimates of the variances of $\overline{X}_{pv}$ and $\overline{X}_{pv'}$, respectively, and $S_{vv'}(p) = \hat{\sigma}_{vv'}(p)$ is the observed covariance. For the $p^{\bullet} \times i^{\circ}$ design,

$$S_v^2(p) = \hat{\sigma}_v^2(p) + \hat{\sigma}_v^2(pi)/n_{iv} \quad \text{and} \quad S_{v'}^2(p) = \hat{\sigma}_{v'}^2(p) + \hat{\sigma}_{v'}^2(pi)/n_{iv'}.$$

For other designs with a single linked facet, different expressions apply for $S_v^2(p)$ and $S_{v'}^2(p)$ in terms of estimated variance components, but Equation 9.55 still applies.

It is evident from the form of Equation 9.53 that estimated standard errors of estimated covariance components tend to be large when the estimate involves many mean products, when the component mean squares and mean products are large, when degrees of freedom are small, and/or when $\pi(\dot{\beta})$ is small. Recall from Section 6.1.1 that estimated standard errors of estimated variance components tend to be large under similar conditions.

In general, the estimated standard error of an estimated covariance component can be smaller or larger than the estimated standard errors for the associated estimated variance components; notationally, $\hat{\sigma}[\hat{\sigma}_{vv'}(\alpha)]$ can be smaller or larger than $\hat{\sigma}[\hat{\sigma}_v^2(\alpha)]$ or $\hat{\sigma}[\hat{\sigma}_{v'}^2(\alpha)]$. However, under some circumstances, relative magnitudes can be predicted. For example, when mean products are small relative to mean squares, estimated standard errors for estimated covariance components (Equation 9.53) tend to be smaller than for estimated variance components (Equation 6.1).

Also, when variance components are equal, standard errors for estimated covariance components tend to be smaller than for estimated variance components. This can be inferred by comparing Equation 9.52 for $\text{var}[\hat{\sigma}(Y, Z)]$ and the corresponding equation for the variance of the estimates of variance (see Kolen, 1985, p. 215):

$$\text{var}[\hat{\sigma}^2(Y)] = 2\sigma^4(Y)/n. \tag{9.56}$$

Suppose that $\sigma^2(Y) = \sigma^2(Z)$. Since $\text{corr}(Y, Z) \leq 1$, it necessarily follows that $\sigma(Y, Z) \leq \sigma^2(Y)$ and, therefore, $\sigma^2(Y, Z) \leq \sigma^4(Y)$. Adding $\sigma^4(Y)$ to both sides of this inequality gives

$$\sigma^4(Y) + \sigma^2(Y, Z) \leq 2\sigma^4(Y).$$

Dividing both sides by n leads to the conclusion that

$$\text{var}[\hat{\sigma}(Y, Z)] \leq \text{var}[\hat{\sigma}^2(Y)].$$

Of course, this result is dependent on normality assumptions.

9.5 Discussion and Other Topics

The *MP* procedure discussed in this chapter for estimating covariance components with balanced designs is directly analogous to the ANOVA procedure for estimating variance components that was extensively discussed in earlier chapters. The *MP* procedure is essentially the procedure proposed by Cronbach et al. (1972) over a quarter century ago. Implementation of the *MP* procedure is relatively straightforward, and the procedure makes no assumptions about distributional form. Other procedures might be considered, however. For example, the "variance of a sum" procedure discussed later in Section 11.1.5 can be used to obtain maximum likelihood and MINQUE estimates of covariance components for some designs. Also, structural equation modeling might be used to estimate at least some of the covariance components in a multivariate design.

9.5.1 Interpreting Covariance Components

As discussed in Section 9.3, a covariance component is the expected value of the product of two effects. This is equivalent to the expected value of the covariance of the two effects, since the expected value of the effects themselves is zero. In interpreting estimates of covariance components, these definitional issues are central.

For multivariate designs with a single linked facet, the interpretation of a covariance component is especially straightforward: it is simply the covariance between universe scores for the v and v' levels of that facet. Furthermore, the observed covariance is an unbiased estimate of the covariance component. For designs with more than one linked facet, however, it is crucial to note that observed covariances are *not* direct estimates of covariance components for the population and universe.

Let us return to the synthetic data example of the multivariate $p^{\bullet} \times i^{\bullet}$ design in Table 9.3. The *MP* estimates of the covariance components are the lower diagonal elements of the following matrices, and the *MS* estimates of the variance components are the diagonal elements:

$$\widehat{\boldsymbol{\Sigma}}_p = \begin{bmatrix} .3682 & .8663 \\ .3193 & .3689 \end{bmatrix} \tag{9.57}$$

$$\widehat{\Sigma}_i \;=\; \begin{bmatrix} .3444 & .2767 \\ .0919 & .3200 \end{bmatrix} \tag{9.58}$$

$$\widehat{\Sigma}_{pi} \;=\; \begin{bmatrix} 1.2522 & .5081 \\ .7048 & 1.5367 \end{bmatrix}. \tag{9.59}$$

One way to interpret estimated covariance components is in terms of the estimated disattenuated correlations based on them. These are reported in italics in the upper-diagonal positions of the three matrices. For example, the estimated correlation between universe scores for persons on v and v' is

$$\hat{\rho}_{vv'}(p) = \frac{\hat{\sigma}_{vv'}(p)}{\sqrt{\hat{\sigma}_v^2(p)\,\hat{\sigma}_{v'}^2(p)}} = \frac{.3193}{\sqrt{.3682 \times .3689}} = .8663. \tag{9.60}$$

In words, ν_p and ξ_p are correlated about .87. (The reader can verify that the observed-score correlation for persons is .73, which is not to be interpreted as an estimate of the disattenuated correlation for persons.)

Similarly, the estimated disattenuated correlation for i is

$$\hat{\rho}_{vv'}(i) = \frac{\hat{\sigma}_{vv'}(i)}{\sqrt{\hat{\sigma}_v^2(i)\,\hat{\sigma}_{v'}^2(i)}} = \frac{.0919}{\sqrt{.3444 \times .3200}} = .2767, \tag{9.61}$$

and the estimated disattenuated correlation for pi is

$$\hat{\rho}_{vv'}(pi) = \frac{\hat{\sigma}_{vv'}(pi)}{\sqrt{\hat{\sigma}_v^2(pi)\,\hat{\sigma}_{v'}^2(pi)}} = \frac{.7048}{\sqrt{1.2522 \times 1.5367}} = .5081. \tag{9.62}$$

The positive values of these two disattenuated correlations suggest that correlated error will be an issue in D study results.

Transforming estimated covariance components into disattenuated correlations occasionally leads to the uncomfortable occurrence of correlations greater than unity. This can occur as a result of sampling error and, of course, it is more likely to occur when sample sizes are small. Also, correlations greater than unity can be indicative of one or more hidden facets in the G study design. For example, suppose occasion is a facet in the universe of admissible observations, but occasion is not explicitly represented in the G study design. Suppose also that the data for estimating variance components were obtained on one occasion, and the data for estimating covariance components were obtained on a different occasion. Then, disattenuated correlations greater than unity may result solely because v and v' effects are more highly correlated for the "covariance-components occasion" than for the "variance-components occasion." In short, a disattenuated correlation greater than unity does not mean the computations are incorrect although, of course, that's always a possibility!

Variance components are necessarily non negative, although ANOVA estimates of them can be negative due to sampling variability. By contrast, covariance components can be negative, so a negative estimate is not necessarily indicative of sampling variability. Even so, in particular studies, there may be good reason to believe that one or more covariance components (i.e., parameters) should be positive, and the occurrence of a negative estimate is probably attributable to sampling variability. In such cases, should a negative estimate be set to zero in the G study? This author thinks not. Even if the estimate is negative, the investigator seldom knows for certain that the parameter is positive. Furthermore, using the procedures discussed in this chapter, the estimate is unbiased, and changing it to zero makes it biased. It is probably not prudent to set a negative estimate to zero, at least not until D study issues are considered.

Throughout this chapter, we have assumed that the structure of the universe of admissible observations and the G study are the same. Of course, this need not be the case. Suppose, for example, that the population and universe of admissible observations have the structure $p^\bullet \times i^\bullet \times h^\bullet$. The G study design might have the structure $p^\bullet \times (i^\circ : h^\bullet)$. If so, in addition to confounded variance components, the G study involves confounded covariance components in the sense that

$$\sigma_{vv'}(i{:}h) = \sigma_{vv'}(i) + \sigma_{vv'}(ih) \tag{9.63}$$

and

$$\sigma_{vv'}(pi{:}h) = \sigma_{vv'}(pi) + \sigma_{vv'}(pih), \tag{9.64}$$

where covariance components to the left of the equalities are for the G study $p^\bullet \times (i^\circ : h^\bullet)$ design, and those to the right are for the $p^\bullet \times i^\bullet \times h^\bullet$ population and universe of admissible observations. These relationships parallel those discussed in Section 3.2.1 for variance components.

As noted above, hidden facets can occur in multivariate G studies. Again, suppose that the population and universe of admissible observations have the structure $p^\bullet \times i^\bullet \times h^\bullet$, where h stands for occasion. The G study might have the $p^\bullet \times i^\bullet$ design and, if so, occasion would be a hidden facet. In this case, assuming the data were all collected on a single occasion,

$$\begin{aligned}
\sigma_{vv'}(p|p^\bullet \times i^\bullet) &= \sigma_{vv'}(p) + \sigma_{vv'}(po) \\
\sigma_{vv'}(i|p^\bullet \times i^\bullet) &= \sigma_{vv'}(i) + \sigma_{vv'}(io) \\
\sigma_{vv'}(pi|p^\bullet \times i^\bullet) &= \sigma_{vv'}(pi) + \sigma_{vv'}(pio),
\end{aligned} \tag{9.65}$$

where covariance components to the left of the equalities are for the G study $p^\bullet \times i^\bullet$ design, and those to the right are for the $p^\bullet \times i^\bullet \times h^\bullet$ population and universe of admissible observations. These relationships parallel those discussed in Section 5.1.4 for variance components.

9.5.2   Computer Programs

In the author's opinion, one very real impediment to the use of multivariate generalizability theory has been that application of the theory requires extensive computations that are often not readily performed with available computer programs/packages. To ameliorate this problem, the computer program mGENOVA discussed in Appendix H was designed specifically for multivariate generalizability theory as discussed in this book. mGENOVA estimates variance and covariance components for each of the designs in Table 9.2, except the last one, the $(p^\circ : c^\bullet) \times i^\circ$ design.

Input for mGENOVA consists of a set of control cards along with a data set. The data layout at the bottom of the figures in Section 9.2 mirrors that employed by mGENOVA. The algorithms used by mGENOVA are not matrix-based; consequently, mGENOVA can process an almost unlimited number of observations very rapidly. mGENOVA provides G study output of the type discussed in this chapter, as well as D study output of the type discussed in subsequent chapters.

For balanced multivariate designs, various computer packages often can be used to perform some or all of the computations. For example, in the appendix to their overview of multivariate generalizability theory, Webb et al. (1983) provide a discussion and example of how the SAS GLM and MATRIX procedures can be used to estimate covariance components based on the MP procedure in Section 9.4.2. SAS IML could be used, as well.

9.6 Exercises

9.1* For the APL Survey discussed in Section 3.5.4, Table 3.7 reports that the estimated variance components for the southern region, for a univariate mixed model with content categories fixed, are:

$$\hat{\sigma}^2(p|V) = .0378, \qquad \hat{\sigma}^2(pv|V) = .0051,$$
$$\hat{\sigma}^2(i{:}v|V) = .0259, \qquad \hat{\sigma}^2(pi{:}v|V) = .1589.$$

(Here we use v and V rather than h and H.) The corresponding variance-covariance matrices for the multivariate $p^\bullet \times i^\circ$ design are:

$$\widehat{\Sigma}_p = \begin{bmatrix} .0342 & .0332 & .0298 & .0353 & .0290 \\ .0332 & .0409 & .0386 & .0463 & .0342 \\ .0298 & .0386 & .0408 & .0348 & .0345 \\ .0353 & .0463 & .0438 & .0548 & .0423 \\ .0290 & .0342 & .0345 & .0423 & .0386 \end{bmatrix}$$

$$\widehat{\Sigma}_i = \begin{bmatrix} .0208 & & & & \\ & .0298 & & & \\ & & .0364 & & \\ & & & .0081 & \\ & & & & .0343 \end{bmatrix}$$

$$\widehat{\Sigma}_{pi} = \begin{bmatrix} .1334 & & & & \\ & .1634 & & & \\ & & .1618 & & \\ & & & .1593 & \\ & & & & .1765 \end{bmatrix}.$$

Use the elements of these matrices with Equations 9.12 to 9.15 to verify the results reported above for the variance components $\hat{\sigma}^2(\alpha|v)$. Why is it that $\hat{\sigma}^2(v|V) = .0013$ in Table 3.7 does not arise as some combination of elements in the multivariate variance-covariance matrices?

9.2* For the $p^\bullet \times i^\circ \times h^\circ$ design, provide the Venn diagram, design matrices, and record layout.

9.3 For the $(p^\bullet : c^\bullet) \times i^\circ$ design, provide the Venn diagram, design matrices, and record layout. For the record layout assume there are two classes with three and two persons per class, respectively; assume there are 6 items in v_1, and 14 items in v_2.

9.4* For the $p^\bullet \times i^\bullet$ design and the synthetic data in Table 9.3, verify the numerical results for the TP terms, the MP terms, and the estimated covariance components.

9.5* If the synthetic data in Table 9.3 were for a $p^\bullet \times i^\circ$ design, what would be $\hat{\sigma}_{vv'}(p)$?

9.6 Suppose the synthetic data in Table 9.3 were for the $i^\bullet : p^\bullet$ design. Determine the mean-squares-and-mean-products matrices and the variance-covariance matrices.

9.7* For the $p^\bullet \times i^\bullet$ design in Table 9.3, estimate the standard errors of the estimated covariance components for i and pi.

9.8 For the $p^\bullet \times i^\bullet$ design in Table 9.3, the estimated variance components for p are nearly equal. It was stated in Section 9.4.3 that the standard error of an estimated covariance component will be smaller than the standard error of the variance components when the variance components are equal. Verify that this statement applies to the estimates in Table 9.3.

9.9* Miller and Kane (2001) discuss a pretest–posttest study in which each of 30 students took the same 15-item test on two occasions. The 15 items were subdivided into five categories (c) of three items (i) each. The pretest, posttest, and difference-score random effects variance components are:

Effect	Pretest	Posttest	Difference
p	.0058	.0334	.0453
h	.1238	.0298	.0498
$i{:}h$	.0279	.0028	.0211
ph	.0177	.0606	.0713
$pi{:}h$	.0432	.0438	.0878

What are the variance-covariance matrices?

10
Multivariate D Studies

This chapter is split into two major parts followed by a few real-data examples. Section 10.1 covers fundamental issues about multivariate universes of generalization and D studies. Section 10.2 treats a selected set of other issues that, in most cases, are probably less central than the issues covered in Section 10.1. All of the issues, and most of the results and procedures, discussed in this chapter apply to both balanced and unbalanced multivariate D study designs. However, attention is focused primarily on balanced designs in the sense that, for each of the fixed levels in the D study, there is a balanced variance components design. Multivariate unbalanced designs are the subject of Chapter 11.[1]

10.1 Universes of Generalization and D Studies

Letting τ designate the objects of measurement "facet," in multivariate generalizability theory, for each τ there are n_v universe scores $\mu_{\tau v}$, each of which is associated with a single level of the fixed facet. In this sense, there is a universe score *profile* for each τ. The universe score (for an object of measurement) for any particular level of the fixed facet is the expected value over *all* facets in the universe of generalization for that level (and that

---

[1] A number of issues covered in this chapter are not treated by Cronbach et al. (1972).

object of measurement). Unless otherwise specified, it is always assumed
here that, for each fixed condition of measurement, all facets are random.

Parts of this section are, in effect, an extension of the Jarjoura and
Brennan (1982, 1983) multivariate $p^\bullet \times I^\circ$ table of specifications model
to more complicated multivariate designs. The initial parts of this section
repeat some of the introduction to multivariate generalizability theory in
Section 9.1.

10.1.1   Variance-Covariance Matrices for D Study Effects

A multivariate D study analysis is, in a sense, a conjunction of n_v univariate
random model analyses that are linked through covariance components. By
way of illustration, let us consider the $p^\bullet \times I^\bullet$ design. The linear model
equations for levels v and v' are:

$$\overline{X}_{pv} = X_{pIv} = \mu_v + \nu_p + \nu_I + \nu_{pI} \tag{10.1}$$

and

$$\overline{X}_{pv'} = X_{pIv'} = \mu_{v'} + \xi_p + \xi_I + \xi_{pI}, \tag{10.2}$$

with the following three D study variance-covariance matrices,

$$\Sigma_p = \begin{bmatrix} \sigma_v^2(p) & \sigma_{vv'}(p) \\ \sigma_{vv'}(p) & \sigma_{v'}^2(p) \end{bmatrix} \tag{10.3}$$

$$\Sigma_I = \begin{bmatrix} \sigma_v^2(I) & \sigma_{vv'}(I) \\ \sigma_{vv'}(I) & \sigma_{v'}^2(I) \end{bmatrix} \tag{10.4}$$

$$\Sigma_{pI} = \begin{bmatrix} \sigma_v^2(pI) & \sigma_{vv'}(pI) \\ \sigma_{vv'}(pI) & \sigma_{v'}^2(pI) \end{bmatrix}. \tag{10.5}$$

As in Chapter 9, the fact that these matrices are represented using only
two rows and two columns does not mean that there are only two levels of
the fixed facet. This compact form simply indicates the notation used to
represent the elements of the matrices.

For balanced designs, multivariate D study analyses in the mean score
metric are performed using procedures that are directly analogous to those
discussed in Chapter 4 for univariate random models. That is, letting $\bar{\alpha}$
designate the indices for a particular D study effect,

$$\sigma_v^2(\bar{\alpha}) = \frac{\sigma_v^2(\alpha)}{\pi_v(\bar{\alpha}|\tau)}, \tag{10.6}$$

where

$$\pi_v(\bar{\alpha}|\tau) = \begin{cases} \text{the product (for } v) \text{ of the D study sam-} \\ \text{ple sizes for all indices in } \bar{\alpha} \text{ except } \tau; \end{cases} \tag{10.7}$$

and

$$\sigma_{vv'}(\bar{\alpha}) = \frac{\sigma_{vv'}(\alpha)}{\pi_{vv'}(\bar{\alpha}|\tau)}, \tag{10.8}$$

where

$$\pi_{vv'}(\bar{\alpha}|\tau) = \begin{cases} \text{the product (for } v \text{ and } v') \text{ of the D study} \\ \text{sample sizes for all indices in } \bar{\alpha} \text{ except } \tau. \end{cases} \quad (10.9)$$

Recall that D study conditions are the same for linked effects, and linked effects produce nonzero covariance components. It follows that the D study sample sizes for v and v' are the same for a nonzero covariance component. So, for example, for the balanced $p^\bullet \times I^\bullet$ design, $\sigma_{vv'}(I) = \sigma_{vv'}(i)/n'_i$ and $\sigma_{vv'}(pI) = \sigma_{vv'}(pi)/n'_i$.

Unless otherwise noted, all results in this chapter are in the mean score metric. To obtain D study variance-covariance matrices for the total score metric, certain rules can be followed. First, to obtain the elements of all D study matrices except Σ_τ, use $\pi_v(\bar{\alpha}|\tau)$ and $\pi_{vv'}(\bar{\alpha}|\tau)$ as *multipliers*, rather than divisors, of G study variance and covariance components, respectively. Second, transform the elements of Σ_τ as follows,

$$\pi_v \pi_{v'} \sigma_{vv'}(\tau),$$

where π_v and $\pi_{v'}$ are the *total* number of D study conditions sampled for v and v', respectively. These sample sizes are the same as those for the highest order interaction, for designs with single observations per cell.

10.1.2 Variance-Covariance Matrices for Universe Scores and Errors of Measurement

Given the variance-covariance matrices for D study effects, it is easy to use them to express variance-covariance matrices for profiles of universe scores (Σ_τ), relative errors (Σ_δ), and absolute errors (Σ_Δ). Rules for doing so are directly analogous to those discussed in Chapter 4 for univariate random models. In particular,

$$\Sigma_\tau = \begin{cases} \text{variance-covariance matrix} \\ \text{for universe scores for } \tau \end{cases} \quad (10.10)$$

$$\Sigma_\delta = \begin{cases} \text{sum of all } \Sigma_{\bar{\alpha}} \text{ such that } \bar{\alpha} \text{ includes } \tau \\ \text{\emph{and} at least one other index} \end{cases} \quad (10.11)$$

$$\Sigma_\Delta = \text{sum of all } \Sigma_{\bar{\alpha}} \text{ except } \Sigma_\tau. \quad (10.12)$$

So, for example, for the $p^\bullet \times I^\bullet$ design, these rules lead to the following matrices

$$\Sigma_\tau = \Sigma_p \quad (10.13)$$

$$\Sigma_\delta = \Sigma_{pI} \quad (10.14)$$

$$\Sigma_\Delta = \Sigma_I + \Sigma_{pI}. \quad (10.15)$$

For any multivariate design, the elements of $\boldsymbol{\Sigma}_\tau$, $\boldsymbol{\Sigma}_\delta$, and $\boldsymbol{\Sigma}_\Delta$ are variance and covariance components, as indicated in the following representations of the matrices.

$$\boldsymbol{\Sigma}_\tau \;=\; \begin{bmatrix} \sigma_v^2(\tau) & \sigma_{vv'}(\tau) \\ \sigma_{vv'}(\tau) & \sigma_{v'}^2(\tau) \end{bmatrix} \qquad (10.16)$$

$$\boldsymbol{\Sigma}_\delta \;=\; \begin{bmatrix} \sigma_v^2(\delta) & \sigma_{vv'}(\delta) \\ \sigma_{vv'}(\delta) & \sigma_{v'}^2(\delta) \end{bmatrix} \qquad (10.17)$$

$$\boldsymbol{\Sigma}_\Delta \;=\; \begin{bmatrix} \sigma_v^2(\Delta) & \sigma_{vv'}(\Delta) \\ \sigma_{vv'}(\Delta) & \sigma_{v'}^2(\Delta) \end{bmatrix}. \qquad (10.18)$$

Again, this compact form for representing the matrices does not necessarily mean that $n_v = 2$.

Note that the relative-error and absolute-error covariance components in $\boldsymbol{\Sigma}_\delta$ and $\boldsymbol{\Sigma}_\Delta$, respectively, are directly analogous to relative-error and absolute-error variance components. For example, for the $p^\bullet \times I^\bullet$ design,

$$\sigma_{vv'}(\delta) = \sigma_{vv'}(pI) = \frac{\sigma_{vv'}(pi)}{n_i'},$$

and

$$\sigma_{vv'}(\Delta) = \sigma_{vv'}(I) + \sigma_{vv'}(pI) = \frac{\sigma_{vv'}(i)}{n_i'} + \frac{\sigma_{vv'}(pi)}{n_i'}.$$

Often, it is convenient to transform relative-error and absolute-error covariance components to their correlated error counterparts; that is,

$$\rho_{vv'}(\delta) = \frac{\sigma_{vv'}(\delta)}{\sqrt{\sigma_v^2(\delta)\sigma_{v'}^2(\delta)}}, \qquad (10.19)$$

and

$$\rho_{vv'}(\Delta) = \frac{\sigma_{vv'}(\Delta)}{\sqrt{\sigma_v^2(\Delta)\sigma_{v'}^2(\Delta)}}. \qquad (10.20)$$

One of the distinct advantages of multivariate generalizability theory is that the off-diagonal elements of $\boldsymbol{\Sigma}_\delta$ and $\boldsymbol{\Sigma}_\Delta$ provide explicit indicators of different types of correlated error. In most other measurement theories, correlated error is assumed not to exist, is confounded with other effects of interest, or lurks in statistical shadows.

As in univariate theory, the expected value of the observed variance for object-of-measurement mean scores for v is

$$ES_v^2(\overline{X}_\tau) = \sigma_v^2(\tau) + \sigma_v^2(\delta), \qquad (10.21)$$

which we sometimes abbreviate $ES_v^2(\tau)$. Similarly, the expected value of the observed covariance between object-of-measurement mean scores for v and v' is

$$ES_{vv'}(\overline{X}_\tau) = \sigma_{vv'}(\tau) + \sigma_{vv'}(\delta), \qquad (10.22)$$

which we sometimes abbreviate $ES_{vv'}(\tau)$. It follows that the observed covariance is an unbiased estimate of the universe score covariance only when relative errors are uncorrelated, as they are for the $p^{\bullet} \times I^{\circ}$ design.

Note that Equation 10.21 is the *expected* value of a variance and Equation 10.22 is the *expected* value of a covariance. Unbiased estimators of these quantities are the observed variance for v, and the observed covariance for v and v', respectively. With persons as the objects of measurement, a matrix display of these estimators is

$$\mathbf{S} = \begin{bmatrix} S_v^2(p) & S_{vv'}(p) \\ S_{vv'}(p) & S_{v'}^2(p) \end{bmatrix}, \tag{10.23}$$

or, more succinctly,

$$\mathbf{S} = \widehat{\mathbf{\Sigma}}_{\tau} + \widehat{\mathbf{\Sigma}}_{\delta}.$$

10.1.3 Composites and A Priori Weights

Occasionally, $\mathbf{\Sigma}_{\tau}$, $\mathbf{\Sigma}_{\delta}$, and/or $\mathbf{\Sigma}_{\Delta}$ contain the primary (or even the sole) statistics of concern. This might occur, for example, when profiles of universe scores are of principal concern. More frequently, however, some composite of universe scores is of interest. In general, we define a composite universe score as

$$\mu_{\tau C} = \sum_{v} w_v \mu_{\tau v}, \tag{10.24}$$

where the w_v are weights defined a priori by an investigator.[2] Sometimes they are called nominal weights.

Usually these a priori, or nominal, weights are defined such that $\sum_v w_v = 1$ and $w_v \geq 0$ for all v. A common example is $w_v = n_{iv}/n_{i+}$, where n_{i+} designates the total number of items over all levels of the fixed facet. In this case, the weights are proportional to the number of items in the measurement procedure that are associated with each v; if items are scored dichotomously, the composite is the proportion of items correct.

The theory discussed here, however, does not *require* that $\sum_v w_v = 1$. For example, when $n_v = 2$, difference scores can be obtained by setting $w_v = 1$ and $w_{v'} = -1$. Also, sometimes all weights are set to unity. This commonly occurs when expressing the composite in the total score metric (see Exercise 10.1).

Composite universe score variance is a weighted sum of the elements in $\mathbf{\Sigma}_{\tau}$:

$$\sigma_C^2(\tau) = \sum_v w_v^2 \sigma_v^2(\tau) + \sum_v \sum_{v \neq v'} w_v w_{v'} \sigma_{vv'}(\tau) \tag{10.25}$$

---

[2]The use of the term "a priori" here is unrelated to the use of that term in the discussion of MINQUE(0) in Section 7.3.1.

$$= \sum_v \sum_{v'} w_v w_{v'} \sigma_{vv'}(\tau). \tag{10.26}$$

Equation 10.26 follows from Equation 10.25 because $\sigma_{vv'}(\tau) = \sigma_v^2(\tau)$ when $v = v'$.

The fact that the w_v are used to *define* composite universe score in no way requires that they be used to *estimate* composite universe score. However, usually the w weights are employed to obtain the following estimator of composite universe score,

$$\overline{X}_{\tau C} = \sum_v w_v \overline{X}_{\tau v}. \tag{10.27}$$

For this estimator, relative error variance for the composite is

$$\sigma_C^2(\delta) = \sum_v \sum_{v'} w_v w_{v'}\, \sigma_{vv'}(\delta), \tag{10.28}$$

where $\sigma_{vv'}(\delta) = \sigma_v^2(\delta)$ when $v = v'$. Similarly, absolute error variance for the composite is

$$\sigma_C^2(\Delta) = \sum_v \sum_{v'} w_v w_{v'}\, \sigma_{vv'}(\Delta), \tag{10.29}$$

where $\sigma_{vv'}(\Delta) = \sigma_v^2(\Delta)$ when $v = v'$.

A multivariate generalizability coefficient can be defined as the ratio of composite universe score variance to itself plus composite relative error variance:

$$E\rho^2 = \frac{\sigma_C^2(\tau)}{\sigma_C^2(\tau) + \sigma_C^2(\delta)} \tag{10.30}$$

$$= \frac{\sum_v \sum_{v'} w_v w_{v'}\, \sigma_{vv'}(\tau)}{\sum_v \sum_{v'} w_v w_{v'}\left[\sigma_{vv'}(\tau) + \sigma_{vv'}(\delta)\right]}. \tag{10.31}$$

Similarly, a multivariate phi coefficient can be defined as the ratio of composite universe score variance to itself plus composite absolute error variance:

$$\Phi = \frac{\sigma_C^2(\tau)}{\sigma_C^2(\tau) + \sigma_C^2(\Delta)} \tag{10.32}$$

$$= \frac{\sum_v \sum_{v'} w_v w_{v'}\, \sigma_{vv'}(\tau)}{\sum_v \sum_{v'} w_v w_{v'}\left[\sigma_{vv'}(\tau) + \sigma_{vv'}(\Delta)\right]}. \tag{10.33}$$

10.1.4 Composites and Effective Weights

It is evident from the form of Equations 10.25 and 10.26 that different levels of the fixed facet can have different contributions to composite universe score variance. In particular, for level v, the proportional contribution is

$$ew_v(\tau) = \frac{w_v \sum_{v'=1}^{n_v} w_{v'}\, \sigma_{vv'}(\tau)}{\sigma_C^2(\tau)}, \qquad (10.34)$$

which means that $\sum_v ew_v(\tau) = 1$.

In terms of distinctions discussed by Wang and Stanley (1970), the $ew_v(\tau)$ play the role of effective weights, whereas the w_v play the role of nominal weights.[3] In the context considered here, nominal weights express the investigator's judgment about the relative importance of the various levels of the fixed facet in specifying the universe of generalization. Effective weights, which are based in part on nominal weights, reflect the relative statistical contribution of a particular level of the fixed facet. It is clear from Equation 10.34 that effective weights and nominal weights are not usually the same.

Effective weights can be defined also with respect to composite relative error variance and/or composite absolute error variance:

$$ew_v(\delta) = \frac{w_v \sum_{v'} w_{v'}\, \sigma_{vv'}(\delta)}{\sigma_C^2(\delta)} \quad \text{and} \quad ew_v(\Delta) = \frac{w_v \sum_{v'} w_{v'}\, \sigma_{vv'}(\Delta)}{\sigma_C^2(\Delta)}.$$

Clearly, the effective weights $ew_v(\tau)$, $ew_v(\delta)$, and $ew_v(\Delta)$ usually will be different. There is no comparable trilogy for w weights. That is, w weights relate to universe scores, only—not errors of measurement.

10.1.5   Composites and Estimation Weights

Usually the nominal w_v weights are used not only to define composite universe score (Equation 10.24) but also to estimate it (Equation 10.27). Sometimes, however, other weights, say a_v, are used for estimation purposes. In this case,

$$\overline{X}_{\tau C|a} = \sum_v a_v \overline{X}_{\tau v}; \qquad (10.35)$$

and mean-squared error involved in using $\overline{X}_{\tau C|a}$ as an estimate of composite universe score is

$$MSE_C(\Delta) \;=\; \boldsymbol{E}[\overline{X}_{\tau C|a} - \mu_{\tau C}]^2$$
$$=\; \left[\sum_v (a_v - w_v)\mu_v\right]^2$$

---

[3]In much of the literature, this distinction is drawn with respect to composite observed score variance, rather than composite universe score variance.

$$+ \sum_v \sum_{v'} (a_v - w_v)(a_{v'} - w_{v'})\sigma_{vv'}(\tau)$$

$$+ \sum_v \sum_{v'} a_v a_{v'} \sigma_{vv'}(\Delta), \tag{10.36}$$

where $\sigma_{vv'}(\Delta)$ is based on the D study sample sizes used with the a_v. Note that there is a nonnegative contribution to $MSE_C(\Delta)$ from the elements of Σ_τ and from the means of the fixed categories (μ_v).

For tests developed according to a table of specifications, a reasonable question to consider is the extent to which the fixed category system, and the assignment of specific numbers of items to categories, leads to a reduction in some measure of error. After all, if no such reduction occurs, it would be prudent to reconsider the need for the particular category system. One way to address this question is to compare $\hat{\sigma}_C^2(\Delta)$ to $MSE_C(\Delta)$ when $MSE_C(\Delta)$ is based on assigning all items to only one of the n_v categories. Obviously, there are n_v such comparisons, and they are likely to give different answers. Even so, this somewhat ad hoc examination of the efficacy of a category system can be useful (see Jarjoura & Brennan, 1982, 1983, for a real-data example).

Also, it can be useful to examine $MSE_C(\Delta)$ for sample sizes slightly different from those actually used with a measurement procedure. This can give an investigator a sense of how important it is that all instances of the measurement procedure (e.g., forms) use exactly the same sample sizes.

Under some circumstances, it may be appropriate to consider mean-square relative error:

$$MSE_C(\delta) \;=\; \sum_v \sum_{v'} (a_v - w_v)(a_{v'} - w_{v'})\sigma_{vv'}(\tau)$$

$$+ \sum_v \sum_{v'} a_v a_{v'} \sigma_{vv'}(\delta). \tag{10.37}$$

For example, if test forms are carefully equated, it might be judged sensible to assume that $\Sigma_\Delta = \Sigma_\delta$ and the profile of μ_v is constant over forms, at least approximately. Under these assumptions, $MSE_C(\Delta)$ in Equation 10.36 becomes $MSE_C(\delta)$ in Equation 10.37.

10.1.6 Synthetic Data Example

Consider, again, the synthetic data example for the $p^\bullet \times i^\bullet$ design in Table 9.3 on page 292. In this data set the G study sample size for items is $n_i = 6$. If the D study sample size is $n_i' = 8$, then

$$\hat{\Sigma}_p \;=\; \begin{bmatrix} .3682 & .3193 \\ .3193 & .3689 \end{bmatrix}$$

$$\hat{\Sigma}_I \;=\; \begin{bmatrix} .0431 & .0115 \\ .0115 & .0400 \end{bmatrix}$$

$$\widehat{\Sigma}_{pI} = \begin{bmatrix} .1565 & .0881 \\ .0881 & .1921 \end{bmatrix}.$$

Recall, as well, that the G study has observed means of $\overline{X}_1 = 4.5167$ and $\overline{X}_2 = 5.0833$.

Using Equations 10.13 to 10.15, the estimates of the universe score, relative error, and absolute error matrices are

$$\widehat{\Sigma}_\tau = \begin{bmatrix} .3682 & .8663 \\ .3193 & .3689 \end{bmatrix} \tag{10.38}$$

$$\widehat{\Sigma}_\delta = \begin{bmatrix} .1565 & .5081 \\ .0881 & .1921 \end{bmatrix} \tag{10.39}$$

$$\widehat{\Sigma}_\Delta = \begin{bmatrix} .1996 & .4627 \\ .0996 & .2321 \end{bmatrix}. \tag{10.40}$$

The upperdiagonal elements in italics are correlations. That is, the estimated correlations between universe scores, relative errors, and absolute errors are $\hat{\rho}_{vv'}(\tau) = .8663$, $\hat{\rho}_{vv'}(\delta) = .5081$, and $\hat{\rho}_{vv'}(\Delta) = .4627$, respectively. Note, in particular, that correlated δ-type error is greater than correlated Δ-type error even though the δ covariance is smaller than the Δ covariance (see Exercise 10.4 for a condition under which this inequality can occur). Note also that the observed-score variance-covariance matrix that would be expected with $n_i' = 8$ is

$$S = \begin{bmatrix} .5247 & .7509 \\ .4074 & .5610 \end{bmatrix}, \tag{10.41}$$

where the upper diagonal element is the expected observed score correlation.

For the sake of specificity, suppose v_1 represented accuracy and v_2 represented speed in solving math word problems administered by computer. In specifying the universe of generalization, an investigator might judge that accuracy is three times more important than speed. This would imply that for this investigator's universe $w_1 = .75$ and $w_2 = .25$. Under these assumptions, composite universe score variance is estimated to be

$$\hat{\sigma}_C^2(\tau) = (.75)(.75)(.3682) + (.25)(.25)(.3689) + 2(.75)(.25)(.3193) = .3499.$$

If examinee reported scores are obtained by weighting accuracy three times as much as speed, then the estimation weights are the same as the w weights, the estimate of composite relative error variance is

$$\hat{\sigma}_C^2(\delta) = (.75)(.75)(.1565) + (.25)(.25)(.1921) + 2(.75)(.25)(.0881) = .1331,$$

the δ-type signal–noise ratio is 2.63, and the multivariate generalizability coefficient is $E\hat{\rho}^2 = .724$.

Under the same assumptions, the estimate of composite absolute error variance is

$$\hat{\sigma}_C^2(\Delta) = (.75)(.75)(.1996) + (.25)(.25)(.2321) + 2(.75)(.25)(.0996) = .1641,$$

the Δ-type signal–noise ratio is 2.31, and $\widehat{\Phi} = .681$. The proportional contribution of accuracy to composite absolute error variance is estimated to be

$$ew_1(\Delta) = \frac{.75[.75(.1996) + .25(.0996)]}{.1641} = .7979,$$

which means that accuracy contributes about 80% and speed contributes about 20% to composite absolute error variance.

These results assume that the a priori weights and the estimation weights are the same. By contrast, suppose a second investigator decided to give equal weight to accuracy and speed in arriving at examinee reported scores, but wanted to interpret these reported scores as estimates of the universe of generalization characterized by the original w weights. In this case, the mean-square error in using $.5\overline{X}_{p1} + .5\overline{X}_{p2}$ as an estimate of $.75\mu_{p1} + .25\mu_{p2}$ is

$$
\begin{aligned}
MSE_C(\Delta) \quad = \quad & [(.50 - .75)4.5167 + (.50 - .25)5.0833]^2 \\
& + [(.50 - .75)^2.3682 + (.50 - .25)^2.3689 \\
& \quad\quad + 2(.50 - .75)(.50 - .25).3193] \\
& + .25[.1996 + .2321 + 2(.0996)] \\
= \quad & .1839,
\end{aligned}
$$

which is about 12% larger than composite absolute error variance (.1641) that is based on using the w weights as estimation weights.

10.2   Other Topics

The topics in Section 10.1 are essential to understanding D study issues in multivariate generalizability theory, but there are numerous other relevant topics. One such topic is multivariate unbalanced designs, which is the subject of Chapter 11. Also, Chapter 12 provides a relatively lengthy treatment of multivariate regressed scores. A few additional topics are covered less extensively in this section. To keep the treatment somewhat less abstract, the notation and discussion here use persons (p) as the objects of measurement. As always, however, any facet could constitute the objects of measurement.

10.2.1   Standard Errors of Estimated Covariance Components

Section 9.4.3 discussed a normality-based approach to estimating standard errors of estimated G study covariance components that are linear combinations of mean products. When the D study covariance components are linear functions of the G study components, the D study components are also linear combinations of the mean products. Under these circumstances, an approximate estimator of the standard error of an estimated covariance component is

$$\hat{\sigma}[\hat{\sigma}_{vv'}(\alpha)] = \sqrt{\sum_{\beta}[f(\beta|\alpha)]^2 \frac{MS_v(\beta)MS_{v'}(\beta) + [MP_{vv'}(\beta)]^2}{df(\beta) + 2}}, \qquad (10.42)$$

where β indexes the mean products that enter $\hat{\sigma}_{vv'}(\alpha)$, and $f(\beta|\alpha)$ is the coefficient of $MP_{vv'}(\beta)$ in the formula for $\hat{\sigma}_{vv'}(\alpha)$. The only difference between Equation 10.42 for D study covariance components and Equation 9.53 for G study covariance components is that $f(\beta|\alpha)$ in Equation 10.42 plays the role of $1/\pi(\dot{\alpha})$ in Equation 9.53.

The more general formulation in Equation 10.42 permits it to apply to any estimator that is a linear function of the mean products. In particular, this equation applies to estimates of universe score covariance components, relative error covariance components, and absolute error covariance components—for levels of a fixed facet and for composites. For example, for the $p^{\bullet} \times I^{\bullet} \times H^{\bullet}$ design

$$\hat{\sigma}_{vv'}(\delta) = \hat{\sigma}_{vv'}(pI) + \hat{\sigma}_{vv'}(pH) + \hat{\sigma}_{vv'}(pIH), \qquad (10.43)$$

where

$$\hat{\sigma}_{vv'}(pI) = \frac{MP_{vv'}(pi) - MP_{vv'}(pih)}{n_h n'_i},$$

$$\hat{\sigma}_{vv'}(pH) = \frac{MP_{vv'}(ph) - MP_{vv'}(pih)}{n_i n'_h},$$

and

$$\hat{\sigma}_{vv'}(pIH) = \frac{MP_{vv'}(pih)}{n'_i n'_h}.$$

It follows that an expression for $\hat{\sigma}_{vv'}(\delta)$ in terms of mean products is

$$\hat{\sigma}_{vv'}(\delta) = \left(\frac{1}{n_h n'_i}\right) MP_{vv'}(pi) + \left(\frac{1}{n_i n'_h}\right) MP_{vv'}(ph)$$
$$+ \left(\frac{1}{n'_i n'_h} - \frac{1}{n_h n'_i} - \frac{1}{n_i n'_h}\right) MP_{vv'}(pih), \qquad (10.44)$$

where the terms in the three sets of parentheses are $f(pi|\delta)$, $f(ph|\delta)$, and $f(pih|\delta)$, respectively.

Therefore, using Equation 10.42, the estimated standard error of $\hat{\sigma}_{vv'}(\delta)$ is

$$\hat{\sigma}[\hat{\sigma}_{vv'}(\delta)] = \left\{ \left(\frac{1}{n_h n'_i}\right)^2 \frac{MS_v(pi)MS_{v'}(pi) + [MP_{vv'}(pi)]^2}{(n_p - 1)(n_i - 1) + 2} \right.$$

$$+ \left(\frac{1}{n_i n_h'} \right)^2 \frac{MS_v(ph)MS_{v'}(ph) + [MP_{vv'}(ph)]^2}{(n_p - 1)(n_h - 1) + 2}$$

$$+ \left(\frac{1}{n_i' n_h'} - \frac{1}{n_h n_i'} - \frac{1}{n_i n_h'} \right)^2$$

$$\left. \frac{MS_v(pih)MS_{v'}(pih) + [MP_{vv'}(pih)]^2}{(n_p - 1)(n_i - 1)(n_h - 1) + 2} \right\}^{1/2}. \quad (10.45)$$

10.2.2 Optimality Issues

To the extent possible, when a composite is under consideration, it is sensible to choose sample sizes that minimize composite error variance, given whatever other constraints may exist. A relatively simple example is choosing D study sample sizes for the $p^\bullet \times I^\circ$ design under the assumption that $n_{i+}' = \sum_v n_{iv}'$ is fixed (see Jarjoura & Brennan, 1982). For this design, assuming the estimation weights equal the a priori weights, composite absolute error variance is

$$\sigma_C^2(\Delta) = \sum_v w_v^2 \left[\frac{\sigma_v^2(i) + \sigma_v^2(pi)}{n_{iv}'} \right].$$

Since the D study sample sizes n_{iv}' are positive integers, iteration might be used to minimize $\sigma_C^2(\Delta)$. This approach to minimization is defensible to the extent that precise estimates of the variance components are available. If the estimates are imprecise, then any minimization claim based on iteration is suspect, at best.

However, if precise estimates are available, a simpler solution involves the following inequality,[4]

$$\sigma_C^2(\Delta) = \sum_v w_v^2 \left[\frac{\sigma_v^2(i) + \sigma_v^2(pi)}{n_{iv}'} \right]$$

$$\geq \frac{1}{n_{i+}'} \left[\sum_v w_v \sqrt{\sigma_v^2(i) + \sigma_v^2(pi)} \right].$$

Minimization of $\sigma_C^2(\Delta)$ occurs when

$$n_{iv}' = \frac{n_{i+}' w_v \sqrt{\sigma_v^2(i) + \sigma_v^2(pi)}}{\sum_v w_v \sqrt{\sigma_v^2(i) + \sigma_v^2(pi)}}, \quad (10.46)$$

although the resulting n_{iv}' are not necessarily integers. To obtain integer values for practical use, it is possible to iterate around the solution in Equation 10.46.

---

[4]This inequality is based on the Cauchy–Schwartz Inequality.

For the synthetic data example introduced in Table 9.3 on page 292, suppose the D study design is $p^{\bullet} \times I^{\circ}$ with $w_1 = .75$ and $w_2 = .25$. If the total number of items is constrained to be $n'_{i+} = 10$, then using Equation 10.46 the optimal sample sizes are

$$n'_{i1} = \frac{10(.75)\sqrt{.3444 + 1.2522}}{.75\sqrt{.3444 + 1.2522} + .25\sqrt{.3200 + 1.5367}} = 7.3559$$

and

$$n'_{i2} = \frac{10(.25)\sqrt{.3200 + 1.5367}}{.75\sqrt{.3444 + 1.2522} + .25\sqrt{.3200 + 1.5367}} = 2.6441.$$

Using $n'_{i1} = 7$ and $n'_{i2} = 3$ gives $\hat{\sigma}^2_C(\Delta) = .167$, and using $n'_{i1} = 8$ and $n'_{i2} = 2$ gives $\hat{\sigma}^2_C(\Delta) = .170$. Consequently, the optimal integer sample sizes are seven and three.

This approach to determining sample sizes that minimize $\sigma^2_C(\Delta)$ can be extended to other, more complicated multivariate designs. Furthermore, constraints such as cost can be employed instead of, or in addition to, fixing total sample sizes (see, e.g., Marcoulides & Goldstein, 1990, 1992). However, the logic of multivariate generalizability theory as presented in this book requires that the w_v weights be specified a priori, and any statistical procedure that modifies these weights a posteriori is at odds with this perspective on the theory. In particular, a statistical procedure described by Joe and Woodward (1976) for choosing weights that maximize a generalizability coefficient is inconsistent with the logic of multivariate generalizability theory presented here. In effect, this procedure produces a set of a posteriori category weights that may have little to do with the relative importance of the categories for the prespecified intended use of the measurement procedure.

Although the w_v need to be specified a priori, it is certainly possible for two investigators to use the same measurement procedure for different purposes, which may well require different weightings of the fixed categories, that is, different values for the w_v. Also, in the early stages of defining a universe of generalization and deciding on the relative importance of the various categories, it may be very helpful to use various statistical procedures, as well as other sources of information, to make an informed choice of the w_v weights. Indeed, at this formative stage, the Joe and Woodward (1976) procedure might be useful. Note, also, that it is possible to specify effective weights and determine the w_v from them (see Dunnette & Hoggatt, 1957, and Wilks, 1938).

The w_v are an integral part of the specification of an investigator's universe of generalization, which provides at least part of the definition of the construct of interest to the investigator. As such, the w_v are fixed, and they apply not only to the measurement procedure that generates a particular set of data, but also to all other randomly parallel instances of the measurement procedure in the universe of generalization. Any optimization procedure that treats the w_v as random is likely to yield different values for the w_v depending on which instance of the measurement procedure is employed; and there is no guarantee that any of these different sets of values for the w_v will accurately reflect the investigator's judgments about the relative importance of the categories. So, any optimization procedure that treats the w_v as random effectively creates a moving validity target for inferences based on the measurement procedure. In the end, the w_c must be specified before any meaningful statements can be made about error variances, generalizability coefficients, and so on; and the investigator must be prepared to provide some type of validity defense for the w_v values.

10.2.3   Conditional Standard Errors of Measurement for Composites

Section 5.4 discussed conditional standard errors of measurement for balanced designs in univariate generalizability theory. As discussed there, $\sigma(\Delta_p)$ is the standard error of the mean for the within-person D study design associated with person p. So, for example, if the across-persons univariate D study design is $p \times I$, then the within-person design for each person is simply I, and $\sigma(\Delta_p)$ is the standard error of the mean for the person's item scores.

This logic is easily extended to multivariate designs. Assuming the estimation weights (a_v) and the a priori weights (w_v) are the same, absolute error for the composite is

$$
\begin{aligned}
\Delta_{pC} &= \overline{X}_{pC} - \mu_{pC} \\
&= \sum_v w_v \left(\overline{X}_{pv} - \mu_{pv} \right) \\
&= \sum_v w_v \, \Delta_{pv}.
\end{aligned}
$$

It follows immediately that the conditional absolute-error SEM is

$$
\sigma_C(\Delta_p) = \sqrt{\sum_v w_v^2 \, \sigma_v^2(\Delta_p) + \sum_{v \neq v'} \sum w_v w_{v'} \, \sigma_{vv'}(\Delta_p)}, \qquad (10.47)
$$

where

$$
\sigma_v^2(\Delta_p) = \mathrm{var}(\overline{X}_{pv} - \mu_{pv}|p)
$$

is the variance of the mean for person p on v, and

$$
\sigma_{vv'}(\Delta_p) = \mathrm{cov}(\overline{X}_{pv} - \mu_{pv}, \overline{X}_{pv'} - \mu_{pv'}|p)
$$

is the covariance of person p's mean scores on v and v'.

TABLE 10.1. Synthetic Data Example of Conditional Absolute Standard Errors of Measurement for Balanced $p^{\bullet} \times I^{\bullet}$ Design with $n'_i = 6$ and $w_1 = w_2 = .5$

p	$\overline{X}_{p1}$	$\overline{X}_{p2}$	$\overline{X}_{pC}$	$\hat{\sigma}^2_1(\Delta_p)$	$\hat{\sigma}^2_2(\Delta_p)$	$\hat{\sigma}_{12}(\Delta_p)$	Actual $\hat{\sigma}_C(\Delta_p)$	Fitted[a] $\hat{\sigma}_C(\Delta_p)$
4	3.1667	3.8333	3.5000	.2278	.2278	.1722	.4472	.4251
2	3.5000	4.6667	4.0833	.3167	.3778	−.1000	.3516	.4451
5	4.3333	4.5000	4.4167	.1778	.3833	.1000	.4362	.4540
1	4.3333	5.0000	4.6667	.1778	.1333	.1000	.3575	.4594
10	4.5000	4.8333	4.6667	.3833	.4278	.2167	.5578	.4594
3	5.0000	4.8333	4.9167	.3333	.6278	.4333	.6760	.4639
7	4.8333	5.1667	5.0000	.2278	.2278	.1722	.4472	.4652
8	4.5000	6.0000	5.2500	.1833	.0667	.0000	.2500	.4685
9	5.5000	5.3333	5.4167	.1833	.3111	.1000	.4167	.4701
6	5.5000	6.6667	6.0833	.4500	.3111	.1333	.5069	.4729
Mn	4.5167	5.0833	4.8000	.2661	.3094	.1328	.4586[b]	.4586[b]

[a] fitted $\hat{\sigma}_C(\Delta_p) = \sqrt{-.0187 + .0822\,\overline{X}_{pC} - .0066\,\overline{X}^2_{pC}}$.

[b] Square root of the average of the variances.

These variances and covariances can be conceptualized in the following manner. Any particular instance of the measurement procedure will result in mean scores on v and v' for the person. The variance of the person's mean scores for v over replications of the measurement procedure is $\sigma^2_v(\Delta_p)$. The covariance of these mean scores for v and v' over replications of the measurement procedure is $\sigma_{vv'}(\Delta_p)$.

The conditional relative-error SEM can be approximated as

$$\sigma_C(\delta_p) \doteq \sqrt{\sigma^2_C(\Delta_p) - [\sigma^2_C(\Delta) - \sigma^2_C(\delta)]}. \qquad (10.48)$$

The correction to $\sigma^2_C(\Delta_p)$ in Equation 10.48 is simply the difference between the *overall* absolute and relative error variances for the composite. The logic behind using this constant correction for all persons mirrors that discussed in Section 5.4 for univariate designs.

Consider again the synthetic data example for the $p^{\bullet} \times I^{\bullet}$ design in Table 9.3 on page 292. Assuming the G and D study sample sizes are the same ($n_i = n'_i = 6$) and the a priori weights for the two variables are equal ($w_1 = w_2 = .5$), Table 10.1 provides conditional absolute-error SEMs for composite scores, for each of the 10 persons, with persons ordered from smallest to largest observed composite score. Let us focus on person number one (the fourth person from the top of Table 10.1). The raw scores, means, variances, and covariance for this person are:

TABLE 10.2. Synthetic Data Within-person $t^\bullet \times r^\circ$ and $T^\bullet \times R^\circ$ Designs

	v_1			v_2						
t	r_1	r_2	r_3	r_1	r_2	r_3				
1	3	2	2	2	3	3	$\widehat{\Sigma}_t$	$=$	$\begin{bmatrix} .0741 & .2235 \\ .2235 & .2926 \end{bmatrix}$	
2	3	3	2	4	4	4				
3	3	3	3	4	4	4	$\widehat{\Sigma}_r$	$=$	$\begin{bmatrix} .0333 & \\ & .2148 \end{bmatrix}$	
4	3	3	3	3	3	4				
5	2	2	1	2	3	1	$\widehat{\Sigma}_{tr}$	$=$	$\begin{bmatrix} .3000 & \\ & .3852 \end{bmatrix}$	
6	1	3	3	2	4	3				
7	2	3	2	2	4	4				
8	3	3	2	2	4	3	$\widehat{\Sigma}_{\Delta_p}$	$=$	$\begin{bmatrix} .0285 & .0224 \\ .0224 & .1137 \end{bmatrix}$	
9	3	3	2	2	4	3				
10	3	3	3	4	4	4				
Mean	2.6	2.8	2.3	2.7	3.7	3.3				

	i_1	i_2	i_3	i_4	i_5	i_6	Mean	Var	Cov
v_1	6	4	3	5	4	4	4.3333	1.0667	
v_2	6	4	5	6	4	5	5.0000	.8000	.6000

It follows that

$$\hat{\sigma}_1^2(\Delta_1) = 1.0667/6 = .1778,$$

$$\hat{\sigma}_2^2(\Delta_1) = .8000/6 = .1333,$$

$$\hat{\sigma}_{12}(\Delta_1) = .6000/6 = .1000,$$

and the estimated conditional absolute-error SEM for person one's composite score is

$$\hat{\sigma}_C(\Delta_1) = \sqrt{.25(.1778) + .25(.1333) + 2(.25)(.1000)} = \sqrt{.1278} = .3575.$$

Table 10.1 also provides fitted values for the conditional absolute-error SEMs, or, more correctly, the square roots of the quadratic-fits for the conditional absolute-error variances. These are obtained in the same manner as discussed in Section 5.4 for univariate designs. Note also that the overall absolute-error SEM can be obtained from the conditional values: specifically,

$$\hat{\sigma}_C(\Delta) = \sqrt{\frac{1}{10} \sum_p \hat{\sigma}_C^2(\Delta_p)} = .4586.$$

As another example, suppose the G study design is $p^\bullet \times t^\bullet \times r^\circ$, the D study design is $p^\bullet \times T^\bullet \times R^\circ$, and the G and D study sample sizes are the same with $n_t = n_t' = 10$ tasks and $n_r = n_r' = 3$ raters. Table 10.2 provides hypothetical data for a particular person in the design. Note that the G study design for this person (and all other persons in this design) is $t^{\bullet} \times r^{\circ}$, and the D study design is $T^{\bullet} \times R^{\circ}$.

Table 10.2 also provides the three G study estimated variance-covariance matrices for this particular person. For these data, the estimated variance of the mean (over 10 tasks and three raters) for v_1 is the conditional absolute-error variance:

$$
\begin{aligned}
\hat{\sigma}_1^2(\Delta_p) &= \frac{\hat{\sigma}_1^2(t)}{n_t'} + \frac{\hat{\sigma}_1^2(r)}{n_r'} + \frac{\hat{\sigma}_1^2(tr)}{n_t' n_r'} \\
&= \frac{.0741}{10} + \frac{.0333}{3} + \frac{.3000}{30} \\
&= .0285.
\end{aligned}
$$

Similarly, for v_2,

$$
\hat{\sigma}_2^2(\Delta_p) = \frac{.2926}{10} + \frac{.2148}{3} + \frac{.3852}{30} = .1137.
$$

The estimated conditional absolute-error covariance is simply

$$
\hat{\sigma}_{12}(\Delta_p) = \frac{\hat{\sigma}_{12}(t)}{n_t'} = \frac{.2235}{10} = .0224.
$$

Note that $\sigma_{12}(r)$ and $\sigma_{12}(tr)$ are not present in $\sigma_{12}(\Delta_p)$ because they are both zero for the $t^{\bullet} \times r^{\circ}$ design.

These conditional absolute-error variance and covariance components are the elements of $\hat{\Sigma}_{\Delta_p}$ in Table 10.2. Using Equation 10.47 with $w_1 = w_2 = .5$, conditional absolute-error variance for the person's composite score is

$$
\hat{\sigma}_C(\Delta_p) = \sqrt{.25(.0285) + .25(.1137) + 2(.25)(.0224)} = \sqrt{.0467} = .2162.
$$

Other examples of conditional SEMs for multivariate composites are provided by Brennan (1996a, 1998).

10.2.4 Profiles and Overlapping Confidence Intervals

It is relatively common practice for test publishers to produce score reports in which a confidence interval is placed around each score in a person's profile of n_v test scores. Almost always the width of each interval is two standard errors of measurement, and the person is usually told something like, "If any two intervals for your scores overlap, then your level of achievement for the two tests is likely the same." This common practice raises a number of questions about overlapping confidence intervals. Here we focus on only two of these questions and develop results based on classical test theory. We then extend and express these results in terms of multivariate generalizability theory. It is especially important to note that the results derived here require normality assumptions.[5]

Let two tests be denoted X and Y with standard errors of measurement of σ_{Ex} and σ_{Ey}, respectively. Under normality assumptions, 68% confidence intervals are $X \pm \sigma_{Ex}$ and $Y \pm \sigma_{Ey}$, with endpoints of (X_L, X_H) and (Y_L, Y_H), respectively. Consider a particular person whose true scores on both tests are the same. For such a person, the probability that the two intervals will *not* overlap is

$$\Pr(\text{no overlap}) = \Pr(X_L > Y_H) + \Pr(X_H < Y_L). \qquad (10.49)$$

Since $\Pr(X_L > Y_H) = \Pr(X_H < Y_L)$, we derive $\Pr(X_L > Y_H)$ and then double it to obtain the result. To begin,

$$
\begin{aligned}
\Pr(X_L > Y_H) &= \Pr(X_L - Y_H > 0) \\
&= \Pr\left[(X - \sigma_{Ex}) - (Y + \sigma_{Ey}) > 0\right] \\
&= \Pr(X - Y > \sigma_{Ex} + \sigma_{Ey}),
\end{aligned}
$$

which essentially translates the original question about "no overlap" to a question about the distribution of observed difference scores. Let us assume that observed difference scores are normally distributed, under the usual classical test theory assumption of uncorrelated errors of measurement. Since we are assuming that $\tau_X = \tau_Y$, the normality assumption is

$$(X - Y) \sim N(0, \sigma_{Ex}^2 + \sigma_{Ey}^2),$$

where $\sigma_{Ex}^2 + \sigma_{Ey}^2$ is the variance of the difference in the error scores, which equals the variance of the observed difference scores. It follows that

$$
\begin{aligned}
\Pr(X_L > Y_H) &= \Pr(X - Y > \sigma_{Ex} + \sigma_{Ey}) \\
&= \Pr\left(z > \frac{\sigma_{Ex} + \sigma_{Ey}}{\sqrt{\sigma_{Ex}^2 + \sigma_{Ey}^2}} \right) \\
&= 1 - \Pr\left(z < \frac{\sigma_{Ex} + \sigma_{Ey}}{\sqrt{\sigma_{Ex}^2 + \sigma_{Ey}^2}} \right).
\end{aligned}
$$

Doubling this result we obtain

$$\Pr(\text{no overlap}) = 2\left[1 - \Pr\left(z < \frac{\sigma_{Ex} + \sigma_{Ey}}{\sqrt{\sigma_{Ex}^2 + \sigma_{Ey}^2}} \right)\right], \qquad (10.50)$$

---

[5]The classical test theory aspects of this section are based largely on an exercise developed by Leonard S. Feldt.

with its complement being

$$\Pr(\text{overlap}) = 2 \times \Pr\left(z < \frac{\sigma_{E_X} + \sigma_{E_Y}}{\sqrt{\sigma_{E_X}^2 + \sigma_{E_Y}^2}}\right) - 1. \qquad (10.51)$$

The probabilities in Equations 10.50 and 10.51 are based on the assumption of uncorrelated error, with intervals formed by adding and subtracting one standard error of measurement to a person's observed scores on X and Y. Also, of course, these probabilities have been derived under the classical test theory assumption of undifferentiated error.

These results are easily extended to designs and situations accommodated in multivariate generalizability theory. For example, for each v, $100\gamma\%$ intervals might be formed with respect to overall absolute error as follows,

$$\overline{X}_{pv} \pm z^*_{(1+\gamma)/2}\,\sigma_v(\Delta), \qquad (10.52)$$

where $z^*_{(1+\gamma)/2}$ is the z-score associated with a two-sided $100\gamma\%$ confidence interval. Under these circumstances, for a person with the same universe score on v and v',

$$\Pr(\text{overlap}) = 2 \times \Pr\left\{z < \frac{z^*_{(1+\gamma)/2}\,[\sigma_v(\Delta) + \sigma_{v'}(\Delta)]}{\sqrt{\sigma_v^2(\Delta) + \sigma_{v'}^2(\Delta) - 2\sigma_{vv'}(\Delta)}}\right\} - 1, \quad (10.53)$$

where the denominator is the standard error of the observed difference scores, which can be denoted $\sigma_D(\Delta)$.

Equation 10.53 has the same form as Equation 10.51, but there are dissimilarities. Most importantly, the standard error of the difference scores incorporates the possibility of correlated absolute error.[6] All other things being equal, large positive values for correlated error lead to larger probabilities of overlap than small positive values. In this sense, if correlated error were present but ignored, the probability of overlap would be understated.

Of course, Equation 10.53 simplifies if correlated error is not present (e.g., the $p^\bullet \times I^\circ$ design), but even then Equation 10.53 differs from the classical test theory result in Equation 10.51, in that Equation 10.53 is based on absolute error, whereas Equation 10.51 uses the error in classical theory. These two types of error may or may not be the same, depending on the design. For example, if the design is $p^\bullet \times I^\circ$, absolute error and the error in classical theory are different; if the design is $I^\circ{:}p^\bullet$, absolute error and classical error are the same.

The probability of overlap result in Equation 10.53 is based on the assumption that the person's universe scores are equal; that is, $\mu_{pv} - \mu_{pv'} = 0$. Without this restriction, we obtain a more general form of Equation 10.53.

---

[6]Less importantly, Equation 10.53 is stated for any confidence coefficient, whereas Equation 10.51 applies as stated for a confidence coefficient of .68 only.

Specifically, for a prespecified value for the difference in universe scores $(\mu_{pv} - \mu_{pv'})$, the probability that two $100\gamma\%$ intervals will overlap is

$$\text{Pr(overlap)} =$$

$$\Pr\left\{ z < \frac{z^*_{(1+\gamma)/2}\,[\sigma_v(\Delta) + \sigma_{v'}(\Delta)] - (\mu_{pv} - \mu_{pv'})}{\sqrt{\sigma_v^2(\Delta) + \sigma_{v'}^2(\Delta) - 2\sigma_{vv'}(\Delta)}} \right\} -$$

$$\Pr\left\{ z < \frac{-z^*_{(1+\gamma)/2}\,[\sigma_v(\Delta) + \sigma_{v'}(\Delta)] - (\mu_{pv} - \mu_{pv'})}{\sqrt{\sigma_v^2(\Delta) + \sigma_{v'}^2(\Delta) - 2\sigma_{vv'}(\Delta)}} \right\}. \quad (10.54)$$

Obviously, intervals such as those given in Equation 10.52 and 10.54 can be expressed in terms of overall relative standard errors of measurement, which leads to probability of overlap equations identical to Equations 10.53 and 10.54 except that δ replaces Δ. Similarly, conditional absolute or relative standard errors of measurement can be used.

As an example, consider the $p^\bullet \times I^\bullet$ design with $n_i' = 6$ using the synthetic data in Table 9.3 on page 292. Dividing each of the elements in $\widehat{\Sigma}_i$ and $\widehat{\Sigma}_{pi}$ by six, and then summing the elements gives

$$\widehat{\Sigma}_\Delta = \begin{bmatrix} .2661 & .1328 \\ .1328 & .3094 \end{bmatrix}.$$

The standard errors for the two levels are $\hat{\sigma}_1(\Delta) = \sqrt{.2661} = .5159$ and $\hat{\sigma}_2(\Delta) = \sqrt{.3094} = .5563$, and the standard error of the difference scores is

$$\hat{\sigma}_D(\Delta) = \sqrt{.2661 + .3094 - 2(.1328)} = .5567.$$

The two 68% confidence intervals are $\overline{X}_{p1} \pm .5159$ and $\overline{X}_{p2} \pm .5563$. Under the assumption that $\mu_{p1} = \mu_{p2}$, the probability that these two intervals will overlap is

$$\begin{aligned} \text{Pr(overlap)} &= 2 \times \Pr\left(z < \frac{.5159 + .5563}{.5567} \right) - 1 \\ &= 2 \times \Pr\left(z < 1.9260 \right) - 1 \\ &= 2 \times (.9730) - 1 \\ &= .946. \end{aligned}$$

10.2.5 Expected Within-Person Profile Variability

When profiles are under consideration, usually interest focuses on individual persons, as discussed in Section 10.2.4. To characterize the entire measurement procedure for a population, however, we can consider expected within-person profile variability, which we denote generically as $\mathcal{V}(*)$. For example, $\mathcal{V}(\overline{X}_p)$ is the expected within-person profile variability for observed scores, and $\mathcal{V}(\mu_p)$ is the expected within-person profile variability for universe scores.

The $\mathcal{V}(*)$ formulas discussed in this section result from the well-known analysis of variance identity. For a row-by-column matrix, the analysis of variance identity guarantees that the average variance within the columns plus the variance of the column means equals the "total" variance, and, similarly, the average variance within the rows plus the variance of the row means equals the "total" variance. Letting Y_{pv} be scores for any variable,

$$\frac{1}{n_v} \sum_v \sigma^2(Y_{pv}) + \text{var}(\overline{Y}_v) = E_p \left[\text{var}(Y_{pv}) \right] + \sigma^2(\overline{Y}_p),$$

where "var" designates variance over the fixed levels of v. The quantity $E_p[\text{var}(Y_{pv})]$ is a measure of the average variability of the profile of Y scores, which we designate simply as $\mathcal{V}(\mathbf{Y}_p)$. It follows from the above equation that

$$\mathcal{V}(\mathbf{Y}_p) = E_p \left[\text{var}(Y_{pv}) \right] = \frac{1}{n_v} \sum_v \sigma^2(Y_{pv}) + \text{var}(\overline{Y}_v) - \sigma^2(\overline{Y}_p). \quad (10.55)$$

Of course, if $\text{var}(Y_{pv})$ is computed for each p in the data, then the average will be an estimate of $\mathcal{V}(\mathbf{Y}_p)$. This direct computational estimate, however, is not always practical, or even possible, because it requires that the entire set of Y_{pv} data be available.

Variance Formulas

Applying Equation 10.55 to raw scores $\overline{X}_{pv}$ for a finite number of persons gives[7]

$$\mathcal{V}(\overline{X}_p) = \frac{n_p - 1}{n_p} [\overline{S}_v^2(p) - \overline{S}_{vv'}(p)] + \text{var}(\overline{X}_v). \quad (10.56)$$

It is important to note that $\overline{S}_v^2(p)$ is the average over the n_v observed variances, and $\overline{S}_{vv'}(p)$ is the average over all n_v^2 elements of the observed variance-covariance matrix, which includes those with $v = v'$. The multiplicative factor $(n_p - 1)/n_p$ has no compelling theoretical relevance or likely practical importance, but it does guarantee that the value obtained using this formula is identical to the value obtained through directly computing the average of the variances of the n_p profiles for a finite number of persons. Of course, this multiplicative factor approaches unity as $n_p \to \infty$.

Consider, again, the synthetic data example for the $p^\bullet \times I^\bullet$ design with $n_i' = 6$. The person mean scores and related statistics and matrices, including the observed score variance-covariance matrix, are provided in Ta-

---

[7]Strictly speaking, this equation is expressed in terms of estimators, since, for example, $S_v^2(p)$ and $S_{vv'}(p)$ have been defined previously as estimators.

TABLE 10.3. Observed Profile Variability for Synthetic Data Example of $p^{\bullet} \times I^{\bullet}$ Design with $n_i' = 6$

p	$\overline{X}_{p1}$	$\overline{X}_{p2}$	Mean	Var[a]			
1	4.3333	5.0000	4.6667	.1111			
2	3.5000	4.6667	4.0833	.3403	$\widehat{\Sigma}_p =$	$\begin{bmatrix} .3682 \\ .3193 \end{bmatrix}$	$\begin{bmatrix} .3193 \\ .3689 \end{bmatrix}$
3	5.0000	4.8333	4.9167	.0069			
4	3.1667	3.8333	3.5000	.1111	$\widehat{\Sigma}_\delta =$	$\begin{bmatrix} .2087 \\ .1175 \end{bmatrix}$	$\begin{bmatrix} .1175 \\ .2561 \end{bmatrix}$
5	4.3333	4.5000	4.4167	.0069			
6	5.5000	6.6667	6.0833	.3403			
7	4.8333	5.1667	5.0000	.0278	$\widehat{\Sigma}_\Delta =$	$\begin{bmatrix} .2661 \\ .1328 \end{bmatrix}$	$\begin{bmatrix} .1328 \\ .3094 \end{bmatrix}$
8	4.5000	6.0000	5.2500	.5625			
9	5.5000	5.3333	5.4167	.0069	$S =$	$\begin{bmatrix} .5769 \\ .4367 \end{bmatrix}$	$\begin{bmatrix} .4367 \\ .6250 \end{bmatrix}$
10	4.5000	4.8333	4.6667	.0278			
Mean	4.5167	5.0833	4.8000	.1542			
Var[a]	.5192	.5625	.4669	.0337			

[a] Biased estimates.

ble 10.3.[8] For these data, the average of the observed variances is

$$\overline{S}_v^2(p) = \frac{.5769 + .6250}{2} = .6010,$$

the average of all elements in the S matrix is

$$\overline{S}_{vv'}(p) = \frac{.5769 + .6250 + 2(.4367)}{4} = .5188,$$

and the variance (biased estimate) of the two means (4.5167 and 5.0833) is .0803. It follows that observed-score profile variability is

$$\hat{\mathcal{V}}(\overline{X}_p) = .9(.6010 - .5188) + .0803 = .1542,$$

which is identical to the value reported in Table 10.3 based on direct computation of the average of the observed profile variances for the 10 persons.

For this synthetic data example, direct computation cannot be employed for any other design or sample size, because the observed scores for other designs and sample sizes are not known. However, Equation 10.56 can be used. So, for example, with $n_i' = 8$, Equation 10.41 provides the expected observed variances and covariances, which leads to

$$\hat{\mathcal{V}}(\overline{X}_p) = .9(.5428 - .4751) + .0803 = .1412.$$

---

[8]Note that the variances in S are unbiased estimates, whereas the "Var" estimates are biased. For example, $(10/9)(.5192) = .5769$.

For universe-score profile variability, applying Equation 10.55 gives

$$\mathcal{V}(\boldsymbol{\mu}_p) = \frac{n_p - 1}{n_p}[\bar{\sigma}_v^2(p) - \bar{\sigma}_{vv'}(p)] + \text{var}(\mu_v). \tag{10.57}$$

For the synthetic data example, the elements of the $\widehat{\boldsymbol{\Sigma}}_p$ matrix are given in Table 10.3. The average of the two variances is .3686, the average of all four variance-covariance elements is .3439, an estimate of var(μ_v) is var($\overline{X}_v$) = .0803, and the universe-score profile variability is

$$\hat{\mathcal{V}}(\boldsymbol{\mu}_p) = .9(.3686 - .3439) + .0803 = .1024.$$

Since universe scores are unknown, this result cannot be obtained through direct computation using persons' universe scores. We retain the multiplicative factor $(n_p - 1)/n_p$ solely for purposes of consistency with other $\mathcal{V}(\overline{\mathbf{X}}_p)$.

The variance of the δ-type errors for a randomly selected person is

$$\mathcal{V}(\boldsymbol{\delta}_p) = \boldsymbol{E}_p\left[\underset{v}{\text{var}}(\delta_{pv})\right] = \bar{\sigma}_v^2(\delta) - \bar{\sigma}_{vv'}(\delta), \tag{10.58}$$

where we have dropped the $(n_p - 1)/n_p$ multiplicative factor. For the synthetic data with $n_i' = 6$, $\hat{\mathcal{V}}(\boldsymbol{\delta}_p) = .058$. Consequently, the standard deviation of the δ-type errors of measurement for a typical person is $\sqrt{.058} = .24$. In a corresponding manner, we can express the variance of the Δ-type errors for a typical person as

$$\mathcal{V}(\boldsymbol{\Delta}_p) = \boldsymbol{E}_p\left[\underset{v}{\text{var}}(\Delta_{pv})\right] = \bar{\sigma}_v^2(\Delta) - \bar{\sigma}_{vv'}(\Delta), \tag{10.59}$$

which is $\hat{\mathcal{V}}(\boldsymbol{\Delta}_p) = .078$ for the synthetic data with $n_i' = 6$.

Relative Variability

As in univariate theory, it is natural to consider functions of variabilities, particularly ratios. One such ratio is

$$\mathcal{G} = \frac{\mathcal{V}(\boldsymbol{\mu}_p)}{\mathcal{V}(\overline{\mathbf{X}}_p)}, \tag{10.60}$$

which is the proportion of the variance in the profile of observed scores for a typical person that is explained by the variance in the profile of universe scores for such a person. If $\mathcal{V}(\boldsymbol{\mu}_p)$ is viewed as a measure of the flatness of the profile of universe scores, and $\mathcal{V}(\overline{\mathbf{X}}_p)$ is viewed as a measure of the flatness of the profile of observed scores, then $\mathcal{G}$ is a measure of the relative flatness of these profiles for a typical person.

$\mathcal{G}$ is also interpretable approximately as a type of generalizability coefficient for a randomly selected person p. For a specified person, we can define a person-level generalizability coefficient as the ratio of $\text{var}(\mu_{pv})$ to $\text{var}(\overline{X}_{pv})$, where the variance is taken over levels of v. Obviously, this ratio is not estimable for a given person, because universe scores are unknown. The expected value, over persons, of this ratio would be the expected generalizability coefficient for a randomly selected person. We approximate this expected value over persons with the ratio of the expected values in Equation 10.60.

For the synthetic data with $n_i' = 6$, $\hat{\mathcal{G}} = .1024/.1542 = .664$, which suggests that, for a typical person, 66% of the variance in observed mean scores for the n_v variables is attributable to variance in universe scores. For the synthetic data with $n_i' = 8$, $\hat{\mathcal{G}} = .1024/.1412 = .725$.

It is important to note that these variability formulas are for profiles— not composites. As such, a priori (w) weights and estimation (a) weights play no role. Also, it is helpful to remember that throughout this section we have been viewing data from the within-person, or "row," perspective. By contrast, when we consider one of the $n_v \times n_v$ matrices such as Σ_p, we are viewing data from a "column" perspective. The measures of (relative) profile variability discussed here provide overall or summative results for the measurement procedure, or average results for a typical person. They should not be construed as capturing all the information in the data. In particular, the variance and covariance components should be examined and reported always.

10.2.6   Hidden Facets

Hidden facets can occur in multivariate generalizability theory, just as they can in univariate generalizability theory. Obviously, the fact that a facet is hidden makes it easy to overlook. Less obviously, the consequences of failing to recognize a hidden facet can be seriously detrimental to interpretations. Often, a hidden facet is effectively fixed, and the likely consequences involve underestimating error variances, overestimating universe score variances, and severely overestimating coefficients. These conclusions are like those discussed previously for univariate analyses, but multivariate analyses involve the additional complexities of covariance components and attention to composites.

Suppose, for example, that the universe of generalization is best represented by the multivariate $p^{\bullet} \times I^{\circ} \times H^{\bullet}$ structure, which means, of course, the H is a random facet in the universe of generalization. However, when the G study data were collected, suppose the same single level of h was employed with all p, i, and v. In this case, undoubtedly the investigator will analyze the G study data according to the $p^{\bullet} \times i^{\circ}$ design, and the estimated variance and covariance components will be estimates of

$$\Sigma_p + \Sigma_{ph} = \left[\begin{array}{cc} \sigma_v^2(p) + \sigma_v^2(ph) & \sigma_{vv'}(p) + \sigma_{vv'}(ph) \\ \sigma_{vv'}(p) + \sigma_{vv'}(ph) & \sigma_{v'}^2(p) + \sigma_{v'}^2(ph) \end{array} \right] \quad (10.61)$$

$$\Sigma_i + \Sigma_{ih} = \left[\begin{array}{cc} \sigma_v^2(i) + \sigma_v^2(ih) & \\ & \sigma_{v'}^2(i) + \sigma_{v'}^2(ih) \end{array} \right] \quad (10.62)$$

$$\Sigma_{pi} + \Sigma_{pih} = \left[\begin{array}{cc} \sigma_v^2(pi) + \sigma_v^2(pih) & \\ & \sigma_{v'}^2(pi) + \sigma_{v'}^2(pih) \end{array} \right]. \quad (10.63)$$

Note that none of the elements of Σ_h is present in any of these three matrices. This is certainly one sense in which h is "hidden."

Now, suppose these $p^\bullet \times i^\circ$ estimated variance and covariance components are used to estimate relative error variances for the categories and the composite. Further, suppose the G and D study sample sizes are the same, which means, among other things, that the investigator really wants $n_h' = 1$ for decisions based on the measurement procedure. Clearly, n_{iv} will divide not only the pi variance components but also the pih variance components, which is what the investigator intends. However, estimates of the ph variance components will be absent from estimates of all relative error variances. This will result in underestimates of relative error variance for each universe score.

If the w weights are all nonnegative, the absence of the ph variance components from the category error variances will also lead to underestimating the composite relative error variance. In addition, the composite relative error variance will be underestimated even more because of the absence of the $\sigma_{vv'}(ph)$, assuming the $\sigma_{vv'}(ph)$ are positive. If some of the $\sigma_{vv'}(ph)$ are negative and/or some of the w weights are negative (e.g., when the composite is a difference score), composite relative error variance is still likely to be biased, but the direction of the bias would have to be determined on a case-by-case basis.

The same conclusions discussed in the previous paragraph apply to absolute error variances, as well. By contrast, universe score variances for the categories and the composite will tend to be overestimated, because they will include the ph variance components. For composite universe score variance there is an additional inflation factor caused by the presence of the $\sigma_{vv'}(ph)$, assuming the $\sigma_{vv'}(ph)$ and the w weights are positive. The conjunction of underestimating error variances and overestimating universe score variance will likely lead to dramatic overestimates of coefficients. Exceptions to these conclusions will occur only if the variance components involving h are negligible; or, in the case of the composite, possibly if some of the $\sigma_{vv'}(ph)$ and/or the w_v are negative.

In beginning this discussion of hidden facets, we assumed that each observation involved the same *single* level of the h facet. This is probably the most frequent way a hidden facet arises. The same logic would apply, however, if each observation in the $p^\bullet \times i^\circ$ design were a mean over the *same* n_h levels of the h facet. Notationally, we would represent this difference by using pH rather than ph in Equations 10.61 through 10.63. We are not suggesting it is necessarily good practice to collapse observations over a facet in this manner. We are merely noting that an analysis of such collapsed data effectively treats H as fixed, which produces biased estimates if the intent is that H be random.

The occurrence of hidden facets is much more common than generally realized. To ascertain whether hidden facets are present, an investigator must clearly specify the intended universe of generalization. If the D study design does not explicitly represent all facets in the universe of generalization, then one or more facets are hidden. Probably the most frequently encountered hidden facet is "occasion." Very often, an investigator's intended universe of generalization involves an occasion facet, in the sense that the investigator intends that the measurement procedure be applicable on different occasions. However, it is relatively rare for G study data to be collected on multiple occasions, and if that is not done, then occasion is a hidden facet. If so, the investigator's conclusions about generalizability need to recognize the biasing consequences of hidden facets.

It is possible for a hidden facet to be random. For example, a hidden occasion facet would be random if a different occasion were used to obtain data for each person. As discussed in Section 5.1.4 for univariate analyses, a hidden random facet induces bias different from that induced by a hidden fixed facet, and the magnitude and direction of bias are not easily ascertained.

10.2.7 Collapsed Fixed Facets in Multivariate Analyses

Suppose a G study is conducted using the $p^{\bullet} \times i^{\circ} \times h^{\bullet}$ design, but subsequently an investigator decides that the n_h levels of the h facet will be considered fixed for some particular D study purpose. That is, for the D study there will be $n_h \times n_v$ levels of the fixed facet. How can the investigator obtain estimates of variance and covariance components to use in this D study? One approach is to reanalyze the G study data. Another approach that *may* be appropriate is to "collapse" the variance-covariance matrices for the $p^{\bullet} \times i^{\circ} \times h^{\bullet}$ design.

Reanalyzing the G study data for the design with $n_h \times n_v$ levels for the two fixed facets is complicated by the fact that i is nested within v but crossed with h. For the simplest case of $n_h = n_v = 2$, the G study variance and covariance components can be arranged in three 4×4 matrices with the structure:

$$
\Sigma_p = \begin{array}{cccc} (v_1 h_1) & (v_1 h_2) & (v_2 h_1) & (v_2 h_2) \end{array}
$$

$$
\Sigma_p = \left[\begin{array}{cccc} X & X & X & X \\ X & X & X & X \\ X & X & X & X \\ X & X & X & X \end{array} \right]
$$

$$\Sigma_i = \begin{bmatrix} X & X & & \\ X & X & & \\ & & X & X \\ & & X & X \end{bmatrix}$$

$$\Sigma_{pi} = \begin{bmatrix} X & X & & \\ X & X & & \\ & & X & X \\ & & X & X \end{bmatrix}.$$

Once these variance-covariance matrices are available, all D study issues can be addressed. In particular, any desired weights can be applied to each of the $n_h n_v$ fixed levels. However, reanalyzing the G study data may not be possible. Or, even if it can be done, the reanalysis may be difficult. In any case, the structure of the resulting matrices is clearly somewhat complicated.

A simpler approach is possible if: (i) the investigator does not need to examine variance and covariance components for each of the fixed levels of h; and (ii) for any composite, the w weights for each level of h are the same, as are the estimation weights. Under these circumstances, the available estimated variance and covariance components for the $p^\bullet \times i^\circ \times h^\bullet$ design can be arrayed as follows.

$$\widehat{\Sigma}_{p|H} = \begin{bmatrix} \hat{\sigma}_v^2(p) + \hat{\sigma}_v^2(pH) & \hat{\sigma}_{vv'}(p) + \hat{\sigma}_{vv'}(pH) \\ \hat{\sigma}_{vv'}(p) + \hat{\sigma}_{vv'}(pH) & \hat{\sigma}_{v'}^2(p) + \hat{\sigma}_{v'}^2(pH) \end{bmatrix} \quad (10.64)$$

$$\widehat{\Sigma}_{i|H} = \begin{bmatrix} \hat{\sigma}_v^2(i) + \hat{\sigma}_v^2(iH) & \\ & \hat{\sigma}_{v'}^2(i) + \hat{\sigma}_{v'}^2(iH) \end{bmatrix} \quad (10.65)$$

$$\widehat{\Sigma}_{pi|H} = \begin{bmatrix} \hat{\sigma}_v^2(pi) + \hat{\sigma}_v^2(piH) & \\ & \hat{\sigma}_{v'}^2(pi) + \hat{\sigma}_{v'}^2(piH) \end{bmatrix}. \quad (10.66)$$

The estimated variance and covariance components that involve H are simply the corresponding h components divided by n_h. These G study matrices are of size $n_v \times n_v$, which is much smaller than the $(n_h n_v) \times (n_h n_v)$ matrices that result from the previously discussed reanalysis of the G study data. The D study variance-covariance matrices for the simplified procedure are easily obtained in the usual manner, that is, by dividing the elements of $\widehat{\Sigma}_{i|H}$ and $\widehat{\Sigma}_{pi|H}$ by the desired values of n'_{iv}.

In essence, this simplified approach treats the fixed h facet in much the same manner that a fixed facet is treated in univariate generalizability theory, whereas the fixed v facet is accorded a complete multivariate treatment. This simplified approach may be sensible, or even necessary, in some circumstances, but simplification has its costs. In particular, when a fixed facet is treated in a univariate manner, the analysis "collapses" over levels of the facet. Consequently, it is not possible to consider or weight any level of the facet differently from any other level.

This simplified treatment of a fixed facet is summarized most succinctly by the three matrices in Equations 10.64 to 10.66, which bear obvious similarities in form to the hidden-facet matrices in Equations 10.61 to 10.63 of Section 10.2.6. There are very important differences, however. First, the matrices in Equations 10.64 to 10.66 result from an intent to treat H as fixed; that is, H is a fixed facet in the universe of generalization. By contrast, in the hidden-facet Equations 10.61 to 10.63, the intent is that h be random in the universe, but actually h is fixed in the design. Second, each of the variance and covariance components in Equations 10.64 to 10.66 is estimable, whereas only the sums are estimable in the hidden-facet Equations 10.61 to 10.63. Obviously, disentangling these complicated matters requires careful attention to the intended universe of generalization and the manner in which a particular design adequately reflects it, or fails to do so.

10.2.8 Computer Programs

The computer program mGENOVA (Brennan, 2001b) that is described in Appendix H can be used to perform all of the analyses discussed in Section 10.1 for the D study versions of the designs in Table 9.2. Also, mGENOVA can assist in performing the analyses discussed in this Section 10.2.

If estimated G study variance and covariance components are available, a spreadsheet program is adequate for obtaining most of the D study results discussed in this chapter. Also, many results discussed here can be formulated in terms of operations on matrices, which are easily programmed using SAS IML.

10.3 Examples

This section discusses three real-data examples of multivariate G and D studies. The first example involves the $p^\bullet \times i^\circ$ and $p^\bullet \times I^\circ$ designs; the second example employs the $p^\bullet \times t^\bullet \times r^\bullet$ and $p^\bullet \times T^\bullet \times R^\bullet$ designs; and the third example uses the $p^\bullet \times t^\bullet \times r^\circ$, $p^\bullet \times T^\bullet \times R^\circ$, and $p^\bullet \times T^\circ \times R^\circ$ designs. Almost all of the theoretical results discussed in this chapter are illustrated in one or more of these examples.

10.3.1 ACT Assessment Mathematics

The ACT Assessment Mathematics Test (*Math*) is a 60-item, 60-minute test (ACT, 1997, p. 9)

> designed to assess the mathematical skills that students have typically acquired in courses taken up to the beginning of grade 12. The test presents multiple-choice items that require students

TABLE 10.4. G Study Variance and Covariance Components for ACT Assessment *Math*

			Persons			
Form	$\hat{\sigma}_1^2(p)$	$\hat{\sigma}_2^2(p)$	$\hat{\sigma}_3^2(p)$	$\hat{\sigma}_{12}(p)$	$\hat{\sigma}_{13}(p)$	$\hat{\sigma}_{23}(p)$
1	.03210	.03645	.03094	.03249	.03015	.03120
2	.03793	.04884	.04050	.03924	.03621	.04158
3	.03104	.03359	.04507	.02982	.03394	.03790
4	.03066	.03741	.04317	.03189	.03351	.03833
5	.03987	.03078	.03048	.03258	.03234	.03025
6	.03334	.03525	.03339	.03215	.03117	.03236
7	.02963	.03822	.02799	.03229	.02640	.03118
8	.02868	.02737	.04205	.02659	.03234	.03205
9	.02884	.03877	.02748	.03180	.02590	.03142
Mean	.03245	.03630	.03567	.03209	.03133	.03403
SE	.00397	.00599	.00698	.00330	.00340	.00410
		Variance	and Covariances	of Estimates[a]		
$\hat{\sigma}_1^2(p)$	1.58e-05	*.252*	*-.054*	*.643*	*.523*	*.189*
$\hat{\sigma}_2^2(p)$	5.99e-06	3.59e-05	*-.054*	*.896*	*.117*	*.595*
$\hat{\sigma}_3^2(p)$	-1.49e-06	-2.27e-06	4.87e-05	*-.147*	*.811*	*.754*
$\hat{\sigma}_{12}(p)$	8.44e-06	1.77e-05	-3.39e-06	1.09e-05	*.263*	*.484*
$\hat{\sigma}_{13}(p)$	7.08e-06	2.38e-06	1.92e-05	2.95e-06	1.16e-05	*.738*
$\hat{\sigma}_{23}(p)$	3.08e-06	1.46e-05	2.16e-05	6.54e-06	1.03e-05	1.68e-05

		Items		Persons × Items		
Form	$\hat{\sigma}_1^2(i)$	$\hat{\sigma}_2^2(i)$	$\hat{\sigma}_3^2(i)$	$\hat{\sigma}_1^2(pi)$	$\hat{\sigma}_2^2(pi)$	$\hat{\sigma}_3^2(pi)$
1	.02532	.03698	.04374	.17681	.17831	.17724
2	.02718	.03469	.03122	.15167	.16831	.17738
3	.03677	.04356	.03482	.14213	.17528	.17081
4	.03096	.04227	.03045	.16712	.17230	.17781
5	.01848	.04967	.03082	.17122	.17232	.18355
6	.04109	.03686	.02912	.15989	.17945	.18912
7	.02916	.02494	.05042	.17172	.18710	.17299
8	.05002	.03760	.02077	.15793	.18693	.18826
9	.02980	.02269	.05008	.17237	.18883	.17423
Mean	.03209	.03658	.03572	.16343	.17876	.17904
SE	.00932	.00856	.01016	.01138	.00744	.00654

[a]Italicized values in upper off-diagonal positions are correlations.

to use their reasoning skills to solve practical problems in mathematics. . . . Four scores are reported for the ACT Mathematics Test: a total test score based on all 60 items, a subscore in Pre-Algebra/Elementary Algebra based on 24 items, a subscore in Intermediate Algebra/Coordinate Geometry based on 18 items, and a subscore in Plane Geometry/Trigonometry based on 18 items.

Here, we view the three subscores as categories in a table of specifications, which means that the G study design is $p^{\bullet} \times i^{\circ}$. This is a real-data version of the synthetic data example used in the introduction to multivariate generalizability theory in Section 9.1.

Table 10.4 provides estimated variance and covariance components for nine forms of *Math*, where each form was administered to over 3000 examinees in a random-groups equating design. Also provided are the means and standard deviations of the estimated variance and covariance components. The standard deviations are empirical estimates of the standard errors, without making normality assumptions. So, for example, the mean of the estimated variance components for the first subscore is .03245 with a standard error of .00397. The relatively small magnitude of this standard error is reflected by the similarity of the nine estimates. The standard errors for the other estimated variance and covariance components are also quite small. For estimated variance and covariance components for persons, the standard errors are on the order of one-tenth of the estimates; for items, the standard errors are on the order of one-fourth of the estimates; and for the interactions, the standard errors are on the order of one-twentieth of the estimates.

Using the means of the estimated variances and covariances in Table 10.4, the G study matrices are

$$\widehat{\Sigma}_p = \begin{bmatrix} .03245 & .03209 & .03133 \\ .03209 & .03630 & .03403 \\ .03133 & .03403 & .03567 \end{bmatrix}$$

$$\widehat{\Sigma}_i = \begin{bmatrix} .03209 & & \\ & .03658 & \\ & & .03572 \end{bmatrix}$$

$$\widehat{\Sigma}_{pi} = \begin{bmatrix} .16343 & & \\ & .17876 & \\ & & .17904 \end{bmatrix}.$$

Using D study sample sizes of $n_{i1} = 24$, $n_{i2} = 18$, and $n_{i3} = 18$, the D study estimated variance-covariance matrices are

$$\widehat{\Sigma}_p = \begin{bmatrix} .03245 & .03209 & .03133 \\ .03209 & .03630 & .03403 \\ .03133 & .03403 & .03567 \end{bmatrix}$$

$$\widehat{\Sigma}_I \;=\; \begin{bmatrix} .00134 & & \\ & .00203 & \\ & & .00198 \end{bmatrix}$$

$$\widehat{\Sigma}_{pI} \;=\; \begin{bmatrix} .00681 & & \\ & .00993 & \\ & & .00995 \end{bmatrix}.$$

It follows that the estimated universe score, relative error, and absolute error matrices are

$$\widehat{\Sigma}_p \;=\; \begin{bmatrix} .03245 & \mathit{.93499} & \mathit{.92088} \\ .03209 & .03630 & \mathit{.94571} \\ .03133 & .03403 & .03567 \end{bmatrix} \tag{10.67}$$

$$\widehat{\Sigma}_\delta \;=\; \begin{bmatrix} .00681 & & \\ & .00993 & \\ & & .00995 \end{bmatrix} \tag{10.68}$$

$$\widehat{\Sigma}_\Delta \;=\; \begin{bmatrix} .00815 & & \\ & .01196 & \\ & & .01193 \end{bmatrix}, \tag{10.69}$$

where the italicized values in the upper diagonal positions of $\widehat{\Sigma}_p$ are disattenuated correlations.

It seems sensible to assume that the numbers of items that contribute to the three subscores are reflective of the relative importance of the categories for the universe of generalization intended by ACT. Under this assumption, the a priori weights are $w_1 = .4$, $w_2 = .3$, and $w_3 = .3$. Using these weights, estimated composite variances are

$$\hat{\sigma}_C^2(p) = .03302, \quad \hat{\sigma}_C^2(\delta) = .00288, \quad \text{and} \quad \hat{\sigma}_C^2(\Delta) = .00345,$$

and the reliability-like coefficients are

$$E\hat{\rho}^2 = .920 \quad \text{and} \quad \widehat{\Phi} = .905.$$

Actually, composite absolute error variance and Φ may be of questionable relevance for *Math*, because the various forms are carefully equated. As a result, the scores that are used operationally are adjusted for differences in the difficulty of the forms, and it is probably sensible to assume that $\Delta_{pC} = \delta_{pC}$.[9]

For *Math*, ACT reports scores for the three subscores as well as the composite. Therefore, it may be relevant to consider characteristics of profiles

---

[9]This is an incomplete explanation because it does not take into account the fact that the transformations of raw scores for *Math* forms to operational (scale) scores are nonlinear (see Kolen & Brennan, 1995). Such transformations have consequences beyond simply removing average form differences. We overlook such complexities here.

of subscores for *Math*. Although subscores for individuals are not available, we can use the variance formulas in Section 10.2.5 to characterize the profile for a typical examinee. For example, using Equation 10.57 the expected variance of the profile of universe scores is

$$\mathcal{V}(\mu_p) = (.03481 - .03326) + .00616 = .0077,$$

and using Equation 10.57 the expected variance of the profile of observed scores is

$$\mathcal{V}(\overline{\mathbf{X}}_p) = (.04370 - .03622) + .00616 = .0136,$$

where .00616 is the average (over forms) of the variance of the three mean scores.[10] It follows that, for a typical person, the proportion of the variance in observed scores explained by universe scores is[11]

$$\mathcal{G} = \frac{.0077}{.0136} \doteq .57.$$

In this sense, we might say that universe-score profiles are a little over 50% flatter than observed-score profiles. Another interpretation is that a generalizability coefficient for a typical person is approximately .57, in the sense discussed in Section 10.2.5.

Importance of Table of Specifications

The disattenuated correlations in $\widehat{\boldsymbol{\Sigma}}_p$ in Equation 10.67 are all quite high, which certainly suggests that the constructs underlying the three subscores are highly correlated. This does not mean, however, that adherence to the table of specifications is of little importance. One perspective on this issue is to consider the consequences of assigning all items to one of the three categories.

For example, the mean-square relative error associated with assigning all 60 items to the first category is obtained using Equation 10.37 with $a_1 = 1$, $a_2 = 0$, and $a_3 = 0$, and with $n_{i1} = 60$, $n_{i2} = 0$, and $n_{i3} = 0$. This gives $MSE_C(\delta) = .00418$. Assigning all items to the second category gives $MSE_C(\delta) = .00442$, and assigning them to the third category gives $MSE_C(\delta) = .00479$. These values are about 50% larger than composite relative error variance $[\hat{\sigma}_C^2(\delta) = .00288]$, which clearly illustrates that adherence to the table of specifications has an impact on measurement precision. Of course, minor deviations from the 24/18/18 specifications would make much less difference.

---

[10]In computing both $\mathcal{V}(\mu_p)$ and $\mathcal{V}(\overline{\mathbf{X}}_p)$, we have dropped the $(n_p - 1)/n_p$ factor. It is unimportant with over 3000 examinees per form.

[11]This result is for illustrative purposes only; it is based on raw scores, not the scale scores actually reported for *Math*.

More about Standard Errors

Because data for multiple forms are available for *Math*, standard errors of
variance and covariance components for the universe of admissible observations are easily estimated without making any normality assumptions,
as discussed previously. However, it is instructive to consider what the estimated standard errors might be under normality assumptions.

For example, using Equation 9.55, the estimated standard error of $\hat{\sigma}_{12}(p)$
for any given form is

$$\hat{\sigma}[\hat{\sigma}_{12}(p)] = \sqrt{\frac{S_1^2(p)\,S_2^2(p) + [S_{12}(p)]^2}{(n_p - 1) + 2}},$$

where

$$S_v^2(p) = \hat{\sigma}_v^2(p) + \hat{\sigma}_v^2(pi)/n_{iv}.$$

Using this formula with data for the first form in Table 10.4 gives

$$\hat{\sigma}[\hat{\sigma}_{12}(p)] = \sqrt{\frac{(.03947)(.04636) + (.03249)^2}{3000}} \doteq .001,$$

where 3000 is an approximate sample size.[12]

The reader can verify that the estimated standard errors of $\hat{\sigma}_{12}(p)$ for all
nine forms are about the same size, namely, .001. This is about three times
smaller than the estimate based on direct computation without normality
assumptions, namely, .003, as provided in Table 10.4. A comparable result
holds for the other estimated covariance components. Also, the standard
errors of the estimated variance components are too small under normality assumptions. Clearly, for this example, the normality-based formulas
lead to underestimates of estimated standard errors. This result may be
influenced by the fact that the underlying data are dichotomous. Also,
the very large sample size for persons pretty much guarantees that the
normality-based estimates will be small for estimated variance and covariance components involving p .

Estimated variance and covariance components are not only fallible—
they are also correlated. For *Math* the variances and covariances (over
forms) of the $\hat{\sigma}_v^2(p)$ and the $\hat{\sigma}_{vv'}(p)$ are provided in the middle of Table 10.4.
The square roots of the variances are the standard errors. The upper diago-
nal elements are correlations. It is evident that some of these estimates are
substantially correlated. The availability of multiple forms makes it possi-
ble to estimate the variances and covariances of the $\hat{\sigma}_v^2(p)$ and the $\hat{\sigma}_{vv'}(p)$
relatively easily, and without making normality assumptions.

D study estimated composite variances such as $\hat{\sigma}_C^2(p)$, $\hat{\sigma}_C^2(\delta)$, and $\hat{\sigma}_C^2(\Delta)$
are subject to sampling variability, too. When multiple forms are available,

---

[12]The actual sample size for the first form and all the other forms is slightly larger
than 3000, but that is immaterial here.

as they are for the *Math* example, standard errors of these quantities can be computed directly. For example, to estimate the standard error of $\hat{\sigma}_C^2(p)$, we can compute $\hat{\sigma}_C^2(p)$ for each form, and then compute the standard deviation of these estimates. Doing so gives $\hat{\sigma}[\hat{\sigma}_C^2(p)] = .0030$.

10.3.2 Painting Assessment

Nuβbaum (1984) describes a real-data example of the G study $p^{\bullet} \times t^{\bullet} \times r^{\bullet}$ design and the D study $p^{\bullet} \times T^{\bullet} \times R^{\bullet}$ design. This is the *Painting* example introduced in Section 9.2.2. In Nuβbaum's study, 60 fourth-grade students (p) were asked to create watercolor paintings on four topics (t). Each painting was evaluated by 25 art students (r); that is, each art student evaluated all 240 paintings. Evaluations were made on three different 10-point scales (v):

1. Are persons and things represented in an objective way?

2. Is the background appropriate?

3. Do relations between the objects become clear to the viewer?

Table 10.5 provides the mean-square and mean-product matrices, along with the G study matrices of estimated variance and covariance components.[13] The rows and columns of the matrices correspond to the three questions about representations, background, and object-relationships, respectively. For example, the estimator of the person covariance component for v and v' is

$$\hat{\sigma}_{vv'}(p) = \frac{MP_{vv'}(p) - MP_{vv'}(pt) - MP_{vv'}(pr) + MP_{vv'}(ptr)}{n_t n_r}.$$

When this formula is applied to v_1 (representations) and v_2 (background), we obtain

$$\hat{\sigma}_{12}(p) = \frac{37.1744 - 7.4388 - 1.0794 + .6638}{4 \times 25} = .2932.$$

Since the $p^{\bullet} \times t^{\bullet} \times r^{\bullet}$ is a fully crossed design, all matrices are full and symmetric, although only the diagonal and lower diagonal elements are explicitly reported in Table 10.5.

It is evident that the elements of $\widehat{\boldsymbol{\Sigma}}_r$ are quite large, which suggests that, for small numbers of raters, absolute error variances are likely to be substantially larger than relative error variances. Note also that the elements of $\widehat{\boldsymbol{\Sigma}}_{pt}$ are substantially larger than the elements of $\widehat{\boldsymbol{\Sigma}}_{pr}$. This suggests that increasing the number of topics will reduce relative error

---

[13]Nuβbaum (1984) does not provide the mean squares and mean products, but they are easily computed using Equation 9.47.

TABLE 10.5. G Study for *Painting* Example

$$\mathbf{M}_p = \begin{bmatrix} 52.9630 & & \\ 37.1744 & 43.2336 & \\ 49.0604 & 42.4304 & 52.4574 \end{bmatrix} \qquad \widehat{\boldsymbol{\Sigma}}_p = \begin{bmatrix} .4156 & & \\ .2932 & .3153 & \\ .3908 & .3295 & .3897 \end{bmatrix}$$

$$\mathbf{M}_t = \begin{bmatrix} 79.5866 & & \\ -37.6932 & 26.3060 & \\ 42.9452 & -21.4924 & 55.9722 \end{bmatrix} \qquad \widehat{\boldsymbol{\Sigma}}_t = \begin{bmatrix} .0424 & & \\ -.0314 & .0053 & \\ .0196 & -.0217 & .0252 \end{bmatrix}$$

$$\mathbf{M}_r = \begin{bmatrix} 332.6465 & & \\ 268.7274 & 288.0886 & \\ 321.8674 & 304.6384 & 384.1964 \end{bmatrix} \qquad \widehat{\boldsymbol{\Sigma}}_r = \begin{bmatrix} 1.3568 & & \\ 1.1070 & 1.1601 & \\ 1.3196 & 1.2549 & 1.5703 \end{bmatrix}$$

$$\mathbf{M}_{pt} = \begin{bmatrix} 10.8626 & & \\ 7.4388 & 10.9400 & \\ 9.5672 & 8.9156 & 13.0002 \end{bmatrix} \qquad \widehat{\boldsymbol{\Sigma}}_{pt} = \begin{bmatrix} .3805 & & \\ .2710 & .3782 & \\ .3518 & .3264 & .4534 \end{bmatrix}$$

$$\mathbf{M}_{pr} = \begin{bmatrix} 1.8905 & & \\ 1.0794 & 2.2486 & \\ 1.1852 & 1.3204 & 2.1492 \end{bmatrix} \qquad \widehat{\boldsymbol{\Sigma}}_{pr} = \begin{bmatrix} .1351 & & \\ .1039 & .1909 & \\ .1033 & .1412 & .1218 \end{bmatrix}$$

$$\mathbf{M}_{tr} = \begin{bmatrix} 6.4741 & & \\ 2.6318 & 8.9010 & \\ 4.7502 & 2.8976 & 6.8372 \end{bmatrix} \qquad \widehat{\boldsymbol{\Sigma}}_{tr} = \begin{bmatrix} .0854 & & \\ .0328 & .1236 & \\ .0663 & .0357 & .0862 \end{bmatrix}$$

$$\mathbf{M}_{ptr} = \begin{bmatrix} 1.3501 & & \\ .6638 & 1.4850 & \\ .7722 & .7556 & 1.6652 \end{bmatrix} \qquad \widehat{\boldsymbol{\Sigma}}_{ptr} = \begin{bmatrix} 1.3501 & & \\ .6638 & 1.4850 & \\ .7722 & .7556 & 1.6652 \end{bmatrix}$$

Note. All matrices are symmetric. Sample sizes are: $n_p = 60$ students, $n_t = 4$ topics, and $n_r = 25$ raters.

variance faster than a corresponding increase in the number of raters. These somewhat abstract statements, of course, should be confirmed by examining specific D study results.

Raters Random

The top half of Table 10.6 reports D study results for the D study random effects $p^{\bullet} \times T^{\bullet} \times R^{\bullet}$ design based on two raters and two topics, as well as two raters and four topics. For both pairs of D study sample sizes, estimated universe-score, relative-error, and absolute-error variance-covariance matrices are reported. For this random effects multivariate design, the estimated universe score matrix is simply $\widehat{\boldsymbol{\Sigma}}_p$. The estimated relative-error matrix is

$$\widehat{\boldsymbol{\Sigma}}_\delta = \frac{\widehat{\boldsymbol{\Sigma}}_{pt}}{n_t'} + \frac{\widehat{\boldsymbol{\Sigma}}_{pr}}{n_r'} + \frac{\widehat{\boldsymbol{\Sigma}}_{ptr}}{n_t' n_r'},$$

TABLE 10.6. Some D Study Results for *Painting* Example

	Raters Random with $n'_r = 2$					
	$n'_t = 2$			$n'_t = 4$		
$\widehat{\Sigma}_\tau$	.4156	*.810*	*.971*	.4156	*.810*	*.971*
	.2932	.3153	*.940*	.2932	.3153	*.940*
	.3908	.3295	.3897	.3908	.3295	.3897
$\widehat{\Sigma}_\delta$	.5953	*.566*	*.650*	.3314	*.574*	*.663*
	.3534	.6558	*.622*	.2027	.3756	*.651*
	.4206	.4227	7039	.2361	.2467	.3824
$\widehat{\Sigma}_\Delta$	1.3163	*.696*	*.782*	1.0311	*.751*	*.823*
	.8994	1.2694	*.754*	.7524	.9725	*.814*
	1.1068	1.0482	1.5232	.9091	.8731	1.1846

Composite Defined Using Weights of $w_1 = .50$, $w_2 = .33$, and $w_3 = .17$

$\hat{\sigma}^2_C(\tau)$	.349		.349
$\hat{\sigma}^2_C(\delta)$	.476		.270
$\hat{\sigma}^2_C(\Delta)$	1.113		.898
$E\hat{\rho}^2$	.423		.564
$\widehat{\Phi}$	.239		.280

	One Fixed Rater					
	$n'_t = 2$			$n'_t = 4$		
$\widehat{\Sigma}_\tau$	.5507	*.752*	*.931*	.5507	*.752*	*.931*
	.3971	.5062	*.925*	.3971	.5062	*.925*
	.4941	.4707	.5115	.4941	.4707	.5115
$\widehat{\Sigma}_\delta$	.8653	*.521*	*.587*	.4327	*.521*	*.587*
	.4674	.9316	*.545*	.2337	.4658	*.545*
	.5620	.5410	1.0593	.2810	.2705	.5297
$\widehat{\Sigma}_\Delta$	.9292	*.487*	*.594*	.4646	*.487*	*.594*
	.4681	.9961	*.520*	.2341	.4980	*.520*
	.6050	.5480	1.1150	.3025	.2740	.5575

Composite Defined Using Weights of $w_1 = .50$, $w_2 = .33$, and $w_3 = .17$

$\hat{\sigma}^2_C(\tau)$	.475		.475
$\hat{\sigma}^2_C(\delta)$	.659		.329
$\hat{\sigma}^2_C(\Delta)$	.692		.346
$E\hat{\rho}^2$	.419		.591
$\widehat{\Phi}$	.407		.579

Note. Italicized values in upper off-diagonal positions are correlations.

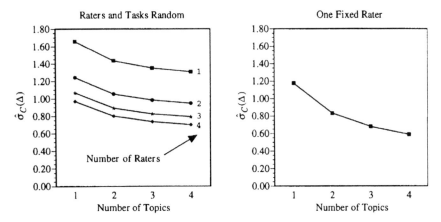

FIGURE 10.1. Estimated absolute-error SEMs for composite for *Painting* example.

where each divisor applies to all elements of the matrix in the numerator.[14]
Similarly,

$$\widehat{\Sigma}_{\Delta} = \frac{\widehat{\Sigma}_t}{n'_t} + \frac{\widehat{\Sigma}_r}{n'_r} + \frac{\widehat{\Sigma}_{pt}}{n'_t} + \frac{\widehat{\Sigma}_{pr}}{n'_r} + \frac{\widehat{\Sigma}_{tr}}{n'_t n'_r} + \frac{\widehat{\Sigma}_{ptr}}{n'_t n'_r}.$$

The upper-diagonal elements of these matrices are correlations. For example, the upper-diagonal elements of $\widehat{\Sigma}_r$ are disattenuated correlations reflecting the extent to which the three criteria measure the same construct. Since these correlations range from about .81 to .97, it is evident that the three criteria are highly correlated but not identical. The correlations in $\widehat{\Sigma}_\delta$ and $\widehat{\Sigma}_\Delta$ are direct indicators of correlated relative error and correlated absolute error, respectively. Note that the correlations in $\widehat{\Sigma}_\delta$ and $\widehat{\Sigma}_\Delta$ *increase* as the number of tasks increases, although the covariances *decrease*, as they must. Clearly, by just about any standard, there are substantial degrees of correlation among both the relative and absolute error components.

Nuβbaum (1984) suggests that when paintings of elementary school children are to be judged, the first criterion (representations) is most important, followed by the second (background) and third (object-relationships). In his study, he employs a priori weights of $w_1 = .50$, $w_2 = .33$, and $w_1 = .17$. For a composite based on these weights, and with the estimation weights equal to them, the middle of Table 10.6 provides composite results. Also, the left-hand graph in Figure 10.1 provides estimated com-

---

[14]This notational convention is applicable in this case because the $p^\bullet \times T^\bullet \times R^\bullet$ design is fully crossed. For any variance-covariance matrix in a fully crossed design, the D study sample sizes are the same for all elements in the matrix. For designs involving nesting, a more complicated matrix notation would be required.

posite absolute-error SEMs for numbers of raters and topics ranging from one to four. If an investigator wanted 68% of the students to have universe scores within one absolute-error SEM of their observed scores, Figure 10.1 suggests that it would be appropriate to use two raters and three topics.

One Fixed Rater

Nuβbaum (1984, p. 227) states,

> A teacher may have good reason not to generalize over judges. He or she may, for example, regard his or her own statement as the only valid one because other teachers have no knowledge of the conditions under which the pictures were painted.

Nuβbaum uses this logic in support of performing a multivariate D study analysis for a single fixed rater, in the sense discussed in Section 10.2.7. That is, his D study analysis collapses over the levels of the rater facet. Or, stated differently, the rater facet is treated as fixed in the univariate sense with $n'_r = 1$.

The bottom half of Table 10.6 reports results for the $p^\bullet \times T^\bullet \times R^\bullet$ D study design with a single fixed rater, for two and four topics. For this design, the estimated universe score matrix is

$$\widehat{\boldsymbol{\Sigma}}_\tau = \widehat{\boldsymbol{\Sigma}}_p + \widehat{\boldsymbol{\Sigma}}_{pr},$$

the estimated relative-error matrix is

$$\widehat{\boldsymbol{\Sigma}}_\delta = \frac{\widehat{\boldsymbol{\Sigma}}_{pt}}{n'_t} + \frac{\widehat{\boldsymbol{\Sigma}}_{ptr}}{n'_t},$$

and the estimated absolute-error matrix is

$$\widehat{\boldsymbol{\Sigma}}_\Delta = \frac{\widehat{\boldsymbol{\Sigma}}_t}{n'_t} + \frac{\widehat{\boldsymbol{\Sigma}}_{pt}}{n'_t} + \frac{\widehat{\boldsymbol{\Sigma}}_{tr}}{n'_t} + \frac{\widehat{\boldsymbol{\Sigma}}_{ptr}}{n'_t}.$$

The bottom of Table 10.6 provides composite results using the a priori weights of $w_1 = .50$, $w_2 = .33$, and $w_1 = .17$. Also, the right-hand graph in Figure 10.1 provides estimated composite absolute-error SEMs for one to four topics. As in univariate theory, fixing raters leads to smaller composite error variances and SEMs, and larger coefficients.

Nuβbaum's (1984) one-fixed-rater D study analysis is probably not ideal. The likely ideal would be to conduct separate D study $p^\bullet \times T^\bullet$ analyses for each specific judge of interest. Alternatively, if the judges of interest were the same 25 judges used in the G study, then a multivariate $p^\bullet \times T^\bullet$ analysis with $25 \times 3 = 75$ levels of v could be employed. Neither of these idealized alternatives is likely to be viable in most circumstances, and Nuβbaum's (1984) one-fixed-rater D study analysis is a reasonable alternative for practical use.

10.3.3 Listening and Writing Assessment

Brennan et al. (1995) describe a Listening and Writing assessment that provides an example of the G study $p^\bullet \times t^\bullet \times r^\circ$ design, and the D study $p^\bullet \times T^\bullet \times R^\circ$ and $p^\bullet \times T^\circ \times R^\circ$ designs. This is the LW example introduced in Section 9.2.2. Each of three preliminary forms (402, 404, 406) of the Listening and Writing tests was administered to one of three groups of 50 examinees. Each examinee (p) listened to 12 tape-recorded messages (t). For each form, the messages were different. Examinees were told to take notes while each message was played. At the conclusion of each message, examinees were told to use their notes to construct a written message. The written messages were scored by trained raters on a five-point holistic scale for both listening skills (L) and writing skills (W). The listening score reflected the accuracy and completeness of the information provided in the written messages. The writing score reflected other aspects of the "quality" of the writing. For each form there were two distinct sets of raters (r). Three raters evaluated the written messages for listening, and a different three raters evaluated the messages for writing. The groups of listening and writing raters were different for each form.

This is a relatively rare example of a true G study in the sense that the primary purpose of gathering the data was to obtain empirical evidence to inform judgments about the design characteristics of measurement procedure that would be used operationally. Table 10.7 provides the G study variance and covariance matrices for each of the forms, as well as the averages over forms. For each matrix, the first row and column is for Listening and the second is for Writing.

Also provided in Table 10.7 are the estimated standard errors of the averages of the estimated variance and covariance components. These are empirical standard errors; that is, they are the standard deviations of the elements divided by the square root of three. As such, they are not based on any normality assumptions. In general, it is clear that the estimated standard errors are small relative to the estimated variance and covariance components themselves.

In the remainder of our discussion of the LW example, we focus on the average estimates for two reasons: they are more stable than the individual form estimates, and the intended inferences to be drawn are for a "general" form (i.e., a randomly parallel form) as opposed to any specific form.

It is evident that $\hat{\sigma}_L^2(p)$ and $\hat{\sigma}_W^2(p)$ are large relative to the other estimated variance components, which suggests that examinees differ considerably with respect to their levels of proficiency in listening and writing. Furthermore, since $\hat{\sigma}_W^2(p) = .691$ is much larger than $\hat{\sigma}_L^2(p) = .324$, examinees appear to be much more variable in writing proficiency than listening proficiency, as these constructs are represented in this measurement procedure.

TABLE 10.7. G Study Variance-Covariance Matrices for *LW* Example of $p^{\bullet} \times t^{\bullet} \times r^{\circ}$ Design

			Form		
	402	404	406	Average[a,b]	SE[c]
$\widehat{\Sigma}_p$	$\begin{bmatrix} .321 & .403 \\ .403 & .740 \end{bmatrix}$	$\begin{bmatrix} .262 & .210 \\ .210 & .481 \end{bmatrix}$	$\begin{bmatrix} .388 & .454 \\ .454 & .854 \end{bmatrix}$	$\begin{bmatrix} .324 & .752 \\ .356 & .691 \end{bmatrix}$	$\begin{bmatrix} .037 & .074 \\ .074 & .110 \end{bmatrix}$
$\widehat{\Sigma}_t$	$\begin{bmatrix} .104 & .041 \\ .041 & .028 \end{bmatrix}$	$\begin{bmatrix} .038 & .036 \\ .036 & .028 \end{bmatrix}$	$\begin{bmatrix} .240 & .039 \\ .039 & .019 \end{bmatrix}$	$\begin{bmatrix} .127 & .962 \\ .039 & .025 \end{bmatrix}$	$\begin{bmatrix} .059 & .001 \\ .001 & .003 \end{bmatrix}$
$\widehat{\Sigma}_r$	$\begin{bmatrix} .030 & \\ & .029 \end{bmatrix}$	$\begin{bmatrix} .0^d & \\ & .0^d \end{bmatrix}$	$\begin{bmatrix} .004 & \\ & .003 \end{bmatrix}$	$\begin{bmatrix} .012 & \\ & .010 \end{bmatrix}$	$\begin{bmatrix} .009 & \\ & .010 \end{bmatrix}$
$\widehat{\Sigma}_{pt}$	$\begin{bmatrix} .398 & .052 \\ .052 & .151 \end{bmatrix}$	$\begin{bmatrix} .315 & .0^e \\ .0^e & .169 \end{bmatrix}$	$\begin{bmatrix} .467 & .040 \\ .040 & .156 \end{bmatrix}$	$\begin{bmatrix} .393 & .120 \\ .030 & .159 \end{bmatrix}$	$\begin{bmatrix} .044 & .016 \\ .016 & .005 \end{bmatrix}$
$\widehat{\Sigma}_{pr}$	$\begin{bmatrix} .008 & \\ & .042 \end{bmatrix}$	$\begin{bmatrix} .008 & \\ & .022 \end{bmatrix}$	$\begin{bmatrix} .027 & \\ & .078 \end{bmatrix}$	$\begin{bmatrix} .014 & \\ & .047 \end{bmatrix}$	$\begin{bmatrix} .006 & \\ & .016 \end{bmatrix}$
$\widehat{\Sigma}_{tr}$	$\begin{bmatrix} .013 & \\ & .017 \end{bmatrix}$	$\begin{bmatrix} .034 & \\ & .004 \end{bmatrix}$	$\begin{bmatrix} .020 & \\ & .003 \end{bmatrix}$	$\begin{bmatrix} .022 & \\ & .008 \end{bmatrix}$	$\begin{bmatrix} .006 & \\ & .005 \end{bmatrix}$
$\widehat{\Sigma}_{ptr}$	$\begin{bmatrix} .307 & \\ & .249 \end{bmatrix}$	$\begin{bmatrix} .314 & \\ & .212 \end{bmatrix}$	$\begin{bmatrix} .328 & \\ & .194 \end{bmatrix}$	$\begin{bmatrix} .317 & \\ & .218 \end{bmatrix}$	$\begin{bmatrix} .006 & \\ & .016 \end{bmatrix}$

[a] Averages of the estimates for the three forms.
[b] Italicized values in upper off-diagonal positions are correlations.
[c] Standard errors of the averages.
[d] Negative estimates of −.003 for Listening and −.001 for Writing set to zero.
[e] Negative estimate of −.002 retained.

The estimated variance components $\hat{\sigma}_L^2(pt)$ and $\hat{\sigma}_W^2(pt)$ are also notably large, which suggests that the rank ordering of examinees differs by task for both Listening and Writing. By contrast, $\hat{\sigma}_L^2(r)$, $\hat{\sigma}_W^2(r)$, $\hat{\sigma}_L^2(pr)$, and $\hat{\sigma}_W^2(pr)$ are all quite small, suggesting that raters are not nearly as large a contributor to total variance as are tasks.

There are positive estimated covariance components for p, t, and pt. The estimated covariance component for persons is particularly large (.356) relative to the estimated variance components (.324 and .691) suggesting that μ_{pL} and μ_{pW} are highly correlated. The estimate of this correlation is

$$\hat{\rho}_{LW}(p) = \frac{\hat{\sigma}_{LW}(p)}{\sqrt{\hat{\sigma}_L^2(p)\,\hat{\sigma}_W^2(p)}} = \frac{.356}{\sqrt{.324 \times .691}} = .752.$$

Also, by the same line of reasoning, the tasks appear to be highly correlated $[\hat{\rho}_{LW}(t) = .682]$ suggesting that the rank ordering of the tasks in terms of difficulty is quite similar for Listening and Writing. By contrast, the pt interaction effects are only slightly correlated $[\hat{\rho}_{LW}(pt) = .120]$.

Brennan et al. (1995) report a number of D study results for one, two, and three raters, but they suggest that cost factors will likely preclude using more than two raters operationally. For this reason, here we restrict consideration to two raters. Table 10.8 provides D study results for two different design structures ($p^{\bullet} \times T^{\bullet} \times R^{\circ}$ and $p^{\bullet} \times T^{\circ} \times R^{\circ}$), two sample sizes for tasks (6 and 12), and two composites—an equally weighted average of the two scores ($w_L = .5$ and $w_W = .5$) and difference scores ($w_L = 1$ and $w_W = -1$).

The top of Table 10.8 provides D study estimated variance-covariance matrices followed by the matrix of estimated absolute-error variance and covariance components. In both $\widehat{\Sigma}_p$ and $\widehat{\Sigma}_{\Delta}$ italicized values are correlations. The estimated correlation between universe scores for Listening and Writing is .752. For the $p^{\bullet} \times T^{\bullet} \times R^{\circ}$ design, the Δ-type errors have an estimated correlation of .114. These errors are necessarily uncorrelated for the $p^{\bullet} \times T^{\circ} \times R^{\circ}$ design.

The $p^{\bullet} \times T^{\bullet} \times R^{\circ}$ design has the same design structure as that in the G study. It assumes that the same tasks would be used to obtain scores on Listening and Writing. By contrast, for the $p^{\bullet} \times T^{\circ} \times R^{\circ}$ design, the tasks used to obtain a Listening score would be different from those used to obtain a Writing score. For both designs, of course, the use of more tasks reduces composite absolute error variance and increases indices of dependability, for both composites.

A principal difference between the $p^{\bullet} \times T^{\bullet} \times R^{\circ}$ and $p^{\bullet} \times T^{\circ} \times R^{\circ}$ designs is that the $p^{\bullet} \times T^{\bullet} \times R^{\circ}$ design has an estimated absolute error covariance of $\hat{\sigma}_{LW}(\Delta) = .011$ for 6 tasks and $\hat{\sigma}_{LW}(\Delta) = .006$ for 12 tasks, whereas there is no absolute error covariance component for the $p^{\bullet} \times T^{\circ} \times R^{\circ}$ design. The role of $\hat{\sigma}_{LW}(\Delta)$ in the two designs has very different consequences for the two composites. Specifically,

- for the composite based on averaging the two scores, the positive covariance component for the $p^{\bullet} \times T^{\bullet} \times R^{\circ}$ design causes $\hat{\sigma}_C^2(\Delta)$ to be *larger* for the $p^{\bullet} \times T^{\bullet} \times R^{\circ}$ design than for the $p^{\bullet} \times T^{\circ} \times R^{\circ}$ design; whereas,

- for the "difference" composite, the positive covariance component for the $p^{\bullet} \times T^{\bullet} \times R^{\circ}$ design causes $\hat{\sigma}_C^2(\Delta)$ to be *smaller* for the $p^{\bullet} \times T^{\bullet} \times R^{\circ}$ design than for the $p^{\bullet} \times T^{\circ} \times R^{\circ}$ design.

All other things being equal, this suggests that the $p^{\bullet} \times T^{\bullet} \times R^{\circ}$ design is preferable when the composite is a difference score, but the $p^{\bullet} \times T^{\circ} \times R^{\circ}$ design is preferable when the composite is an average score—provided $\sigma_C^2(\Delta)$ is positive.

Positively correlated error decreases $\sigma_C^2(\Delta)$ when the composite is a difference score, because $\sigma_{LW}(\Delta)$ has a negative contribution to absolute error

TABLE 10.8. D Study Variance-Covariance Matrices and Composite Results for *LW* Example with Two Raters

	$p^\bullet \times T^\bullet \times R^\circ$ Design		$p^\bullet \times T^\circ \times R^\circ$ Design	
	$n'_t = 6$	$n'_t = 12$	$n'_t = 6$	$n'_t = 12$
$\widehat{\Sigma}_p$	$\begin{bmatrix} .324 & .752 \\ .356 & .691 \end{bmatrix}$	$\begin{bmatrix} .324 & .752 \\ .356 & .691 \end{bmatrix}$	$\begin{bmatrix} .324 & .752 \\ .356 & .691 \end{bmatrix}$	$\begin{bmatrix} .324 & .752 \\ .356 & .691 \end{bmatrix}$
$\widehat{\Sigma}_T$	$\begin{bmatrix} .021 & .006 \\ .006 & .004 \end{bmatrix}$	$\begin{bmatrix} .011 & .003 \\ .003 & .002 \end{bmatrix}$	$\begin{bmatrix} .021 & \\ & .004 \end{bmatrix}$	$\begin{bmatrix} .011 & \\ & .002 \end{bmatrix}$
$\widehat{\Sigma}_R$	$\begin{bmatrix} .006 & \\ & .005 \end{bmatrix}$	$\begin{bmatrix} .006 & \\ & .005 \end{bmatrix}$	$\begin{bmatrix} .006 & \\ & .005 \end{bmatrix}$	$\begin{bmatrix} .006 & \\ & .005 \end{bmatrix}$
$\widehat{\Sigma}_{pT}$	$\begin{bmatrix} .066 & .005 \\ .005 & .026 \end{bmatrix}$	$\begin{bmatrix} .033 & .003 \\ .003 & .013 \end{bmatrix}$	$\begin{bmatrix} .066 & \\ & .026 \end{bmatrix}$	$\begin{bmatrix} .033 & \\ & .013 \end{bmatrix}$
$\widehat{\Sigma}_{pR}$	$\begin{bmatrix} .007 & \\ & .024 \end{bmatrix}$	$\begin{bmatrix} .007 & \\ & .024 \end{bmatrix}$	$\begin{bmatrix} .007 & \\ & .024 \end{bmatrix}$	$\begin{bmatrix} .007 & \\ & .024 \end{bmatrix}$
$\widehat{\Sigma}_{TR}$	$\begin{bmatrix} .002 & \\ & .001 \end{bmatrix}$	$\begin{bmatrix} .001 & \\ & .000 \end{bmatrix}$	$\begin{bmatrix} .002 & \\ & .001 \end{bmatrix}$	$\begin{bmatrix} .001 & \\ & .000 \end{bmatrix}$
$\widehat{\Sigma}_{pTR}$	$\begin{bmatrix} .026 & \\ & .018 \end{bmatrix}$	$\begin{bmatrix} .013 & \\ & .009 \end{bmatrix}$	$\begin{bmatrix} .026 & \\ & .018 \end{bmatrix}$	$\begin{bmatrix} .013 & \\ & .009 \end{bmatrix}$
$\widehat{\Sigma}_\Delta$	$\begin{bmatrix} .128 & .114 \\ .011 & .078 \end{bmatrix}$	$\begin{bmatrix} .070 & .093 \\ .006 & .053 \end{bmatrix}$	$\begin{bmatrix} .128 & \\ & .078 \end{bmatrix}$	$\begin{bmatrix} .070 & \\ & .053 \end{bmatrix}$

Composite Defined Using Weights of $w_L = .5$ and $w_W = .5$

$\hat{\sigma}^2_C(p)$	.432	.432	.432	.432
$\hat{\sigma}^2_C(\Delta)$	.057	.034	.051	.031
$\widehat{\Phi}$	.883	.928	.894	.933

Composite Defined Using Weights of $w_L = 1$ and $w_W = -1$

$\hat{\sigma}^2_C(p)$	.304	.304	.304	.304
$\hat{\sigma}^2_C(\Delta)$	.183	.112	.206	.124
$\widehat{\Phi}$	.624	.730	.596	.711

Note. Italicized values in upper off-diagonal positions are correlations.

variance for difference scores:

$$\sigma_C^2(\Delta) = \sigma_L^2(\Delta) + \sigma_L^2(\Delta) - 2\,\sigma_{LW}(\Delta).$$

By contrast, for the "average" composite,

$$\sigma_C^2(\Delta) = .25\,\sigma_L^2(\Delta) + .25\,\sigma_L^2(\Delta) + .50\,\sigma_{LW}(\Delta).$$

10.4 Exercises

10.1* For the Rajaratnam et al. (1965) synthetic data example considered in Section 9.1 (see, especially, Table 9.1 on page 270), provide $\widehat{\mathbf{\Sigma}}_\tau$, $\widehat{\mathbf{\Sigma}}_\delta$, $\widehat{\mathbf{\Sigma}}_\Delta$ for the total score metric. For this metric, specify the w weights and use them to obtain $\hat{\sigma}_C^2(\tau)$, $\hat{\sigma}_C^2(\delta)$, and $\hat{\sigma}_C^2(\Delta)$. Verify that $E\hat{\rho}^2$ and $\widehat{\Phi}$ are unchanged by this change in metric.

10.2* For the Rajaratnam et al. (1965) synthetic data example in Section 9.1, if the a priori weights were all one-third, what would be the sample sizes that minimize absolute error variance under the constraint that $n'_{i+} = 10$?

10.3 For the Rajaratnam et al. (1965) synthetic data example in Section 9.1, what is $MSE_C(\Delta)$ if all eight items were assigned to the second category, but the w weights stay unchanged (i.e., .25, .50, and .25, respectively)?

10.4 Consider a multivariate $p^\bullet \times I^\bullet$ design with two levels of a fixed facet, and suppose that $\hat{\sigma}_1^2(I) = \hat{\sigma}_2^2(I)$, $\hat{\sigma}_1^2(pI) = \hat{\sigma}_2^2(pI)$, and both estimated D study covariance components are positive. Under what circumstances is $\hat{\rho}_{12}(\delta) > \hat{\rho}_{12}(\Delta)$?

10.5* For the synthetic data example in Section 10.1.6, what is $\hat{\sigma}_C^2(\Delta)$ for the D study $I^\bullet : p^\bullet$ design assuming $n'_i = 8$, $w_1 = .75$, and $w_2 = .25$?

10.6* For the example in Section 10.1.6, what is $E\hat{\rho}^2$ for the D study $p^\bullet \times I^\circ$ design assuming $n'_i = 8$, $w_1 = .75$, and $w_2 = .25$? Provide a verbal description of the $p^\bullet \times I^\circ$ design for the accuracy/speed measurement procedure. If items were administered by computer, is this a likely design for this hypothetical example?

10.7 For the synthetic data example in Section 10.1.6 (see, also, Table 9.3 on page 292), suppose the investigator considered accuracy and speed to be equally important in characterizing the universe of generalization. Further suppose she had to do scoring by hand, and her scoring resources were limited, so much so that she could afford to score only two items for each student. However, she can administer as many items as she wants, and she can get a measure of speed for each student on each item. She decides that students will take different items, but they will all take the same number of items. It is important, she decides, that the SEM for the composite be no larger than 2/3. What number of items should be administered to each student?

10.8* For the synthetic data example discussed in Section 10.2.3, what are $\hat{\sigma}_C(\Delta_1)$ and $\hat{\sigma}_C(\delta_1)$ for the $p^\bullet \times I^\bullet$ design with $n_i' = 8$, $w_1 = .75$, and $w_2 = .25$?

10.9 For the synthetic data example in Section 9.1 (raw data in Table 9.1 on page 270) with a priori weights proportional to sample size, what is the conditional absolute standard error of measurement for the first person, using the mean-score metric?

10.10 Derive Equation 10.54 based on the classical test theory model with uncorrelated error and assuming that $\sigma_{EX} = \sigma_{EY} = \sigma_E$.

10.11 Section 10.2.3 provided conditional SEMs for the first person in the synthetic data in Table 9.3 on page 292. What is the probability that 68% confidence intervals for this person will overlap if the true difference in this person's universe scores is $\mu_{11} - \mu_{12} = .5$?

10.12* Suppose a test called *Science Assessment* consists of two separately timed parts. Part 1 contains 24 dichotomously scored items. Part 2 contains two open-ended items with responses coded 0 to 4. The technical manual for *Science Assessment* provides the following information,

	Part 1	Part 2
Mean	16.00	4.00
Standard Deviation	3.64	1.82
Coefficient alpha	.70	.50

where the means and standard deviations are for total scores. The correlation between total scores on the two parts is .40.

(a) Provide $\hat{\Sigma}_p$ and $\hat{\Sigma}_{pi}$ for both the mean-score metric and the total-score metric.

(b) Provide $\hat{\Sigma}_\tau$ and $\hat{\Sigma}_\delta$ for both the mean-score metric and the total-score metric.

(c) Procedures in classical test theory for estimating the reliability of a battery composite (see Feldt & Brennan, 1989, p. 117) yield a coefficient of .742. Show that this value can be obtained easily from the total-score $\hat{\Sigma}_\tau$ and $\hat{\Sigma}_\delta$ matrices.

(d) Using the mean-score $\widehat{\Sigma}_\tau$ and $\widehat{\Sigma}_\delta$ matrices, what are the w weights that are required to give the classical reliability coefficient of .742 in (c)?

(e) Provide at least one advantage and one disadvantage of the classical test theory procedure in (c) compared to the procedure in (d).

(f) What is the contribution of Part 1 to composite universe (or true) score variance? Show that the answer is the same for both the mean-score and total-score metrics.

(g) The technical manual for *Science Assessment* states that, "Coefficient alpha for Part 2 is based on a single rating of each examinee's response to each open-ended item." Discuss the appropriateness of Coefficient alpha under these circumstances.

10.13 For Nuβbaum's (1984) *Painting* example, Table 10.6 reports that $\hat{\sigma}_{12}(\delta) = .2027$. Under normality assumptions, what is the estimated standard error of this covariance component?

10.14 Consider the G study variance-covariance matrices for the LW example in Table 10.7. Provide an estimate of the standard error of $\hat{\sigma}_{LW}(p)$ if Form 402 were the only form available. Discuss this estimate relative to the standard error of .074 reported in Table 10.7.

10.15* Recall Exercise 9.9 based on the study by Miller and Kane (2001). Suppose categories are considered fixed, and the D study sample sizes are the same as those in the G study. Provide $\widehat{\Sigma}_\tau$, $\widehat{\Sigma}_\delta$, and $\widehat{\Sigma}_\Delta$. For the difference-score composite, determine $\hat{\sigma}_C^2(\tau)$, $\hat{\sigma}_C^2(\delta)$, and $\hat{\sigma}_C^2(\Delta)$. If $\overline{X}_1 = .2578$ and $\overline{X}_2 = .7955$ in the proportion-correct metric, estimate the error-tolerance ratio (see Section 2.5.1) for absolute interpretations of difference scores; that is, estimate

$$E/T = \sqrt{\frac{\sigma_C^2(\Delta)}{\sigma^2(\mu_{p2} - \mu_{p1})}} = \sqrt{\frac{\sigma_C^2(\Delta)}{\sigma_C^2(\tau) + (\mu_2 - \mu_1)^2}}.$$

11
Multivariate Unbalanced Designs

The power and flexibility of multivariate generalizability are purchased at the price of complex conceptual and statistical issues that become even more challenging in unbalanced situations. Recall that a covariance components design is called *unbalanced* in this book if either of the corresponding variance components designs is unbalanced. Many real-data applications of multivariate generalizability theory involve unbalanced situations.[1]

For unbalanced multivariate designs, perhaps the most challenging statistical issue is the estimation of covariance components. That is the subject of Section 11.1. D study issues are considered subsequently in Section 11.2 in the context of several real-data examples. Most of the D study formulas and theoretical results for balanced designs discussed in Chapter 10 also apply in unbalanced situations.

11.1 Estimating G Study Covariance Components

For balanced covariance components designs, the expected-mean-product equations are relatively simple, and it is easy to use them to estimate the covariance components. For unbalanced covariance components designs, however, complexities arise. The fundamental theoretical problem is that there

---

[1] As noted in Chapter 8, randomly discarding data to achieve a balanced univariate design is often problematic. In most cases, there is good reason to believe that this ad hoc strategy is likely to be even more problematic when applied to multivariate designs.

TABLE 11.1. Some Procedures for Estimating Covariance Components

Mult Design	Observed Cov.	MP Terms	Analogous TP Terms	CP Terms	Comp Means	Var. of Sum[a]
$p^\bullet \times i^\bullet$		nb		nb		nb
$p^\bullet \times i^\circ$	nb					
$i^\bullet : p^\bullet$		b	u	b/u		b/u
$i^\circ : p^\bullet$	b/u					
$p^\bullet \times i^\bullet \times h^\bullet$		nb		nb		nb
$p^\bullet \times i^\bullet \times h^\circ$		nb		nb		
$p^\bullet \times i^\circ \times h^\bullet$		nb		nb		
$p^\bullet \times i^\circ \times h^\circ$	nb					
$p^\bullet \times (i^\bullet : h^\bullet)$		b	u	b/u		b/u
$p^\bullet \times (i^\circ : h^\bullet)$		b		b/u	u	
$p^\bullet \times (i^\circ : h^\circ)$	b/u				u	
$(p^\bullet : c^\bullet) \times i^\bullet$		b	u	b/u		b/u
$(p^\bullet : c^\bullet) \times i^\circ$		b		b/u	u	
$(p^\circ : c^\bullet) \times i^\circ$	b/u				u	

Note. An entry in a cell indicates that the procedure is appropriate for the indicated design. The notation "nb" means the covariance components design is necessarily balanced; "b" means the procedure is appropriate for balanced designs only; "b/u" means the procedure is appropriate for balanced or unbalanced designs; and "u" means the procedure might be considered for unbalanced designs.

[a] Provided v and v' have the same pattern of missing data.

are numerous estimators, and no unambiguously clear basis for choosing among them. This is the same problem encountered in estimating variance components with unbalanced designs, but estimation issues for covariance components are even more complex, in several respects. First, much less is known about estimators of covariance components. Second, a particular procedure for estimating covariance components is not necessarily applicable to all designs. Third, estimating covariance components often requires more complex computations than estimating variance components.

Table 11.1 provides a summary of some procedures that can be used to estimate covariance components for both balanced and unbalanced designs. The procedure identified as "*MP* terms" in Table 11.1 refers to the mean-products procedure discussed in Section 9.4 for balanced designs. The procedures in Table 11.1 for unbalanced designs are the focus of this section.

A general discussion of each procedure is provided along with some illustrative results. Although the illustrations are restricted to designs in Table 11.1, the procedures themselves apply more widely. Unless otherwise noted, the term "unbalanced" is to be understood as unbalanced with respect to nesting only—not missing data. Some of the procedures discussed are applicable to designs that involve missing data, but this topic is not treated explicitly until Section 11.1.6.

Estimating covariance components for unbalanced designs is one of the most complicated statistical issues covered in this book. Some readers may wish to skim this section initially and return to it after studying the examples in Section 11.2.

11.1.1 Observed Covariance

When the only linked facet is that for the objects of measurement [e.g., p in the $p^\bullet \times (i^\circ : h^\circ)$ design], the covariance of the observed mean scores for that facet is an unbiased estimator of the covariance component. For unbalanced designs, the observed mean is defined here as the simple average over all levels of all facets. So, for example, for the multivariate unbalanced $p^\bullet \times (i^\circ : h^\circ)$ design

$$\overline{X}_{pv} = \frac{\sum_h \sum_i X_{pihv}}{\sum_h n_{i:h}}, \tag{11.1}$$

a similar expression applies for $\overline{X}_{pv'}$, and the observed covariance is given by Equation 9.7. We call the expression for $\overline{X}_{pv}$ in Equation 11.1 a *simple mean* to distinguish it from the mean (over levels of h) of the mean (over levels of i). We consider covariance components based on such *compound means* later. Unless otherwise noted, all references to mean scores imply simple means. Of course, simple means and compound means are equivalent for balanced designs.

11.1.2 Analogous TP Terms

Recall from Section 7.1.1 that analogous T terms can be used to estimate variance components when the variance components design is unbalanced. Similarly, for *some* unbalanced covariance components designs, analogous TP terms can be used to estimate covariance components. Specifically, analogous TP terms can be used to estimate covariance components for unbalanced multivariate designs in which all levels of all facets are linked, which implies that the variance components designs and the covariance components designs have the same structure. We say that such designs are "full"; that is, each facet has a superscript *filled* circle in the representation of the multivariate design.

The definition and use of analogous TP terms parallels the definition and use of analogous T terms, as discussed in Section 7.1.1. Specifically, analogous TP terms are defined as

$$TP_{vv'}(\alpha) = \sum_\alpha \tilde{n}_\alpha \overline{X}_{\alpha v} \overline{X}_{\alpha v'}, \tag{11.2}$$

where $\tilde{n}_\alpha$ is the total number of observations for a given level of α, and the mean scores ($\overline{X}_{\alpha v}$ and $\overline{X}_{\alpha v'}$) are sums of observations divided by the total number of observations summed (e.g., Equation 11.1).

The coefficient of $\mu_v \mu_{v'}$ in the expected value of every TP term is simply

$$k[\mu_v \mu_{v'}, \boldsymbol{E}\,TP_{vv'}(\alpha)] = n_+, \tag{11.3}$$

where n_+ is the total number of observations for v (or v') in the design.[2] The coefficient of $\sigma_{vv'}(\alpha)$ in the expected value of the TP term for β is

$$k[\sigma_{vv'}(\alpha), \boldsymbol{E}\,TP_{vv'}(\beta)] = \sum_\beta \left(\sum_\gamma \frac{\tilde{n}_{\beta\gamma}^2}{\tilde{n}_\beta} \right), \tag{11.4}$$

where

> γ is the set of all indices in α that are not in β (if $\beta = \mu$, then $\gamma = \alpha$);
>
> $\tilde{n}_{\beta\gamma}$ is the total number of observations in the *variance components* design that is associated with a given combination of levels of β and γ; and
>
> $\tilde{n}_\beta$ is the total number of observations for a given level of β (note that $\tilde{n}_\beta = \sum_\gamma \tilde{n}_{\beta\gamma}$).

One useful special case of Equation 11.4 is

$$k[\sigma_{vv'}(\alpha), \boldsymbol{E}\,TP_{vv'}(\mu_v \mu_{v'})] = \sum_\alpha \frac{\tilde{n}_\alpha^2}{n_+}, \tag{11.5}$$

where $\tilde{n}_\alpha$ is the total number of observations for a given level of α. Note also that

$$k[\sigma_{vv'}(\alpha), \boldsymbol{E}\,TP_{vv'}(\alpha)] = n_+, \tag{11.6}$$

and

$$k[\sigma_{vv'}(\alpha), \boldsymbol{E}\,TP_{vv'}(\omega)] = n_+, \tag{11.7}$$

where ω is the effect associated with all the indices in the covariance components design.

Consider, for example, the unbalanced $i^\bullet : p^\bullet$ design in which $n_{i:p}$ is not a constant for all p. The analogous TP terms are

$$
\begin{aligned}
TP_{vv'}(p) &= \sum_p n_{i:p} \overline{X}_{pv} \overline{X}_{pv'} \\
TP_{vv'}(i{:}p) &= \sum_p \sum_i X_{piv} X_{piv'} \\
TP_{vv'}(\mu) &= n_+ \overline{X}_v \overline{X}_{v'},
\end{aligned}
\tag{11.8}
$$

---

[2]For full multivariate designs, the total number of observations is the same for both v and v'.

where $n_+ = \sum_p n_{i:p}$.

It is relatively easy to show that the expected values of the analogous TP terms in Equation Set 11.8 are

$$E\,TP_{vv'}(p) = n_+\mu_v\mu_{v'} + n_+\sigma_{vv'}(p) + n_p\sigma_{vv'}(i{:}p) \qquad (11.9)$$

$$E\,TP_{vv'}(i{:}p) = n_+\mu_v\mu_{v'} + n_+\sigma_{vv'}(p) + n_+\sigma_{vv'}(i{:}p) \qquad (11.10)$$

$$E\,TP_{vv'}(\mu) = n_+\mu_v\mu_{v'} + r_i\sigma_{vv'}(p) + \sigma_{vv'}(i{:}p), \qquad (11.11)$$

where

$$r_i = \sum_p n_{i:p}^2/n_+. \qquad (11.12)$$

Equations 11.9 to 11.11 are easily solved for estimators of the covariance components:

$$\hat{\sigma}_{vv'}(i{:}p) = \frac{TP_{vv'}(i{:}p) - TP_{vv'}(p)}{n_+ - n_p} \qquad (11.13)$$

$$\hat{\sigma}_{vv'}(p) = \frac{TP_{vv'}(p) - TP_{vv'}(\mu) - (n_p - 1)\hat{\sigma}_{vv'}(i{:}p)}{n_+ - r_i}. \qquad (11.14)$$

Table 11.2 provides a synthetic data example of the unbalanced $i^\bullet{:}p^\bullet$ design. In this data set, there are four, five, or six items nested within each of 10 persons, with responses evaluated with respect to two levels of a fixed facet. Using analogous TP terms with Equations 11.13 and 11.14, the estimates of the covariance components are

$$\hat{\sigma}_{12}(i{:}p) = \frac{931.0000 - 927.1667}{50 - 10} = .0958 \qquad (11.15)$$

and

$$\hat{\sigma}_{12}(p) = \frac{927.1667 - 910.52000 - 9(.0958)}{50 - 5.12} = .3517. \qquad (11.16)$$

Other aspects of this example are considered in subsequent sections.

The analogous TP terms in Equation Set 11.8 parallel the analogous T terms in Table 7.1 on page 220 used to estimate variance components for the unbalanced $i{:}p$ design. There is a similar relationship between the expected TP terms in Equations 11.9 to 11.11 and the expected T terms in Equation Set 7.7. Also, the estimators of the covariance components in Equations 11.13 and 11.14 parallel the estimators of the variance components in Equations 7.9 and 7.10. Analogous mean products can be defined also, but computations are usually easier with analogous TP terms.

There are similar relationships for other unbalanced multivariate designs that are full. For example, the equations in Section 7.1.3 for estimating variance components for the unbalanced $p \times (i{:}h)$ design can be transformed to obtain equations for estimating covariance components for the $p^\bullet \times (i^\bullet{:}h^\bullet)$ design. The process is simple: replace analogous T terms with analogous

TABLE 11.2. Synthetic Data Example for Unbalanced $i^{\bullet}:p^{\bullet}$ Design

p	Scores for v_1						Scores for v_2						$n_{i:p}$	$\overline{X}_{p1}$	$\overline{X}_{p2}$	$\overline{X}_{p1}\overline{X}_{p2}$
1	4 3 3 4						5 5 4 5						4	3.5000	4.7500	16.6250
2	3 3 3 3 3						4 5 5 4 5						5	3.0000	4.6000	13.8000
3	4 4 4 3 4						3 3 3 3 3						5	3.8000	3.0000	11.4000
4	3 4 4 4 4 4						4 4 3 3 4 4						6	3.8333	3.6667	14.0556
5	5 4 4 3						4 2 3 2						4	4.0000	2.7500	11.0000
6	5 5 5 5 5						6 6 5 6 6						5	5.0000	5.8000	29.0000
7	5 4 5 5 5 5						5 5 5 6 5 5						6	4.8333	5.1667	24.9722
8	5 5 5 4						6 6 5 5						4	4.7500	5.5000	26.1250
9	5 5 4 4 5						5 6 5 4 5						5	4.6000	5.0000	23.0000
10	4 4 4 4 3 4						4 4 4 4 4 4						6	3.8333	4.0000	15.3333

$n_p = 10$ $\qquad\qquad$ $n_+ = 50$ $\qquad\qquad$ $\ddot{n}_i = 4.8781$ $\qquad\qquad$ $r_i = 5.12$

$\overline{X}_1 = 4.1200$ $\qquad\qquad$ $\overline{X}_2 = 4.4200$ $\qquad\qquad$ $\overline{X}_1\overline{X}_2 = 18.2104$

$$\begin{aligned}
TP_{12}(p) &= 927.1667 & CP_{12}(p) &= 185.3111 \\
TP_{12}(i{:}p) &= 931.0000 & CP_{12}(i{:}p) &= 931.0000 \\
TP_{12}(\mu) &= 910.5200 & CP_{12}(\mu) &= 18.2104
\end{aligned}$$

TP terms, and replace variance components with covariance components. Likewise, the equations in Appendix E for estimating variance components for the unbalanced $(p{:}c) \times i$ design can be transformed to obtain equations for estimating covariance components for the $(p^{\bullet}{:}c^{\bullet}) \times i^{\bullet}$ design.

For any unbalanced variance components design, there is a set of analogous T terms (see Equation 7.1) that can be used to estimate the variance components. By contrast, for unbalanced multivariate designs that are *not* full, typically some analogous TP terms do not exist. Note from Equation 11.2 that analogous TP terms involve the product of two means multiplied by the *common* number of observations used to determine each mean. Whenever a different number of observations is used for each mean, no analogous TP term exists. Consider, for example, the unbalanced $p^{\bullet} \times (i^{\circ}{:}h^{\bullet})$ design with two levels (v and v') for the fixed facet, and assume the $n_{i:h}$ for v do not equal the $m_{i:h}$ for v'. For this design, analogous TP terms do not exist for p, h, and ph. Next, we consider an alternative to analogous TP terms that can be used with any covariance components design.

11.1.3 CP Terms

Another approach for estimating covariance components for an unbalanced design involves direct use of the sums-of-cross-products terms and their expected values. These lead to a set of simultaneous linear equations that can be solved for the estimated covariance components. This approach is quite general in that it can be used for any unbalanced covariance components design, but expressions for the resulting estimated covariance components are often complex.

A sum-of-cross-products term for a component α in a covariance components design is

$$
CP_{vv'}(\alpha) = \begin{cases} \sum_\alpha \overline{X}_{\alpha v} \overline{X}_{\alpha v'} & \text{if } \alpha \neq \mu \\[2ex] \overline{X}_v \overline{X}_{v'} & \text{if } \alpha = \mu \end{cases}, \tag{11.17}
$$

where the mean scores in Equation 11.17 are sums of observations divided by the total number of observations summed (e.g., Equation 11.1). Comparing Equations 9.45 and 11.17, it is obvious that for balanced designs TP terms and CP terms differ only by a multiplicative factor, $\pi(\dot{\alpha})$. For balanced designs, TP terms have the advantage of being related to MP terms which, in turn, are easily used to estimate covariance components. For unbalanced designs, however, usually it is easier to work directly with CP terms (except, perhaps, when analogous TP terms exist).

The expected value of each $CP_{vv'}(\alpha)$ term involves the product of the means, $\mu_v \mu_{v'}$, as well as each of the covariance components. The coefficient of the product of the means is

$$
k[\mu_v \mu_{v'}, \boldsymbol{E}\, CP_{vv'}(\alpha)] = \begin{cases} \sum_\alpha(1) & \text{if } \alpha \neq \mu \\[2ex] 1 & \text{if } \alpha = \mu, \end{cases} \tag{11.18}
$$

where $\sum_\alpha(1)$ is the number of levels associated with α.

The coefficient of each of the covariances included in the expected value of the sum-of-cross-products term for β is much more complicated. For any particular covariance component $\sigma_{vv'}(\alpha)$, the coefficient is

$$
k[\sigma_{vv'}(\alpha), \boldsymbol{E}\, CP_{vv'}(\beta)] = \sum_\beta \left(\sum_\gamma \frac{\tilde{n}_{\beta\gamma} \tilde{m}_{\beta\gamma}}{\tilde{n}_\beta \tilde{m}_\beta} \right), \tag{11.19}
$$

where

> γ is the set of all indices in α that are not in β (if $\beta = \mu$, then $\gamma = \alpha$); note that γ plus α includes all indices in the covariance components design;

if β or γ is the null set (i.e., no indices), the associated summation operator disappears;

$\tilde{n}_{\beta\gamma}$ is the total number of observations in the *variance components* design for v that are associated with a given combination of levels of β and γ;

$\tilde{m}_{\beta\gamma}$ is the total number of observations in the *variance components* design for v' that are associated with a given combination of levels of β and γ;

$\tilde{n}_{\beta}$ is the total number of observations for a given level of β for variable v ($\tilde{n}_{\beta} = \sum_{\gamma} \tilde{n}_{\beta\gamma}$); and

$\tilde{m}_{\beta}$ is the total number of observations for a given level of β for variable v' ($\tilde{m}_{\beta} = \sum_{\gamma} \tilde{m}_{\beta\gamma}$).

One useful special case of Equation 11.19 is

$$k[\sigma_{vv'}(\alpha), \boldsymbol{E}\,CP_{vv'}(\mu)] = \sum_{\alpha} \frac{\tilde{n}_{\alpha}\tilde{m}_{\alpha}}{n_+ m_+}, \tag{11.20}$$

where $\tilde{n}_{\alpha}$ is the total number of observations for v for a given level of α, $\tilde{m}_{\alpha}$ is the total number of observations for v' for a given level of α, n_+ is the total number of observations in the variance components design for v, and m_+ is the total number of observations in the variance components design for v'. Note also that

$$k[\sigma_{vv'}(\alpha), \boldsymbol{E}\,CP_{vv'}(\alpha)] = \sum_{\alpha} (1), \tag{11.21}$$

which is the number of levels associated with α; and letting ω be the effect associated with all indices in the *covariance* components design,

$$k[\sigma_{vv'}(\alpha), \boldsymbol{E}\,CP_{vv'}(\omega)] = \sum_{\omega} (1), \tag{11.22}$$

which is the total number of levels for all the indices in the covariance components design.

$i^{\bullet}{:}p^{\bullet}$ Design

Consider, for example, the unbalanced $i^{\bullet}{:}p^{\bullet}$ design. The three CP terms are

$$
\begin{aligned}
CP_{vv'}(p) &= \sum_{p} \overline{X}_{pv}\overline{X}_{pv'} \\
CP_{vv'}(i{:}p) &= \sum_{p}\sum_{i} X_{piv}X_{piv'} \\
CP_{vv'}(\mu) &= \overline{X}_{v}\overline{X}_{v'}.
\end{aligned}
\tag{11.23}
$$

For the coefficient of $\sigma_{vv'}(p)$ in $\boldsymbol{E}CP_{vv'}(\mu)$, Equation 11.20 applies, which gives

$$k[\sigma_{vv'}(p), \boldsymbol{E}CP_{vv'}(\mu)] = \sum_p \frac{\tilde{n}_p \tilde{m}_p}{n_+ m_+} = \sum_p \frac{n_{i:p}^2}{n_+^2} = \frac{r_i}{n_+},$$

where r_i is given by Equation 11.12. Using Equation 11.20 again, the coefficient of $\sigma_{vv'}(i{:}p)$ in $\boldsymbol{E}CP_{vv'}(\mu)$ is

$$k[\sigma_{vv'}(i{:}p), \boldsymbol{E}CP_{vv'}(\mu)] = \sum_p \sum_i \frac{\tilde{n}_{pi} \tilde{m}_{pi}}{n_+ m_+} = \sum_p \sum_i \frac{1}{n_+^2} = \frac{1}{n_+}.$$

Note that both $\tilde{n}_{pi}$ and $\tilde{m}_{pi}$ are unity because there is only one observation for each combination of levels of p and i. The coefficient of $\sigma_{vv'}(i{:}p)$ in $\boldsymbol{E}CP_{vv'}(p)$ is obtained using Equation 11.19 with $\beta = p$ and $\gamma = i$,

$$k[\sigma_{vv'}(i{:}p), \boldsymbol{E}CP_{vv'}(p)] = \sum_p \left(\sum_i \frac{1}{\tilde{n}_p \tilde{m}_p} \right) = \sum_p \left(\frac{1}{n_{i:p}} \right) = \frac{n_p}{\ddot{n}_i},$$

where $\ddot{n}_i$ is the harmonic mean of the $n_{i:p}$; namely,

$$\ddot{n}_i = \left[\frac{1}{n_p} \sum_p \left(\frac{1}{n_{i:p}} \right) \right]^{-1}.$$

In a similar manner, the other k terms in the three $\boldsymbol{E}CP$ equations can be obtained. Doing so gives the following set of three equations in terms of the three unknowns (the product of the means and the two covariance components):

$$\boldsymbol{E}CP_{vv'}(\mu) = \mu_v \mu_{v'} + \left(\frac{r_i}{n_+} \right) \sigma_{vv'}(p) + \left(\frac{1}{n_+} \right) \sigma_{vv'}(i{:}p) \quad (11.24)$$

$$\boldsymbol{E}CP_{vv'}(p) = n_p \mu_v \mu_{v'} + n_p \sigma_{vv'}(p) + \left(\frac{n_p}{\ddot{n}_i} \right) \sigma_{vv'}(i{:}p) \quad (11.25)$$

$$\boldsymbol{E}CP_{vv'}(i{:}p) = n_+ \mu_v \mu_{v'} + n_+ \sigma_{vv'}(p) + n_+ \sigma_{vv'}(i{:}p). \quad (11.26)$$

After replacing the parameters in Equations 11.24 to 11.26 with estimators, algebraic procedures can be used to obtain the following estimators of the covariance components:

$$\hat{\sigma}_{vv'}(i{:}p) = \frac{n_p CP_{vv'}(i{:}p) - n_+ CP_{vv'}(p)}{n_p n_+} \left(\frac{\ddot{n}_i}{\ddot{n}_i - 1} \right) \quad (11.27)$$

$$\hat{\sigma}_{vv'}(p) = \frac{CP_{vv'}(p) - n_p CP_{vv'}(\mu)}{n_p (1 - r_i/n_+)} - \frac{(n_+ - \ddot{n}_i)\hat{\sigma}_{vv'}(i{:}p)}{\ddot{n}_i(n_+ - r_i)}. \quad (11.28)$$

Using these equations with the synthetic data in Table 11.2 gives

$$\hat{\sigma}_{12}(i{:}p) = \frac{10(931.0000) - 50(185.3111)}{10(50)} \left(\frac{4.8781}{1 - 4.8781} \right) = .1118, \quad (11.29)$$

TABLE 11.3. Coefficients of $\mu_v \mu_{v'}$ and Covariance Components in Expected Values of CP Terms for Unbalanced $p^{\bullet} \times (i^{\circ} : h^{\bullet})$ Design

ECP term	Coefficients			
	$\mu_v \mu_{v'}$	$\sigma_{vv'}(p)$	$\sigma_{vv'}(h)$	$\sigma_{vv'}(ph)$
$ECP_{vv'}(\mu)$	1	$1/n_p$	t	t/n_p
$ECP_{vv'}(p)$	n_p	n_p	tn_p	tn_p
$ECP_{vv'}(h)$	n_h	n_h/n_p	n_h	n_h/n_p
$ECP_{vv'}(ph)$	$n_p n_h$	$n_p n_h$	$n_p n_h$	$n_p n_h$

and

$$\hat{\sigma}_{12}(p) = \frac{185.3111 - 10(18.2104)}{10(1 - 5.12/50)} - \frac{(50 - 4.8781)(.1118)}{4.8781(50 - 5.12)} = .3343. \quad (11.30)$$

Recall that the estimates based on analogous TP terms, given by Equations 11.15 and 11.16 are $\hat{\sigma}_{12}(i:p) = .0958$ and $\hat{\sigma}_{12}(p) = .3517$, respectively. Obviously, the two procedures give different estimates for this small set of synthetic data.

$p^{\bullet} \times (i^{\circ} : h^{\bullet})$ Design

As a more complicated example, consider the $p \times h$ covariance components design in the $p^{\bullet} \times (i^{\circ} : h^{\bullet})$ multivariate design. For the coefficient of $\sigma_{vv'}(p)$ in $ECP_{vv'}(ph)$, $\alpha = p$, $\beta = ph$, $\tilde{n}_{\beta} = \tilde{n}_{\beta\gamma} = n_{i:h}$, $\tilde{m}_{\beta} = \tilde{m}_{\beta\gamma} = m_{i:h}$, and the coefficient is $\sum_p \sum_h (1) = n_p n_h$. In a similar manner, using Equations 11.18 to 11.22, each of the coefficients in all four ECP terms can be obtained. They are provided in Table 11.3, where

$$t = \frac{\sum_h n_{i:h} m_{i:h}}{\left(\sum_h n_{i:h}\right)\left(\sum_h m_{i:h}\right)}, \quad (11.31)$$

with $n_{i:h}$ designating the number of levels of $i:h$ for v and $m_{i:h}$ designating the number of levels of $i:h$ for v'.

After replacing parameters with estimators in the ECP equations, algebraic procedures or matrix operations can be used to obtain the following estimators of the covariance components in terms of sums of cross-products of observed mean scores.

$$\hat{\sigma}_{vv'}(ph) = \frac{CP_{vv'}(ph) - n_h CP_{vv'}(p) - n_p CP_{vv'}(h) + n_p n_h CP_{vv'}(\mu)}{n_h(1 - t)(n_p - 1)} \quad (11.32)$$

$$\hat{\sigma}_{vv'}(p) = \frac{CP_{vv'}(p) - n_p CP_{vv'}(\mu)}{n_p - 1} - t\,\hat{\sigma}_{vv'}(ph) \quad (11.33)$$

$$\hat{\sigma}_{vv'}(h) = \frac{CP_{vv'}(h) - n_h CP_{vv'}(\mu)}{n_h(1 - t)} - \frac{\hat{\sigma}_{vv'}(ph)}{n_p}. \quad (11.34)$$

TABLE 11.4. Synthetic Data Example for Unbalanced $p^\bullet \times (i^\circ : h^\bullet)$ Design

	v_1			v_2		
p	h_1	h_2	h_3	h_1	h_2	h_3
1	2 6 4	2 7 5 4	5 9 8 7 8	1 3 1 5	4 3 5 3	4 4 4 4
2	6 6 4	1 2 4 2	8 6 6 6 9	7 7 5 6	3 3 4 3	5 7 4 5
3	2 6 9	2 6 4 4	7 9 5 7 9	6 5 4 4	3 4 6 5	6 6 6 2
4	5 5 6	4 3 3 2	5 5 6 6 7	7 7 5 8	6 4 4 5	7 9 7 8
5	5 6 9	6 6 4 3	7 7 8 6 8	9 6 5 9	0 0 0 1	6 7 1 7
6	4 4 5	2 3 2 3	3 5 5 5 2	4 3 5 5	1 2 2 0	4 2 1 1
7	4 6 6	0 5 1 3	6 6 3 5 5	5 4 1 5	0 1 1 2	1 2 3 4
8	6 5 7	2 2 4 4	7 9 6 9 6	7 6 2 3	6 7 4 6	7 7 9 6
9	4 7 6	6 5 6 4	4 5 4 3 4	5 6 2 3	5 3 6 5	4 5 4 5
10	2 4 5	2 5 3 2	5 6 6 3 6	7 8 3 7	0 0 1 2	4 2 2 4

	v_1			v_2			Simple Means[a]		
p	h_1	h_2	h_3	h_1	h_2	h_3	$\overline{X}_{p1}$	$\overline{X}_{p2}$	$\overline{X}_{p1}\overline{X}_{p2}$
1	4.00	4.50	7.40	2.50	3.75	4.00	5.5833	3.4167	19.0764
2	5.33	2.25	7.00	6.25	3.25	5.25	5.0000	4.9167	24.5833
3	5.67	4.00	7.40	4.75	4.50	5.00	5.8333	4.7500	27.7083
4	5.33	3.00	5.80	6.75	4.75	7.75	4.7500	6.4167	30.4792
5	6.67	4.75	7.20	7.25	0.25	5.25	6.2500	4.2500	26.5625
6	4.33	2.50	4.00	4.25	1.25	2.00	3.5833	2.5000	8.9583
7	5.33	2.25	5.00	3.75	1.00	2.50	4.1667	2.4167	10.0694
8	6.00	3.00	7.40	4.50	5.75	7.25	5.5833	5.8333	32.5694
9	5.67	5.25	4.00	4.00	4.75	4.50	4.8333	4.4167	21.3472
10	3.67	3.00	5.20	6.25	0.75	3.00	4.0833	3.3333	13.6111
Sum									214.9653
Mean	5.200	3.450	6.040	5.025	3.000	4.650	4.9667	4.2250	

$$CP_{12}(p) = 214.9653$$
$$CP_{12}(h) = (5.200 \times 5.205) + (3.450 \times 3.000)$$
$$+ (6.040 \times 4.650) = 64.5660$$
$$CP_{12}(ph) = (4.00 \times 2.50) + \cdots + (5.20 \times 3.00) = 666.8708$$
$$CP_{12}(\mu) = 4.9667 \times 4.2250 = 20.9842$$

[a]See Equation 11.1.

Table 11.4 provides a synthetic data example of the $p^{\bullet} \times (i^{\circ} : h^{\bullet})$ design with two levels of a fixed facet, for 10 persons and three levels of h. There are three, four, and five items nested within the levels of h for v_1, and four items nested within each of the levels of h for v_2. For these data, t in Equation 11.31 is

$$t = \frac{3(4) + 4(4) + 5(4)}{12(12)} = .3333.$$

Replacing the CP terms in Table 11.4 in Equations 11.32 to 11.34 gives

$$\hat{\sigma}_{12}(ph) = \frac{666.8708 - 3(214.9653) - 10(64.5660) + 30(20.9842)}{3(1 - .3333)(10 - 1)}$$

$$= .3244$$

$$\hat{\sigma}_{12}(p) = \frac{214.9653 - 10(20.9842)}{10 - 1} - .3333(.3244) = .4611 \quad (11.35)$$

$$\hat{\sigma}_{12}(h) = \frac{64.5660 - 3(20.9842)}{3(1 - .3333)} - \frac{.3333}{10} = .7743.$$

C Terms

An obvious special case of CP terms is to set $v = v'$, which results in a set of quadratic forms that can be used to estimate variance components for unbalanced designs. We call these quadratic forms C terms. The coefficients of μ^2 and the variance components in the expected value of any C term are easily obtained as special cases of Equations 11.18 to 11.22. For example, the C-terms version of Equation 11.18 is

$$k[\mu^2, \boldsymbol{E}C(\alpha)] = \begin{cases} \sum_{\alpha}(1) & \text{if } \alpha \neq \mu \\ \\ 1 & \text{if } \alpha = \mu, \end{cases} \quad (11.36)$$

and the C-terms version of Equation 11.19 is

$$k[\sigma^2(\alpha), \boldsymbol{E}C(\beta)] = \sum_{\beta} \sum_{\gamma} \left(\frac{\tilde{n}^2_{\beta\gamma}}{\tilde{n}^2_{\beta}} \right), \quad (11.37)$$

where, as before, γ is the set of all indices in α that are not in β.

Note that C terms are different from the analogous T terms discussed in Chapter 7 for estimating variance components for unbalanced designs. The basic difference is that analogous T terms have a multiplier, whereas C terms do not. With unbalanced designs, C terms and analogous T terms usually give different estimates of variance components, as illustrated later in Section 11.1.5. With balanced designs, C terms and analogous T terms give the same estimates.

11.1.4   Compound Means

Products of simple means are the basis for the CP-terms procedure for estimating covariance components for unbalanced designs. Simple means are sums of observations divided by the number of observations summed (e.g., Equation 11.1). For example, $\overline{X}_{p1}$ and $\overline{X}_{p2}$ in Table 11.4 are simple means. Also, simple means were discussed as the basis for computing observed covariances that estimate covariance components for designs with only one linked facet.

By contrast, *compound* means are defined here as averages of averages. For example, for the $p^{\bullet} \times (i^{\circ} : h^{\circ})$ design, the compound mean for persons for v is defined as

$$\overline{X}^{*}_{pv} \equiv \frac{1}{n_h} \sum_h \left(\frac{1}{n_{i:h}} \sum_i X_{pihv} \right), \qquad (11.38)$$

which is the average over levels of h of the average over levels of i. For designs with only one linked facet, covariances based on compound means are unbiased estimates of the covariance components. So, for example, for the $p^{\bullet} \times (i^{\circ} : h^{\circ})$ design,

$$S_{vv'}(\overline{X}^{*}_{p}) = \frac{n_p}{n_p - 1} \left(\frac{\sum_p \overline{X}^{*}_{pv} \overline{X}^{*}_{pv'}}{n_p} - \overline{X}^{*}_{v} \overline{X}^{*}_{v'} \right) \qquad (11.39)$$

is an unbiased estimator of $\hat{\sigma}_{vv'}(p)$, but not the same estimator as the observed covariance based on simple means (see Section 11.1.1).

Compound means can be used also to obtain unbiased estimates of covariance components for the $p \times h$ design in the multivariate $p^{\bullet} \times (i^{\circ} : h^{\bullet})$ design. Simple means over levels of i are used as the $n_p \times n_h$ cell entries in the $p \times h$ design, and the mean-product procedure for balanced designs in Section 9.4 is applied to these means. In doing so, the marginals for p and h are compound means.

Table 11.5 illustrates some of the computations for this procedure using the synthetic data in Table 11.4. The TP terms at the bottom of Table 11.5 are indeed TP terms in the sense of Equation 9.45, except that the means are compound means (e.g., $\overline{X}^{*}_{pv}$) rather than simple means (e.g., $\overline{X}_{pv}$). Using Equation 9.46, the mean-product equations are

$$MP_{12}(p) = \frac{635.5000 - 620.6525}{9} = 1.6497$$

$$MP_{12}(h) = \frac{645.6600 - 620.6525}{2} = 12.5038 \qquad (11.40)$$

$$MP_{12}(ph) = \frac{666.8708 - 635.5000 - 645.6600 + 620.6525}{9 \times 2} = .3535.$$

TABLE 11.5. Compound Means for Synthetic Data Example for Unbalanced $p^{\bullet} \times (i^{\circ} : h^{\bullet})$ Design in Table 11.4

	v_1			v_2			Compound Means		
p	h_1	h_2	h_3	h_1	h_2	h_3	$\overline{X}_{p1}$	$\overline{X}_{p2}$	$\overline{X}_{p1}\overline{X}_{p2}$
1	4.00	4.50	7.40	2.50	3.75	4.00	5.3000	3.4167	18.1083
2	5.33	2.25	7.00	6.25	3.25	5.25	4.8611	4.9167	23.9005
3	5.67	4.00	7.40	4.75	4.50	5.00	5.6889	4.7500	27.0222
4	5.33	3.00	5.80	6.75	4.75	7.75	4.7111	6.4167	30.2296
5	6.67	4.75	7.20	7.25	0.25	5.25	6.2056	4.2500	26.3736
6	4.33	2.50	4.00	4.25	1.25	2.00	3.6111	2.5000	9.0278
7	5.33	2.25	5.00	3.75	1.00	2.50	4.1944	2.4167	10.1366
8	6.00	3.00	7.40	4.50	5.75	7.25	5.4667	5.8333	31.8889
9	5.67	5.25	4.00	4.00	4.75	4.50	4.9722	4.4167	21.9606
10	3.67	3.00	5.20	6.25	0.75	3.00	3.9556	3.3333	13.1852
Sum									211.8333
Mean	5.200	3.450	6.040	5.025	3.000	4.650	4.8967	4.2250	

$$
\begin{aligned}
TP_{12}(p) &= 3 \times 211.8333 = 635.5000 \\
TP_{12}(h) &= 10[(5.200 \times 5.205) + (3.450 \times 3.000) \\
&\quad + (6.040 \times 4.650)] = 645.6600 \\
TP_{12}(ph) &= (4.00 \times 2.50) + \cdots + (5.20 \times 3.00) = 666.8708 \\
TP_{12}(\mu) &= 10 \times 3 \times 4.8966666 \times 4.2250000 = 620.6525
\end{aligned}
$$

Using Equations 9.39 to 9.41, the estimates of the covariance components are

$$
\begin{aligned}
\hat{\sigma}_{12}(p) &= \frac{1.6497 - .3535}{3} = .4321 \\
\hat{\sigma}_{12}(h) &= \frac{12.5038 - .3535}{10} = 1.2150 \qquad (11.41) \\
\hat{\sigma}_{12}(ph) &= .3535.
\end{aligned}
$$

These can be compared with the CP-terms estimates of .4611, .7743, and .3244, respectively, given by Equation Set 11.35.

11.1.5 Variance of a Sum

Searle et al. (1992, Chap. 11) discuss a procedure for estimating covariance components based on the well-known result that the variance of a sum of two variables equals the sum of the two variances plus twice the covariance.

TABLE 11.6. Continuation of Synthetic Data Example for Unbalanced $i^\bullet : p^\bullet$ Design in Table 11.2

p	Scores for v_1	Scores for v_2	Scores for $v_1 + v_2$	$n_{i:p}$	$\overline{X}_{p(v_1+v_2)}$
1	4 3 3 4	5 5 4 5	9 8 7 9	4	8.2500
2	3 3 3 3 3	4 5 5 4 5	7 8 8 7 8	5	7.6000
3	4 4 4 3 4	3 3 3 3 3	7 7 7 6 7	5	6.8000
4	3 4 4 4 4 4	4 4 3 3 4 4	7 8 7 7 8 8	6	7.5000
5	5 4 4 3	4 2 3 2	9 6 7 5	4	6.7500
6	5 5 5 5 5	6 6 5 6 6	11 11 10 11 11	5	10.8000
7	5 4 5 5 5 5	5 5 5 6 5 5	10 9 10 11 10 10	6	10.0000
8	5 5 5 4	6 6 5 5	11 11 10 9	4	10.2500
9	5 5 4 4 5	5 6 5 4 5	10 11 9 8 10	5	9.6000
10	4 4 4 4 3 4	4 4 4 4 4 4	8 8 8 8 7 8	6	7.8333

	v_1	v_2	$v_1 + v_2$
$\overline{X}$	4.1200	4.4200	8.5400
$T(p)$	867.7500	1022.3334	3744.4167
$T(i:p)$	876.0000	1033.0000	3771.0000
$T(\mu)$	848.7200	976.8200	3646.5800
$\hat{\sigma}^2(p)$	.3827	.9606	2.0467
$\hat{\sigma}^2(i:p)$	.2063	.2667	.6646
$C(p)$	173.1625	205.3139	749.0986
$C(i:p)$	876.0000	1033.0000	3771.0000
$C(\mu)$	16.9744	19.5364	72.9316
$\hat{\sigma}^2(p)$	.3280	1.0752	2.0717
$\hat{\sigma}^2(i:p)$	.2563	.1618	.6417

In our notation,

$$\sigma^2_{v+v'}(\alpha) = \sigma^2_v(\alpha) + \sigma^2_{v'}(\alpha) + 2\sigma_{vv'}(\alpha). \tag{11.42}$$

Replacing parameters with estimates gives

$$\hat{\sigma}_{vv'}(\alpha) = \frac{\hat{\sigma}^2_{v+v'}(\alpha) - \hat{\sigma}^2_v(\alpha) - \hat{\sigma}^2_{v'}(\alpha)}{2}. \tag{11.43}$$

This means that a covariance component for v and v' can be estimated using the variance components for v, v', and $v + v'$, provided the covariance components design is the same as both variance components designs. That is, in the terminology introduced earlier, the multivariate design must be full.

Consider, again, the synthetic data for the $i^\bullet : p^\bullet$ design in Table 11.2. The raw data are repeated in Table 11.6, which also provides estimated variance components for v_1, v_2, and $v_1 + v_2$. Actually, two sets of estimated variance components are provided: one based on T terms and another based on C terms. The T terms are the analogous T terms discussed in Section 7.1.2. C terms are discussed in Section 11.1.3.

Using Equation 11.43 with the analogous T-terms estimates gives

$$\hat{\sigma}_{12}(i{:}p) = \frac{.6646 - .2063 - .2667}{2} = .0958,$$

and

$$\hat{\sigma}_{12}(p) = \frac{2.0467 - .3827 - .9606}{2} = .3517,$$

which are necessarily identical to the results in Equations 11.15 and 11.16, respectively. Similarly, using Equation 11.43 with the C-terms estimates gives

$$\hat{\sigma}_{12}(i{:}p) = \frac{.6416 - .2563 - .1618}{2} = .1118,$$

and

$$\hat{\sigma}_{12}(p) = \frac{2.0717 - .3280 - 1.0752}{2} = .3343,$$

which are necessarily identical to the results in Equations 11.29 and 11.30, respectively.

Strictly speaking, Equation 11.43 is silent about whether the same procedures are used to estimate the three variance components. For example, variance components for v and v' could be estimated using analogous T terms, and variance components for $v + v'$ could be estimated using C terms. However, if the same types of estimators are not used for all three, it seems reasonable to assume that the estimated covariance component may have unusual (and, surely, unknown) statistical characteristics.

Provided the same procedure is used to estimate variance components for v, v', and $v+v'$, the variance-of-a-sum procedure leads to the same estimates as those that could be obtained otherwise. That is, if analogous T terms are used for all three, the estimates will be identical to those obtained using analogous TP terms, and if C terms are used for all three, the estimates will be identical to those obtained using CP terms. Although the variance-of-a-sum procedure is restricted to full multivariate designs, it has the distinct advantages of being flexible, elegant, and simple to understand.

The variance-of-a-sum procedure permits any procedure that can be used to estimate variance components to be tailored to estimate covariance components—provided the multivariate design is full. This means, for example, that the procedures in urGENOVA, SAS, SPSS, and S-Plus for estimating variance components (see Section 7.3.2) can all be used to estimate covariance components for full multivariate designs. For such designs, the variance-of-a-sum procedure affords the investigator an extraordinary variety of choices in procedures and computer programs/packages for estimation.

11.1.6 Missing Data

To this point, the discussion of estimating covariance components for unbalanced designs has given explicit consideration to designs that are unbalanced with respect to nesting only. Actually, however, the procedures that have been discussed can be used to obtain unbiased estimates of covariance components when covariance components designs are crossed but unbalanced with respect to missing data, provided both v and v' have the same pattern of missing data. Rarely, however, does the same pattern of missing data occur by chance. Usually, some data must be eliminated.

Consider, for example, the synthetic data for the $p^{\bullet} \times i^{\bullet}$ design in Table 11.7. This design is unbalanced in the sense that observations are missing for p_2 and i_4 on v_1, for p_3 and i_5 on v_1, and for p_7 and i_4 on v_2. That is, there are two missing observations for v_1, one missing observation for v_2, and neither of the missing observations for v_1 corresponds to the missing observation for v_2. Clearly, the patterns of missing data are different for v_1 and v_2. To make the missing data patterns the same, we can eliminate the observations that correspond to the missing data. These are indicated by an asterisk in Table 11.7. The reduced data set has 57 observations for both v_1 and v_2.

One way to estimate the covariance components is to use the variance-of-a-sum procedure in Section 11.1.5. That is,

- use the analogous T terms procedure in Section 7.1.4 to estimate the variance components for the unbalanced $p \times i$ design for v_1, v_2, and $v_1 + v_2$, and

- use Equation 11.43 to estimate the covariance components.

Table 11.7 provides the analogous T terms and estimates of the variance components. Using the latter in Equation 11.43 gives

$$
\hat{\sigma}_{12}(p) = \frac{1.6053 - .3977 - .4204}{2} = .3936
$$

$$
\hat{\sigma}_{12}(i) = \frac{.6713 - .3225 - .2675}{2} = .0407 \qquad (11.44)
$$

$$
\hat{\sigma}_{12}(pi) = \frac{3.9982 - 1.2244 - 1.4906}{2} = .6416.
$$

Alternatively, the covariance components can be estimated using analogous TP terms (see Section 11.1.2), which are provided at the bottom of Table 11.7. The estimators of the covariance components are obtained by replacing analogous T terms in Equation Set 7.17 with analogous TP terms. The resulting estimated covariance components are identical to those computed above (see Exercise 11.8).

Eliminating data to obtain the same pattern of missing data for v and v' is potentially problematic. For example, doing so could result in a reduced

TABLE 11.7. Synthetic Data Example for Unbalanced $p^\bullet \times i^\bullet$ Design

p	v_1						v_2						$v_1 + v_2$						$\tilde{n}_p$
	i_1	i_2	i_3	i_4	i_5	i_6	i_1	i_2	i_3	i_4	i_5	i_6	i_1	i_2	i_3	i_4	i_5	i_6	
1	6	4	3	5	4	4	6	4	5	6	4	5	12	8	8	11	8	9	6
2	3	2	2		5	5	6	4	5	*	2	5	9	6	7		7	10	5
3	6	5	7	5		3	6	5	8	4	*	3	12	10	15	9		6	5
4	4	2	2	3	3	5	5	4	2	4	3	5	9	6	4	7	6	10	6
5	4	4	3	5	4	6	4	5	5	3	3	7	8	9	8	8	7	13	6
6	8	5	4	7	5	4	9	6	6	5	7	7	17	11	10	12	12	11	6
7	5	4	5	*	4	4	4	5	6		4	5	9	9	11		8	9	5
8	4	5	3	4	5	6	6	7	6	6	5	6	10	12	9	10	10	12	6
9	7	5	4	6	6	5	6	7	3	5	5	6	13	12	7	11	11	11	6
10	5	3	3	7	4	5	6	5	4	6	2	6	11	8	7	13	6	11	6
$\tilde{n}_i$	10	10	10	8	9	10	10	10	10	8	9	10	10	10	10	8	9	10	
Sum	52	39	36	42	40	47	58	52	50	39	35	55	110	91	86	81	75	102	

	v_1	v_2	$v_1 + v_2$
$\overline{X}$	4.4912	5.0702	9.5614
$T(p)$	1181.3000	1500.3667	5329.5333
$T(i)$	1171.2778	1485.5361	5263.2250
$T(pi)$	1254.0000	1583.0000	5549.0000
$T(\mu)$	1149.7544	1465.2807	5210.9649
$\hat{\sigma}^2(p)$	.3977	.4204	1.6053
$\hat{\sigma}^2(i)$	.3225	.2675	.6713
$\hat{\sigma}^2(pi)$	1.2244	1.4906	3.9982

n_p	=	10	r_p	=	9.5614	$TP_{12}(p)$ =	1323.9333
n_i	=	6	r_i	=	5.7368	$TP_{12}(i)$ =	1303.2056
n_+	=	57	λ_p	=	1.0093	$TP_{12}(pi)$ =	1356.0000
			λ_i	=	1.0052	$TP_{12}(\mu)$ =	1279.9649

Note. In the top part of the table, a blank entry indicates missing data, and an asterisk indicates data that are disregarded.

TABLE 11.8. Coefficients of $\mu_v\mu_{v'}$ and Covariance Components in Expected Values of CP Terms for $p^{\bullet} \times i^{\bullet}$ Design with Missing Data

ECP term	Coefficients			
	$\mu_v\mu_{v'}$	$\sigma_{vv'}(p)$	$\sigma_{vv'}(i)$	$\sigma_{vv'}(pi)$
$ECP_{vv'}(\mu)$	1	$\displaystyle\sum_p \frac{\tilde{n}_p\tilde{m}_p}{n_+m_+}$	$\displaystyle\sum_i \frac{\tilde{n}_i\tilde{m}_i}{n_+m_+}$	$\dfrac{q_+}{n_+m_+}$
$ECP_{vv'}(p)$	n_p	n_p	$\displaystyle\sum_p \frac{q_p}{\tilde{n}_p\tilde{m}_p}$	$\displaystyle\sum_p \frac{q_p}{\tilde{n}_p\tilde{m}_p}$
$ECP_{vv'}(i)$	n_i	$\displaystyle\sum_i \frac{q_i}{\tilde{n}_i\tilde{m}_i}$	n_i	$\displaystyle\sum_i \frac{q_i}{\tilde{n}_i\tilde{m}_i}$
$ECP_{vv'}(pi)$	q_+	q_+	q_+	q_+

Note. q_+ is the total number of products of X_{piv} and $X_{piv'}$ such that neither observation is missing; q_p is the number of items for person p such that neither the response for v nor the response for v' is missing; and q_i is the number of persons for item i such that neither the response for v nor the response for v' is missing.

data set that is considerably smaller than the original one. Also, for the estimators to be unbiased, the missing data must be missing at random; that is, the pattern of missing data must be uncorrelated with the model effects. This statement applies both to data that are truly missing (e.g., the three observations not present in Table 11.7) and to any data that are eliminated (e.g., the three observations with an asterisk in Table 11.7). This assumption may be strained in many real data contexts.

It is possible to derive the expected values of CP terms without assuming that missing data patterns are the same. For the $p^{\bullet} \times i^{\bullet}$ design, the expected values are provided in Table 11.8. Although they are quite complex, they have the advantage of using more of the data to estimate covariance components than is used when data are eliminated to create the same pattern of missing data.

There are missing data situations that cannot be handled easily by application of the procedures discussed here. Consider, for example, the synthetic data example of the $p^{\bullet} \times (i^{\circ}:h^{\bullet})$ design in Table 11.4, and suppose the response for p_1, h_1, and v_1 was missing, as well as the response for p_2, h_3, and v_2. The compound-means procedure could still be used in the manner indicated in Section 11.1.4 to estimate the covariance components. However, for the CP-terms procedure, the coefficients in Table 11.3 no longer apply, and the expected values of the CP terms would have to be derived. Note also that in estimating the variance components, the analogous T terms would change, which means that the coefficients in Table 7.4 and the estimators in Equation Set 7.14 no longer apply. Of course, the expected values of the analogous T terms can be derived and used to estimate the variance components, but doing so is tedious.

11.1.7   Choosing a Procedure

For multivariate designs with a single linked facet, the covariances of observed mean scores are unbiased estimates of the covariance components. If the design is unbalanced, the investigator must choose between simple means or compound means, but otherwise there are no particular estimation ambiguities. For multivariate unbalanced designs with more than one linked facet, however, ambiguities abound in selecting an estimation procedure.

There are several statements that can be made about the principal procedures that have been discussed in this chapter: the MP, TP, CP, and compound-means procedures. First, they all give the same estimates for balanced designs. Second, they require no assumptions about distributional form. Third, they all give unbiased estimates. Fourth, other statistical characteristics of the estimates for the various procedures are generally unknown.[3] It follows that there is no obvious statistical basis for choosing among the procedures, and only ad hoc suggestions can be offered.

For balanced multivariate designs, there is historical precedent favoring the MP procedure. It is essentially the procedure proposed by Cronbach et al. (1972) over a quarter century ago, and it is closely related to the ANOVA procedure discussed in Chapter 3 for estimating variance components for balanced designs. For unbalanced designs, the TP procedure is not always applicable, but if it does apply, this procedure has the advantage of being closely related to the analogous ANOVA procedure for estimating variance components that was discussed in Chapter 7. The CP procedure is somewhat more complicated than the TP procedure, but the CP procedure has the distinct advantage of being applicable to any unbalanced multivariate design. The compound-means procedure is a MP-like procedure that can be viewed as a covariance-components version of an unweighted analysis-of-means.

Of course, other procedures might be considered as well. Indeed, there are probably an unlimited number of quadratic-forms procedures that might be employed. In particular, the variance-of-a-sum procedure makes numerous procedures applicable to estimating covariance components for full multivariate designs, including maximum likelihood and MINQUE procedures. It seems likely that most of the strengths and weaknesses of these procedures for estimating variance components extend to covariance components, as well.

There is no theoretical requirement that the procedure used to estimate covariance components must parallel that used to estimate variance components. For example, the CP-terms procedure could be used to estimate covariance components and the analogous T-terms procedure could be used to estimate variance components.

---

[3]This is a slight exaggeration, but it is true enough for the discussion here.

For unbalanced multivariate designs, there is always the alternative of randomly discarding data to obtain a balanced design. Indeed, for designs that are unbalanced with respect to missing data, eliminating data is one approach that was explicitly discussed in Section 11.1.6 for estimating covariance components. It was noted there, however, that this approach can be problematic if it results in eliminating large amounts of data. Furthermore, Chapter 8 illustrated that randomly discarding data to achieve a balanced design for estimating variance components can be a questionable practice. Since a multivariate design is a kind of conjunction of univariate random designs that are linked through covariance components, the strategy of randomly eliminating data seems likely to be even more problematic when applied to multivariate designs. Still, undoubtedly there are times when this ad hoc approach is sensible, at least relative to other available alternatives.

11.2   Examples of G Study and D Study Issues

The procedures for estimating covariance components discussed in Section 11.1 are illustrated here using real-data examples of several unbalanced multivariate designs. This section also illustrates numerous D study issues, including the estimation of D study covariance components. Two of the examples discussed here involve two-facet table of specifications models. These examples employ the G study $p^\bullet \times (i^\circ : h^\circ)$ and $p^\bullet \times (i^\circ : h^\bullet)$ designs and the corresponding D study $p^\bullet \times (I^\circ : H^\circ)$ and $p^\bullet \times (I^\circ : H^\bullet)$ designs. The third example uses the G study $p^\bullet : (d^\bullet \times o^\bullet)$ design and the D study $P^\bullet : (d^\bullet \times O^\bullet)$ design to examine measurement characteristics of reading-minus-math difference scores for school districts.

11.2.1   ITBS Maps and Diagrams Test

In describing the $p^\bullet \times (i^\circ : h^\circ)$ design, Section 9.2.3 introduced the *MD* example, the *Maps and Diagrams* test of the *Iowa Tests of Basic Skills* (ITBS) (Hoover et al., 1993). For this test, there are two distinct types of stimulus materials, maps and diagrams, that are fixed conditions of measurement in the sense that every form of the *Maps and Diagrams* test contains both types of stimulus material. Specifically, for Forms K and L, Level 10, there are two maps and two diagrams. For each of the two maps there are six and seven items, respectively; similarly, there are six and seven items, respectively, for each of the two diagrams. The covariance components design is unbalanced because both of the variance components designs are unbalanced; that is, there are different numbers of items nested within each of the two maps and different numbers of items nested within each of the two diagrams.

TABLE 11.9. G Study for *Maps and Diagrams*

	Variance and Covariance Components	
Mean Squares	Estimates	Standard Errors[a]

$$\mathbf{M}_p = \begin{bmatrix} .7299 & .0354^b \\ .0354^b & .6727 \end{bmatrix} \quad \widehat{\boldsymbol{\Sigma}}_p = \begin{bmatrix} .0404 & .0354 \\ .0354 & .0319 \end{bmatrix} \quad \begin{bmatrix} .0030 & .0028 \\ .0028 & .0027 \end{bmatrix}$$

$$\mathbf{M}_h = \begin{bmatrix} 161.6959 & \\ & 68.4253 \end{bmatrix} \quad \widehat{\boldsymbol{\Sigma}}_h = \begin{bmatrix} .0062 & \\ & .0001 \end{bmatrix} \quad \begin{bmatrix} .0060 & \\ & .0018 \end{bmatrix}$$

$$\mathbf{M}_{i:h} = \begin{bmatrix} 43.9569 & \\ & 66.8994 \end{bmatrix} \quad \widehat{\boldsymbol{\Sigma}}_{i:h} = \begin{bmatrix} .0148 & \\ & .0226 \end{bmatrix} \quad \begin{bmatrix} .0036 & \\ & .0047 \end{bmatrix}$$

$$\mathbf{M}_{ph} = \begin{bmatrix} .2041 & \\ & .2578 \end{bmatrix} \quad \widehat{\boldsymbol{\Sigma}}_{ph} = \begin{bmatrix} .0032 & \\ & .0116 \end{bmatrix} \quad \begin{bmatrix} .0001 & \\ & .0001 \end{bmatrix}$$

$$\mathbf{M}_{pi:h} = \begin{bmatrix} .1832 & \\ & .1831 \end{bmatrix} \quad \widehat{\boldsymbol{\Sigma}}_{pi:h} = \begin{bmatrix} .1832 & \\ & .1831 \end{bmatrix} \quad \begin{bmatrix} .0062 & \\ & .0093 \end{bmatrix}$$

Note. Sample sizes are: $n_p = 2951$ students, $n_h = 2$ maps with 6 and 7 items in v_1, and $n_h = 2$ diagrams with 6 and 7 items in v_2.

[a]Based on two replications.

[b]Observed covariance.

For this test, Table 11.9 provides mean-square and mean-product matrices $(\mathbf{M})$, as well as estimated variance-covariance matrices $(\widehat{\boldsymbol{\Sigma}})$, based on the responses of 2951 students who took Form K in a standardization study. In each matrix, the first row/column is for maps and the second is for diagrams.

Since the only linked "facet" is for the objects of measurement (students), all entries in the $\mathbf{M}$ matrices are mean squares except for the off-diagonal elements of $\mathbf{M}_p$, which contain the observed covariance for students' maps and diagrams scores based on simple means (see Equation 11.1). This observed covariance is an unbiased estimate of the covariance component for maps and diagrams. Since the two variance component designs are *unbalanced* $p \times (i:h)$ designs, the mean squares are actually "analogous" mean squares, and formulas for estimating the variance components are given in Section 7.1.3.

When the standardization was conducted, both Forms K and L were administered to randomly equivalent groups of students, which permits us to obtain two estimates for each variance and covariance component. This, in turn, enables us to estimate standard errors for each of the components, as reported in the last column of Table 11.9. The standard error for a single estimate is simply the standard deviation of the estimates, and when there are only two replicates (say X_1 and X_2), the standard error (SE) formula is especially simple:

$$SE = \frac{|X_1 - X_2|}{\sqrt{2}}. \tag{11.45}$$

Consider, for example, the covariance component for persons. The Forms K and L estimates are $\hat{\sigma}_{vv'}(p) = .0354$ and $\hat{\sigma}_{vv'}(p) = .0315$, respectively. Therefore, the estimated standard error for any single estimate is:

$$\hat{\sigma}[\hat{\sigma}_{vv'}(p)] = \frac{|.0354 - .0315|}{\sqrt{2}} = .0028,$$

which is reported in Table 11.9. This estimate does not make any normality assumptions. It can be compared to the normality-based estimate for Form K that is obtained using Equation 9.55:

$$\hat{\sigma}[\hat{\sigma}_{vv'}(p)] = \sqrt{\frac{(.0561)(.0517) + (.0354)^2}{2952}} = .0012,$$

which is almost 60% smaller than the estimate of the standard error that does not make normality assumptions. For this dichotomous-data example, the empirically based estimate of .0028 seems much more credible.

Table 11.10 provides results for a D study that uses the same sample sizes as the G study, and that assumes that the w weights are proportional to the sample sizes. The only complexity caused by the unbalanced nature of this $p^{\bullet} \times (I^{\circ} : H^{\circ})$ design involves the divisors of the elements of $\widehat{\Sigma}_h$ and $\widehat{\Sigma}_{ph}$ required to obtain $\widehat{\Sigma}_H$ and $\widehat{\Sigma}_{pH}$, respectively. Because these matrices are diagonal, the elements under consideration are variance components, and the divisors are those discussed in Section 7.2.2 for the unbalanced univariate $p \times (I : H)$ design:

$$\breve{n}'_h = \frac{n'^2_{i+}}{\sum_h n'^2_{i:h}}, \tag{11.46}$$

where $h = 1, \ldots, n'_h$ and $n'_{i+} = \sum_h n'_{i:h}$. If all of the $n'_{i:h}$ are the same constant for a given level of v, then $\breve{n}'_h = n'_h$ for that level of v. Equation 11.46 is the same as Equation 7.28, except for the use of D study sample sizes. With $n'_{i:h}$ sample sizes of six and seven for both v_1 (maps) and v_2 (diagrams), $\breve{n}'_h = 1.988$ for both levels of the fixed facet.

Once the D study estimated variance-covariance matrices are obtained, almost all of the D study methodology discussed in Chapter 10 can be applied directly, which leads to the results at the bottom of Table 11.10. Two-replicate estimated standard errors for the composite results are provided in parentheses. These provide direct estimates of the stability of quantities such as $E\hat{\rho}^2$.

TABLE 11.10. D Study for *Maps and Diagrams*

					D Study	
G Study Components			Divisors[a]		Components	

$$\widehat{\boldsymbol{\Sigma}}_p = \begin{bmatrix} .0404 & .0354 \\ .0354 & .0319 \end{bmatrix} \qquad\qquad \widehat{\boldsymbol{\Sigma}}_p = \begin{bmatrix} .0404 & .0354 \\ .0354 & .0319 \end{bmatrix}$$

$$\widehat{\boldsymbol{\Sigma}}_h = \begin{bmatrix} .0062 & \\ & .0001 \end{bmatrix} \qquad \begin{matrix} 1.988 \\ 1.988 \end{matrix} \qquad \widehat{\boldsymbol{\Sigma}}_H = \begin{bmatrix} .0031 & \\ & .0000 \end{bmatrix}$$

$$\widehat{\boldsymbol{\Sigma}}_{i:h} = \begin{bmatrix} .0143 & \\ & .0226 \end{bmatrix} \qquad \begin{matrix} 13.000 \\ 13.000 \end{matrix} \qquad \widehat{\boldsymbol{\Sigma}}_{I:H} = \begin{bmatrix} .0011 & \\ & .0017 \end{bmatrix}$$

$$\widehat{\boldsymbol{\Sigma}}_{ph} = \begin{bmatrix} .0032 & \\ & .0116 \end{bmatrix} \qquad \begin{matrix} 1.988 \\ 1.988 \end{matrix} \qquad \widehat{\boldsymbol{\Sigma}}_{pH} = \begin{bmatrix} .0016 & \\ & .0058 \end{bmatrix}$$

$$\widehat{\boldsymbol{\Sigma}}_{pi:h} = \begin{bmatrix} .1832 & \\ & .1831 \end{bmatrix} \qquad \begin{matrix} 13.000 \\ 13.000 \end{matrix} \qquad \widehat{\boldsymbol{\Sigma}}_{pI:H} = \begin{bmatrix} .0141 & \\ & .0141 \end{bmatrix}$$

Composite Using $w_1 = w_2 = .5$

$$\hat{\sigma}_C^2(\tau) = .0358 \ (.0028^b)$$
$$\hat{\sigma}_C^2(\delta) = .0089 \ (.0004^b)$$
$$\hat{\sigma}_C^2(\Delta) = .0104 \ (.0006^b)$$
$$E\hat{\rho}^2 = .801 \ (.020^b)$$
$$\hat{\Phi} = .775 \ (.003^b)$$

$$\widehat{\boldsymbol{\Sigma}}_\tau = \begin{bmatrix} .0404 & .9855^c \\ .0354 & .0319 \end{bmatrix}$$

$$\widehat{\boldsymbol{\Sigma}}_\delta = \begin{bmatrix} .0157 & \\ & .0199 \end{bmatrix}$$

$$\widehat{\boldsymbol{\Sigma}}_\Delta = \begin{bmatrix} .0200 & \\ & .0217 \end{bmatrix}$$

[a] D study sample sizes are: $n'_h = 2$ maps with 6 and 7 items in v_1, and $n'_h = 2$ diagrams with 6 and 7 items in v_2.

[b] Estimated standard errors based on two forms.

[c] Disattenuated correlation.

Since the disattenuated correlation between the universe scores for maps and diagrams is quite high (.986), it is natural to consider what might happen to measurement precision for composite scores if a form of the test contained only maps or only diagrams, even though the universe of generalization remained unchanged with an equal weighting of maps and diagrams. In principle, we could answer this question using either $MSE_C(\Delta)$ or $MSE_C(\delta)$ in Section 10.1.5. Here we use $MSE_C(\delta)$, given by Equations 10.37, because the various forms of MD are carefully equated, which should diminish, if not eliminate, the contribution of the variance components in the $\widehat{\boldsymbol{\Sigma}}_H$ and $\widehat{\boldsymbol{\Sigma}}_{I:H}$ matrices.

With $w_1 = w_2 = .5$, $MSE_C(\delta)$ based on using two diagrams with six items and two diagrams with seven items is

$$MSE_C(\delta) = \frac{\hat{\sigma}_1^2(p) + \hat{\sigma}_2^2(p) - 2\,\hat{\sigma}_{12}(p)}{4} + \left[\frac{\hat{\sigma}_2^2(ph)}{\breve{n}'_h} + \frac{\hat{\sigma}_2^2(pi:h)}{n'_{i+}} \right]$$

$$= \frac{.0403 + .0319 - 2(.0354)}{4} + \left(\frac{.0116}{3.9765} + \frac{.1831}{26} \right)$$

$$= .0103,$$

which is about 15% larger than $\hat{\sigma}_C^2(\delta) = .0089$, although the absolute magnitude of the difference is quite small. The corresponding result based on using four maps is virtually indistinguishable from $\hat{\sigma}_C^2(\delta) = .0089$.[4]

It certainly appears that there would not be much change in overall measurement precision if a form of the *Maps and Diagrams* test had unequal numbers of maps and diagrams, provided the total number of stimuli was four and the total number of items was about 26. Of course, there are other reasons for using equal numbers of maps and diagrams, not the least of which is that doing so faithfully reflects the intended universe of generalization.

Another matter that might be considered is the measurement precision of a half-length form of the *Maps and Diagrams* test. This is not as simple an issue as it may appear, however, because there are a number of possible interpretations of "half-length" for this test, such as

1. half of the passages, in the sense of one map with six items and one diagram with seven items; and

2. half of the items per passage, in the sense of two maps with three items each, and two diagrams with three and four items.

These would be half-length forms in that they both involve half (13) of the number of items in a full-length form (26).

Results for these two possibilities are provided in the left and right sides, respectively, of Table 11.11. The divisors that give the $\widehat{\Sigma}_H$ and $\widehat{\Sigma}_{pH}$ matrices are the $\breve{n}_h'$ given by Equation 11.46, but otherwise the computations are straightforward. It is evident that using all four passages with half of the items per passage leads to somewhat greater measurement precision than using half of the passages. Roughly, error variances are 10 to 12% lower, coefficients are 3 to 4% higher, and signal–noise ratios are 13 to 14% higher using all of the passages with half of the items per passage. The crux of the explanation for this difference is that using all of the passages leads to lower values for the estimated variance components in the $\widehat{\Sigma}_H$ and $\widehat{\Sigma}_{pH}$ matrices. The differences associated with these two perspectives on "half-length" are not very large because the elements of $\widehat{\Sigma}_h$ and $\widehat{\Sigma}_{ph}$ are relatively small.

There are other interpretations of "half-length" that might be considered. In particular, length might be viewed with respect to the amount of time needed to complete the test. From this perspective, using half of the

---

[4]Computations using Form L give similar results.

TABLE 11.11. Half-Length D Studies for *Maps and Diagrams*

Half of the Passages[a]			Half of the Items per Passage[b]		
Divisors	Components		Divisors	Components	
	$\widehat{\Sigma}_p = \begin{bmatrix} .0404 & .0354 \\ .0354 & .0319 \end{bmatrix}$			$\widehat{\Sigma}_p = \begin{bmatrix} .0404 & .0354 \\ .0354 & .0319 \end{bmatrix}$	
1.000 1.000	$\widehat{\Sigma}_H = \begin{bmatrix} .0062 & \\ & .0001 \end{bmatrix}$		2.000 1.960	$\widehat{\Sigma}_H = \begin{bmatrix} .0031 & \\ & .0000 \end{bmatrix}$	
6.000 7.000	$\widehat{\Sigma}_{I:H} = \begin{bmatrix} .0025 & \\ & .0032 \end{bmatrix}$		6.000 7.000	$\widehat{\Sigma}_{I:H} = \begin{bmatrix} .0025 & \\ & .0032 \end{bmatrix}$	
1.000 1.000	$\widehat{\Sigma}_{pH} = \begin{bmatrix} .0032 & \\ & .0116 \end{bmatrix}$		2.000 1.960	$\widehat{\Sigma}_{pH} = \begin{bmatrix} .0016 & \\ & .0059 \end{bmatrix}$	
6.000 7.000	$\widehat{\Sigma}_{pI:H} = \begin{bmatrix} .0305 & \\ & .0262 \end{bmatrix}$		6.000 7.000	$\widehat{\Sigma}_{pI:H} = \begin{bmatrix} .0305 & \\ & .0262 \end{bmatrix}$	

$$\widehat{\Sigma}_\tau = \begin{bmatrix} .0404 & .9855^c \\ .0354 & .0319 \end{bmatrix} \qquad \widehat{\Sigma}_\tau = \begin{bmatrix} .0404 & .9855^c \\ .0354 & .0319 \end{bmatrix}$$

$$\widehat{\Sigma}_\delta = \begin{bmatrix} .0338 & \\ & .0377 \end{bmatrix} \qquad \widehat{\Sigma}_\delta = \begin{bmatrix} .0322 & \\ & .0321 \end{bmatrix}$$

$$\widehat{\Sigma}_\Delta = \begin{bmatrix} .0424 & \\ & .0410 \end{bmatrix} \qquad \widehat{\Sigma}_\Delta = \begin{bmatrix} .0377 & \\ & .0353 \end{bmatrix}$$

Composite Defined Using Weights of $w_1 = w_2 = .50$

$$\hat{\sigma}_C^2(\tau) = 0.0358 \qquad\qquad \hat{\sigma}_C^2(\tau) = 0.0358$$
$$\hat{\sigma}_C^2(\delta) = 0.0179 \qquad\qquad \hat{\sigma}_C^2(\delta) = 0.0161$$
$$\hat{\sigma}_C^2(\Delta) = 0.0209 \qquad\qquad \hat{\sigma}_C^2(\Delta) = 0.0183$$

$$E\hat{\rho}^2 = 0.667 \qquad\qquad E\hat{\rho}^2 = 0.690$$
$$\widehat{\Phi} = 0.632 \qquad\qquad \widehat{\Phi} = 0.662$$

$$\widehat{S/N}(\delta) = 2.001 \qquad\qquad \widehat{S/N}(\delta) = 2.228$$
$$\widehat{S/N}(\Delta) = 1.714 \qquad\qquad \widehat{S/N}(\Delta) = 1.958$$

[a]D study sample sizes are: $n'_h = 1$ map with six items in v_1, and $n'_h = 1$ diagram with seven items in v_2.

[b]D study sample sizes are: $n_h = 2$ maps with three and three items in v_1, and $n_h = 2$ diagrams with three and four items in v_2.

[c]Disattenuated correlations.

passages likely results in a considerably "shorter" test than using all of the passages with half of the items per passage.

Strictly speaking, for the composite results in Table 11.11, it is assumed that $w_1 = w_2 = .5$ and the estimation weights equal the w weights, which means that the reported score for a person is an equally weighted average of the person's average map(s) score and average diagram(s) score; that is, $\overline{X}_p = .5\,\overline{X}_{p1} + .5\,\overline{X}_{p2}$. If the reported score were the simple proportion of items correct over the 13 items, the results in Table 11.11 would not be quite right, because there are unequal numbers of items associated with the map(s) and diagram(s) scores (6 and 7, respectively). Under these circumstances, $\overline{X}_{pC} = (6/13)\,\overline{X}_{p1} + (7/13)\,\overline{X}_{p2}$, and the estimation weights do not exactly equal the a priori weights.

11.2.2 ITED Literary Materials Test

To illustrate the $p^{\bullet} \times (i^{\circ} \!:\! h^{\bullet})$ design, Section 9.2.3 introduced the *LM* example, the *Ability to Interpret Literary Materials* test of the *Iowa Tests of Educational Development* (ITED) (Feldt et al., 1994). Form L, Level 17/18, of this test contains five passages (levels of h) with 9, 8, 9, 8, and 10 items, respectively, for a total of 44 items. In addition to being nested within a passage, each item also can be viewed as being nested within one of two fixed process categories. The numbers of items in each of the passages associated with the first category are 4, 4, 7, 2, and 6, respectively; the numbers of items in each of the passages associated with the second category are 5, 4, 2, 6, and 4, respectively. Since each passage contributes items to both categories, and each person responds to items in both categories, the covariance components design is $p \times h$. This design is unbalanced since the variance components designs for both categories are unbalanced.

For the *LM* example, Table 11.12 provides analogous mean-square and mean-product matrices ($\mathbf{M}$), as well as estimated variance-covariance matrices ($\widehat{\boldsymbol{\Sigma}}$), based on the responses of 2450 students who took Form L in a standardization study. The off-diagonal terms in the $\mathbf{M}_p$, $\mathbf{M}_h$, and $\mathbf{M}_{ph}$ matrices are the mean products for the compound means procedure discussed in Section 11.1.4.

When the standardization was conducted, two forms (K and L) were administered to somewhat different groups of students. Based on the estimated variance and covariance components for these two forms, the last column of matrices in Table 11.12 provides two-replicate estimated standard errors of the type discussed in the previous section for the *MD* example. Note, however, that Forms K and L have different patterns of $n_{i:h}$ sample sizes. For Form K the sample sizes are 7/2/2/6/6 for v_1 and 2/6/6/3/4 for v_2; for Form L the sample sizes are 4/4/7/2/6 for v_1 and 5/4/2/6/4 for v_2. To be true replicates, the patterns would have to be the same (and the groups would have to be randomly equivalent). The differences are sub-

TABLE 11.12. G Study for *Literary Materials* Based on Mean Squares and Mean Products

Analogous Mean Squares and Mean Products[a]	Variance and Covariance Components	
	Estimates	Standard Errors[b]
$\mathbf{M}_p = \begin{bmatrix} 1.2473 & .2109 \\ .2109 & 1.0630 \end{bmatrix}$	$\widehat{\boldsymbol{\Sigma}}_p = \begin{bmatrix} .0443 & .0405 \\ .0405 & .0401 \end{bmatrix}$	$\begin{bmatrix} .0010 & .0028 \\ .0028 & .0030 \end{bmatrix}$
$\mathbf{M}_h = \begin{bmatrix} 105.4833 & 2.1558 \\ 2.1558 & 43.7483 \end{bmatrix}$	$\widehat{\boldsymbol{\Sigma}}_h = \begin{bmatrix} .0087 & .0009 \\ .0009 & .0000 \end{bmatrix}$	$\begin{bmatrix} .0047 & .0010 \\ .0010 & .0020 \end{bmatrix}$
$\mathbf{M}_{i:h} = \begin{bmatrix} 11.3881 & \\ & 44.2438 \end{bmatrix}$	$\widehat{\boldsymbol{\Sigma}}_{i:h} = \begin{bmatrix} .0046 & \\ & .0180 \end{bmatrix}$	$\begin{bmatrix} .0009 & \\ & .0013 \end{bmatrix}$
$\mathbf{M}_{ph} = \begin{bmatrix} .2167 & .0083 \\ .0083 & .2163 \end{bmatrix}$	$\widehat{\boldsymbol{\Sigma}}_{ph} = \begin{bmatrix} .0143 & .0083 \\ .0083 & .0089 \end{bmatrix}$	$\begin{bmatrix} .0018 & .0009 \\ .0009 & .0011 \end{bmatrix}$
$\mathbf{M}_{pi:h} = \begin{bmatrix} .1531 & \\ & .1798 \end{bmatrix}$	$\widehat{\boldsymbol{\Sigma}}_{pi:h} = \begin{bmatrix} .1531 & \\ & .1798 \end{bmatrix}$	$\begin{bmatrix} .0059 & \\ & .0057 \end{bmatrix}$

Note. Sample sizes for Form L are: $n_p = 2450$ students and $n_h = 5$ passages with 4, 4, 7, 2, and 6 items in v_1, and 5, 4, 2, 6, and 4 items in v_2.
[a]Off-diagonal elements are mean products for compound means procedure.
[b]Based on two replications.

stantial enough to cast some doubt on the estimated standard errors. Still, they are the only readily available estimates.[5]

In Table 11.12, analogous mean squares are used to estimate variance components (see Equation Set 7.14), and the compound means procedure is employed to estimate covariance components. By contrast, in Table 11.13 C terms are used to estimate variance components, and the CP-terms procedure is employed to estimate covariance components (see Section 11.1.3). The estimated standard errors in Table 11.13 are analogous to those in Table 11.12.

Using the estimated standard errors as benchmarks, we observe that almost all of the Form L estimates in $\widehat{\boldsymbol{\Sigma}}_h$, $\widehat{\boldsymbol{\Sigma}}_{i:h}$, and $\widehat{\boldsymbol{\Sigma}}_{ph}$ based on C and CP terms are substantially different from those based on analogous MS and MP terms. Also, it is evident that the estimated standard errors of the elements of $\widehat{\boldsymbol{\Sigma}}_h$, $\widehat{\boldsymbol{\Sigma}}_{i:h}$, and $\widehat{\boldsymbol{\Sigma}}_{ph}$ based on C and CP terms are noticeably

---

[5]The previously discussed normality-based procedures for estimating standard errors of estimated variance components (see Section 6.1.1) and estimated covariance components (see Section 9.4.3) do not apply to unbalanced designs.

TABLE 11.13. G Study for *Literary Materials* Based on C Terms and CP Terms

C Terms and CP Terms	Variance and Covariance Components	
	Estimates	Standard Errors[a]
$\mathbf{C}_p = \begin{bmatrix} 1216.10 & 1027.41 \\ 1027.41 & 904.70 \end{bmatrix}$	$\widehat{\boldsymbol{\Sigma}}_p = \begin{bmatrix} .0452 & .0430 \\ .0430 & .0409 \end{bmatrix}$	$\begin{bmatrix} .0003 & .0018 \\ .0018 & .0035 \end{bmatrix}$
$\mathbf{C}_h = \begin{bmatrix} 2.27693 & 1.84766 \\ 1.84766 & 1.53788 \end{bmatrix}$	$\widehat{\boldsymbol{\Sigma}}_h = \begin{bmatrix} .0180 & -.007 \\ -.007 & -.026 \end{bmatrix}$	$\begin{bmatrix} .0132 & .0111 \\ .0111 & .0333 \end{bmatrix}$
$\mathbf{C}_{i:h} = \begin{bmatrix} 10.4256 & \\ & 7.0525 \end{bmatrix}$	$\widehat{\boldsymbol{\Sigma}}_{i:h} = \begin{bmatrix} .0029 & \\ & .0388 \end{bmatrix}$	$\begin{bmatrix} .0035 & \\ & .0301 \end{bmatrix}$
$\mathbf{C}_{ph} = \begin{bmatrix} 6753.88 & 5124.37 \\ 5124.37 & 4937.70 \end{bmatrix}$	$\widehat{\boldsymbol{\Sigma}}_{ph} = \begin{bmatrix} .0097 & .0059 \\ .0059 & .0045 \end{bmatrix}$	$\begin{bmatrix} .0042 & .0016 \\ .0016 & .0028 \end{bmatrix}$
$\mathbf{C}'_{pi:h} = \begin{bmatrix} 37470 & \\ & 29044 \end{bmatrix}$	$\widehat{\boldsymbol{\Sigma}}_{pi:h} = \begin{bmatrix} .1568 & \\ & .1834 \end{bmatrix}$	$\begin{bmatrix} .0023 & \\ & .0089 \end{bmatrix}$

Note. Sample sizes for Form L are: $n_p = 2450$ students and $n_h = 5$ passages with 4, 4, 7, 2, and 6 items in v_1, and 5, 4, 2, 6, and 4 items in v_2.

[a]Based on two replications.

larger those based on analogous *MS* and *MP* terms. It seems plausible that these differences in estimated standard errors are attributable, at least partly, to the rather severely unbalanced nature of the design, coupled with the fact that the sample size patterns for Forms K and L are quite different. There is at least the hint of a suggestion that estimates based on *C* and *CP* terms tend to be more variable than those based on analogous *MS* and *MP* terms.

These observations do not provide unequivocal guidance for a choice of estimates, but there are pragmatic reasons for choosing the estimates based on the analogous *MS* terms and *MP* terms:

- the variance-component estimates based on analogous *MS* terms have more precedent than those based on *C* terms; and

- the covariance-component estimates based on the compound-means procedure are simpler to conceptualize and compute.

Note also that the compound-means procedure does not explicitly incorporate the $n'_{i:h}$ sample sizes. This may be somewhat defensible for this example in that Forms K and L of *Literary Materials* have substantially different sample size patterns, even though the two forms are used interchangeably (after equating).

TABLE 11.14. D Study for *Literary Materials* Using G Study Estimates from Table 11.12

		D Study[a]	
G Study Components	Divisors[b]	Components	
$\widehat{\Sigma}_p = \begin{bmatrix} .0443 & .0405 \\ .0405 & .0401 \end{bmatrix}$		$\widehat{\Sigma}_p = \begin{bmatrix} .0443 & .0405 \\ .0405 & .0401 \end{bmatrix}$	
$\widehat{\Sigma}_h = \begin{bmatrix} .0087 & .0009 \\ .0009 & .0000 \end{bmatrix}$	4.372 4.546	$\widehat{\Sigma}_H = \begin{bmatrix} .0020 & .0002 \\ .0002 & .0000 \end{bmatrix}$	
$\widehat{\Sigma}_{i:h} = \begin{bmatrix} .0046 & \\ & .0180 \end{bmatrix}$	23.000 21.000	$\widehat{\Sigma}_{I:H} = \begin{bmatrix} .0002 & \\ & .0009 \end{bmatrix}$	
$\widehat{\Sigma}_{ph} = \begin{bmatrix} .0143 & .0083 \\ .0083 & .0089 \end{bmatrix}$	4.372 4.546	$\widehat{\Sigma}_{pH} = \begin{bmatrix} .0033 & .0015 \\ .0015 & .0020 \end{bmatrix}$	
$\widehat{\Sigma}_{pi:h} = \begin{bmatrix} .1531 & \\ & .1798 \end{bmatrix}$	23.000 21.000	$\widehat{\Sigma}_{pI:H} = \begin{bmatrix} .0067 & \\ & .0086 \end{bmatrix}$	

Composite: $w_1 = .523$; $w_2 = .477$

$\hat{\sigma}_C^2(\tau) = .0415 \ (.0033^c)$

$\hat{\sigma}_C^2(\delta) = .0059 \ (.0003^c)$

$\hat{\sigma}_C^2(\Delta) = .0067 \ (.0007^c)$

$E\hat{\rho}^2 = .876 \ (.012^c)$

$\widehat{\Phi} = .861 \ (.021^c)$

$\widehat{\Sigma}_\tau = \begin{bmatrix} .0443 & .9614^d \\ .0405 & .0401 \end{bmatrix}$

$\widehat{\Sigma}_\delta = \begin{bmatrix} .0099 & .1444^d \\ .0015 & .0105 \end{bmatrix}$

$\widehat{\Sigma}_\Delta = \begin{bmatrix} .0121 & .1391^d \\ .0016 & .0114 \end{bmatrix}$

[a]D study sample sizes are: $n_h = 5$ passages with 4, 4, 7, 2, and 6 items in v_1, and 5, 4, 2, 6, and 4 items in v_2.

[b]For covariance components in $\widehat{\Sigma}_H$ and $\widehat{\Sigma}_{pH}$, the divisor is 5.616.

[c]Estimated standard errors based on two forms.

[d]Disattenuated correlations.

The G study estimated variance and covariance components based on analogous MS and MP terms in Table 11.12 are used in Table 11.14 to estimate results for a D study with the same sample sizes as the G study, with w weights proportional to sample sizes, and with estimation weights equal to w weights. To obtain the D study variance component estimates, the G study estimates are divided by $\breve{n}_h'$ given in Equation 11.46. The D study covariance component estimates in $\widehat{\Sigma}_H$ and $\widehat{\Sigma}_{pH}$ are obtained using the divisor:

$$\breve{n}_{h_{vv'}}' = \frac{n_{i+}' m_{i+}'}{\sum_h n_{i:h}' m_{i:h}'}, \tag{11.47}$$

where $h = 1, \ldots, n'_h$, $n'_{i:h}$ are the D study sample sizes for v, $m'_{i:h}$ are the D study sample sizes for v', $n'_{i+} = \sum_h n'_{i:h}$, and $m'_{i+} = \sum_h m'_{i:h}$. (Note that $\breve{n}'_{h_{vv'}}$ would be n'_h if all of the $n'_{i:h}$ and the $m'_{i:h}$ were the same constant.)

To prove that Equation 11.47 is the divisor of $\sigma_{vv'}(h)$, recall that the convention adopted in this book is that ν designates effects associated with level v, and ξ designates effects associated with level v'. Also, to simplify notation, suppose the G and D study sample sizes are the same. Then

$$\sigma_{vv'}(H) = E\left[\left(\frac{\sum_h n_{i:h}\nu_h}{n_{i+}}\right)\left(\frac{\sum_h m_{i:h}\xi_h}{m_{i+}}\right)\right]$$

$$= \frac{1}{n_{i+}m_{i+}}E\left[\sum_h n_{i:h}m_{i:h}(\nu_h\xi_h) + \sum\sum_{h\neq h'} n_{i:h}m_{i:h'}(\nu_h\xi_{h'})\right].$$

Since $E\nu_h\xi_{h'} = 0$ for $h \neq h'$, it follows that

$$\sigma_{vv'}(H) = \frac{1}{n_{i+}m_{i+}}E\left[\sum_h n_{i:h}m_{i:h}(\nu_h\xi_h)\right]$$

$$= \left(\frac{\sum_h n_{i:h}m_{i:h}}{n_{i+}m_{i+}}\right)\sigma_{vv'}(h).$$

An entirely analogous proof applies for $\sigma_{vv'}(pH)$, and similar proofs apply when the G and D study sample sizes are different.

For the LM example, recall that the $n_{i:h}$ were 4, 4, 7, 2, and 6 items for v_1, and the corresponding $m_{i:h}$ were 5, 4, 2, 6, and 4 items for v_2. This gives

$$\breve{n}'_{h_{12}} = \frac{23 \times 21}{20 + 16 + 14 + 12 + 24} = 5.616.$$

Once the D study estimated variance and covariance matrices are determined, the remaining D study results in Table 11.14 are easily obtained in the usual manner. Both relative and absolute errors for the two categories are slightly correlated (about .14), but universe scores are very highly correlated (about .96). Such a high correlation indicates that universe scores for the two categories are nearly linearly related, but this does not mean that the categories should be ignored in test development. For example, if the universe of generalization remained unchanged, but all items were selected from the first category and the estimation weights were set at $a_1 = 1$ and $a_2 = 0$, then $MSE_C(\delta) = .0071$, which is about 20% larger than $\hat{\sigma}_C^2(\delta) = .0059$.

For the LM example, overall absolute error variance for the $p^\bullet \times (I^\circ : H^\bullet)$ design is $\hat{\sigma}_C^2(\Delta) = .0067$. Both conceptually and computationally, this is equal to the average of the conditional absolute error variances for each of the 2450 persons in the sample. For any person, conditional absolute error variance is the variance of the mean for the within-person $I^\circ : H^\bullet$ design, in

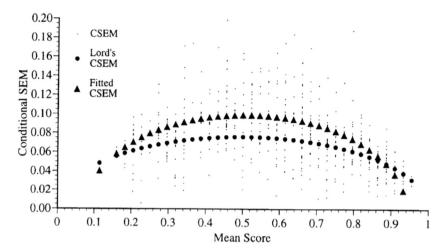

FIGURE 11.1. Conditional absolute standard errors of measurement for *Literary Materials* for 494 examinees.

the sense discussed in Section 10.2.3 (see also Exercise 11.12). Figure 11.1 provides a plot of conditional SEMs, $\hat{\sigma}_C(\Delta_p)$, for nearly 500 examinees, along with a quadratic fit to the conditional SEMs for all 2450 examinees.[6] Also plotted are Lord's conditional SEMs in which all 44 items in *Literary Materials* are assumed to be sampled from an undifferentiated universe; that is,

$$\hat{\sigma}_C(\Delta_p) = \sqrt{\frac{\overline{X}_{pC}(1 - \overline{X}_{pC})}{43}}.$$

It is evident from Figure 11.1 that this simplistic model substantially understates the magnitude of the conditional SEMs.

Thus far in this section, results have been presented only for Form L of *Literary Materials*. There are other forms, one of which is Form K. These forms are developed according to the same specifications, to the extent possible. However, Forms K and L have different patterns of numbers of items within passage for the first process category and for the second process category. Such dissimilar patterns are common—indeed, almost inevitable—in passage-based tests with process categories constituting part of a table of specifications for forms. Test developers may be able to achieve some degree of balance across categories, but seldom is the pool of available passages so rich that the same patterns of sample sizes can be used across forms.

---

[6]The quadratic fit to the error variances for all 2450 examinees is given by the polynomial equation $\hat{\sigma}_C^2(\Delta_p) = -.0038 + .0537\,\overline{X}_{pC} - .0529\,\overline{X}_{pC}^2$. The square roots of these fitted values are given by the solid triangles in Figure 11.1.

Using $n'_{i:h}$ and $m'_{i:h}$ to designate sample sizes for Categories 1 and 2, respectively, sample-size statistics for Forms K and L are given in the following table.

Form	Category 1			Category 2			$\breve{n}'_{h_{12}}$
	$n'_{i:h}$	n'_{i+}	$\breve{n}'_h$	$m'_{i:h}$	m'_{i+}	$\breve{n}'_h$	
K	7 2 2 6 6	23	4.101	2 6 6 3 4	21	4.366	6.037
L	4 4 7 2 6	23	4.372	5 4 2 6 4	21	4.546	5.616

Both forms have 23 items in Category 1 and 21 items in Category 2, which means that w weights will be the same for the two forms when the weights are proportional to the numbers of items within categories. However, the dissimilar sample-size patterns cause the two $\breve{n}'_h$ and $\breve{n}'_{h_{12}}$ to be different across forms. For this reason, Forms K and L in the LM example might have somewhat different psychometric characteristics even if, in all other respects, the forms were perfectly parallel.

Dissimilar sample size patterns are perhaps inevitable, but they are still a reflection of less than ideal experimental control in test development. Furthermore, differences in sample size patterns highlight an issue in conceptualizing replications of the measurement procedure. Strictly speaking, randomly parallel instances of the measurement procedure have the same sample size patterns. That is, sample sizes are not random effects. Rather, both in estimation procedures and in conceptualizing replications, sample sizes are fixed effects.

Let us consider three possible sample-size patterns that might be used for the LM example:

1. sample-size patterns used in Form K;

2. sample-size patterns used in Form L; and

3. approximately constant sample sizes (e.g., 4, 4, 5, 5, and 5 items for v_1, and 4, 4, 4, 4, and 5 items for v_2).

Let us suppose, as well, that "ideal" forms would be built according to the third set of specifications.

For all three sample-size patterns, composite universe score variance is necessarily the same, because universe scores are not affected by D study sample sizes. Error variances, however, are different. The D study results in Table 11.14 assume that the universe of generalization corresponds to the first set of sample-size patterns, although we are arguing hypothetically that "ideal" forms would be created according to the third set. Consequently, for some purposes, it might be sensible to consider the expected correlation between a randomly selected form with Form K sample sizes (say X) and a randomly selected form with "approximately constant" sample sizes (say $\mathcal{X}$):

$$E\rho_{X\mathcal{X}} \doteq \frac{\sigma_C^2(\tau)}{\sqrt{ES_{C_X}^2(\tau)\, ES_{C_{\mathcal{X}}}^2(\tau)}},$$

where the first expected-observed-score-variance is for X and the second is for $\mathcal{X}$. The value of this correlation is necessarily somewhere between $E\rho^2$ for the Form K sample sizes and $E\rho^2$ for the "ideal" sample sizes.

For the LM example, this theoretical argument is moot because there is virtually no difference in composite relative error variances for the universes (assuming the G study estimated variance and covariance components in Table 11.12 are the parameters). Still, in other contexts, consequential differences might be observed. In any case, it bears repeating that sample sizes are fixed conditions of measurement in generalizability theory.

11.2.3   District Mean Difference Scores

Table 11.15 provides raw score summary statistics for Reading Comprehension (R) and Math Total (M) grade equivalent scores on the *Iowa Tests of Basic Skills (ITBS)* for fourth-grade students in a sample of 103 Iowa school districts that tested in four consecutive years in the 1990s. For each year, the summary statistics reported are for the distribution of unweighted district mean scores. For example, the mean of 4.5961 for Reading Comprehension in 1995 to 1996 was obtained by computing the mean grade equivalent score for students in each of the 103 districts, and then computing the mean (unweighted) of these means. The number of students in the districts ranges from 10 to over 600, with arithmetic means for each of the four years in the low 70s, and harmonic means in the 30s (see Table 11.15).

For any given year, the data provided in Table 11.15 do not provide a basis for generalization in that the data were collected on the same occasion using the same form of the *ITBS*.[7] Across years, however, different students were involved each year and two different forms were employed. So, variability in the Reading Comprehension and Math Total statistics over years reflects variability in both a person facet and an item facet.

G Study Design

One approach to examining the issue of the generalizability of these district (d) mean scores over forms and years begins with an analysis employing a G study multivariate $p^{\bullet}:(d^{\bullet} \times o^{\bullet})$ design, where p designates students

---

[7]Districts can choose when they administer the *ITBS*; some test in the Fall and some in the Spring. The data reported in Table 11.15 are for districts that chose to test in the Fall.

TABLE 11.15. Grade-Equivalent Scores for District Means Example

Year	Form	Test	Mean	S.D.	r	S_{diff}	HM^a	n_{tot}
1995–96	K	R	4.5961	.5231	.7011	.3828	36.26	7656
		M	4.3823	.4528				
1996–97	L	R	4.5715	.5339	.7728	.3499	32.31	7252
		M	4.4039	.5000				
1997–98	K	R	4.4511	.5911	.8181	.3427	33.99	7397
		M	4.2758	.5272				
1998–99	L	R	4.4384	.5946	.8413	.3228	32.41	7386
		M	4.3324	.5303				
Over Years		R	4.5106	.4541	.8610	.2338	33.67^b	29691
		M	4.3483	.4274				

^{a}Harmonic mean of number of students within each district.
^{b}Harmonic mean over districts of harmonic means over occasions within district.

within districts (d), with each student tested in one of four years or occasions (o). Note that this representation of the design involves a complicated confounding of years and forms in what is being called the "occasion" facet.

For this design, the levels of the fixed facet are Reading Comprehension and Math Total grade-equivalent scores. This is an unbalanced multivariate design in that there are different numbers of students within districts, and the design is full in the sense that there are R and M scores for all districts, for all students within districts, and for all four years.

The second and third columns of Table 11.16 provide estimated variance components for R and M, respectively, using the analogous-ANOVA procedure in Section 11.1.2. Since this is a full multivariate design, covariance components can be obtained using the variance-of-a-sum procedure discussed in Section 11.1.5. To obtain them, the fourth column of Table 11.16 provides estimated variance components for the sum of R and M.[8] The estimated covariance components based on Equation 11.43 are in the fifth column.

The patterns are very similar for the estimated variance components for R, the estimated variance components for M, and the estimated covariance components. Students account for most of the variance and covariance, occasion differences (i.e., year/form) are of little consequence, and there is evidence of variability and covariability in district mean scores.

At the bottom of Table 11.16, the estimated variance and covariance matrices are provided, with the upper-diagonal elements being disattenuated

---

[8]The estimated variance components for R, M, and $R + M$ were obtained using urGENOVA.

TABLE 11.16. G Study for District Means Example of $p^{\bullet}:(d^{\bullet} \times o^{\bullet})$ Design for Reading Comprehension and Math Total Grade-Equivalent Scores

Effect	$\hat{\sigma}_R^2$	$\hat{\sigma}_M^2$	$\hat{\sigma}_{R+M}^2$	$\hat{\sigma}_{RM}$	$\hat{\rho}_{RM}$	$\hat{\sigma}_R^2 + \hat{\sigma}_M^2 - 2\,\hat{\sigma}_{RM}$
d	.1668	.1398	.5791	.1363	.8925	.0340
o	.0038	.0014	.0097	.0022	.9529	.0008
do	.0178	.0180	.0601	.0121	.6772	.0116
$p{:}do$	3.3461	2.1765	9.0997	1.7886	.6628	1.9455

$$\widehat{\Sigma}_d = \begin{bmatrix} .1668 & .8925 \\ .1363 & .1398 \end{bmatrix} \qquad \widehat{\Sigma}_o = \begin{bmatrix} .0038 & .9529 \\ .0022 & .0014 \end{bmatrix}$$

$$\widehat{\Sigma}_{do} = \begin{bmatrix} .0178 & .6772 \\ .0121 & .0180 \end{bmatrix} \qquad \widehat{\Sigma}_{p:do} = \begin{bmatrix} 3.3461 & .6628 \\ 1.7886 & 2.1765 \end{bmatrix}$$

Note. Italicized values are disattenuated correlations.

correlations—all of which are quite large.[9] Interpreting these correlations requires some thought. For example, consider $\hat{\rho}_{RM}(d) = .89$. This is the estimated correlation between Reading and Math Total grade-equivalent mean scores for districts, where each such mean is over students, forms, and years. Also, consider $\hat{\rho}_{RM}(o) = .95$. This is the estimated correlation between Reading and Math Total grade-equivalent mean scores for occasions, where each such mean is over students and districts. The fact that $\hat{\rho}_{RM}(d)$ is somewhat smaller than $\hat{\rho}_{RM}(o)$ suggests that district mean scores for R and M are not as highly correlated as occasion (i.e., year/form) mean scores.

The student correlation $\hat{\rho}_{RM}(p{:}do) = .66$ is *not* a disattenuated correlation in the usual sense of that term. It is interpretable as an estimate of the correlation between R and M for students in a randomly selected district for a randomly selected occasion. This correlation is essentially an observed score correlation, not an estimate of the correlation between R and M universe scores for students. At the student level, there is only one observation for R and one observation for M in these data, so there is no statistical basis that permits generalizing student observed scores to universe scores for R and M. Generalization is possible for district mean scores because there are district scores for different forms and different years (and, therefore, different samples of students, too).

Note also that $\hat{\rho}_{RM}(do)$ and $\hat{\rho}_{RM}(p{:}do)$ are almost identical (about .67). Apparently, the district–occasion interaction effects for R and M are about as highly correlated as the observed student scores for R and M, but these

---

[9]The size of these correlations is partly attributable to scaling issues, but not entirely.

correlations are both considerably smaller than the district means disattenuated correlation of .89.

D Study Design and Sample Sizes

With districts as the objects of measurement, the D study multivariate design is $P^{\bullet} : (d^{\bullet} \times O^{\bullet})$. Since the design is full, for each of the four matrices, the D study sample-size divisor of the G study variance and covariance components is the same. It is obvious that the divisor is n'_o for the elements of $\widehat{\Sigma}_o$ and $\widehat{\Sigma}_{do}$. Less obviously, we use $n'_o \ddot{n}'_p$ as the divisor for the elements of $\widehat{\Sigma}_{p:do}$, where $\ddot{n}'_p$ is the harmonic mean (over districts) of the $\ddot{n}'_{p:d}$ (harmonic mean over occasions of the $n'_{p:d}$ within a district), as discussed next.

Consider, for example, $\sigma_{RM}(P{:}dO)$, under the simplifying assumption that G and D study sample sizes are the same. By definition,

$$
\begin{aligned}
\sigma_{RM}(P{:}dO) &= E(\nu_{P:dO})(\xi_{P:dO}) \\
&= E\left(\frac{1}{n_o}\sum_o^{n_o}\sum_p^{n_{p:do}}\frac{\nu_{p:do}}{n_{p:do}}\right)\left(\frac{1}{n_o}\sum_o^{n_o}\sum_p^{n_{p:do}}\frac{\xi_{p:do}}{n_{p:do}}\right) \\
&= \frac{1}{n_o^2}\sum_o^{n_o}\sum_p^{n_{p:do}}\frac{E\nu_{p:do}\xi_{p:do}}{n_{p:do}^2} \\
&= \left(\frac{1}{n_o^2}\sum_o\frac{1}{n_{p:do}}\right)\sigma_{RM}(p{:}do) \\
&= \frac{\sigma_{RM}(p{:}do)}{n_o\ddot{n}_{p:d}}, \tag{11.48}
\end{aligned}
$$

where $\ddot{n}_{p:d}$ is the harmonic mean of the $n_{p:do}$ within a district:

$$
\ddot{n}_{p:d} = \left[\frac{1}{n_o}\sum_o\left(\frac{1}{n_{p:do}}\right)\right]^{-1}.
$$

The result in Equation 11.48 is for a randomly selected district, and it is conditional on the harmonic mean $\ddot{n}_{p:d}$. The average value over all districts is

$$
\frac{1}{n_d}\sum_d\frac{\sigma_{RM}(p{:}do)}{n_o\ddot{n}_{p:d}} = \frac{\sigma_{RM}(p{:}do)}{n_o\ddot{n}_p},
$$

where $\ddot{n}_p$ is the harmonic mean (over districts) of the $\ddot{n}_{p:d}$. Therefore, for many purposes we use

$$
\sigma_{RM}(P{:}dO) = \frac{\sigma_{RM}(p{:}do)}{n_o\ddot{n}_p}. \tag{11.49}
$$

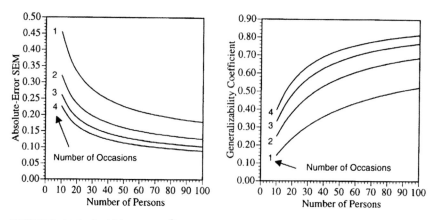

FIGURE 11.2. $\hat{\sigma}_C(\Delta)$ and $E\hat{\rho}^2$ for district mean difference scores for Reading Comprehension and Math Total.

Mean Difference Scores

Suppose an investigator is interested in the the measurement characteristics of district mean difference scores for R and M. Typical questions might focus on a generalizability coefficient and standard error of measurement of such difference scores, where the universe of generalization involves a person (i.e., student) facet and an occasion (i.e., year/form) facet, and the objects of measurement are districts. For such questions, the a priori weights are $w_R = 1$ and $w_M = -1$.

Estimated composite absolute error variance is

$$\hat{\sigma}_C^2(\Delta) = \frac{\hat{\sigma}_R^2(o) + \hat{\sigma}_M^2(o) - 2\,\hat{\sigma}_{RM}(o)}{n'_o}$$
$$+ \frac{\hat{\sigma}_R^2(do) + \hat{\sigma}_M^2(do) - 2\,\hat{\sigma}_{RM}(do)}{n'_o}$$
$$+ \frac{\hat{\sigma}_R^2(p{:}do) + \hat{\sigma}_M^2(p{:}do) - 2\,\hat{\sigma}_{RM}(p{:}do)}{n'_o \ddot{n}'_p}. \qquad (11.50)$$

The numerators of each of the fractions in this equation are provided in the last column of Table 11.16. For the G study data, $\ddot{n}_p = 33.67$. Using this value with $n'_o = 2$ gives an estimated absolute-error SEM of

$$\hat{\sigma}_C(\Delta) = \sqrt{\frac{.0008}{2} + \frac{.0116}{2} + \frac{1.9455}{2(33.67)}} = .1873. \qquad (11.51)$$

The left panel of Figure 11.2 reports $\hat{\sigma}_C(\Delta)$ for $n'_o = 1, 2, 3, 4$ with $\ddot{n}'_p$ ranging from 10 to 100. Using $n'_o = 2$ is consistent with the usual practice in Iowa of reporting averages over two years (and, hence, two forms). Doing so has a number of advantages, including offsetting the effects of any errors that may be present in the equating of Forms K and L, and partially offsetting unusual results for districts with small numbers of students. It is clear from Figure 11.2 that using two years' worth of data substantially reduces the absolute-error SEM for the difference in mean scores over what it would be if only one year's data were used.

Estimated composite universe score variance for districts is

$$\hat{\sigma}_C^2(d) = \hat{\sigma}_R^2(d) + \hat{\sigma}_M^2(d) - 2\,\hat{\sigma}_{RM}(d) = .0340,$$

where it is to be emphasized that d stands for "district" not "difference." Estimated relative error variance is

$$\hat{\sigma}_C^2(\delta) \quad = \quad \frac{\hat{\sigma}_R^2(do) + \hat{\sigma}_M^2(do) - 2\,\hat{\sigma}_{RM}(do)}{n_o'}$$
$$+ \frac{\hat{\sigma}_R^2(p{:}do) + \hat{\sigma}_M^2(p{:}do) - 2\,\hat{\sigma}_{RM}(p{:}do)}{n_o'\ddot{n}_p'}. \tag{11.52}$$

For $n_o' = 2$ and $\ddot{n}_p' = 33.67$ (the value in the G study data), $\hat{\sigma}_C^2(\delta) = .0347$ and

$$E\hat{\rho}^2 = \frac{.0340}{.0340 + .0347} = .495. \tag{11.53}$$

The right panel of Figure 11.2 reports $E\hat{\rho}^2$ for $n_o' = 1, 2, 3, 4$ and $\ddot{n}_p'$ ranging from 10 to 100.

At least some approaches to estimating the reliability of district mean difference scores would assume uncorrelated errors. For this example, given the magnitude of the covariances between the effects for R and M (see Table 11.16), such an assumption would lead to seriously *under*estimating reliability. For example, assuming $\sigma_{RM}(do) = \sigma_{RM}(p{:}do) = 0$, $n_o' = 2$, and $\ddot{n}_p' = 33.67$, it is easily verified that $E\hat{\rho}^2 = .254$, which is about half as large as the value that incorporates correlated effects (.495).

As another perspective on this example, note that the observed-score variances for each of the four occasions (the square of the S_{diff} values in Table 11.15) are .1466, .1224, .1175, and .1042. The average of these observed variances, .1227, is clearly an estimate of expected observed score variance for a single occasion. It follows that one estimate of a generalizability coefficient for $n_o' = 1$ and the $n_{p{:}do}$ sample sizes in the G study data is

$$E\hat{\rho}^2 = \frac{.0340}{.1227} = .2771, \tag{11.54}$$

which is an estimated generalizability coefficient for a single occasion. For this example, the Spearman–Brown formula applies (see Exercise 11.13) and can be used with Equation 11.54 to obtain an estimated generalizability coefficient for mean scores over two occasions:

$$E\hat{\rho}^2 = \frac{2(.2771)}{1.2771} = .434, \tag{11.55}$$

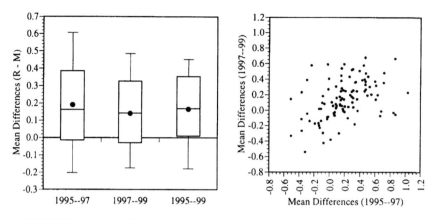

FIGURE 11.3. Box plots and scatterplot for district mean difference scores.

which is associated with $\hat{\sigma}_C^2(\delta) = .0443$. The discrepancy between the estimated generalizability coefficients in Equations 11.53 and 11.55 is attributable primarily to the fact that the error variance in Equation 11.53 is based directly on analogous-ANOVA estimates of variance and covariance components, whereas the error variance in Equation 11.55 is not.

As an additional perspective, note that the correlation (over districts) between the average of the difference scores for the first two years (1995–96 and 1996–97) and the average of the difference scores for the last two years (1997–98 and 1998–99) is

$$r(1995\text{--}97, 1997\text{--}99) = .434. \tag{11.56}$$

Figure 11.3 provides Box plots for 1995–97, 1997–99, and 1995–99, along with a scatterplot for the 1995–97 and 1997–99 district mean difference scores. The correlation in Equation 11.56 is an estimate of reliability for district mean difference scores based on two years' worth of data. Clearly, this estimate is virtually identical to the estimate in Equation 11.55. Note also that the covariance associated with Equation 11.56 is .0343, which is an estimate of universe score variance for district mean difference scores. This estimate is quite close to the estimate derived previously, namely, .0340.

Which approach is preferable for estimating reliability for two occasions—the approach that led to Equation 11.53, the approach that led to Equation 11.55, or the two-year correlation in Equation 11.56? That depends. The two-year correlation provides a very direct estimate of reliability for the unweighted distribution of district mean difference scores, using averages over two years, although both pairs of years use the same forms. The second approach gives a similar result, but it is less direct, and it uses multivariate generalizability theory to estimate universe score variance for the district mean difference scores. However, neither of these approaches gives estimates of all of the variance and covariance components, and neither of them permits the kind of analysis reported in Figure 11.2, based on multi-

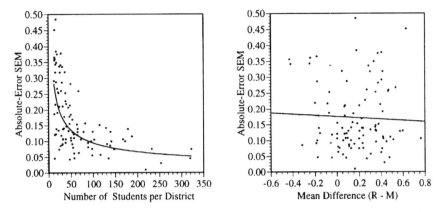

FIGURE 11.4. Absolute-error conditional standard errors of measurement for district means $\hat{\sigma}_C(\Delta_d)$.

variate generalizability theory (the first approach). Also, the first approach is easily extended to obtain conditional standard errors of measurement for districts, as discussed next.

Conditional SEMs

Within each district, the data conform to an unbalanced $p^\bullet : o^\bullet$ multivariate design. It follows that, for any district, the absolute error variance for the mean difference score is the variance of the mean for the $P^\bullet : O^\bullet$ design. That is, the estimated absolute-error SEM is the square root of

$$\hat{\sigma}_C^2(\Delta_d) = \frac{\hat{\sigma}_R^2(o|d) + \hat{\sigma}_M^2(o|d) - 2\,\hat{\sigma}_{RM}(o|d)}{n_o'}$$
$$+ \frac{\hat{\sigma}_R^2(p{:}o|d) + \hat{\sigma}_M^2(p{:}o|d) - 2\,\hat{\sigma}_{RM}(p{:}o|d)}{n_o'\ddot{\bar{n}}_{p:d}'}, \quad (11.57)$$

where the estimated variance components are conditional on the district.

These conditional absolute SEMs for $n_o' = 2$ are plotted in two ways in Figure 11.4.[10] As illustrated in the right panel, $\hat{\sigma}_C(\Delta_d)$ does not appear to vary too much as a function of mean difference scores. By contrast, the left panel illustrates that $\hat{\sigma}_C(\Delta_d)$ is much larger for small districts than for large ones.[11] The arithmetic mean of the conditional absolute SEMs is .0444, which is quite a bit larger than the overall absolute error variance .0351 (the square of the value in Equation 11.51). This discrepancy is directly

---

[10]These data are for 101 districts. Two very large districts were disregarded that have about 600 students per year.

[11]The curve in the left panel is a power function $[1/(e^{.015}x^{.507})$, where x is number of persons] that appears to represent the data reasonably well. In the right panel, a linear function is fit to the data.

attributable to the fact that the lack of balance in the design is with respect to nesting within the objects of measurement (districts).

11.3   Discussion and Other Topics

This chapter has provided extensive discussions of issues associated with multivariate unbalanced designs. Still, there are numerous other statistical issues that have not been covered. In particular, standard errors and confidence intervals for estimated covariance components have not been treated, except in the limited sense of Equation 11.45, which requires replications. Unfortunately, the published literature provides very little information about the variability of estimated covariance components for unbalanced designs.

Much of the power and flexibility of multivariate generalizability theory is directly related to the many estimated variance and covariance components available to the investigator. However, because there are many parameters to be estimated, a G study typically requires large amounts of data to achieve satisfactorily stable estimates.[12] This is an additional reason to be less than sanguine about any approach to estimation that involves eliminating data.

The need for large amounts of data was one motivating factor in the development of the computer program mGENOVA that was first mentioned in Chapter 9 and is discussed in Appendix H. mGENOVA estimates variance and covariance components for each of the designs in Table 9.2, except the last one [the $(p^\circ : c^\bullet) \times i^\circ$ design]. Although mGENOVA cannot handle missing data designs such as those discussed in Section 11.1.6, it can estimate covariance components using analogous TP and CP terms, and it can estimate variance components using T and C terms. The algorithms used by mGENOVA are not matrix-based; consequently, mGENOVA can process an almost unlimited number of observations very rapidly. mGENOVA provides both G study and D study results.

11.4   Exercises

11.1* Derive $E\,TP_{vv'}(p)$ in Equation 11.9.

11.2 For the balanced $i^\bullet : p^\bullet$ design, show that the estimators of the covariance components based on CP terms (Equations 11.27 and

---

[12] In the absence of known formulas for standard errors of estimated covariance components for unbalanced designs, this statement is technically speculative, but it seems almost certainly true.

11.28) are equivalent to those based on TP terms (Equations 11.13 and 11.14).

11.3* For the balanced $p^\bullet \times (i^\circ : h^\bullet)$ design, show that the estimators of the covariance components based on CP terms (Equations 11.32 to 11.34) are equivalent to those based on MP terms (Equations 9.39 to 9.41). Note that for the balanced $p^\bullet \times (i^\circ : h^\bullet)$ design, $n_{i:h}$ is a constant for all levels of h associated with v, and $m_{i:h}$ is a constant for all levels of h associated with v'.

11.4 For the unbalanced $p^\bullet \times (i^\circ : h^\bullet)$ design prove that

$$\boldsymbol{E}\, CP_{vv'}(p) = n_p \mu_v \mu_{v'} + n_p \sigma_{vv'}(p) + t n_p \sigma_{vv'}(h) + t n_p \sigma_{vv'}(ph),$$

as provided in Table 11.3. Use a level of detail comparable to that used to prove Equation 9.34.

11.5* Verify the C-terms estimates of the variance components at the bottom of Table 11.6 for the unbalanced $i^\bullet : p^\bullet$ design.

11.6 For the $p^\bullet \times (i^\circ : h^\circ)$ design, prove that the expected value of the compound-means covariance in Equation 11.39 is $\sigma_{vv'}(p)$.

11.7 For the $p^\bullet \times (i^\circ : h^\bullet)$ design, the compound-means estimates of co-variance components in Equation Set 11.41 result from the expected mean-products in Equations 9.39 to 9.41. Prove that the expected mean product for p given by Equation 9.39 applies with compound means.

11.8* For the synthetic data example of the $p^\bullet \times i^\bullet$ design with missing data in Table 11.7, verify that the estimates of the covariance components based on TP terms are identical to those provided in Equation Set 11.44, which are based on the variance-of-a-sum procedure.

11.9 For the $p^\bullet \times i^\bullet$ design with missing data, verify the coefficients for $\boldsymbol{E}\, CP_{vv'}(p)$ reported in Table 11.8.

11.10* Using the expected CP equations in Table 11.8 for the $p^\bullet \times i^\bullet$ design with missing data, determine estimates of the covariance components for the data in Table 11.7 assuming the observations with an asterisk are all five.

11.11* Table 11.10 reports that $\boldsymbol{E}\hat{\rho}^2 = .801$ for the MD example of the multivariate $p^\bullet \times (i^\circ : h^\circ)$ design. How much smaller or larger might $\boldsymbol{E}\hat{\rho}^2$ be if both the random h facet and the fixed v facet were ignored, that is, if $\boldsymbol{E}\hat{\rho}^2$ were based on a univariate analysis for the undifferentiated set of 26 items? Why is the difference so small?

11.12 For the *LM* example, Figure 11.1 provides a plot of conditional
SEMs, $\hat{\sigma}_C(\Delta_p)$, for nearly 500 examinees, along with a quadratic
fit to the conditional SEMs. [The quadratic fit to the error variances
for all 2450 examinees is $\hat{\sigma}_C^2(\Delta_p) = -.0038 + .0537\overline{X}_{pC} - .0529\overline{X}_{pC}^2$.]
The item-level scores for one of the examinees can be summarized
as

	h_1	h_2	h_3	h_4	h_5
v_1	4/4	3/4	7/7	1/2	5/6
v_2	5/5	3/4	1/2	4/6	1/4

where the notation a/b means a items correct out of a total of b
items. What is $\hat{\sigma}_C(\Delta_p)$ for this examinee? What is the fitted value?

11.13 Why is it reasonable to apply the Spearman–Brown formula to the
estimated generalizability coefficient in Equation 11.54?

12
Multivariate Regressed Scores

Recall from Section 5.5.2 that Kelley's regressed score estimates of true score are easily extended to estimates of universe score in univariate generalizability theory. The basic equation is

$$\hat{\mu}_p = (1 - \boldsymbol{E}\rho^2)\mu + \boldsymbol{E}\rho^2\overline{X}_p, \tag{12.1}$$

where p stands for an object of measurement. The principal way in which univariate generalizability theory extends Kelley's work is through the many different D study designs and universes of generalization that may be employed to characterize $\boldsymbol{E}\rho^2$ and arrive at an estimate of it.

Multivariate generalizability theory permits us to extend further these basic ideas to the estimation of universe score profiles and universe score composites. Kelley's original development was based on applying results from simple linear regression to the classical test theory model. We obtain regressed score estimates of profiles by applying results from multiple linear regression to models of the type used in generalizability theory. Doing so is relatively straightforward conceptually, but notational complexities can easily obscure matters. For this reason, we begin with a review of basic results for multiple linear regression with two independent variables using relatively standard terminology and notation. Then we translate these results into the terminology of generalizability theory and the notation used in this book. These developments permit us to obtain a multiple regression estimate of each universe score. We call this set of estimates the estimated universe score profile.

After considering profiles, we show how the same basic multiple regression theory can be used to obtain least squares estimates of composite universe scores. The theory applies to any composite universe score defined in terms of a priori weights. It is important to note that *predictions* of composite universe scores using multiple regression employ *statistical* weights for the observed scores, rather than the a priori (w) weights used to define the composite. After discussing and illustrating the theory of predicted composites, relationships between predicted composites and estimated profiles are considered.

In this chapter, the notation used does not explicitly distinguish between parameters and estimates, with one important exception, namely, a hat ˆ over a symbol designates a regressed score estimate. To maintain the parameter/estimate distinction everywhere would likely add more confusion than clarity to already complicated notational conventions. Also, we often use classical test theory notation when doing so results in simpler expressions without limiting the generality of results. In addition, for simplicity, the discussion and notation assume that the objects of measurement are persons, although the theory permits any facet to play that role.

Much of the discussion of profiles and predicted composites is in terms of single-facet multivariate designs with $n_v = 2$. The theory per se has no such restriction, however, and various parts of this chapter specify procedures and equations for treating more complicated designs. One synthetic data example is used throughout to illustrate computations and discuss results. The results are hypothetical, of course, and not intended to be illustrative of what is likely to happen with any particular real data set.

12.1   Multiple Linear Regression

The linear model for the regression of one dependent variable on two independent variables is

$$\hat{Y} = b_0 + b_1 X_1 + b_2 X_2 + e_y, \qquad (12.2)$$

where

$$b_0 = \overline{Y} - b_1 \overline{X}_1 - b_2 \overline{X}_2. \qquad (12.3)$$

Sometimes this is referred to as the raw score regression equation, to distinguish it from the standard score regression equation:

$$\hat{Z}_Y = \beta_1 Z_{X_1} + \beta_2 Z_{X_2} + e_{z_y}, \qquad (12.4)$$

where, for example, $Z_Y = (Y - \overline{Y})/S_Y$ with S_Y designating the observed score standard deviation for Y.[1] The standard score regression equation is

---

[1]A model-fit error term e is included in both Equations 12.2 and 12.4, which is consistent with traditional statistical notation. By contrast, it is usual in measurement texts

related to the raw score regression equation through the equalities

$$b_1 = \beta_1 \frac{S_Y}{S_{X_1}} \tag{12.5}$$

and

$$b_2 = \beta_2 \frac{S_Y}{S_{X_2}}, \tag{12.6}$$

where S_{X_1} and S_{X_2} are observed score standard deviations.

The principal advantage of the standard score regression equation is that the βs have simpler expressions than the bs in the raw score regression equation. The βs can be obtained from the normal equations:

$$\begin{aligned}
\beta_1 &+ r_{X_1 X_2}\beta_2 &= r_{Y X_1} \\
r_{X_1 X_2}\beta_1 &+ \beta_2 &= r_{Y X_2},
\end{aligned} \tag{12.7}$$

where the rs are Pearson product-moment correlations. Using matrix procedures it is relatively easy to determine that

$$\beta_1 = \frac{r_{Y X_1} - r_{Y X_2} r_{X_1 X_2}}{1 - r_{X_1 X_2}^2} \tag{12.8}$$

and

$$\beta_2 = \frac{r_{Y X_2} - r_{Y X_1} r_{X_1 X_2}}{1 - r_{X_1 X_2}^2}. \tag{12.9}$$

Using Equations 12.5 and 12.6, it is easy to show that

$$b_1 = \frac{S_{Y X_1}/S_{X_1}^2 - r_{X_1 X_2}[S_{Y X_2}/(S_{X_1} S_{X_2})]}{1 - r_{X_1 X_2}^2} \tag{12.10}$$

and

$$b_2 = \frac{S_{Y X_2}/S_{X_2}^2 - r_{X_1 X_2}[S_{Y X_1}/(S_{X_1} S_{X_2})]}{1 - r_{X_1 X_2}^2}. \tag{12.11}$$

The so-called multiple R^2 is the squared correlation between Y and $\hat{Y}$, which is the proportion of the variance in Y explained by the multiple regression. It can be shown that

$$R^2 = \beta_1 r_{Y X_1} + \beta_2 r_{Y X_2} = \frac{b_1 S_{Y X_1} + b_2 S_{Y X_2}}{S_Y^2}. \tag{12.12}$$

The standard deviation of dependent-variable observations about the regression line is called the standard error of estimate. For standard scores it is

$$\sigma_{Z_Y | Z_{\mathbf{x}}} = \sqrt{1 - R^2}, \tag{12.13}$$

---

not to include the e term in regressed score estimation equations such as Equation 12.1. This is not entirely a matter of tradition, however. In Equation 12.1, individual es are not observable because the individual universe scores μ_p are not observable. Still, the theory of regressed score estimates implicitly recognizes that es exist.

or, letting $\mathcal{E}_Z = Z_Y - \hat{Z}_Y$, it can be denoted $\sigma_{\mathcal{E}_Z}$. For raw scores, the standard error of estimate is

$$\sigma_{Y|\mathbf{X}} = \sigma_Y \sqrt{1 - R^2}, \tag{12.14}$$

or, letting $\mathcal{E} = Y - \hat{Y}$, it can be denoted $\sigma_{\mathcal{E}}$. The variance of the errors of estimate for raw scores can be expressed in a number of different ways, including

$$\sigma_{\mathcal{E}}^2 = \sigma_{Y|\mathbf{X}}^2 = \sigma_{(Y-\hat{Y})}^2 = \sigma_Y^2 - \sigma_{\hat{Y}}^2 = \sigma_Y^2(1 - R^2).$$

Some of these equalities rely on the fact that $\sigma_{\hat{Y}}^2 = \sigma_{Y\hat{Y}}$.

The extension to more than two independent variables, say k of them, involves solving the normal equations

$$
\begin{array}{ccccccccc}
\beta_1 & + & r_{X_1 X_2}\beta_2 & + & \cdots & + & r_{X_1 X_k}\beta_k & = & r_{YX_1} \\
r_{X_2 X_1}\beta_1 & + & \beta_2 & + & \cdots & + & r_{X_2 X_k}\beta_k & = & r_{YX_2} \\
 & & & & & & & \vdots & \\
r_{X_k X_1}\beta_1 & + & r_{X_k X_2}\beta_2 & + & \cdots & + & \beta_k & = & r_{YX_k}.
\end{array}
\tag{12.15}
$$

Although the solutions of these equations are rather complicated expressions for the βs, the basic notions and interpretations are unchanged from what they are for the two-independent-variables case.

It is also possible, and sometimes easier, to obtain the b weights directly using the variance-covariance matrix of observed scores. Specifically, the normal equations for the bs are:

$$
\begin{array}{ccccccccc}
S_{X_1}^2 b_1 & + & S_{X_1 X_2} b_2 & + & \cdots & + & S_{X_1 X_k} b_k & = & S_{YX_1} \\
S_{X_2 X_1} b_1 & + & S_{X_2}^2 b_2 & + & \cdots & + & S_{X_2 X_k} b_k & = & S_{YX_2} \\
 & & & & & & & \vdots & \\
S_{X_k X_1} b_1 & + & S_{X_k X_2} b_2 & + & \cdots & + & S_{X_k}^2 b_k & = & S_{YX_k},
\end{array}
\tag{12.16}
$$

where $S_{X_i X_j}$ is the observed score covariance for X_i and X_j, and

$$b_0 = \overline{Y} - \sum_v b_v \overline{X}_v.$$

These normal equations are often especially appealing in generalizability theory because expressions for variances and covariances are usually simpler than those for correlations. The β weights for the standard score regression are simply

$$\beta_i = b_i \frac{S_{X_i}}{S_Y}. \tag{12.17}$$

A general formula for the variance of the predicted Y values is

$$S_{\hat{Y}}^2 = \sum_{j=1}^{k} b_j S_{YX_j}. \tag{12.18}$$

The squared correlation between Y and $\hat{Y}$ or, equivalently, the proportion of the variance in Y explained by the multiple regression, is

$$R^2 = \frac{S_{\hat{Y}}^2}{S_Y^2} = \sum_{j=1}^{k} \beta_j r_{YX_j} = \frac{1}{S_Y^2} \sum_{j=1}^{k} b_j S_{YX_j}. \qquad (12.19)$$

Equations 12.13 and 12.14 provide the standard error of estimate for standard scores and raw scores, respectively, for any number of variables.

The application of multiple linear regression to multivariate generalizability theory takes two primary forms, both of which result in certain specifications of the r_{YX_j} correlations and the S_{YX_j} covariances. First, when we consider profiles, Y is a universe score for one of the variables, which we denote T_v. In this case, as illustrated and discussed in Section 12.2, $S_{T_vX_j} = \sigma_{vj}(p)$ and $r_{T_vX_j} = \rho_{vj}(p)$. Second, as discussed in Section 12.3, if Y represents a universe score composite, then S_{YX_j} is a function of universe score variances and covariances, and, similarly, r_{YX_j} is a function of universe score correlations. Note that, in this chapter, often j is used to index the independent variables in the regression equation, while v is used to index the dependent variables.

12.2 Estimating Profiles Through Regression

In a typical multiple regression, observed values are available for both the dependent and independent variables. That is *not* the case, however, for the application of multiple regression to the estimation of universe scores, because universe scores play the role of Y and are obviously unknown. At first blush, this may seem to present an insurmountable problem, because without having values for Y we cannot directly compute the correlations involving Y in the normal equations. With the models we are using, however, we can determine the correlations even though we do not know the individual scores. Next, we illustrate the process of doing so when there are only two independent variables. Extending the process to $n_v > 2$ is straightforward.

12.2.1 Two Independent Variables

When $n_v = 2$, we need to obtain two prediction equations: one for v_1 and one for v_2. For raw scores, these equations can be represented as

$$\hat{Y}_v = b_{0v} + b_{1v}X_{1v} + b_{2v}X_{2v}.$$

For standard scores, a representation is

$$\hat{Z}_{Y_v} = \beta_{1v}Z_{X_{1v}} + \beta_{2v}Z_{X_{2v}}$$

or, more simply,

$$\hat{Z}_{Y_v} = \beta_{1v} Z_{1v} + \beta_{2v} Z_{2v}.$$

Suppose Y represents universe scores for v_1. This means that

$$Y = \mu_{p1} = \mu_v + \nu_{p1},$$

which we abbreviate $Y = T_1$. Then, the covariance of Y with X_1 is

$$S_{YX_1} = S_{T_1 X_1} = S_{T_1(T_1+E_1)} = S_{T_1}^2 + S_{T_1 E_1} = S_{T_1}^2 = \sigma_1^2(p). \tag{12.20}$$

That is, the covariance between universe scores and observed scores for a single level of the fixed facet (v_1 in this case) equals universe score variance. It follows that r_{YX_1} in Equation Set 12.7 is

$$r_{YX_1} = r_{T_1 X_1} = \frac{S_{T_1 X_1}}{S_{T_1} S_{X_1}} = \frac{S_{T_1}^2}{S_{T_1} S_{X_1}} = \frac{S_{T_1}}{S_{X_1}} = \sqrt{E\rho_1^2}. \tag{12.21}$$

Several aspects of the derivations in Equations 12.20 and 12.21 should be noted. First, the classical test theory representation of X_1 as $T_1 + E_1$ is used in Equation 12.20 for simplicity of notation only. The term E_1 stands for all terms in the linear model for v_1 except $T_1 = \mu_v + \nu_{p1}$. The covariance of all such terms with T_1 is zero. Second, the derivation essentially says that the correlation between universe scores and observed scores is the square root of the generalizability coefficient, which is consistent with the fact that a generalizability coefficient is the squared correlation between universe scores and observed scores. Third, the generalizability coefficient in Equation 12.21 is the univariate coefficient for v_1, not the coefficient for some composite of the two variables. That is why the coefficient has a subscript of 1.

When Y represents universe scores for v_1, the covariance of Y with X_2 is

$$S_{YX_2} = S_{T_1 X_2} = S_{T_1(T_2+E_2)} = S_{T_1 T_2} = \sigma_{12}(p). \tag{12.22}$$

It follows that r_{YX_2} in Equation Set 12.7 is

$$r_{YX_2} = r_{T_1 X_2} = \frac{S_{T_1 T_2}}{S_{T_1} S_{X_2}} = \frac{S_{T_1 T_2}}{S_{T_1} S_{T_2}} \frac{S_{T_2}}{S_{X_2}} = \rho_{12}(p)\sqrt{E\rho_2^2}. \tag{12.23}$$

When Y represents universe scores for v_2, derivations similar to those in Equations 12.20 to 12.23 give

$$S_{YX_2} = S_{T_2 X_2} = \sigma_2^2(p)$$

$$r_{YX_2} = r_{T_2 X_2} = \sqrt{E\rho_2^2}$$

$$S_{YX_1} = S_{T_2 X_1} = S_{T_1 T_2} = \sigma_{12}(p)$$

$$r_{YX_1} = r_{T_2 X_1} = \rho_{12}(p)\sqrt{E\rho_1^2}.$$

Subsequently, to simplify notation, usually we use these abbreviations for observed score quantities:

$$S_1 = S_{X_1}, \qquad S_2 = S_{X_2},$$

$$S_{12} = S_{X_1 X_2}, \qquad \text{and} \qquad r_{12} = r_{X_1 X_2}. \qquad (12.24)$$

Also, usually we abbreviate certain generalizability theory parameters as

$$\sigma_1 = \sigma_1(p), \qquad \sigma_2 = \sigma_2(p), \qquad \sigma_{12} = \sigma_{12}(p) \qquad \rho_{12} = \rho_{12}(p),$$

$$\rho_1 = \sqrt{E\rho_1^2}, \qquad \text{and} \qquad \rho_2 = \sqrt{E\rho_2^2}. \qquad (12.25)$$

Note that we are using r and S for observed score statistics and ρ and σ for the parameters of interest in the generalizability theory model.

Using Equations 12.8 and 12.9, the beta coefficients for the standardized regression of universe scores for v_1 on observed scores for both variables are

$$\beta_{1|T_1} = \frac{\rho_1 - \rho_2 \rho_{12} r_{12}}{1 - r_{12}^2} \qquad \text{and} \qquad \beta_{2|T_1} = \frac{\rho_2 \rho_{12} - \rho_1 r_{12}}{1 - r_{12}^2}. \qquad (12.26)$$

Similarly, the beta coefficients for the standardized regression of universe scores for v_2 are

$$\beta_{1|T_2} = \frac{\rho_1 \rho_{12} - \rho_2 r_{12}}{1 - r_{12}^2} \qquad \text{and} \qquad \beta_{2|T_2} = \frac{\rho_2 - \rho_1 \rho_{12} r_{12}}{1 - r_{12}^2}. \qquad (12.27)$$

These coefficients are relatively simple, but there is a subtle issue about the definition of the generalizability coefficients in standardized regressions. For any level of v, the normal equations leading to the standardized regression coefficients involve standard scores for the independent (observed) variables *and* the dependent variable, that is, the universe scores for the level of v. This means that generalizability coefficients are not interpretable as ratios of variances for standardized universe scores and standardized raw scores; such ratios are always unity. Rather, we must interpret generalizability coefficients as squared correlations between universe scores and observed scores; such squared correlations are unaltered by the linear transformations involved in standardization.

Using Equations 12.10 and 12.11, the b coefficients for the raw score regression of universe scores for v_1 on observed scores for both variables are

$$b_{1|T_1} = \frac{\rho_1^2 - \rho_1 \rho_2 \rho_{12} r_{12}}{1 - r_{12}^2} \qquad \text{and} \qquad b_{2|T_1} = \frac{\sigma_1}{S_2} \left(\frac{\rho_2 \rho_{12} - \rho_1 r_{12}}{1 - r_{12}^2} \right), \qquad (12.28)$$

with

$$b_{0|T_1} = \overline{X}_1 - b_{1|T_1} \overline{X}_1 - b_{2|T_1} \overline{X}_2 = (1 - b_{1|T_1}) \overline{X}_1 - b_{2|T_1} \overline{X}_2. \qquad (12.29)$$

Similarly, the b coefficients for the raw score regression of universe scores for v_2 are

$$b_{1|T_2} = \frac{\sigma_2}{S_1}\left(\frac{\rho_1\rho_{12} - \rho_2 r_{12}}{1 - r_{12}^2}\right) \quad \text{and} \quad b_{2|T_2} = \frac{\rho_2^2 - \rho_1\rho_2\rho_{12}r_{12}}{1 - r_{12}^2}, \quad (12.30)$$

with

$$b_{0|T_2} = (1 - b_{2|T_2})\overline{X}_2 - b_{1|T_2}\overline{X}_1. \quad (12.31)$$

Note that the observed score statistics can be expressed in terms of the generalizability theory model parameters as follows.

$$S_1 = \sqrt{\sigma_1^2(p) + \sigma_1^2(\delta)}, \quad (12.32)$$

$$S_2 = \sqrt{\sigma_2^2(p) + \sigma_2^2(\delta)}, \quad (12.33)$$

$$S_{12} = \sigma_{12}(p) + \sigma_{12}(\delta), \quad (12.34)$$

and

$$r_{12} = \frac{\sigma_{12}(p) + \sigma_{12}(\delta)}{\sqrt{\sigma_1^2(p) + \sigma_1^2(\delta)}\sqrt{\sigma_2^2(p) + \sigma_2^2(\delta)}}. \quad (12.35)$$

It is particularly important to note that the observed covariance and correlation involve the correlated error component $\sigma_{12}(\delta)$. When $\sigma_{12}(\delta) \neq 0$, the classical test theory formula for obtaining the disattenuated correlation for the objects of measurement is *not* valid. In our abbreviated notational system, when $\sigma_{12}(\delta) \neq 0$,

$$\rho_{12} = \frac{\sigma_{12}}{\sigma_1\sigma_2} \neq \frac{r_{12}}{\rho_1\rho_2}.$$

For expository purposes, let us focus, now, on the standardized regressions for the two universe scores. Using Equations 12.26 and 12.27, the regressed score estimate of the universe score profile is given by the following two equations.

$$\hat{Z}_{p1} = \left[\frac{\rho_1 - \rho_2\rho_{12}r_{12}}{1 - r_{12}^2}\right]Z_{p1} + \left[\frac{\rho_2\rho_{12} - \rho_1 r_{12}}{1 - r_{12}^2}\right]Z_{p2} \quad (12.36)$$

$$\hat{Z}_{p2} = \left[\frac{\rho_1\rho_{12} - \rho_2 r_{12}}{1 - r_{12}^2}\right]Z_{p1} + \left[\frac{\rho_2 - \rho_1\rho_{12}r_{12}}{1 - r_{12}^2}\right]Z_{p2}, \quad (12.37)$$

where $\hat{Z}_{pv}$ is to be understood in the sense of the Z score associated with the predicted universe score for variable v (i.e., $\hat{Z}_{T_v}$ or, even more specifically, $\hat{Z}_{\mu_{pv}}$), and Z_{pv} is to be understood in the sense of the Z score associated with the vth raw score $Z_{X_{pv}}$. Using Equation 12.12, the proportions of variance explained by the two regressions are given by

$$R_1^2 = (\beta_{1|T_1})\rho_1 + (\beta_{2|T_1})\rho_2\rho_{12} \quad (12.38)$$

and
$$R_2^2 = (\beta_{1|T_2})\rho_1\rho_{12} + (\beta_{2|T_2})\rho_2. \tag{12.39}$$

Using Equation 12.13, the corresponding standard errors of estimate are $\sqrt{1 - R_1^2}$ and $\sqrt{1 - R_2^2}$, respectively.

Regression to the Mean and Profile Flatness

Equations 12.36 and 12.37 constitute a regressed score estimate of the profile in two senses. First, both variables are regressed toward their respective means (zero for these standardized regressions). This is guaranteed by the fact that $R^2 \leq 1$ for both variables. In addition, the profile of predicted standardized universe scores tends to be flatter than the profile of observed standardized scores.

To illustrate these points, and to comment further on them, let us assume that $\rho_1 = \rho_2$ in the standardized regression Equations 12.36 and 12.37. Under this assumption, using Equations 12.38 and 12.39,

$$R_1^2 = R_2^2 = R^2 = \rho^2 \left(\frac{1 + \rho_{12}^2 - 2\rho_{12}r_{12}}{1 - r_{12}^2} \right).$$

Clearly, R^2 gets smaller as ρ^2 decreases. Also, under classical test theory assumptions (or, equivalently, the $p^\bullet \times I^\circ$ design), $r_{12} = \rho^2\rho_{12}$, and it can be shown that R^2 gets smaller as $|\rho_{12}|$ decreases. Furthermore, for a constant positive value of ρ_{12}, R^2 gets smaller as r_{12} *increases*, which can occur, for example, when there is positively correlated relative error (see Equation 12.35) in the $p^\bullet \times I^\bullet$ design.

With $\rho_1 = \rho_2$ in Equations 12.36 and 12.37, it is easy to determine that

$$\hat{Z}_{p2} - \hat{Z}_{p1} = \rho \left(\frac{1 - \rho_{12}}{1 - r_{12}} \right) (Z_{p2} - Z_{p1}). \tag{12.40}$$

Whenever $r_{12} \leq \rho_{12}$, the coefficient of $Z_{p2} - Z_{p1}$ is less than or equal to unity,[2] which means that

$$\hat{Z}_{p2} - \hat{Z}_{p1} \leq Z_{p2} - Z_{p1}.$$

That is, the profile of predicted standardized universe scores tends to be flatter than the profile of standardized observed scores.

These statements about regression to the mean and profile flatness are admittedly complex, but the basic ideas apply to any number of variables, and to both standard and raw scores. These matters are discussed and

---

[2]Under classical test theory assumptions, which correspond to the assumptions for the $p^\bullet \times I^\circ$ design, $r_{12} = \rho_1\rho_2\rho_{12}$, which necessarily means that $r_{12} \leq \rho_{12}$. For the $p^\bullet \times I^\bullet$ design, it is not necessarily true that $r_{12} \leq \rho_{12}$, but it is very likely to be true with real data.

illustrated further in Sections 12.2.2 and especially 12.2.6. For now, we simply note that the degree of regression to the mean and the flatness of the predicted profiles are affected by generalizability coefficients, universe score correlations, and the amount of correlated δ-type correlated error; and all three factors influence the intercorrelations of the raw scores.

Extending the Theory

It is conceptually easy to extend the theory discussed thus far to obtain regression coefficients, R^2 values, and standard errors of estimate for $n_v > 2$, for both standard scores and raw scores, but the algebra is more complicated. The crux of the matter is to determine the normal equations (Equation 12.15 for standard scores, or Equation 12.16 for raw scores), and then solve them for the regression weights. For raw scores, the left side of the normal equations involves the observed score variances and covariances for all n_v variables. For variable v, the right side involves the covariances

$$S_{YX_j} = S_{T_vX_j} = \sigma_{vj}. \tag{12.41}$$

Using these covariances in Equation 12.19, along with the derived b weights gives

$$R_v^2 = \frac{\sigma_v^2(\hat{\mu}_p)}{\sigma_v^2} \tag{12.42}$$

$$= \frac{\sum_j b_{jv}\sigma_{vj}}{\sigma_v^2}. \tag{12.43}$$

With $S_Y^2 = \sigma_v^2$ in Equation 12.14, the associated standard error of estimate is:

$$\sigma_v(\mathcal{E}) = \sigma_v\sqrt{1 - R_v^2}. \tag{12.44}$$

For standard scores, the left side of the normal equations involves the observed score correlations for all n_v variables. For variable v, the right side involves the correlations

$$r_{YX_j} = r_{T_vX_j} = \begin{cases} \rho_v & \text{if } v = j \\ \rho_{vj}\rho_j & \text{if } v \neq j. \end{cases} \tag{12.45}$$

Using these correlations in Equation 12.19 along with the derived β weights, gives R_v^2. The associated standard error of estimate $\sigma_v(\mathcal{E})$ is given by Equation 12.13.

Profile equations and their associated statistics [e.g., R^2 and $\sigma(\mathcal{E})$] are D study results. We have discussed estimating them primarily under the assumptions that the G and D study sample sizes are the same, and the D study design structure mirrors that used in the G study (e.g., $p^\bullet \times i^\bullet$ and $p^\bullet \times I^\bullet$, or $p^\bullet \times i^\circ$ and $p^\bullet \times I^\circ$). It is possible, however, to estimate profile equations for sample sizes and/or designs different from those used in a G study. That is, we can obtain answers to "What if ..." questions for regressed score estimates of profiles, just as we can for other D study statistics. To do so, the normal equations must be specified for the D study sample sizes and design. For example, for raw score profile equations, instead of using the observed variances and covariances from the G study, the investigator would use their expected values for the D study sample sizes and design. For any particular level of v, the right side of the normal equations is unchanged, because the right side involves only universe score variances or covariances. This process is illustrated at the end of Section 12.2.2 and then again in Section 12.2.5.

12.2.2 Synthetic Data Example

Consider the synthetic data example for the $p^{\bullet} \times i^{\bullet}$ design in Table 9.3 for 10 persons, six items, and two fixed variables, in which each item contributes scores for each variable. Assuming $n'_i = n_i = 6$, the reader can easily verify that

$$\Sigma_p = \begin{bmatrix} .3682 & .8663 \\ .3193 & .3689 \end{bmatrix} \tag{12.46}$$

$$\Sigma_\delta = \begin{bmatrix} .2087 & .5081 \\ .1175 & .2561 \end{bmatrix} \tag{12.47}$$

$$S = \begin{bmatrix} .5769 & .7273 \\ .4367 & .6250 \end{bmatrix}, \tag{12.48}$$

where the entries in italics in the upper-diagonal cells are correlations. Note in particular that the observed score correlation $r_{12} = .7273$ is *not* equal to the universe score correlation $\rho_{12} = .8663$, because of the δ-type correlated-error component $\sigma_{12}(\delta) = .1175$ (see Equation 12.35). Also, it is easily verified that

$$\rho_1^2 = \frac{.3682}{.5769} = .6382 \quad \text{and} \quad \rho_2^2 = \frac{.3689}{.6250} = .5902. \tag{12.49}$$

Using these results in Equations 12.36 and 12.37 gives the regressed score estimates of the Z-score profile equations:

$$\hat{Z}_{p1} = .6684\, Z_{p1} + .1794\, Z_{p2}$$
$$\hat{Z}_{p2} = .2830\, Z_{p1} + .5624\, Z_{p2}$$

and

$$\hat{Z}_{p2} - \hat{Z}_{p1} = .3830\, Z_{p2} - .3853\, Z_{p1}.$$

This suggests that the predicted profile is about 38% flatter than the observed profile.

TABLE 12.1. Regressed Score Estimates for Synthetic Data Example for Balanced $p^{\bullet} \times I^{\bullet}$ Design in Table 9.3

	Standard Scores				Regressed-Score Estimates			
p	Z_{p1}	Z_{p2}	Mean	Var[a]	$\hat{Z}_{p1}$	$\hat{Z}_{p2}$	Mean	Var[a]
1	−.2544	−.1111	−.1828	.0051	−.1900	−.1345	−.1623	.0008
2	−1.4110	−.5556	−.9833	.1829	−1.0427	−.7118	−.8773	.0274
3	.6708	−.3333	.1687	.2521	.3885	.0024	.1955	.0373
4	−1.8736	−1.6667	−1.7701	.0107	−1.5513	−1.4676	−1.5095	.0018
5	−.2544	−.7778	−.5161	.0685	−.3096	−.5094	−.4095	.0100
6	1.3647	2.1111	1.7379	.1393	1.2910	1.5736	1.4323	.0200
7	.4395	.1111	.2753	.0270	.3137	.1869	.2503	.0040
8	−.0231	1.2222	.5996	.3877	.2039	.6808	.4423	.0569
9	1.3647	.3333	.8490	.2660	.9720	.5737	.7728	.0396
10	−.0231	−.3333	−.1782	.0241	−.0753	−.1940	−.1346	.0035
Mean	.0000	.0000	.0000	.1363	.0000	.0000	.0000	.0201
Var[a]	1.0000	1.0000	.8637	.0158	.6534	.6280	.6205	.0003

$$\hat{Z}_{p1} = .6684\,Z_{p1} + .1794\,Z_{p2} \qquad R_1^2 = .6534$$
$$\hat{Z}_{p2} = .2830\,Z_{p1} + .5624\,Z_{p2} \qquad R_2^2 = .6280$$

	Raw Scores				Regressed-Score Estimates			
p	$\overline{X}_{p1}$	$\overline{X}_{p2}$	Mean	Var[a]	$\hat{\mu}_{p1}$	$\hat{\mu}_{p2}$	Mean	Var[a]
1	4.3333	5.0000	4.6667	.1111	4.4073	5.0058	4.7066	.0896
2	3.5000	4.6667	4.0833	.3403	3.9165	4.6732	4.2948	.1432
3	5.0000	4.8333	4.9167	.0069	4.7403	5.0847	4.9125	.0297
4	3.1667	3.8333	3.5000	.1111	3.6237	4.2377	3.9307	.0943
5	4.3333	4.5000	4.4167	.0069	4.3384	4.7898	4.5641	.0509
6	5.5000	6.6667	6.0833	.3403	5.2598	5.9900	5.6249	.1333
7	4.8333	5.1667	5.0000	.0278	4.6972	5.1910	4.9441	.0610
8	4.5000	6.0000	5.2500	.5625	4.6340	5.4756	5.0548	.1771
9	5.5000	5.3333	5.4167	.0069	5.0761	5.4139	5.2450	.0285
10	4.5000	4.8333	4.6667	.0278	4.4733	4.9716	4.7224	.0621
Mean	4.5167	5.0833	4.8000	.1542	4.5167	5.0833	4.8000	.0870
Var[a]	.5192	.5625	.4669	.0337	.2165	.2085	.2058	.0023

$$\hat{\mu}_{p1} = 1.4050 + .5339\,\overline{X}_{p1} + .1377\,\overline{X}_{p2} \qquad \sigma_1(\mathcal{E}) = .3572$$
$$\hat{\mu}_{p2} = 1.8647 + .2263\,\overline{X}_{p1} + .4321\,\overline{X}_{p2} \qquad \sigma_2(\mathcal{E}) = .3705$$

[a] Biased estimates.

The top half of Table 12.1 provides the observed Z-scores and regressed score estimates for each of the 10 persons.[3] Also provided are means and variances for all variables. For both v_1 and v_2, regression to the mean occurs in the sense that the variance of the regressed score estimates is smaller than the variance of the standardized universe scores. This fact is evident from the results reported in Table 12.1. For example, for v_1, the variance of $\hat{Z}_{p1}$ is .6534, and the variance of the standardized universe scores is necessarily one. Since R^2 is the proportion of universe score variance explained by the regression, it follows that $R_1^2 = .6534$. Alternatively, R_1^2 can be obtained using Equations 12.12, 12.21, and 12.23:

$$
\begin{aligned}
R_1^2 &= \beta_1 r_{YX_1} + \beta_2 r_{YX_2} \\
&= \beta_1 \rho_1 + \beta_2 \rho_{12} \rho_2 \\
&= .6684\sqrt{.6382} + .1794(.8663)\sqrt{.5902} \\
&= .6534.
\end{aligned}
$$

The top half of Table 12.1 also illustrates that regression occurs in the sense that the profiles of predicted standardized universe scores are usually flatter than the profiles of observed standardized scores. Specifically, the average within-person variance for the regressed standardized scores (.0201) is less than the average within-person variance for the observed standardized scores (.1363), which means that the regressed standardized scores are about 85% less variable than the observed standardized scores.

For this synthetic data example, the bottom half of Table 12.1 provides results in the raw score metric. Using Equations 12.28 to 12.31, the profile equations are

$$
\begin{aligned}
\hat{\mu}_{p1} &= 1.4050 + .5339\,\overline{X}_{p1} + .1377\,\overline{X}_{p2} \\
\hat{\mu}_{p2} &= 1.8647 + .2263\,\overline{X}_{p1} + .4321\,\overline{X}_{p2}.
\end{aligned}
$$

Alternatively, b_1 and b_2 for v_1 can be obtained by solving the normal equations

$$
\begin{aligned}
.5769b_1 &+ .4367b_2 &= .3682 \\
.4367b_1 &+ .6250b_2 &= .3193,
\end{aligned}
\tag{12.50}
$$

with b_0 determined using Equation 12.29. Similarly, b_1 and b_2 for v_2 can be obtained by solving the normal equations

$$
\begin{aligned}
.5769b_1 &+ .4367b_2 &= .3193 \\
.4367b_1 &+ .6250b_2 &= .3689,
\end{aligned}
\tag{12.51}
$$

with b_0 determined using Equation 12.31.

---

[3]The observed Z scores were computed using the so-called "biased" variances (i.e., using a denominator of 10). Also, all within-person variances are "biased." That is, in Table 12.1 "Var" indicates "biased" variances.

The R^2 values are necessarily the same for raw scores as they are for standard scores, because the two types of scores are linear transformations of each other. It follows that, for v_1, the proportion of universe score variance explained by the regression is $R_1^2 = .6534$. This value can also be obtained using Equation 12.42:

$$R_1^2 = \frac{\sigma_1^2(\hat{\mu}_p)}{\sigma_1^2} = \frac{.2404}{.3682} = .6534,$$

where $\sigma_1^2 = .3682$ is provided in Equation 12.46, and $\sigma_1^2(\hat{\mu}_p) = .2404$ is the "unbiased" version of the variance (.2165) reported at the bottom of Table 12.1. Using Equation 12.44, the standard error of estimate for v_1 is

$$\sigma_1(\mathcal{E}) = \sigma_1\sqrt{1 - R_1^2} = \sqrt{.3682}\sqrt{1 - .6534} = .3572.$$

Similarly, for v_2, the proportion of universe score variance explained by the regression is

$$R_2^2 = \frac{\sigma_2^2(\hat{\mu}_p)}{\sigma_2^2} = \frac{.2317}{.3689} = .6280,$$

and the standard error of estimate is

$$\sigma_2(\mathcal{E}) = \sigma_2\sqrt{1 - R_2^2} = \sqrt{.3689}\sqrt{1 - .6280} = .3705.$$

It is also evident from the bottom half of Table 12.1 that regression occurs in the sense that the profiles of predicted universe scores are generally flatter than the profiles of observed (raw) scores. Specifically, the average within-person variance for the regressed scores (.0870) is less than the average within-person variance for the observed scores (.1542), which means that the regressed scores are about 44% less variable than the observed scores.

12.2.3   Variances and Covariances of Regressed Scores

Table 12.2 provides formulas for variances and covariances of regressed score estimates for any number of variables. Although the formulas are quite general, many of them are rather unusual compared to traditional regression results. For this reason, some of them are explained here.

Formulas for variances and covariances are reported in several ways. For each such formula, the first expression is in terms of n_v^2 sums of products of β weights and observed correlations (for standard scores), or n_v^2 sums of products of b weights and observed variances and covariances (for raw scores). These are "traditional" expressions from multiple regression theory, although they are usually stated in matrix terms. Because they involve n_v^2 sums of products, they are tedious to compute.

The remaining expressions for variances and covariances have only n_v products, each of which involves a universe score correlation (for standard

TABLE 12.2. Formulas for Multivariate Regressed Score Estimates of Universe Scores

---

Standard Scores

$$R_v^2 = \sigma_v^2(\hat{Z}_p) = \sum_{j=1}^{n_v} \beta_{jv}\rho_j\rho_{vj} \qquad\qquad R_{vv'} = \frac{\sigma_{vv'}(\hat{Z}_p)}{\rho_{vv'}}$$

$$\sigma_v(\mathcal{E}_Z) = \sqrt{1 - R_v^2} \qquad\qquad \sigma_{vv'}(\mathcal{E}_Z) = \rho_{vv'}(1 - R_{vv'})$$

$$\sigma_v^2(\hat{Z}_p) = \sum_{j=1}^{n_v}\sum_{j'=1}^{n_v} \beta_{jv}\beta_{j'v}r_{jj'} = \sum_{j=1}^{n_v} \beta_{jv}\rho_j\rho_{vj} = R_v^2$$

$$\sigma_{vv'}(\hat{Z}_p) = \sum_{j=1}^{n_v}\sum_{j'=1}^{n_v} \beta_{jv}\beta_{j'v'}r_{jj'} = \sum_{j=1}^{n_v} \beta_{jv'}\rho_j\rho_{vj} = \sum_{j=1}^{n_v} \beta_{jv}\rho_j\rho_{v'j}$$

$$\mathcal{V}(\mathbf{Z}_{X_p}) = 1 - \overline{r}_{vv'}$$

$$\mathcal{V}(\mathbf{Z}_{\mu_p}) = 1 - \overline{\sigma}_{vv'}$$

$$\mathcal{V}(\hat{\mathbf{Z}}_{\mu_p}) = \overline{R}_v^2 - \overline{\sigma}_{vv'}(\hat{Z}_p) = \overline{R}_v^2 - \overline{R}_{vv'}$$

Raw Scores

$$R_v^2 = \frac{\sigma_v^2(\hat{\mu}_p)}{\sigma_v^2} = \frac{1}{\sigma_v^2}\sum_{j=1}^{n_v} b_{jv}\sigma_{vj} \qquad\qquad R_{vv'} = \frac{\sigma_{vv'}(\hat{\mu}_p)}{\sigma_{vv'}}$$

$$\sigma_v(\mathcal{E}) = \sigma_v\sqrt{1 - R_v^2} \qquad\qquad \sigma_{vv'}(\mathcal{E}) = \sigma_{vv'}(1 - R_{vv'})$$

$$\sigma_v^2(\hat{\mu}_p) = \sum_{j=1}^{n_v}\sum_{j'=1}^{n_v} b_{jv}b_{j'v}S_{jj'} = \sum_{j=1}^{n_v} b_{jv}\sigma_{vj} = R_v^2\sigma_v^2$$

$$\sigma_{vv'}(\hat{\mu}_p) = \sum_{j=1}^{n_v}\sum_{j'=1}^{n_v} b_{jv}b_{j'v'}S_{jj'} = \sum_{j=1}^{n_v} b_{jv'}\sigma_{vj} = \sum_{j=1}^{n_v} b_{jv}\sigma_{v'j}$$

$$\mathcal{V}(\overline{\mathbf{X}}_p) = \frac{n_p - 1}{n_p}[\overline{S}_v^2 - \overline{S}_{vv'}] + \mathrm{var}(\overline{X}_v)$$

$$\mathcal{V}(\boldsymbol{\mu}_p) = \frac{n_p - 1}{n_p}[\overline{\sigma}_v^2 - \overline{\sigma}_{vv'}] + \mathrm{var}(\mu_v)$$

$$\mathcal{V}(\hat{\boldsymbol{\mu}}_p) = \frac{n_p - 1}{n_p}[\overline{\sigma}_v^2(\hat{\mu}_p) - \overline{\sigma}_{vv'}(\hat{\mu}_p)] + \mathrm{var}(\hat{\mu}_v)$$

Note. b_{jv} is an abbreviation for $b_{j|T_v}$ or, even more specifically, $b_{j|\mu_{pv}}$. Similarly, β_{jv} is an abbreviation for $\beta_{j|T_v}$. The notations $\overline{\sigma}_{vv'}$, $\overline{S}_{vv'}$, and $\overline{R}_{vv'}$ mean the average over all n_v^2 elements; $\overline{\sigma}_v^2$, $\overline{S}_v^2$, and $\overline{R}_v^2$ mean the average of the squares over n_v elements.

scores), or a universe score variance or covariance (for raw scores). These simplified expressions do not apply in multiple regression generally. Rather, they are a consequence of the fact that, for regressed score estimates, the dependent variable is the universe score for one of the independent variables. A unique and useful feature of these expressions is that they depend only on the regression weights and the elements of Σ_p.

Consider, for example, the formula for the variance of the regressed score estimates for v:

$$\sigma_v^2(\hat{\mu}_p) = \sum_{j=1}^{n_v} b_{jv}\sigma_{vj}. \tag{12.52}$$

Note that b_{jv} is an abbreviation for $b_{j|T_v}$ or, even more specifically, $b_{j|\mu_{pv}}$. Applied to the synthetic data example for v_1, this equation gives

$$\sigma_1^2(\hat{\mu}_p) = b_{11}\sigma_{11} + b_{21}\sigma_{12} = .5339(.3682) + .1377(.3193) = .2405, \tag{12.53}$$

which is the unbiased version of the result reported at the bottom of Table 12.1[4]; that is, $(10/9) \times .2405 = .2165$. Essentially, then, $\sigma_1^2(\hat{\mu}_p)$ is the sum of the products of the b terms and σ terms for v_1. For v_2, the variance of raw score regressed score estimates is

$$\sigma_2^2(\hat{\mu}_p) = b_{12}\sigma_{21} + b_{22}\sigma_{22} = .2263(.3193) + .4321(.3689) = .2316, \tag{12.54}$$

which is the sum of the products of the b terms and σ terms for v_2.

There are two simplified formulas for the covariance of the regressed score estimates:

$$\sigma_{vv'}(\hat{\mu}_p) = \sum_{j=1}^{n_v} b_{jv'}\sigma_{vj} = \sum_{j=1}^{n_v} b_{jv}\sigma_{v'j}. \tag{12.55}$$

That is, $\sigma_{vv'}(\hat{\mu}_p)$ is the sum of the products of the b terms for v' and the σ terms for v or, equivalently, the sum of the products of the b terms for v and the σ terms for v'. For the synthetic data, using the first formula,

$$\sigma_{12}(\hat{\mu}_p) = b_{12}\sigma_{11} + b_{22}\sigma_{12} = .2263(.3682) + .4321(.3193) = .2213, \tag{12.56}$$

and using the second formula,

$$\sigma_{12}(\hat{\mu}_p) = b_{11}\sigma_{21} + b_{21}\sigma_{22} = .5339(.3193) + .1377(.3689) = .2213.$$

It follows that the estimated correlation between regressed score estimates for v_1 and v_2 is

$$\rho_{12}(\hat{\mu}_p) = \frac{\sigma_{12}(\hat{\mu}_p)}{\sqrt{\sigma_1^2(\hat{\mu}_p)}\sqrt{\sigma_1^2(\hat{\mu}_p)}} = \frac{.2213}{\sqrt{.2405}\sqrt{.2316}} = .937.$$

---

[4]For the purposes of the discussion here, obviously it would be better to report the unbiased estimates in Table 12.1. However, for the subsequent discussion of profile variability, the so-called "biased" estimates are more convenient. In any case, with data sets of typical size, this is a trivial matter.

By definition, R^2 is the proportion of universe score variance explained by the regression (see Equation 12.42); that is,

$$R_v^2 = \frac{\sigma_v^2(\hat{\mu}_p)}{\sigma_v^2}.$$

Similarly, the proportion of universe score covariance for v and v' explained by the corresponding regressed-score estimate equations is:

$$R_{vv'} = \frac{\sigma_{vv'}(\hat{\mu}_p)}{\sigma_{vv'}}, \tag{12.57}$$

which is $.2213/.3193 = .693$ for the synthetic data. That is, about 69% of the universe-score covariance is explained by the the covariance between regressed score estimates for the two variables.

Note that, for the synthetic data, the correlation between regressed score estimates ($.937$) is larger than the correlation between observed scores ($.727$). This result suggests that the two-variable profile of regressed score estimates is generally flatter than the profile of observed scores. This intuitive notion is formalized at the end of Section 12.2.6.

12.2.4 Standard Errors of Estimate and Tolerance Intervals

As reported in Table 12.2 and Equation 12.44, the standard error of estimate for a universe score is

$$\sigma_v(\mathcal{E}) = \sigma_v \sqrt{1 - R_v^2}.$$

This formula can be viewed simply as a translation of the well-known multiple regression Equation 12.14 to the notation for multivariate regressed score estimates of universe scores. An alternative formula is

$$\sigma_v(\mathcal{E}) = \sqrt{\sigma_v^2 - \sigma_v^2(\hat{\mu}_p)}. \tag{12.58}$$

A common use of the standard error of estimate is to establish a tolerance interval. Under normality assumptions, a $100\gamma\%$ tolerance interval for a variable in the profile is

$$\hat{\mu}_{pv} \pm z^*_{(1+\gamma)/2}\, \sigma_v(\mathcal{E}), \tag{12.59}$$

where $z^*_{(1+\gamma)/2}$ is the normal deviate corresponding to the upper $(1 + \gamma)/2$ percentile point. This interval is centered around the person's regressed score estimate for the variable. A tolerance interval is conditional on observed scores, and, as such, it is meaningful to say that the probability is $100\gamma\%$ that the interval includes the universe score. For the synthetic data example of the $p^\bullet \times I^\bullet$ design with $n_i' = 6$, we have already determined in Section 12.2.2 that $\sigma_1(\mathcal{E}) = .3572$ and $\sigma_2(\mathcal{E}) = .3705$. Under normality assumptions, it follows that 68% tolerance intervals for v_1 and v_2 are $\hat{\mu}_{p1} \pm .3572$ and $\hat{\mu}_{p2} \pm .3705$, respectively.

As reported in Table 12.2, the covariance of the errors of estimate is

$$\sigma_{vv'}(\mathcal{E}) = \sigma_{vv'}(1 - R_{vv'}). \tag{12.60}$$

An alternative formula is

$$\sigma_{vv'}(\mathcal{E}) = \sigma_{vv'} - \sigma_{vv'}(\hat{\mu}_p). \tag{12.61}$$

For the synthetic data example of the $p^\bullet \times I^\bullet$ design with $n'_i = 6$,

$$\sigma_{12}(\mathcal{E}) = .3193 \left(1 - \frac{.2213}{.3193}\right) = .3193 - .2213 = .0980.$$

Since the standard errors of estimate for the two variables are .3572 and .3705, the correlation between the errors of estimates for the two variables is

$$\rho_{12}(\mathcal{E}) = \frac{.0980}{.3572 \times .3705} = .740.$$

The covariance between correlated $\mathcal{E}$-type errors, $\sigma_{vv'}(\mathcal{E})$, plays an important role in making probability statements about overlapping tolerance intervals for universe scores. For example, under normality assumptions, the probability that $100\gamma\%$ tolerance intervals for v and v' will overlap when $\mu_{pv} = \mu_{pv'}$ is

$$\Pr(\text{overlap}) = 2 \times \Pr\left\{z < \frac{z^*_{(1+\gamma)/2}\,[\sigma_v(\mathcal{E}) + \sigma_{v'}(\mathcal{E})]}{\sqrt{\sigma_v^2(\mathcal{E}) + \sigma_{v'}^2(\mathcal{E}) - 2\sigma_{vv'}(\mathcal{E})}}\right\} - 1, \tag{12.62}$$

provided, of course, that the intervals are formed in the multivariate manner discussed in this chapter.[5] The logic leading to Equation 12.62 mirrors that discussed in Section 10.2.4 for confidence intervals.

For the synthetic data example of the $p^\bullet \times I^\bullet$ design with $n'_i = 6$, using 68% tolerance intervals,

$$
\begin{aligned}
\Pr(\text{overlap}) &= 2 \times \Pr\left\{z < \frac{.3572 + .3705}{\sqrt{(.3572)^2 + (.3705)^2 - 2(.0980)}}\right\} - 1 \\
&= 2 \times \Pr(z < 2.7727) - 1 \\
&= 2(.9972) - 1 \\
&= .994,
\end{aligned}
$$

---

[5] Equation 12.62 would not apply if the two tolerance intervals were obtained using the univariate Equation 12.1.

which means it is virtually certain that two 68% tolerance intervals will overlap when $\mu_{pv} = \mu_{pv'}$. This probability-of-overlap result is conditional on the D study $p^\bullet \times I^\bullet$ design with $n_i' = 6$ for both variables. If the design and/or sample sizes change, then the standard errors of estimate *and* the covariance of the errors of estimate will change (see Table 12.3 for a number of examples). Note, in particular, that it is *not* necessarily true that $\sigma_{vv'}(\mathcal{E}) = 0$ for the $p^\bullet \times I^\circ$ design, even though $\sigma_{vv'}(\delta) = 0$ for this design.

If tolerance intervals are constructed for an examinee's scores on different tests in a battery, then a relative strength or weakness for the examinee might be declared if the intervals for two test scores do not overlap. This is the same kind of logic discussed in Section 10.2.4 for confidence intervals. For any specific confidence coefficient, tolerance intervals will be shorter than confidence intervals, because the standard error of estimate is smaller than the standard error of measurement. This may appear to suggest that there is likely to be a greater probability that a relative strength or weakness will be declared using tolerance intervals than using confidence intervals. Actually, however, matters are more complicated than this ad hoc reasoning suggests, because tolerance intervals are centered around regressed score estimates of universe scores, whereas confidence intervals are centered around observed mean scores.

Recall from Section 10.2.4 that for the synthetic data example of the $p^\bullet \times I^\bullet$ design with $n_i' = 6$, using 68% confidence intervals, the probability of overlap was .946. By contrast, in this section we obtained .994 for tolerance intervals. This means that a relative strength or weakness is less likely to be declared using tolerance intervals in this example.

12.2.5 Different Sample Sizes and/or Designs

Profile equations are D study results. In univariate generalizability theory, it is a relatively simple matter to estimate results for a D study that has a different set of sample sizes and/or design structure from that used in a G study. This can be done for profiles, too, but the process is not as simple. The most general approach is to state the normal equations for the revised design, solve them for the regression coefficients, and then use the equations in Table 12.2 to obtain the desired results. Of course, if $n_v = 2$, solving the normal equations can be circumvented by using the equations in Section 12.2.1 directly.

It is important to realize that the D study design and sample sizes affect observed score variances and covariances and, of course, any quantities based on them. Since observed score variances and covariances are incorporated in the left side of the normal equations, regression weights are affected by changes in design and/or sample sizes. Universe score variances and covariances are not affected.

Let us now reconsider the synthetic data example for the $p^\bullet \times I^\bullet$ design using $n_i' = 8$ rather than $n_i' = 6$. Statistics that are sample-size dependent are altered. For v_1,

$$\sigma_1^2(\delta) = \frac{1.2522}{8} = .1565, \qquad \sigma_2^2(\delta) = \frac{1.5367}{8} = .1921,$$

$$S_1^2 = .3682 + .1565 = .5247, \qquad S_2^2 = .3689 + .1921 = .5610,$$

$$\rho_1^2 = \frac{.3682}{.5247} = .7017, \qquad \text{and} \qquad \rho_2^2 = \frac{.3689}{.5610} = .6576.$$

Also, using Equation 12.34, the expected observed covariance (i.e., the value of the observed covariance that is expected for $n_i' = 8$) becomes

$$S_{12} = \sigma_{12}(p) + \sigma_{12}(\delta) = .3193 + \frac{.7048}{8} = .4074,$$

and using Equation 12.35 the corresponding observed correlation is

$$r_{12} = \frac{.4074}{\sqrt{.5247}\sqrt{.5610}} = .7508.$$

The normal equations are

$$\begin{array}{rcrcl}
.5247b_1 & + & .4074b_2 & = & .3682 \\
.4074b_1 & + & .5610b_2 & = & .3193.
\end{array} \tag{12.63}$$

These equations can be solved for the bs or, more simply, we can use the closed-form expressions in Equation 12.28. Note that these normal equations involve smaller observed score variances and covariances than those in Equation 12.50 for $n_i' = 6$.

The resulting regressed score estimation equation for v_1 is

$$\hat{\mu}_{p1} = 1.1321 + .5956\,\overline{X}_{p1} + .1366\,\overline{X}_{p2}.$$

It follows that

$$R_1^2 = \frac{.5956(.3682) + .1366(.3193)}{.3682} = .7141$$

and the standard error of estimate is

$$\sigma_1(\mathcal{E}) = \sqrt{.3682}\sqrt{1 - .7141} = .3244.$$

Now, suppose both the design and sample size are different from that used in the G study. Specifically, suppose the design is $p^\bullet \times I^\circ$ with $n_i' = 8$ for both variables. For this design, there is no correlated δ-type or Δ-type error. All the statistics are the same as for the $p^\bullet \times I^\bullet$ design with $n_i' = 8$ except for S_{12} and r_{12}. Since $\sigma_{12}(\delta) = 0$, Equations 12.34 and 12.35 give

$$S_{12} = \sigma_{12}(p) = .3193 \qquad \text{and} \qquad r_{12} = \frac{.3193}{\sqrt{.5247}\sqrt{.5610}} = .5885.$$

TABLE 12.3. Changes in R^2, Standard Errors of Estimate, and Other Statistics for Different Sample Sizes and Designs with the Synthetic Data in Table 9.3

Design	v	n_i'	ρ^2	b_1	b_2	R^2	$\sigma(\mathcal{E})$	R_{12}	$\sigma_{12}(\mathcal{E})$	$\mathcal{V}(\hat{\mu}_p)$	r_{12}
$p^{\bullet} \times I^{\bullet}$	1	6	.6382	.5339	.1377	.6534	.3572				
	2	6	.5902	.2263	.4321	.6280	.3705	.6931	.0980	.0870	.7274
$p^{\bullet} \times I^{\bullet}$	1	8	.7017	.5956	.1366	.7141	.3244				
	2	8	.6576	.2245	.4946	.6889	.3388	.7535	.0787	.0884	.7509
$p^{\bullet} \times I^{\circ}$	1	6	.6382	.4956	.2577	.7190	.3216				
	2	6	.5902	.3162	.4287	.7024	.3314	.7933	.0660	.0842	.5317
$p^{\bullet} \times I^{\circ}$	1	8	.7017	.5436	.2597	.7689	.2917				
	2	8	.6576	.3187	.4762	.7520	.3024	.8437	.0499	.0852	.5885
$I^{\bullet}{:}p^{\bullet}$	1	6	.5804	.4666	.1597	.6051	.3813				
	2	6	.5438	.2205	.3969	.5877	.3900	.6511	.1114	.0856	.6892
$I^{\bullet}{:}p^{\bullet}$	1	8	.6485	.5280	.1632	.6696	.3488				
	2	8	.6138	.2254	.4568	.6518	.3584	.7166	.0905	.0869	.7171
$I^{\circ}{:}p^{\bullet}$	1	6	.5804	.4502	.2588	.6746	.3461				
	2	6	.5438	.3009	.4022	.6626	.3528	.7492	.0801	.0835	.4867
$I^{\circ}{:}p^{\bullet}$	1	8	.6485	.4987	.2663	.7296	.3155				
	2	8	.6138	.3097	.4493	.7173	.3229	.8064	.0618	.0844	.5466

Note. For v_1, $b_0 = (1 - b_1)(4.5167) + b_2(5.0833)$;
for v_2, $b_0 = (1 - b_2)(5.0833) + b_1(4.5167)$.

The resulting regressed score estimation equation for v_1 is

$$\hat{\mu}_{p1} = .7410 + .5436\,\overline{X}_{p1} + .2597\,\overline{X}_{p2},$$

which leads to $R_1^2 = .7689$ and $\sigma_1(\mathcal{E}) = .2917$.

The top half of Table 12.3 summarizes the results for both crossed-design structures ($p^{\bullet} \times I^{\bullet}$ and $p^{\bullet} \times I^{\circ}$), both sample sizes ($n_i' = 6$ and $n_i' = 8$), and both variables. Note that, for both variables, R^2 increases and $\sigma(\mathcal{E})$ decreases when:

- the design stays the same ($p^{\bullet} \times I^{\bullet}$), but sample size ($n_i'$) increases, leading to larger generalizability coefficients; or

- sample size stays the same, but the design changes from one with linked conditions ($p^{\bullet} \times I^{\bullet}$) and positively correlated δ-type errors to the corresponding design with uncorrelated δ-type errors ($p^{\bullet} \times I^{\circ}$).

These conclusions generalize to other sample sizes and designs. See, for example, the results in Table 12.3 for the $I^{\bullet}{:}p^{\bullet}$ and $I^{\circ}{:}p^{\bullet}$ nested designs.

Note that positively correlated δ-type error leads to larger raw score intercorrelations (r_{12}) for the $p^{\bullet} \times I^{\bullet}$ design. Therefore, larger raw score intercorrelations that result from joint sampling, as opposed to independent sampling, are associated with smaller values for R^2 and larger standard errors of estimate for individual universe scores.[6]

As indicated by the results in Table 12.3, $R_{vv'}$ and $\sigma_{vv'}(\mathcal{E})$ are also affected by changes in the D study design structure and/or sample sizes. Consider, for example, the $p^{\bullet} \times I^{\circ}$ design with $n'_i = 6$. The normal equations are the same as those in Equation 12.50 except that the covariance .4367 is replaced by .3193:

$$
\begin{array}{ccccc}
.5769b_1 & + & .3193b_2 & = & .3682 \\
.3193b_1 & + & .6250b_2 & = & .3193.
\end{array}
$$

Solving these equations for the regression coefficients gives

$$
b_1 = .4956 \qquad \text{and} \qquad b_2 = .2577.
$$

Using Equation 12.55

$$
\sigma_{12}(\hat{\mu}_p) = (.4956)(.3193) + (.2577)(.3689) = .2533,
$$

which leads to

$$
\sigma_{12}(\mathcal{E}) = \sigma_{12} - \sigma_{12}(\hat{\mu}_p) = .3193 - .2533 = .0660.
$$

Note, in particular, that $\sigma_{vv'}(\mathcal{E})$ is *not* zero for the $p^{\bullet} \times I^{\circ}$ design with $n'_i = 6$, although it is smaller than $\sigma_{vv'}(\mathcal{E}) = .0980$ for the $p^{\bullet} \times I^{\bullet}$ design.

12.2.6 Expected Within-Person Profile Variability

When profiles are under consideration, usually interest focuses on individual persons, as discussed in Sections 10.2.4 and 12.2.4. To characterize the entire measurement procedure for a population, however, we can consider expected within-person profile variability using the $\mathcal{V}(*)$ formulas in Table 12.2. For both standard and raw scores, three formulas are provided— one for observed scores, one for universe scores, and one for regressed score estimates of universe scores. The logic behind the derivation of these formulas was developed in Section 10.2.5 (see especially Equation 10.55), which also discussed $\mathcal{V}(\overline{\mathbf{X}}_p)$ and $\mathcal{V}(\boldsymbol{\mu}_p)$.

The expected regressed score profile variability is

$$
\mathcal{V}(\hat{\boldsymbol{\mu}}_p) = \frac{n_p - 1}{n_p} [\overline{\sigma}_v^2(\hat{\mu}_p) - \overline{\sigma}_{vv'}(\hat{\mu}_p)] + \text{var}(\hat{\mu}_v). \tag{12.64}
$$

---

[6]In the context of the discussion here, changes in raw score intercorrelations are a consequence of changes in the amount of δ-type correlated error, not changes in universe score covariances or correlations.

We take this equation as a definition of expected profile flatness or, more specifically, lack of flatness. For the synthetic data with $n_i' = 6$ for the $p^\bullet \times I^\bullet$ design, the computational results in Equations 12.53, 12.54, and 12.56 give

$$\mathcal{V}(\hat{\boldsymbol{\mu}}_p) = .9(.2361 - .2287) + .0803 = .0870.$$

This is the value reported in Table 12.1 based on direct computation of the average (over persons) of the variances of the regressed score estimates.[7]

The flatness of the profiles of predicted scores is affected by generalizability coefficients, universe score correlations, and δ-type correlated errors—all three of which affect raw score intercorrelations. In general,

- larger generalizability coefficients,

- larger values of $|\rho_{12}|$, and

- decreases in δ-type correlated error

tend to be associated with less regression to the mean, smaller standard errors of estimate, and flatter profiles of predicted scores. The reader can verify that these statements are consistent with results reported in Table 12.3 for the synthetic data.

Using the formulas in Table 12.2, it can be shown that

$$\mathcal{V}(\hat{\boldsymbol{\mu}}_p) \leq \mathcal{V}(\boldsymbol{\mu}_p) \leq \mathcal{V}(\overline{\mathbf{X}}_p),$$

which mirrors the inequality relationships in univariate theory[8]; that is,

$$\sigma^2(\hat{\mu}_p) \leq \sigma^2(\mu_p) \leq \sigma^2(\overline{X}_p).$$

As in univariate theory, it is natural to consider various functions of variabilities, particularly ratios. One such ratio is given by Equation 10.60 in Section 10.2.5:

$$\mathcal{G} = \frac{\mathcal{V}(\boldsymbol{\mu}_p)}{\mathcal{V}(\overline{\mathbf{X}}_p)},$$

which is interpretable approximately as a type of generalizability coefficient for a randomly selected person p.

Another obvious ratio to consider is

$$\mathcal{R}^2 = \frac{\mathcal{V}(\hat{\boldsymbol{\mu}}_p)}{\mathcal{V}(\boldsymbol{\mu}_p)}, \tag{12.65}$$

---

[7]For this example, $\overline{\sigma}_{vv'}(\hat{\mu}_p)$ is the average over four terms, not two. Also, var($\hat{\mu}_v$) is the biased estimate of the variance of the two means (4.5167 and 5.0833).

[8]$\sigma^2(\overline{X}_p)$ is the variance of observed persons' mean scores for the population, which was denoted $ES^2(p)$ in earlier chapters.

which we designate $\mathcal{R}^2$ because of its obvious similarity to the squared correlation in a multiple regression.[9] This ratio can be interpreted approximately as the proportion of expected variability in universe scores for a "typical" person that is explained by the regressions. For the synthetic data with $n'_i = 6$ for the $p^\bullet \times I^\bullet$ design, $\mathcal{R}^2 = .0870/.1024 = .849$. The corresponding approximation to the variance of the errors of estimate for a typical person is

$$\mathcal{V}(\boldsymbol{\mu}_p) - \mathcal{V}(\hat{\boldsymbol{\mu}}_p) = \mathcal{V}(\boldsymbol{\mu}_p)(1 - \mathcal{R}^2), \qquad (12.66)$$

which is .015 for the synthetic data. This means that the standard error of the regressed score estimates for a typical person is roughly $\sqrt{.015} = .12$. Recall that the standard deviation of the δ-type errors of measurement for a typical person is $\sqrt{.058} = .24$, which is twice as large as the standard error of estimate. Such statements are sometimes used as an argument in favor of using regressed score estimates.

Finally, the proportional reduction in profile variability attributable to using regressed score estimates is

$$\mathcal{R}\mathcal{V} = 1 - \frac{\mathcal{V}(\hat{\boldsymbol{\mu}}_p)}{\mathcal{V}(\overline{\mathbf{X}}_p)}, \qquad (12.67)$$

which is $1 - .0870/.1542 = .436$ for the synthetic data with $n'_i = 6$ for the $p^\bullet \times I^\bullet$ design. That is, for a typical person, the use of regressed score estimates reduces profile variability by about 44%. This particular value can be computed directly (although less efficiently) using the persons' scores in Table 12.1, whereas $\mathcal{G}$ and $\mathcal{R}^2$ cannot, since universe scores for any particular person are unknown.

Also, direct computation of $\mathcal{R}\mathcal{V}$ (and its component parts) is not possible for different sample sizes and/or designs, because, of course, persons' observed scores are unavailable for different sample sizes and/or designs. The formulas in Table 12.2 can be used, however. For example, for the $p^\bullet \times I^\circ$ design with $n'_i = 6$, these formulas give $\mathcal{V}(\overline{\mathbf{X}}_p) = .2070$ and $\mathcal{V}(\hat{\boldsymbol{\mu}}_p) = .0842$, which leads to $\mathcal{R}\mathcal{V} = .593$. Comparing this result to $\mathcal{R}\mathcal{V} = .436$ for the $p^\bullet \times I^\bullet$ design, it is evident that, for a typical person, the use of the $p^\bullet \times I^\circ$ design leads to greater reduction in profile variability (i.e., a flatter profile) than does the $p^\bullet \times I^\bullet$ design.

Although measures of relative variability for profiles have conceptual similarities with traditional statistics in univariate generalizability theory, there are differences, too. In particular, in univariate theory with a single dependent variable,

$$\left[\rho^2 = \frac{\sigma^2(\mu_p)}{\sigma^2(\overline{X}_p)} \right] = \left[\frac{\sigma^2(\hat{\mu}_p)}{\sigma^2(\mu_p)} = R^2 \right],$$

---

[9]Previously, $\mathcal{R}$ was used to designate a set of random facets. Obviously, $\mathcal{R}^2$ is *not* the square of a set of random facets.

but the corresponding equality does not hold for the relative variability indices based on multiple dependent variables; that is,

$$\left[\mathcal{G} = \frac{\mathcal{V}(\mu_p)}{\mathcal{V}(\overline{X}_p)} \right] \neq \left[\frac{\mathcal{V}(\hat{\mu}_p)}{\mathcal{V}(\mu_p)} = \mathcal{R}^2 \right].$$

12.3   Predicted Composites

The multiple linear regression procedure described in Section 12.1 can be applied also to composites of $k = n_v$ universe scores. For example, the prediction equation for raw scores is

$$\hat{Y} = \hat{\mu}_{pC} = b_0 + b_1 \overline{X}_{p1} + \cdots + b_k \overline{X}_{pk},\qquad (12.68)$$

and there is a corresponding equation for standard scores in terms of β weights. It is important to recognize that the left-hand side of Equation 12.68 is an estimate of

$$Y = \mu_{pC} = w_1 \mu_{p1} + \cdots + w_k \mu_{pk}.\qquad (12.69)$$

Since both the left- and right-hand sides of Equation 12.68 are linear composites, the equation is reminiscent of a canonical correlation, but there is a very important difference. Namely, the left-hand side is a composite based on a priori (i.e., investigator-specified) weights, and only the right-hand side is based on statistically optimal (in a least-squares sense) weights. (For a canonical correlation, both sides are based on statistical weights.)

To estimate the b weights in Equation 12.68, the crux of the matter is to express the right side of the normal equations (Equation 12.16) with respect to estimable model parameters. Once that is done, the equations can be solved for the b weights. When the independent variables $(\overline{X}_{pv})$ are observed score versions of the same variables that enter the composite (μ_{pv}), the right-hand sides of the normal equations are the covariances

$$S_{YX_j} = \sum_{v=1}^{n_v} w_v \sigma_{vj},\qquad (12.70)$$

where $Y = \mu_{pC}$, $\sigma_{vj} = \sigma_{vj}(p)$ in the simplified notation introduced in Equation 12.25, and j indexes the observed scores for the n_v variables.

Replacing Equation 12.70 in Equation 12.18 gives the following formula for the variance of the regressed score estimates of the composite,

$$\sigma_C^2(\hat{\mu}_p) = \sum_{j=1}^{n_v} b_j \left(\sum_{v=1}^{n_v} w_v \sigma_{vj} \right).\qquad (12.71)$$

It follows that R^2 for the predicted composite scores is

$$R_C^2 \;=\; \frac{\sigma_C^2(\hat{\mu}_p)}{\sigma_C^2(\mu_p)} \tag{12.72}$$

$$=\; \frac{\sum_j b_j \left(\sum_v w_v \sigma_{vj} \right)}{\sum_j w_j \left(\sum_v w_v \sigma_{vj} \right)}, \tag{12.73}$$

where all summations range from 1 to n_v.

With error defined as $\mathcal{E}_{pC} = \mu_{pC} - \hat{\mu}_{pC}$, the standard error of estimate is

$$\sigma_C(\mathcal{E}) \;=\; \sigma_C(\mu_p)\sqrt{1 - R_C^2} \tag{12.74}$$

$$=\; \sqrt{\sum_j (w_j - b_j) \left(\sum_v w_v \sigma_{vj} \right)}, \tag{12.75}$$

where, again, both summations range from 1 to n_v. Under normality assumptions, a two-sided $100\gamma\%$ tolerance interval for universe composite scores is

$$\hat{\mu}_{pC} \;\pm\; z^*_{(1+\gamma)/2}\, \sigma_C(\mathcal{E}), \tag{12.76}$$

where $z^*_{(1+\gamma)/2}$ is the $[100(1+\gamma)/2]$th percentile point of the normal distribution. The tolerance interval is centered around the regressed score estimate of the composite score. Such an interval is conditional on observed scores and, as such, it is meaningful to say that the true composite score is contained within the interval $100\gamma\%$ of the time.

As indicated by Equation 12.73, R_C^2 is interpretable as the proportion of universe composite scores predicted by the regression. Equivalently, R_C^2 is the squared correlation between universe composite scores and predicted composite scores. As such, R_C^2 is a type of reliability or generalizability coefficient.

Both R_C^2 in Equation 12.73 and $\sigma_C(\mathcal{E})$ in Equation 12.75 are expressed in terms of b weights, w weights, and elements of Σ_p. No other parameters or statistics are required, although there are equivalent expressions for R_C^2 and $\sigma_C(\mathcal{E})$ that make it appear that other parameters and statistics are involved. Error variances associated with the individual variables are absorbed into these equations through the b weights.

R_C^2 and $\sigma_C(\mathcal{E})$ apply for any universe score composite, even trivial ones. For example, the formulas for R_v^2 and $\sigma_v(\mathcal{E})$ for profiles (see Table 12.2) are special cases of these two equations in which $w_v = 1$ and the other ws are zero.

The theory outlined above for predicted composites is illustrated next by considering difference scores. This is the simplest example of a multivariate predicted composite in that it requires considering only $n_v = 2$ variables, which permits us to make direct use of the results in Section 12.1 that introduced multiple linear regression. Consequently, we can examine algebraic expressions for the b weights, which is not easily done when $n_v > 2$. This generalizability theory perspective on difference scores is considerably more flexible than traditional perspectives.

12.3.1 Difference Scores

When there are two independent variables, an obvious example of a composite is difference, change, or growth scores. In our notation, this composite is $\mu_{pC} = \mu_{p2} - \mu_{p1}$, which means that $w_2 = 1$, and $w_1 = -1$. We wish to obtain the prediction equation

$$\hat{\mu}_{pC} = b_0 + b_1 \overline{X}_{p1} + b_2 \overline{X}_{p2}.$$

To do so we use Equations 12.10 and 12.11.

In classical test theory notation, the universe score composite is typically denoted $Y = T_2 - T_1$. The covariance of universe difference scores with X_1 is

$$S_{YX_1} = S_{(T_2-T_1)(T_1+E_1)} = S_{T_1T_2} - S_{T_1}^2 = \sigma_{12} - \sigma_1^2. \tag{12.77}$$

Similarly, the covariance of universe difference scores with X_2 is

$$S_{YX_2} = S_{(T_2-T_1)(T_2+E_2)} = S_{T_2}^2 - S_{T_1T_2} = \sigma_2^2 - \sigma_{12}. \tag{12.78}$$

Replacing these results in Equations 12.10 and 12.11 gives the raw score regression coefficients for the prediction of $T_2 - T_1$, namely,

$$b_1 = \frac{1}{1 - r_{12}^2}\left[\frac{\sigma_{12}}{S_1^2} - \rho_1^2 - r_{12}\left(\frac{\sigma_2^2 - \sigma_{12}}{S_1 S_2}\right)\right] \tag{12.79}$$

and

$$b_2 = \frac{1}{1 - r_{12}^2}\left[\rho_2^2 - \frac{\sigma_{12}}{S_2^2} - r_{12}\left(\frac{\sigma_{12} - \sigma_1^2}{S_1 S_2}\right)\right], \tag{12.80}$$

where S_1 and S_2 are the abbreviations for observed score standard deviations given in Equation 12.24. It follows from Equation 12.3 that

$$b_0 = (1 - b_2)\overline{X}_2 - (1 + b_1)\overline{X}_1. \tag{12.81}$$

Using Equation 12.12, the squared correlation between universe difference scores and their predicted values is

$$R_C^2 = \frac{b_1(\sigma_{12} - \sigma_1^2) + b_2(\sigma_2^2 - \sigma_{12})}{\sigma_1^2 + \sigma_2^2 - 2\sigma_{12}}, \tag{12.82}$$

where the denominator is the variance of the universe difference scores, $\sigma_C^2(\mu_p)$, and the numerator is the variance of their regressed score estimates. The standard error of estimate based on using the two observed scores to predict the universe difference score can be obtained using Equation 12.74 with R_C^2 given by Equation 12.82, or using Equation 12.75 directly. Using either approach,

$$\sigma_C(\mathcal{E}) = \sqrt{(1 - b_2)(\sigma_2^2 - \sigma_{12}) - (1 + b_1)(\sigma_{12} - \sigma_1^2)}. \qquad (12.83)$$

Under the assumptions of classical test theory, the covariance between true scores equals the covariance between observed scores. From the perspective of multivariate generalizability theory, this statement holds if the only linked "facet" is that associated with the objects of measurement (e.g., the $p^\bullet \times I^\circ$ design). Whenever these assumptions are true,

$$\sigma_{12} = S_{12},$$

and Equations 12.79 and 12.80 become the Lord–McNemar regression coefficients (see Lord, 1956, 1958; McNemar, 1958; Feldt & Brennan, 1989).

Without making the classical test theory assumptions,

$$\sigma_{12} = S_{12} - \sigma_{12}(\delta),$$

as is the case for the $p^\bullet \times I^\bullet$ design. It follows that the prediction of universe difference scores based on Equations 12.79 to 12.81 is more general than the Lord–McNemar prediction. That is, the derivation provided here permits the consideration of designs in which correlated δ-type error may affect the prediction, whereas the Lord–McNemar prediction effectively assumes such correlated errors do not exist.

Indeed, Equations 12.79–12.81 are applicable to difference scores based on *any* multivariate design with $n_v = 2$. Different designs lead to different values for S_1, S_2, ρ_1^2, ρ_2^2, and r_{12}, but the equations themselves are unaltered. From another perspective, the bs in Equations 12.79 and 12.80 are simply the solutions to the normal equations:

$$
\begin{aligned}
S_1^2\, b_1 + S_{12}\, b_2 &= \sigma_{12} - \sigma_1^2 \\
S_{12}\, b_1 + S_2^2\, b_2 &= \sigma_2^2 - \sigma_{12}.
\end{aligned}
$$

Instead of using X_1 and X_2 as distinct independent variables to predict universe difference scores, there is a long tradition in measurement of employing the observed difference $X_{p2} - X_{p1}$ to predict $\mu_{pC} = \mu_{p2} - \mu_{p1}$ using the simple (i.e., single independent variable) linear regression

$$\hat{\mu}_{pC} = (1 - \boldsymbol{E}\rho^2)(\overline{X}_2 - \overline{X}_1) + \boldsymbol{E}\rho^2(\overline{X}_{p2} - \overline{X}_{p1}). \qquad (12.84)$$

In Equation 12.84, $\boldsymbol{E}\rho^2$ is the reliability of the difference scores, which equals R^2 for this simple linear regression. Conceptually, this is the Kelley regressed score estimate of the universe difference score. The usual expression for the reliability of the difference scores (see Feldt & Brennan, 1989, p. 118) is

$$E\rho^2 = 1 - \frac{\sigma_1^2(\delta) + \sigma_2^2(\delta)}{S_1^2 + S_1^2 - 2S_{12}} = \frac{\sigma_1^2 + \sigma_2^2 - 2S_{12}}{S_1^2 + S_1^2 - 2S_{12}}, \qquad (12.85)$$

which effectively assumes uncorrelated δ-type errors, as is the case for the $p^{\bullet} \times I^{\circ}$ design. For the $p^{\bullet} \times I^{\bullet}$ design, however, this assumption is *not* made, and the reliability of the difference scores is

$$E\rho^2 = 1 - \frac{\sigma_1^2(\delta) + \sigma_2^2(\delta) - 2\sigma_{12}(\delta)}{S_1^2 + S_1^2 - 2S_{12}} = \frac{\sigma_1^2 + \sigma_2^2 - 2\sigma_{12}}{S_1^2 + S_1^2 - 2S_{12}}. \qquad (12.86)$$

Since $R^2 = E\rho^2$ for the single-variable prediction, the standard error of estimate is the standard error of measurement. From Equation 12.85 for $E\rho^2$, it is evident that the standard error of estimate for the single-variable prediction for the $p^{\bullet} \times I^{\circ}$ design is

$$\sigma_{C1}(\mathcal{E}) = \sqrt{\sigma_1^2(\delta) + \sigma_2^2(\delta)}.$$

From Equation 12.86, the standard error of estimate for the single-variable prediction for the $p^{\bullet} \times I^{\bullet}$ design is

$$\sigma_{C1}(\mathcal{E}) = \sqrt{\sigma_1^2(\delta) + \sigma_2^2(\delta) - 2\sigma_{12}(\delta)},$$

which is obviously smaller than for the $p^{\bullet} \times I^{\circ}$ design when $\sigma_{12}(\delta)$ is positive. The same conclusion applies to the two-variable prediction. For both prediction equations, then, when correlated error is positive,

- R_C^2 for the $p^{\bullet} \times I^{\bullet}$ design is greater than or equal to R_C^2 for the $p^{\bullet} \times I^{\circ}$ design, and

- $\sigma_C(\mathcal{E})$ for the $p^{\bullet} \times I^{\bullet}$ design is less than or equal to $\sigma_C(\mathcal{E})$ for the $p^{\bullet} \times I^{\circ}$ design.

In this sense, positively correlated δ-type error (in the $p^{\bullet} \times I^{\bullet}$ design) leads to *more* dependable predictions of universe difference scores, all other things being equal.

12.3.2 Synthetic Data Example

For the synthetic data example in Table 9.3, the two-variable prediction equation is

$$\hat{\mu}_{pC} = .4597 - .3076\,\overline{X}_{p1} + .2944\,\overline{X}_{p2}, \qquad (12.87)$$

TABLE 12.4. Predicted Difference Scores for Synthetic Data Example for Balanced $p^{\bullet} \times I^{\bullet}$ Design in Table 9.3

p	Two Indep Vars			One Indep Var	
	$\overline{X}_{p1}$	$\overline{X}_{p2}$	$\hat{\mu}_{pC}$	$\overline{X}_{p2} - \overline{X}_{p1}$	$\hat{\mu}_{pC}$
1	4.3333	5.0000	.5985	.6667	.5967
2	3.5000	4.6667	.7568	1.1667	.7467
3	5.0000	4.8333	.3444	−.1667	.3467
4	3.1667	3.8333	.6140	.6667	.5967
5	4.3333	4.5000	.4514	.1667	.4467
6	5.5000	6.6667	.7302	1.1667	.7467
7	4.8333	5.1667	.4938	.3333	.4967
8	4.5000	6.0000	.8416	1.5000	.8467
9	5.5000	5.3333	.3378	−.1667	.3467
10	4.5000	4.8333	.4982	.3333	.4967
Mean	4.5167	5.0833	.5667	.5667	.5667
Var	.5769	.6250	.02965	.3284	.02956
R_C^2			.30092		.30001

which is easily obtained using the numerical results in Equations 12.46 to 12.49 and the value of r_{12} obtained from Equation 12.35:

$$r_{12} = \frac{\sigma_{12}(p) + \sigma_{12}(\delta)}{S_1 S_2} = \frac{.3193 + .1175}{\sqrt{.5769}\sqrt{.6250}} = .7273. \tag{12.88}$$

For the single-variable prediction, using Equation 12.86 the estimated reliability of the difference scores is $E\hat{\rho}^2 = .3000$, which leads to

$$\hat{\mu}_{pC} = .3967 + .3000\,(\overline{X}_{p2} - \overline{X}_{p1}). \tag{12.89}$$

The estimated variance of the universe difference scores is

$$\sigma_C^2(\mu_p) = .3682 + .3689 - 2(.3193) = .0985,$$

which applies, of course, to both prediction equations. For the two-variable prediction, using Equation 12.82,

$$R_C^2 = \frac{-.3076(.3193 - .3682) + .2944(.3689 - .3193)}{.0985} = .3009.$$

For the single-variable prediction, using Equation 12.86,

$$R_{C1}^2 = E\rho^2 = \frac{.3682 + .3689 - 2(.3193)}{.5769 + .6250 - 2(.4367)} = .3000,$$

where we append the 1 to the subscript of R^2 to emphasize that it is for a regression equation with a single independent variable.

Table 12.4 provides these two statistics, as well as the observed scores and predicted values for all persons, for both prediction equations. It is evident from the values of R_C^2 and R_{C1}^2 that the two equations are nearly identical in terms of their ability to explain the variability in the universe difference scores. This near identity is related to the fact that $b_1 = -.3076$ is nearly equal in absolute value to $b_2 = .2944$, both of which are very close in absolute value to $E\hat{\rho}^2 = .3000$ for the single-variable regression. Indeed, it can be shown that R^2 for the two prediction equations will be equal when $b_1 = -b_2$, which will occur when $\rho_1 = \rho_2$ and $S_1 = S_2$ (see Exercise 12.10).

For the two-variable prediction, using Equation 12.75, the standard error of estimate is

$$\sigma_C(\mathcal{E}) = \sqrt{.0985}\sqrt{1 - .3009} = .2624.$$

For the single-variable prediction, the standard error of estimate is

$$\sigma_{C1}(\mathcal{E}) = \sqrt{.0985}\sqrt{1 - .3000} = .2626.$$

The near equality of these two estimates is a direct consequence, of course, of the fact that R_C^2 and R_{C1}^2 are nearly identical. For either prediction equation, using Equation 12.76, under normality assumptions a 68% tolerance interval for universe difference scores is approximately $\hat{\mu}_{pC} \pm .26$. This is an interval that is conditional on observed scores, and, as such, it is meaningful to say that 68% of the true difference scores lie within the interval.

A cursory view of the two sets of predicted values suggests that they are more different than might be expected, given that the R_C^2 and R_{C1}^2 are nearly identical. In particular, note that persons with the *same* observed difference score can have *different* predicted scores for the two-variable regression, although they must have the same predicted score for the single-variable regression. This difference is evident, for example, for the second and sixth persons who have an observed difference of 1.1667. For the second person the two-variable prediction is larger than the one-variable prediction, but this inequality is reversed for the sixth person. Among other things, this means that the tolerance intervals (Equation 12.76) for the two predictions will be different for persons with the same observed difference score. By most statistical perspectives (except, perhaps, parsimony) there is much to recommend the two-variable prediction over its one-variable counterpart, even when their R^2 values are similar.

12.3.3 Different Sample Sizes and/or Designs

Prediction equations also can be developed for different sample sizes and/or designs. If the design is the same but sample sizes differ, then $\sigma_1^2(\delta)$, $\sigma_2^2(\delta)$, and $\sigma_{12}(\delta)$ will all be affected, as will any statistics based on them, that is, S_1, S_2, ρ_1^2, ρ_2^2, S_{12}, and r_{12}. Otherwise, however, the same equations apply. Prediction equations for the $I^\bullet{:}p^\bullet$ design can be estimated by replacing

δ-type variances and covariances with Δ-type variances and covariances which, of course, will affect S_1, S_2, ρ_1^2, ρ_2^2, S_{12}, and r_{12}. For any choice of sample sizes or design, however, the estimated universe score variances and covariances still apply, as do the means for the levels of v.

When sample sizes are the same, prediction equations for the $p^{\bullet} \times I^{\circ}$ design can be estimated using the same equations and statistics used for the $p^{\bullet} \times I^{\bullet}$ design, with one important exception; namely, r_{12} in Equation 12.88 must be replaced by

$$ r_{12} = \frac{\sigma_{12}(p)}{S_1 S_2} = \frac{.3193}{\sqrt{.5769}\sqrt{.6250}} = .5317, \tag{12.90}$$

which is an estimate of the observed correlation when there are no δ-type correlated errors. This leads to the two-variable prediction

$$ \hat{\mu}_{pC} = .5075 - .1794\,\overline{X}_{p1} + .1711\,\overline{X}_{p2}, $$

with $R_C^2 = .1752$. Using Equation 12.85, the one-variable prediction is

$$ \hat{\mu}_{pC} = .4676 + .1749\,(\overline{X}_{p2} - \overline{X}_{p1}), $$

with $R_{C1}^2 = E\rho^2 = .1749$.

Whether the two-variable or one-variable prediction equation is used, R^2 for the $p^{\bullet} \times I^{\circ}$ design is about .17, and R^2 for the $p^{\bullet} \times I^{\bullet}$ design is about .30. For this synthetic data example, then, the percent of the variance in universe difference scores explained by the regression nearly doubles using the design that explicitly recognizes correlated errors. This occurs because, for the $p^{\bullet} \times I^{\bullet}$ design, positively correlated δ-type error decreases the standard error of estimate. It follows that, if interest focuses on using regression to explain the variance in universe difference scores, then the $p^{\bullet} \times I^{\bullet}$ design is preferable to the $p^{\bullet} \times I^{\circ}$ design if errors are positively correlated. Of course, this conclusion assumes that the universe of generalization is one for which I can be "scored" with respect to each level of v. If not, then the $p^{\bullet} \times I^{\bullet}$ design is not sensible. When the two scores are pre- and posttest measures, this conclusion is consistent with the conventional wisdom that says, "All other things being equal, if you want to measure change, don't change the measure."

12.3.4   Relationships with Estimated Profiles

For a universe difference score composite, the previous section demonstrated that, if errors are positively correlated, the $p^{\bullet} \times I^{\bullet}$ design yields a larger R^2 value and a smaller standard error of estimate than the $p^{\bullet} \times I^{\circ}$ design. By contrast, it was demonstrated in Section 12.2.5 that for regressed score estimates of individual universe scores (i.e., predicted profiles) the

$p^{\bullet} \times I^{\bullet}$ design yields *smaller* R^2 values and larger standard error of estimates than the $p^{\bullet} \times I^{\circ}$ design. In this sense, the "best" design for measuring status may not be the same as the "best" design for measuring differences or change. This is an example of what might be called an optimal-design-structure paradox, meaning that the "best" design structure for making decisions based on regressed score estimates for individual variables may not be the "best" design structure for making decisions based on regressed score estimates for a composite. This is not a statement that generalizes to all composites, but it definitely can happen.

Let us reexamine the synthetic data results for the two-variable profile and the predicted difference score composite based on two variables:

$$\hat{\mu}_{p1} = 1.4050 + .5339\,\overline{X}_{p1} + .1377\,\overline{X}_{p2}$$
$$\hat{\mu}_{p2} = 1.8647 + .2263\,\overline{X}_{p1} + .4321\,\overline{X}_{p2}$$
$$\hat{\mu}_{pC} = .4597 - .3076\,\overline{X}_{p1} + .2944\,\overline{X}_{p2}.$$

It is immediately evident that $\hat{\mu}_{pC} = \hat{\mu}_{p2} - \hat{\mu}_{p2}$. This is a specific example of a more general result:

$$\hat{\mu}_{pC} = \sum_{v=1}^{n_v} w_v \hat{\mu}_{pv}, \tag{12.91}$$

provided the independent variables for $\hat{\mu}_{pC}$ and the $\hat{\mu}_{pv}$ are the observed score variables corresponding to the n_v universe scores (and only these variables). One consequence of Equation 12.91 is that regression coefficients for $\hat{\mu}_{pC}$ are easily determined using the regression coefficients for the n_v predictions of μ_{pv}. Specifically,

$$b_k = \sum_{v=1}^{n_v} w_v b_{kv}, \tag{12.92}$$

for $k = 0, \ldots, n_v$. Note that b with one subscript refers to regression coefficients in $\hat{\mu}_{pC}$, and b with two subscripts refers to regression coefficients in $\hat{\mu}_{pv}$.

The standard error of estimate for a composite can be determined from the variances and covariances of the errors of estimate for the individual variables:

$$\sigma_C(\mathcal{E}) = \sqrt{\sigma^2 \left(\sum_v w_v \, \mathcal{E}_v \right)}$$
$$= \sqrt{\sum_v w_v^2 \, \sigma_v^2(\mathcal{E}) + \sum_v \sum_{v'} w_v w_{v'} \, \sigma_{vv'}(\mathcal{E})}, \tag{12.93}$$

where $v \neq v'$ for the double summation. For the synthetic data, using the standard errors of estimate for the two variables computed in Section 12.2.2 and the covariance of the errors of estimate computed in Section 12.2.4,

$$\sigma_C(\mathcal{E}) = \sqrt{(.3572)^2 + (.3705)^2 - 2(.0980)} = .2624,$$

as was obtained through "direct" computation in Section 12.3.2.

R^2 for the composite also can be determined using statistics for profile equations. Two such expressions are:

$$R_C^2 = \frac{\sum_v w_v^2\, \sigma_v^2(\hat{\mu}_p) + \sum_v \sum_{v'} w_v w_{v'}\, \sigma_{vv'}(\hat{\mu}_p)}{\sum_v w_v^2\sigma_v^2 + \sum_v \sum_{v'} w_v w_{v'}\, \sigma_{vv'}} \qquad (12.94)$$

$$= 1 - \frac{\sum_v w_v^2\sigma_v^2(\mathcal{E}) + \sum_v \sum_{v'} w_v w_{v'}\, \sigma_{vv'}(\mathcal{E})}{\sum_v w_v^2\, \sigma_v^2 + \sum_v \sum_{v'} w_v w_{v'}\, \sigma_{vv'}}, \qquad (12.95)$$

where $v \neq v'$ in the double summations. For the synthetic data (five-digit versions of the results in Equations 12.53, 12.54, and 12.56), Equation 12.94 gives

$$R_C^2 = \frac{.24053 + .23164 - 2(.22127)}{.0985} = .3009,$$

and, equivalently, Equation 12.95 gives

$$R_C^2 = 1 - \frac{(.2624)^2}{.0985} = .3009,$$

as was obtained through "direct" computation in Section 12.3.2.

12.3.5   Other Issues

As discussed in the introduction to Section 12.3, R_C^2 in Equation 12.73 is interpretable as the proportion of universe composite scores predicted by the regression or, equivalently, the squared correlation between universe composite scores and predicted composite scores. As such, it is a type of reliability or generalizability coefficient for predicted composites.

Another type of generalizability coefficient for predicted composites is

$$\rho_C(\hat{\mu}_p, \hat{\mu}_p') = \frac{\sum_v \sum_{v'} b_v b_{v'} \sigma_{vv'}}{\sum_v \sum_{v'} b_v b_{v'} S_{vv'}} \qquad (12.96)$$

$$= 1 - \frac{\sum_v \sum_{v'} b_v b_{v'} \sigma_{vv'}(\delta)}{\sum_v \sum_{v'} b_v b_{v'} S_{vv'}} \qquad (12.97)$$

$$= 1 - \frac{\sum_v \sum_{v'} b_v b_{v'} \sigma_{vv'}(\delta)}{\sum_j b_j \left(\sum_v w_v \sigma_{vj}\right)}, \qquad (12.98)$$

where all subscripts range from 1 to n_v. $\rho_C(\hat{\mu}_p, \hat{\mu}_p')$ is interpretable as an estimate of the correlation between predicted composites for two randomly parallel instances of the measurement procedure. That is, letting $\hat{Y}$ and $\hat{Y}'$ be predicted composites for the two parallel procedures, $\rho_C(\hat{\mu}_p, \hat{\mu}_p')$ is the ratio of their covariance to their variance. Equation 12.97 is a formula from classical test theory. Equation 12.98 expresses $\rho_C(\hat{\mu}_p, \hat{\mu}_p')$ in terms of variance and covariance components from a multivariate analysis. The derivation of Equations 12.96 to 12.98 is provided in the answer to Exercise 12.13.

The coefficient $\rho_C(\hat{\mu}_p, \hat{\mu}_p')$ is more closely aligned with tradition in measurement theory (see, e.g., Feldt & Brennan, 1989, p. 119) than is R_C^2, but both $\rho_C(\hat{\mu}_p, \hat{\mu}_p')$ and R_C^2 are based on widely accepted, although different, definitions of reliability. When $n_v = 1$, the different definitions give equivalent results.

For purposes of simplicity, and to illustrate conceptual and statistical issues, most of our treatment of predicted composites has focused on the simplest case: difference scores and $n_v = 2$ variables. The theory per se, however, has no such restriction. The general equations provided at the beginning of Section 12.3 apply to any number of variables and any universe score composite. The principal complexity is that the normal equations (Equations 12.16) with the right-side elements given by Equation 12.70 must be solved for the b weights. The b weights are specific to the D study design and sample sizes. Changes in design or sample sizes generally alter the *observed* score variances and covariances on the left side of the normal equations, which leads to changes in the b weights.

The theory of predicted composites has been developed here using raw score regression weights, b. Occasionally, however, predicted composites based on β weights are appropriate. Suppose, for example, that each universe score and raw score is transformed to a T score (not to be confused with a true score T) according to the usual formula: T $= 50 + 10\, Z$. The appropriate regression weights are the β weights in the prediction equation:

$$\hat{T}_{pC} = \beta_1 T_{X_1} + \cdots + \beta_k T_{X_k},$$

where $k = n_v$. These β weights can be obtained directly through solving the normal equations in Equation Set 12.15 or, if the b weights are available, the β weights are easily determined using Equation 12.17. Since the T scores are a linear transformation of the Z scores, R_C^2 remains unchanged. Using Equation 12.75, the standard error of estimate is

$$\sigma_C(\mathcal{E}) = 10\sqrt{\sum_j (w_j - \beta_j)\left(\sum_v w_v \sigma_{vj}\right)}.$$

Jarjoura (1983) describes a method for obtaining so-called "best" linear predictions (BLPs) for composite universe scores for the $p^\bullet \times I^\circ$ design. Jarjoura's BLP equation has a form similar to the predicted-composite equation for the $p^\bullet \times I^\circ$ design discussed in this chapter, although the theory underlying the BLP equation is considerably different. The BLP equation has additional terms related to the category means, which gives it the appearance of a Δ-like version of the predicted-composite equation for the $p^\bullet \times I^\circ$ design. However, the measurement error variance for the BLP equation is less than $\sigma_C^2(\Delta)$. Searle et al. (1992, Chap. 7, especially

Sect. 7.4) provide an extensive overview of the statistical theory for BLPs. Applying this theory to the various types of multivariate designs typically used in generalizability theory is possible, but not easy.

12.4 Comments

The multivariate perspective on profiles and regressed score estimates discussed in this chapter is very powerful and flexible. The illustrations discussed here have focused on single-facet multivariate designs with $n_v = 2$ for simplicity of presentation, but the theory extends readily to any multivariate design with any number of variables. For raw scores, the essentials of the process are:

- specify the normal equations in Equation Set 12.16 based on the multivariate design under consideration, using Equation 12.41 for the rightside of the normal equations when profiles are under consideration, or using Equation 12.70 for the right side when predicted composites are desired;

- solve the normal equations for the bs; and

- use the formulas in Table 12.2 and/or Section 12.2.4 if profiles are under consideration, or use Equations 12.71 to 12.76 for predicted composites.

A similar set of steps applies to standard scores.

Many of the most important results in this chapter have been expressed solely in terms of universe score variance and covariance components, regression (b or β) weights, and a priori (w) weights (when composites are under consideration). See, especially, the formulas in Table 12.2 for profiles and Equations 12.70 to 12.75 for predicted composites. This is purposeful. Once universe score variance and covariance components are estimated, regression weights are obtained, and a priori weights are specified, just about everything else of interest for multivariate regressed scores can be obtained.

The entire discussion of multivariate regressed scores in this chapter has avoided any consideration of the complexities that arise with unbalanced designs in generalizability theory. However, in order to use the theory discussed in this chapter, the variance and covariance components on the right side of the normal equations must be estimated. So, the complexities brought about by unbalanced designs are as much an issue here as elsewhere.

Regressed score estimates in multivariate theory have the same potential for misuse or misunderstanding that was mentioned in Section 5.5.2 in discussing regressed score estimates in univariate theory. In particular, the theory per se does not tell an investigator whether it is sensible to regress scores to the mean of any particular population. In some cases, therefore, applying the theory could have unintended and/or unwanted consequences. It is especially important to be sensitive to such matters when persons' scores are regressed to the mean for their race, gender, or ethnicity. Doing so is not wrong from a statistical perspective, per se, but the investigator has the responsibility to provide a substantive, theory-based rationale for choosing the population(s) used as the referent(s) for scores.

Since the theory discussed in this chapter is essentially an application of multiple linear regression analyses, it is natural to think that any computer package that performs such analyses can be used in multivariate generalizability theory. That is not correct, however, because general purpose packages require observed values for the dependent variable. For the theory discussed here, the dependent variable is universe scores for each of the separate variables (in the case of regressed score estimates of profiles) or a function of universe scores (in the case of predicted composites) and, of course, universe scores are unknown. For the D study versions of the illustrative designs in Table 9.2, mGENOVA can be used to perform most of the analyses discussed in this chapter, or at least obtain output that facilitates hand computation of desired results. Also, if estimated variance and covariance components are available, the matrix procedures in SAS IML can be used relatively easily to obtain the results in this chapter.

12.5 Exercises

12.1* If Y is the universe score for variable v (i.e., $Y = T_v$), and the independent variables are the observed scores for all n_v variables, show that the relationship between the b and β weights discussed in Section 12.1 is

$$b_{j|T_v} = \beta_{j|T_v} \left[\frac{\sigma_v(p)}{s_{X_j}} \right],$$

for $j = 1, \ldots, n_v$.

12.2   Equation 12.23 states that $r_{T_1 X_2} = \rho_{12}(p)\sqrt{E\rho_2^2}$. In classical test theory, one disattenuation formula is $r_{T_1 X_2} = r_{X_1 X_2}/\sqrt{E\rho_1^2}$. What is the principal source of the difference in these two results?

12.3* Using Equation 12.18, derive Equation 12.52 for $\sigma_v^2(\hat{\mu}_p)$.

12.4   Derive Equation 12.55 for $\sigma_{vv'}(\hat{\mu}_p)$.

12.5* For $n_v = 2$, prove that $R_v^2 \geq \rho_v^2$.

12.6   For the synthetic data, Section 12.2.3 derived the result $R_{12} = .693$ for raw scores. Verify that this result is the same for standard scores.

12.7* Verify the results reported in Table 12.3 for the $p^{\bullet} \times I^{\bullet}$ and $p^{\bullet} \times I^{\circ}$ designs for v_2.

12.8 For the nested-design results in Table 12.3, provide the normal equations for:

(a) v_1 in the $I^{\bullet}:p^{\bullet}$ design with $n'_i = 6$;

(b) v_2 in the $I^{\bullet}:p^{\bullet}$ design with $n'_i = 8$;

(c) v_1 in the $I^{\circ}:p^{\bullet}$ design with $n'_i = 6$; and

(d) v_2 in the $I^{\circ}:p^{\bullet}$ design with $n'_i = 8$.

12.9* For the synthetic data, Table 12.3 reports that $\sigma_{12}(\mathcal{E}) = .0660$ for the $p^{\bullet} \times I^{\circ}$ design with $n'_i = 6$. This result was verified at the end of Section 12.2.5 using normal equations to obtain the bs for the first variable. Verify that $\sigma_{12}(\mathcal{E}) = .0660$ based on the bs for the second variable obtained using Equation 12.30.

12.10 For the two-variable and single-variable predictions of universe difference scores for the $p^{\bullet} \times I^{\bullet}$ design discussed in Section 12.3, show that R^2 is the same when $\rho_1 = \rho_2$ and $S_1 = S_2$.

12.11* For the synthetic data example in Section 12.3 (see also Table 9.3 on page 292) determine the two-variable and single-variable predictions of universe difference scores for the $I^{\circ}:p^{\bullet}$ design assuming $n'_i = 6$ for both v_1 and v_2. Also, determine R^2 for both predictions.

12.12* Consider the Rajaratnam et al. (1965) synthetic data in Table 9.1 that are discussed in Section 9.1.

(a) For the composite based on w weights proportional to the numbers of items for the three variables, the regressed score prediction equation for the composite is

$$\hat{\mu}_{pC} = 0.2331 + 0.2446\,\overline{X}_{p1} + 0.4608\,\overline{X}_{p2} + 0.2433\,\overline{X}_{p3}.$$

What proportion of the variance in true composite scores is explained by this regression? What is a 90% tolerance interval for the first person?

(b) Suppose it were sensible to consider the three levels of v separately. The regressed score estimates of the universe scores (as raw scores, not standard scores) for the three levels are

$$
\begin{aligned}
\hat{\mu}_{p1} &= .0501 + .7487\,\overline{X}_{p1} + .1202\,\overline{X}_{p2} + .0057\,\overline{X}_{p3} \\
\hat{\mu}_{p2} &= .1339 + .1127\,\overline{X}_{p1} + .8439\,\overline{X}_{p2} + .0441\,\overline{X}_{p3} \\
\hat{\mu}_{p3} &= .6144 + .0043\,\overline{X}_{p1} + .0353\,\overline{X}_{p2} + .8794\,\overline{X}_{p3}.
\end{aligned}
$$

What are the three values of R_v^2 and $R_{vv'}$?

(c) Suppose the raw scores for the three levels were converted to so-called T scores (not to be confused with true scores T) according to the usual formula: $\mathrm{T}_{pv} = 50 + 10\,Z_{pv}$. What is $\mathcal{RV}$?

12.13 Derive the results in Equations 12.96 to 12.98 for $\rho_C(\hat{\mu}_p, \hat{\mu}_p')$ for a predicted composite with two randomly parallel instances of the measurement procedure.

Appendix A
Degrees of Freedom and Sums of Squares for Selected Balanced Designs

See Section 3.2 for notational conventions employed in the following tables as well as general formulas for degrees of freedom and sums of squares for any complete balanced design.

A.1 Single-Facet Designs

TABLE A.1. Degrees of Freedom and Sums of Squares for the $p \times i$ Design

α	$df(\alpha)$	$T(\alpha)$	$SS(\alpha)$
p	$n_p - 1$	$n_i \sum \overline{X}_p^2$	$T(p) - T(\mu)$
i	$n_i - 1$	$n_p \sum \overline{X}_i^2$	$T(i) - T(\mu)$
pi	$(n_p - 1)(n_i - 1)$	$\sum \sum X_{pi}^2$	$T(pi) - T(p) - T(i) + T(\mu)$
Mean(μ)		$n_p n_i \overline{X}^2$	
Total	$n_p n_i - 1$		$T(pi) - T(\mu)$

TABLE A.2. Degrees of Freedom and Sums of Squares for the $i\!:\!p$ Design

α	$df(\alpha)$	$T(\alpha)$	$SS(\alpha)$
p	$n_p - 1$	$n_i \sum \overline{X}_p^2$	$T(p) - T(\mu)$
$i\!:\!p$	$n_p(n_i - 1)$	$\sum \sum X_{pi}^2$	$T(i\!:\!p) - T(p)$
Mean(μ)		$n_p n_i \overline{X}^2$	
Total	$n_p n_i - 1$		$T(i\!:\!p) - T(\mu)$

A.2   Selected Two-Facet Designs

For each of these designs $T(\mu) = \sum\sum\sum X_{pih}^2$, and the total number of degrees of freedom is $n_p n_i n_h - 1$.

TABLE A.3. Degrees of Freedom and Sums of Squares for the $p \times i \times h$ Design

α	$df(\alpha)$	$T(\alpha)$	$SS(\alpha)$
p	$n_p - 1$	$n_i n_h \sum \overline{X}_p^2$	$T(p) - T(\mu)$
i	$n_i - 1$	$n_p n_h \sum \overline{X}_i^2$	$T(i) - T(\mu)$
h	$n_h - 1$	$n_p n_i \sum \overline{X}_h^2$	$T(h) - T(\mu)$
pi	$(n_p - 1)(n_i - 1)$	$n_h \sum\sum \overline{X}_{pi}^2$	$T(pi) - T(p) - T(i) + T(\mu)$
ph	$(n_p - 1)(n_h - 1)$	$n_i \sum\sum \overline{X}_{ph}^2$	$T(ph) - T(p) - T(h) + T(\mu)$
ih	$(n_i - 1)(n_h - 1)$	$n_p \sum\sum \overline{X}_{ih}^2$	$T(ih) - T(i) - T(h) + T(\mu)$
pih	$(n_p - 1)(n_i - 1)$ $\times (n_h - 1)$	$\sum\sum\sum X_{pih}^2$	$T(pih) - T(pi) - T(ph) - T(ih)$ $+ T(p) + T(i) + T(h) - T(\mu)$

TABLE A.4. Degrees of Freedom and Sums of Squares for the $p \times (i\!:\!h)$ Design

α	$df(\alpha)$	$T(\alpha)$	$SS(\alpha)$
p	$n_p - 1$	$n_i n_h \sum \overline{X}_p^2$	$T(p) - T(\mu)$
h	$n_h - 1$	$n_p n_i \sum \overline{X}_h^2$	$T(h) - T(\mu)$
$i\!:\!h$	$n_h(n_i - 1)$	$n_p \sum\sum \overline{X}_{i:h}^2$	$T(i\!:\!h) - T(h)$
ph	$(n_p - 1)(n_h - 1)$	$n_i \sum\sum \overline{X}_{ph}^2$	$T(ph) - T(p) - T(h) + T(\mu)$
$pi\!:\!h$	$n_h(n_p - 1)(n_i - 1)$	$\sum\sum\sum X_{pih}^2$	$T(pi\!:\!h) - T(ph)$ $- T(i\!:\!h) + T(h)$

TABLE A.5. Degrees of Freedom and Sums of Squares for the $(i:p) \times h$ Design

α	$df(\alpha)$	$T(\alpha)$	$SS(\alpha)$
p	$n_p - 1$	$n_i n_h \sum \overline{X}_p^2$	$T(p) - T(\mu)$
h	$n_h - 1$	$n_p n_i \sum \overline{X}_h^2$	$T(h) - T(\mu)$
$i{:}p$	$n_p(n_i - 1)$	$n_h \sum\sum \overline{X}_{i:p}^2$	$T(i{:}p) - T(p)$
ph	$(n_p - 1)(n_h - 1)$	$n_i \sum\sum \overline{X}_{ph}^2$	$T(ph) - T(p) - T(h) + T(\mu)$
$ih{:}p$	$n_p(n_i - 1)(n_h - 1)$	$\sum\sum\sum X_{pih}^2$	$T(ih{:}p) - T(ph)$ $- T(i{:}p) + T(p)$

TABLE A.6. Degrees of Freedom and Sums of Squares for the $i:(p \times h)$ Design

α	$df(\alpha)$	$T(\alpha)$	$SS(\alpha)$
p	$n_p - 1$	$n_i n_h \sum \overline{X}_p^2$	$T(p) - T(\mu)$
h	$n_h - 1$	$n_p n_i \sum \overline{X}_h^2$	$T(h) - T(\mu)$
ph	$(n_p - 1)(n_h - 1)$	$n_i \sum\sum \overline{X}_{ph}^2$	$T(ph) - T(p) - T(h) + T(\mu)$
$i{:}ph$	$n_p n_h(n_i - 1)$	$\sum\sum\sum X_{pih}^2$	$T(i{:}ph) - T(ph)$

TABLE A.7. Degrees of Freedom and Sums of Squares for the $(i \times h):p$ Design

α	$df(\alpha)$	$T(\alpha)$	$SS(\alpha)$
p	$n_p - 1$	$n_i n_h \sum \overline{X}_p^2$	$T(p) - T(\mu)$
$i{:}p$	$n_p(n_i - 1)$	$n_h \sum\sum \overline{X}_{i:p}^2$	$T(i{:}p) - T(p)$
$h{:}p$	$n_p(n_h - 1)$	$n_i \sum\sum \overline{X}_{h:p}^2$	$T(h{:}p) - T(h)$
$ih{:}p$	$n_p(n_i - 1)(n_h - 1)$	$\sum\sum\sum X_{pih}^2$	$T(ih{:}p) - T(i{:}p)$ $- T(h{:}p) + T(p)$

TABLE A.8. Degrees of Freedom and Sums of Squares for the $i:h:p$ Design

α	$df(\alpha)$	$T(\alpha)$	$SS(\alpha)$
p	$n_p - 1$	$n_i n_h \sum \overline{X}_p^2$	$T(p) - T(\mu)$
$h{:}p$	$n_p(n_h - 1)$	$n_i \sum\sum \overline{X}_{h:p}^2$	$T(h{:}p) - T(h)$
$i{:}h{:}p$	$n_p n_h(n_i - 1)$	$\sum\sum\sum X_{pih}^2$	$T(i{:}h{:}p) - T(h{:}p)$

Appendix B

Expected Mean Squares and Estimators of Random Effects Variance Components for Selected Balanced Designs

See Section 3.4 for notational conventions employed in the following tables as well as general formulas for expected mean squares and estimators of variance components for any complete balanced design.

B.1 Single-Facet Designs

TABLE B.1. Expected Mean Squares and $\hat{\sigma}^2(\alpha)$ for the $p \times i$ Design

α	$EMS(\alpha)$	$\hat{\sigma}^2(\alpha)$
p	$\sigma^2(pi) + n_i\sigma^2(p)$	$[MS(p) - MS(pi)]/n_i$
i	$\sigma^2(pi) + n_p\sigma^2(i)$	$[MS(i) - MS(pi)]/n_p$
pi	$\sigma^2(pi)$	$MS(pi)$

TABLE B.2. Expected Mean Squares and $\hat{\sigma}^2(\alpha)$ for the $i\!:\!p$ Design

α	$EMS(\alpha)$	$\hat{\sigma}^2(\alpha)$
p	$\sigma^2(i\!:\!p) + n_i\sigma^2(p)$	$[MS(p) - MS(i\!:\!p)]/n_i$
$i\!:\!p$	$\sigma^2(i\!:\!p)$	$MS(i\!:\!p)$

B.2 Selected Two-Facet Designs

TABLE B.3. Expected Mean Squares and $\hat{\sigma}^2(\alpha)$ for the $p \times i \times h$ Design

α	$EMS(\alpha)$
p	$\sigma^2(pih) + n_i\sigma^2(ph) + n_h\sigma^2(pi) + n_in_h\sigma^2(p)$
i	$\sigma^2(pih) + n_p\sigma^2(ih) + n_h\sigma^2(pi) + n_pn_h\sigma^2(i)$
h	$\sigma^2(pih) + n_p\sigma^2(ih) + n_i\sigma^2(ph) + n_pn_i\sigma^2(h)$
pi	$\sigma^2(pih) + n_h\sigma^2(pi)$
ph	$\sigma^2(pih) + n_i\sigma^2(ph)$
ih	$\sigma^2(pih) + n_p\sigma^2(ih)$
pih	$\sigma^2(pih)$

α	$\hat{\sigma}^2(\alpha)$
p	$[MS(p) - MS(pi) - MS(ph) + MS(pih)]/n_in_h$
i	$[MS(i) - MS(pi) - MS(ih) + MS(pih)]/n_pn_h$
h	$[MS(h) - MS(ph) - MS(ih) + MS(pih)]/n_pn_i$
pi	$[MS(pi) - MS(pih)]/n_h$
ph	$[MS(ph) - MS(pih)]/n_i$
ih	$[MS(ih) - MS(pih)]/n_p$
pih	$MS(pih)$

TABLE B.4. Expected Mean Squares and $\hat{\sigma}^2(\alpha)$ for the $p \times (i{:}h)$ Design

α	$EMS(\alpha)$	$\hat{\sigma}^2(\alpha)$
p	$\sigma^2(pi{:}h) + n_i\sigma^2(ph) + n_in_h\sigma^2(p)$	$[MS(p) - MS(ph)]/n_in_h$
h	$\sigma^2(pi{:}h) + n_i\sigma^2(ph)$	$[MS(h) - MS(i{:}h)$
	$\quad + n_p\sigma^2(i{:}h) + n_pn_i\sigma^2(h)$	$\quad - MS(ph) + MS(pi{:}h)]/n_pn_i$
$i{:}h$	$\sigma^2(pi{:}h) + n_p\sigma^2(i{:}h)$	$[MS(i{:}h) - MS(pi{:}h)]/n_p$
ph	$\sigma^2(pi{:}h) + n_i\sigma^2(ph)$	$[MS(ph) - MS(pi{:}h)]/n_i$
$pi{:}h$	$\sigma^2(pi{:}h)$	$MS(pi{:}h)$

TABLE B.5. Expected Mean Squares and $\hat{\sigma}^2(\alpha)$ for the $(i{:}p) \times h$ Design

α	$EMS(\alpha)$	$\hat{\sigma}^2(\alpha)$
p	$\sigma^2(ih{:}p) + n_i\sigma^2(ph)$ $+ n_h\sigma^2(i{:}p) + n_in_h\sigma^2(p)$	$[MS(p) - MS(i{:}p)$ $- MS(ph) + MS(ih{:}p)]/n_in_h$
h	$\sigma^2(ih{:}p) + n_i\sigma^2(ph) + n_pn_i\sigma^2(h)$	$[MS(h) - MS(ph)]/n_pn_i$
$i{:}p$	$\sigma^2(ih{:}p) + n_h\sigma^2(i{:}p)$	$[MS(i{:}p) - MS(ih{:}p)]/n_h$
ph	$\sigma^2(ih{:}p) + n_i\sigma^2(ph)$	$[MS(ph) - MS(ih{:}p)]/n_i$
$ih{:}p$	$\sigma^2(ih{:}p)$	$MS(ih{:}p)$

TABLE B.6. Expected Mean Squares and $\hat{\sigma}^2(\alpha)$ for the $i{:}(p \times h)$ Design

α	$EMS(\alpha)$	$\hat{\sigma}^2(\alpha)$
p	$\sigma^2(i{:}ph) + n_i\sigma^2(ph) + n_in_h\sigma^2(p)$	$[MS(p) - MS(ph)]/n_in_h$
h	$\sigma^2(i{:}ph) + n_i\sigma^2(ph) + n_pn_i\sigma^2(h)$	$[MS(h) - MS(ph)]/n_pn_i$
ph	$\sigma^2(i{:}ph) + n_i\sigma^2(ph)$	$[MS(ph) - MS(i{:}ph)]/n_i$
$i{:}ph$	$\sigma^2(i{:}ph)$	$MS(i{:}ph)$

TABLE B.7. Expected Mean Squares and $\hat{\sigma}^2(\alpha)$ for the $(i \times h){:}p$ Design

α	$EMS(\alpha)$	$\hat{\sigma}^2(\alpha)$
p	$\sigma^2(ih{:}p) + n_i\sigma^2(h{:}p)$ $+ n_h\sigma^2(i{:}p) + n_in_h\sigma^2(p)$	$[MS(p) - MS(i{:}p)$ $- MS(h{:}p) + MS(ih{:}p)]/n_in_h$
$i{:}p$	$\sigma^2(ih{:}p) + n_h\sigma^2(i{:}p)$	$[MS(i{:}p) - MS(ih{:}p)]/n_h$
$h{:}p$	$\sigma^2(ih{:}p) + n_i\sigma^2(h{:}p)$	$[MS(h{:}p) - MS(ih{:}p)]/n_i$
$ih{:}p$	$\sigma^2(ih{:}p)$	$MS(ih{:}p)$

TABLE B.8. Expected Mean Squares and $\hat{\sigma}^2(\alpha)$ for the $i{:}h{:}p$ Design

α	$EMS(\alpha)$	$\hat{\sigma}^2(\alpha)$
p	$\sigma^2(ih{:}p) + n_i\sigma^2(h{:}p) + n_in_h\sigma^2(p)$	$[MS(p) - MS(h{:}p)]/n_in_h$
$h{:}p$	$\sigma^2(ih{:}p) + n_i\sigma^2(h{:}p)$	$[MS(h{:}p) - MS(i{:}h{:}p)]/n_i$
$i{:}h{:}p$	$\sigma^2(ih{:}p)$	$MS(i{:}h{:}p)$

Appendix C

Matrix Procedures for Estimating Variance Components and Their Variability

Parts of the following discussion are based on Searle (1971, pp. 406, 415–417) and Searle et al. (1992, pp. 128–129, 137–138).

C.1 Estimated Variance Components

Let $j = 1, 2, \ldots, k$ designate the effects in a design (main effects and interaction effects, but not μ). Also, let

$$\mathbf{P} = \begin{cases} k \times k \text{ upper-triangular matrix of coefficients of the variance components in the } EMS \text{ equations for the model} \\ \text{(random, mixed, or sampling from a finite universe),} \end{cases}$$

and

$$\mathbf{a} = k \times 1 \text{ column vector of mean squares for the design.}$$

Then,

$$\hat{\sigma}^2 = \mathbf{P}^{-1}\mathbf{a} \tag{C.1}$$

is a $k \times 1$ column vector whose elements are unbiased estimates of the variance components. For a random model, these estimates are identical to those resulting from using the algorithm in Section 3.4.4. For any model, the estimates in Equation C.1 are the same as those obtained using the algorithm in Section 3.4.4 in conjunction with Equation 3.34. For a mixed model, the "variances" associated with fixed effects are called quadratic forms in traditional statistical literature.

C.2   Variability of Estimated Variance Components

Estimated variance components are themselves subject to sampling variability. Furthermore, two *estimated* variance components are generally *not* uncorrelated, unless there are no common mean squares used in estimating the two variance components.

C.2.1   Single Conditions

Assuming a multivariate normal distribution for the score effects, the symmetric variance-covariance matrix associated with the estimated variance components in Equation C.1 is

$$\hat{\mathbf{V}} = \mathbf{P}^{-1}\mathbf{D}_1(\mathbf{P}^{-1})', \tag{C.2}$$

where

$$\mathbf{D}_1 = \begin{cases} k \times k \text{ diagonal matrix containing the} \\ \text{diagonal elements } 2(MS_j)^2/(df_j + 2), \end{cases} \tag{C.3}$$

and

$$(\mathbf{P}^{-1})' \quad \text{means the transpose of} \quad \mathbf{P}^{-1}.$$

The diagonal elements of $\hat{\mathbf{V}}$ are unbiased estimators of the variances of the estimated variance components, under the normality assumption. Their square roots have been denoted previously as $\hat{\sigma}[\hat{\sigma}^2(\alpha)]$ or $\hat{\sigma}[\hat{\sigma}^2(\alpha|M)]$.

C.2.2   Mean Scores

The symmetric matrix $\hat{\mathbf{V}}$ provides the estimated variances and covariances associated with the $\hat{\sigma}^2(\alpha|M)$, which are estimated variance components for single conditions. There is a corresponding variance-covariance matrix associated with the $\hat{\sigma}^2(\bar{\alpha}|M')$, that is, with the estimated variance components for means over n' conditions from facets in the universe of generalization (see Equation 5.1). If $N' = N$ for each facet and the G and D study designs have the same structure, then this variance-covariance matrix is

$$\widehat{\mathbf{W}} = \mathbf{D}_2\hat{\mathbf{V}}\mathbf{D}_2, \tag{C.4}$$

where

$$\mathbf{D}_2 = \begin{cases} k \times k \text{ diagonal matrix with elements } C(\bar{\alpha}|\tau)/d(\bar{\alpha}|\tau), \text{ where } d(\bar{\alpha}|\tau) \\ \text{and } C(\bar{\alpha}|\tau) \text{ are defined by Equations 5.2 and 5.3, respectively.} \end{cases}$$

The square roots of the diagonal elements in $\widehat{\mathbf{W}}$ are the estimated standard errors of the variance components associated with mean scores. They have been denoted previously as $\hat{\sigma}[\hat{\sigma}^2(\bar{\alpha})]$ and $\hat{\sigma}[\hat{\sigma}^2(\bar{\alpha}|M)]$.

Using the symmetric matrix $\widehat{\mathbf{W}}$, the estimated standard error of $\hat{\sigma}^2(\Delta)$ is

$$\hat{\sigma}[\hat{\sigma}^2(\Delta)] = \begin{cases} \text{square root of the sum of the elements in } \widehat{\mathbf{W}} \\ \text{excluding those rows and columns of } \widehat{\mathbf{W}} \text{ as-} \\ \text{sociated with effects that do not enter } \Delta. \end{cases} \quad (\text{C.5})$$

As indicated below, the standard errors of $\hat{\sigma}^2(\delta)$ and $Est[ES^2(p)]$ can be estimated in a similar manner.

C.3   Example

Table C.1 provides the vectors and matrices in Equations C.1 and C.2 for the $p \times (r\!:\!t)$ design using Synthetic Data Set No. 4 in Table 3.2 on page 73. Row and column $j = 1, 2, 3, 4$, and 5 correspond to $\alpha = p$, t, $r\!:\!t$, pt, and $pr\!:\!t$, respectively. The square roots of the diagonal elements of $\widehat{\mathbf{V}}$ are the estimated standard errors of the G study estimated random effects variance components.

Table C.1 also provides the $\mathbf{D}_2$ and $\widehat{\mathbf{W}}$ matrices in Equation C.4 assuming $n'_t = 3$ and $n'_r = 4$. The square roots of the diagonal elements in $\widehat{\mathbf{W}}$ are the estimated standard errors denoted previously as $\hat{\sigma}[\hat{\sigma}^2(\bar{\alpha})]$. For example,

$$\hat{\sigma}[\hat{\sigma}^2(pR)] = \sqrt{.0158} = .126.$$

Since $\hat{\sigma}^2(\Delta)$ is the sum of all the $\hat{\sigma}^2(\bar{\alpha})$ except $\hat{\sigma}^2(p)$, the estimated standard error of $\hat{\sigma}^2(\Delta)$ is the square root of the sum of all entries in $\widehat{\mathbf{W}}$ except those in the first row and column:

$$\hat{\sigma}[\hat{\sigma}^2(\Delta)] = \sqrt{.0322} = .179.$$

Since $\hat{\sigma}^2(\delta) = \hat{\sigma}^2(pR) + \hat{\sigma}^2(pI\!:\!R)$, the estimated standard error of $\hat{\sigma}^2(\delta)$ is the square root of the sum of all entries in the fourth and fifth rows and columns of $\widehat{\mathbf{W}}$:

$$\hat{\sigma}[\hat{\sigma}^2(\delta)] = \sqrt{.0149} = .122.$$

Since $Est[ES^2(p)]$ includes all the $\hat{\sigma}^2(\bar{\alpha})$ except $\hat{\sigma}^2(T)$ and $\hat{\sigma}^2(R\!:\!T)$, the estimated standard error of $Est[ES^2(p)]$ is the square root of the sum of all entries in $\widehat{\mathbf{W}}$ except those in the second and third rows and columns:

$$\hat{\sigma}[Est[ES^2(p)]] = \sqrt{.1340} = .366.$$

C.4   Complex Cases

Equation C.4 is appropriate when $N' = N$ for each facet. If this condition is not fulfilled, then there is a transformation matrix $\mathbf{T}$ that converts the

TABLE C.1. Matrices for Obtaining Estimated Variance Components and Standard Errors for Synthetic Data Set No. 4

$$\mathbf{P} = \begin{bmatrix} 12 & 0 & 0 & 4 & 1 \\ & 40 & 10 & 4 & 1 \\ & & 10 & 0 & 1 \\ & & & 4 & 1 \\ & & & & 1 \end{bmatrix}$$

$$\mathbf{P}^{-1} = \begin{bmatrix} .083 & .0000 & .0000 & -.0833 & .0000 \\ & .0250 & -.0250 & -.0250 & .0250 \\ & & .1000 & .0000 & -.1000 \\ & & & .0250 & -.0250 \\ & & & & 1.0000 \end{bmatrix}$$

$$\mathbf{a} = \begin{bmatrix} 10.2963 \\ 24.1000 \\ 8.8556 \\ 4.6185 \\ 2.3802 \end{bmatrix} \qquad \hat{\sigma}^2 = \begin{bmatrix} .4731 \\ .3252 \\ .6475 \\ .5596 \\ 2.3802 \end{bmatrix}$$

$$\mathbf{D}_1 = \begin{bmatrix} 19.2752 & & & & \\ & 290.4050 & & & \\ & & 14.2583 & & \\ & & & 2.1331 & \\ & & & & .1365 \end{bmatrix}$$

$$\hat{\mathbf{V}} = \begin{bmatrix} .1487 & .0044 & .0000 & -.0444 & .0000 \\ .0044 & .1918 & -.0360 & -.0142 & .0034 \\ .0000 & -.0360 & .1439 & .0034 & -.0137 \\ -.0444 & -.0142 & .0034 & .1418 & -.0341 \\ .0000 & .0034 & -.0137 & .0341 & .1365 \end{bmatrix}$$

$$\mathbf{D}_2 = \begin{bmatrix} 1.0000 & & & & \\ & .3333 & & & \\ & & .0833 & & \\ & & & .3333 & \\ & & & & .0833 \end{bmatrix}.$$

$$\widehat{\mathbf{W}} = \begin{bmatrix} .1487 & .0015 & .0000 & -.0148 & .0000 \\ .0015 & .0213 & -.0010 & -.0016 & .0001 \\ .0000 & -.0010 & .0010 & .0001 & -.0001 \\ -.0148 & -.0016 & .0001 & .0158 & -.0009 \\ .0000 & .0001 & -.0001 & -.0009 & .0009 \end{bmatrix}.$$

$\hat{\sigma}^2(\alpha|M)$ associated with facets of size N to the $\hat{\sigma}^2(\alpha|M')$ associated with facets of size N'. For example, to convert the $\hat{\sigma}^2(\alpha)$ for Synthetic Data Set No. 4 to the $\hat{\sigma}^2(\alpha|T)$ in Table 5.2, the transformation matrix is

$$
\mathbf{T} = \begin{bmatrix} 1 & 0 & 0 & .33 & 0 \\ & 1 & 0 & 0 & 0 \\ & & 1 & 0 & 0 \\ & & & 1 & 0 \\ & & & & 1 \end{bmatrix},
$$

and the $\hat{\sigma}^2(\alpha|M')$ are the elements of the vector $\mathbf{T}\hat{\boldsymbol{\sigma}}^2$. Given $\mathbf{T}$, the variance-covariance matrix associated with the $\hat{\sigma}^2(\bar{\alpha}|M')$ is

$$
\widehat{\mathbf{W}} = \mathbf{D}_2(\mathbf{T}\widehat{\mathbf{V}}\mathbf{T}')\mathbf{D}_2. \tag{C.6}
$$

When a fixed facet occurs as a primary index in $\bar{\alpha}$, all entries in the corresponding row and column of $\widehat{\mathbf{W}}$ will be zero.

Also, Equation C.4 assumes that the G and D study designs have the same structure. Consider, however, the possibility that the G study design is $p \times (r{:}t)$, the D study design is $R{:}T{:}p$, and both models are random. In this case, the transformation matrix is:

$$
\mathbf{T} = \begin{bmatrix} 1 & 0 & 0 & 0 & 0 \\ & 1 & 0 & 1 & 0 \\ & & 1 & 0 & 1 \end{bmatrix},
$$

where the sequence of the columns is $\alpha = p,\, t,\, r{:}t,\, pt,\, pr{:}t$, and the sequence of the rows is $\alpha = p,\, t{:}p,\, r{:}t{:}p$. For this transformation matrix, both $\mathbf{D}_2$ and $\widehat{\mathbf{W}}$ are 3×3 matrices.

Appendix D
Table for Simplified Use of Satterthwaite's Procedure

Section 6.2.2 describes Satterthwaite's (1941, 1946) procedure for placing confidence intervals on variance components. A simplified version of that procedure employs the multiplicative factors:

$$\text{multiplier for lower limit} = \frac{2r^2}{\chi_U^2(2r^2)}$$

and

$$\text{multiplier for upper limit} = \frac{2r^2}{\chi_L^2(2r^2)}$$

in Equations 6.20 and 6.21, respectively, where $\chi_U^2(\nu)$ and $\chi_L^2(\nu)$ are the *lower* $U = (1+\gamma)/2$ and $L = (1-\gamma)/2$ percentile points of the chi-squared distribution with $\nu = 2r^2$ effective degrees of freedom. Table D.1 tabulates these multiplicative factors for 66.67, 80, 90, and 95% confidence intervals.[1]

---

[1] The values reported here are a subset of those provided in Brennan (1992a).

TABLE D.1. Multiplicative Factors for $100(\gamma)\%$ Confidence Intervals on Variance Components

ν	r	66.67% Lower	66.67% Upper	80% Lower	80% Upper	90% Lower	90% Upper	95% Lower	95% Upper
8	2.000	0.6855	1.8801	0.5987	2.2927	0.5157	2.9282	0.4560	3.6716
8.5	2.062	0.6915	1.8356	0.6060	2.2217	0.5240	2.8100	0.4647	3.4903
9	2.121	0.6970	1.7960	0.6128	2.1595	0.5318	2.7076	0.4729	3.3350
9.5	2.179	0.7022	1.7612	0.6193	2.1046	0.5392	2.6181	0.4806	3.2004
10	2.236	0.7071	1.7299	0.6254	2.0558	0.5461	2.5392	0.4879	3.0827
10.5	2.291	0.7118	1.7018	0.6312	2.0120	0.5527	2.4678	0.4950	2.9763
11	2.345	0.7162	1.6763	0.6366	1.9726	0.5589	2.4048	0.5016	2.8837
11.5	2.398	0.7203	1.6531	0.6418	1.9368	0.5649	2.3481	0.5079	2.8008
12	2.449	0.7243	1.6318	0.6469	1.9037	0.5706	2.2967	0.5139	2.7262
12.5	2.500	0.7281	1.6122	0.6516	1.8738	0.5760	2.2499	0.5198	2.6586
13	2.550	0.7317	1.5942	0.6561	1.8463	0.5812	2.2071	0.5253	2.5971
13.5	2.598	0.7352	1.5775	0.6604	1.8210	0.5862	2.1677	0.5307	2.5409
14	2.646	0.7385	1.5619	0.6646	1.7975	0.5909	2.1315	0.5358	2.4893
14.5	2.693	0.7417	1.5474	0.6685	1.7756	0.5955	2.0980	0.5408	2.4402
15	2.739	0.7447	1.5339	0.6723	1.7553	0.6000	2.0661	0.5455	2.3961
15.5	2.784	0.7477	1.5212	0.6760	1.7362	0.6042	2.0371	0.5500	2.3553
16	2.828	0.7506	1.5093	0.6795	1.7184	0.6083	2.0100	0.5544	2.3173
16.5	2.872	0.7533	1.4979	0.6829	1.7017	0.6123	1.9846	0.5587	2.2818
17	2.915	0.7559	1.4873	0.6863	1.6860	0.6161	1.9608	0.5629	2.2487
17.5	2.958	0.7585	1.4773	0.6895	1.6711	0.6198	1.9384	0.5669	2.2176
18	3.000	0.7609	1.4678	0.6925	1.6568	0.6233	1.9173	0.5707	2.1884
18.5	3.041	0.7633	1.4588	0.6955	1.6435	0.6269	1.8974	0.5745	2.1599
19	3.082	0.7656	1.4503	0.6984	1.6309	0.6302	1.8786	0.5781	2.1339
19.5	3.122	0.7678	1.4421	0.7012	1.6189	0.6335	1.8608	0.5817	2.1094
20	3.162	0.7699	1.4344	0.7039	1.6075	0.6366	1.8434	0.5851	2.0861
21	3.240	0.7741	1.4199	0.7090	1.5863	0.6426	1.8120	0.5917	2.0432
22	3.317	0.7779	1.4067	0.7139	1.5669	0.6483	1.7834	0.5980	2.0044
23	3.391	0.7816	1.3946	0.7185	1.5492	0.6538	1.7574	0.6038	1.9683
24	3.464	0.7851	1.3834	0.7229	1.5329	0.6590	1.7335	0.6094	1.9359
25	3.536	0.7884	1.3730	0.7271	1.5178	0.6638	1.7115	0.6149	1.9063
26	3.606	0.7915	1.3633	0.7310	1.5037	0.6685	1.6908	0.6200	1.8790
27	3.674	0.7945	1.3543	0.7348	1.4907	0.6730	1.6719	0.6249	1.8537
28	3.742	0.7974	1.3458	0.7384	1.4785	0.6772	1.6544	0.6296	1.8296
29	3.808	0.8002	1.3379	0.7418	1.4671	0.6813	1.6380	0.6340	1.8078
30	3.873	0.8028	1.3305	0.7451	1.4565	0.6852	1.6226	0.6384	1.7874
31	3.937	0.8053	1.3235	0.7483	1.4465	0.6890	1.6082	0.6425	1.7683
32	4.000	0.8077	1.3169	0.7514	1.4370	0.6926	1.5945	0.6465	1.7499
33	4.062	0.8100	1.3106	0.7543	1.4281	0.6961	1.5817	0.6504	1.7331
34	4.123	0.8123	1.3047	0.7571	1.4197	0.6994	1.5696	0.6541	1.7172
35	4.183	0.8144	1.2991	0.7598	1.4117	0.7027	1.5582	0.6576	1.7022
36	4.243	0.8165	1.2938	0.7624	1.4039	0.7058	1.5474	0.6611	1.6880
37	4.301	0.8185	1.2887	0.7650	1.3967	0.7088	1.5372	0.6645	1.6742
38	4.359	0.8204	1.2839	0.7674	1.3898	0.7117	1.5275	0.6677	1.6614
39	4.416	0.8222	1.2793	0.7698	1.3833	0.7145	1.5180	0.6709	1.6493
40	4.472	0.8242	1.2745	0.7723	1.3767	0.7174	1.5090	0.6740	1.6378
41	4.528	0.8260	1.2703	0.7745	1.3707	0.7201	1.5005	0.6770	1.6267
42	4.583	0.8277	1.2662	0.7766	1.3650	0.7227	1.4924	0.6798	1.6161
43	4.637	0.8293	1.2623	0.7787	1.3595	0.7251	1.4846	0.6826	1.6059
44	4.690	0.8309	1.2586	0.7807	1.3542	0.7276	1.4772	0.6853	1.5962
45	4.743	0.8324	1.2550	0.7827	1.3491	0.7299	1.4700	0.6879	1.5869

TABLE D.2. Multiplicative Factors for $100(\gamma)\%$ Confidence Intervals on Variance Components (continued)

ν	r	66.67% Lower	66.67% Upper	80% Lower	80% Upper	90% Lower	90% Upper	95% Lower	95% Upper
46	4.796	0.8339	1.2516	0.7845	1.3442	0.7322	1.4632	0.6905	1.5780
47	4.848	0.8354	1.2483	0.7864	1.3396	0.7344	1.4566	0.6930	1.5694
48	4.899	0.8368	1.2451	0.7882	1.3350	0.7366	1.4503	0.6954	1.5612
49	4.950	0.8382	1.2420	0.7899	1.3307	0.7387	1.4442	0.6978	1.5533
50	5.000	0.8395	1.2390	0.7916	1.3265	0.7407	1.4383	0.7001	1.5457
51	5.050	0.8408	1.2361	0.7933	1.3224	0.7427	1.4326	0.7023	1.5383
52	5.099	0.8421	1.2333	0.7949	1.3185	0.7447	1.4271	0.7045	1.5312
53	5.148	0.8433	1.2306	0.7965	1.3147	0.7466	1.4219	0.7066	1.5244
54	5.196	0.8445	1.2280	0.7980	1.3111	0.7485	1.4167	0.7087	1.5178
55	5.244	0.8457	1.2255	0.7995	1.3075	0.7503	1.4118	0.7107	1.5114
56	5.292	0.8469	1.2230	0.8010	1.3041	0.7520	1.4070	0.7127	1.5052
57	5.339	0.8480	1.2207	0.8025	1.3008	0.7538	1.4024	0.7147	1.4993
58	5.385	0.8491	1.2184	0.8039	1.2975	0.7555	1.3979	0.7166	1.4935
59	5.431	0.8502	1.2161	0.8052	1.2944	0.7571	1.3935	0.7185	1.4879
60	5.477	0.8512	1.2139	0.8066	1.2913	0.7587	1.3893	0.7203	1.4824
61	5.523	0.8522	1.2118	0.8079	1.2884	0.7603	1.3852	0.7221	1.4772
62	5.568	0.8532	1.2098	0.8092	1.2855	0.7619	1.3812	0.7238	1.4720
63	5.612	0.8542	1.2078	0.8104	1.2827	0.7634	1.3773	0.7255	1.4671
64	5.657	0.8552	1.2058	0.8116	1.2800	0.7649	1.3736	0.7272	1.4622
65	5.701	0.8561	1.2039	0.8128	1.2773	0.7664	1.3699	0.7289	1.4575
66	5.745	0.8570	1.2021	0.8140	1.2748	0.7678	1.3663	0.7305	1.4530
67	5.788	0.8579	1.2003	0.8152	1.2722	0.7692	1.3628	0.7321	1.4485
68	5.831	0.8588	1.1985	0.8163	1.2698	0.7706	1.3595	0.7336	1.4442
69	5.874	0.8597	1.1968	0.8174	1.2674	0.7719	1.3562	0.7351	1.4400
70	5.916	0.8606	1.1951	0.8185	1.2651	0.7732	1.3529	0.7366	1.4359
71	5.958	0.8614	1.1935	0.8196	1.2628	0.7745	1.3498	0.7381	1.4319
72	6.000	0.8622	1.1919	0.8206	1.2606	0.7758	1.3468	0.7396	1.4280
73	6.042	0.8630	1.1903	0.8217	1.2584	0.7771	1.3438	0.7410	1.4242
74	6.083	0.8638	1.1888	0.8227	1.2563	0.7783	1.3408	0.7424	1.4205
75	6.124	0.8646	1.1873	0.8237	1.2542	0.7795	1.3380	0.7437	1.4168
76	6.164	0.8654	1.1859	0.8247	1.2522	0.7807	1.3352	0.7451	1.4133
77	6.205	0.8661	1.1844	0.8256	1.2502	0.7819	1.3325	0.7464	1.4098
78	6.245	0.8668	1.1830	0.8266	1.2483	0.7830	1.3298	0.7477	1.4064
79	6.285	0.8676	1.1817	0.8275	1.2464	0.7842	1.3272	0.7490	1.4031
80	6.325	0.8683	1.1803	0.8284	1.2445	0.7853	1.3247	0.7502	1.3999
81	6.364	0.8690	1.1790	0.8293	1.2427	0.7864	1.3222	0.7515	1.3967
82	6.403	0.8697	1.1777	0.8302	1.2409	0.7874	1.3198	0.7527	1.3937
83	6.442	0.8703	1.1765	0.8310	1.2392	0.7885	1.3174	0.7539	1.3906
84	6.481	0.8710	1.1753	0.8319	1.2375	0.7895	1.3150	0.7551	1.3877
85	6.519	0.8717	1.1740	0.8327	1.2358	0.7906	1.3128	0.7563	1.3848
86	6.557	0.8723	1.1729	0.8336	1.2342	0.7916	1.3105	0.7574	1.3819
87	6.595	0.8729	1.1717	0.8344	1.2326	0.7926	1.3083	0.7585	1.3791
88	6.633	0.8736	1.1706	0.8352	1.2310	0.7935	1.3062	0.7596	1.3764
89	6.671	0.8742	1.1694	0.8360	1.2294	0.7945	1.3040	0.7607	1.3737
90	6.708	0.8748	1.1683	0.8367	1.2279	0.7955	1.3020	0.7618	1.3711
95	6.892	0.8777	1.1631	0.8405	1.2207	0.8000	1.2922	0.7670	1.3587
100	7.071	0.8804	1.1584	0.8439	1.2142	0.8043	1.2832	0.7718	1.3474
105	7.246	0.8829	1.1540	0.8472	1.2081	0.8082	1.2750	0.7763	1.3371
110	7.416	0.8853	1.1499	0.8502	1.2025	0.8119	1.2674	0.7806	1.3275
115	7.583	0.8875	1.1462	0.8531	1.1973	0.8154	1.2604	0.7846	1.3187

TABLE D.3. Multiplicative Factors for $100(\gamma)\%$ Confidence Intervals on Variance Components (continued)

		66.67%		80%		90%		95%	
ν	r	Lower	Upper	Lower	Upper	Lower	Upper	Lower	Upper
120	7.746	0.8896	1.1427	0.8558	1.1925	0.8188	1.2539	0.7884	1.3105
130	8.062	0.8934	1.1363	0.8607	1.1838	0.8248	1.2421	0.7953	1.2958
140	8.367	0.8969	1.1307	0.8651	1.1761	0.8303	1.2317	0.8016	1.2828
150	8.660	0.9000	1.1258	0.8692	1.1693	0.8353	1.2226	0.8073	1.2714
160	8.944	0.9028	1.1213	0.8729	1.1632	0.8398	1.2144	0.8125	1.2612
170	9.220	0.9054	1.1173	0.8762	1.1577	0.8440	1.2070	0.8173	1.2520
180	9.487	0.9078	1.1136	0.8793	1.1527	0.8478	1.2002	0.8217	1.2436
190	9.747	0.9101	1.1102	0.8822	1.1481	0.8514	1.1941	0.8259	1.2360
200	10.000	0.9121	1.1071	0.8849	1.1439	0.8547	1.1885	0.8297	1.2291
210	10.247	0.9140	1.1043	0.8874	1.1400	0.8578	1.1833	0.8332	1.2227
220	10.488	0.9158	1.1017	0.8897	1.1364	0.8607	1.1785	0.8366	1.2167
230	10.724	0.9175	1.0992	0.8919	1.1331	0.8634	1.1741	0.8397	1.2113
240	10.954	0.9191	1.0969	0.8940	1.1300	0.8660	1.1700	0.8427	1.2061
250	11.180	0.9206	1.0948	0.8959	1.1271	0.8684	1.1661	0.8455	1.2014
300	12.247	0.9270	1.0858	0.9042	1.1150	0.8788	1.1500	0.8574	1.1815
350	13.229	0.9320	1.0790	0.9107	1.1057	0.8869	1.1377	0.8669	1.1664
400	14.142	0.9361	1.0735	0.9161	1.0983	0.8936	1.1279	0.8747	1.1545
450	15.000	0.9395	1.0690	0.9205	1.0922	0.8992	1.1199	0.8812	1.1447
500	15.811	0.9425	1.0652	0.9243	1.0871	0.9040	1.1132	0.8868	1.1365
550	16.583	0.9450	1.0620	0.9276	1.0828	0.9081	1.1075	0.8916	1.1296
600	17.321	0.9472	1.0592	0.9305	1.0790	0.9117	1.1026	0.8958	1.1236
650	18.028	0.9492	1.0567	0.9331	1.0757	0.9150	1.0982	0.8996	1.1183
700	18.708	0.9509	1.0546	0.9354	1.0728	0.9178	1.0944	0.9030	1.1136
750	19.365	0.9525	1.0526	0.9375	1.0702	0.9205	1.0910	0.9060	1.1095
800	20.000	0.9539	1.0509	0.9393	1.0678	0.9228	1.0879	0.9088	1.1057
850	20.616	0.9552	1.0493	0.9411	1.0657	0.9250	1.0851	0.9113	1.1023
900	21.213	0.9564	1.0478	0.9426	1.0637	0.9270	1.0825	0.9137	1.0992
950	21.794	0.9576	1.0465	0.9441	1.0619	0.9288	1.0802	0.9158	1.0964
1000	22.361	0.9586	1.0452	0.9454	1.0603	0.9305	1.0781	0.9178	1.0938
1500	27.386	0.9659	1.0366	0.9550	1.0488	0.9427	1.0630	0.9321	1.0756
2000	31.623	0.9703	1.0316	0.9609	1.0420	0.9501	1.0542	0.9408	1.0650
2500	35.355	0.9734	1.0281	0.9649	1.0374	0.9551	1.0483	0.9468	1.0578
3000	38.730	0.9756	1.0256	0.9678	1.0341	0.9589	1.0439	0.9513	1.0526
3500	41.833	0.9774	1.0237	0.9702	1.0315	0.9619	1.0406	0.9548	1.0486
4000	44.721	0.9788	1.0221	0.9720	1.0294	0.9643	1.0379	0.9576	1.0453
4500	47.434	0.9800	1.0208	0.9736	1.0277	0.9663	1.0356	0.9599	1.0426
5000	50.000	0.9810	1.0197	0.9749	1.0262	0.9679	1.0338	0.9619	1.0404
5500	52.440	0.9819	1.0188	0.9761	1.0250	0.9694	1.0322	0.9637	1.0385
6000	54.772	0.9827	1.0180	0.9771	1.0239	0.9707	1.0308	0.9652	1.0368
6500	57.009	0.9833	1.0173	0.9780	1.0229	0.9718	1.0295	0.9665	1.0353
7000	59.161	0.9839	1.0166	0.9787	1.0221	0.9728	1.0284	0.9677	1.0340
7500	61.237	0.9845	1.0161	0.9795	1.0213	0.9737	1.0274	0.9688	1.0328
8000	63.246	0.9849	1.0155	0.9801	1.0206	0.9745	1.0266	0.9697	1.0317
8500	65.192	0.9854	1.0151	0.9807	1.0200	0.9753	1.0257	0.9706	1.0308
9000	67.082	0.9858	1.0146	0.9812	1.0194	0.9759	1.0250	0.9714	1.0299
9500	68.920	0.9862	1.0142	0.9817	1.0189	0.9766	1.0243	0.9722	1.0291
10000	70.711	0.9865	1.0139	0.9822	1.0184	0.9772	1.0237	0.9729	1.0283
15000	86.603	0.9890	1.0113	0.9854	1.0150	0.9813	1.0193	0.9778	1.0230
20000	100.000	0.9904	1.0098	0.9873	1.0130	0.9838	1.0167	0.9807	1.0199

Appendix E
Formulas for Selected Unbalanced Random Effects Designs

Section 7.1 provides analogous-ANOVA formulas for estimating random effects variance components for the $i{:}p$ and $p \times (i{:}h)$ designs that are unbalanced with respect to nesting, as well as the $p \times i$ design that is unbalanced with respect to missing data. Here, formulas are provided for the unbalanced $i{:}h{:}p$ and $(p{:}c) \times i$ designs.

Unbalanced $i{:}h{:}p$ Design

For the unbalanced $i{:}h{:}p$ design $n_{h{:}p}$ is the number of levels of h for each p, and $n_{i{:}h{:}p}$ is the number of levels of i for each level of h and p. We also define

$$n_+ = \text{total number of observations in design,}$$

$$n_{h+} = \textstyle\sum_p n_{h{:}p} = \text{total number of levels of } h,$$

$$\widetilde{n}_p = \text{the number of observations for a particular } p, \text{ and}$$

$$\widetilde{n}_{ph} = n_{i{:}h{:}p}.$$

Given these notational conventions, the degrees of freedom, T terms, and sums of squares are given in the following table.

Effect	df	T	SS
p	$n_p - 1$	$\sum_p \tilde{n}_p \overline{X}_p^2$	$T(p) - T(\mu)$
$h{:}p$	$n_{h+} - n_p$	$\sum_p \sum_h \tilde{n}_{ph} \overline{X}_{ph}^2$	$T(h{:}p) - T(p)$
$i{:}h{:}p$	$n_+ - n_{h+}$	$\sum_p \sum_h \sum_i X_{pih}^2$	$T(i{:}h{:}p) - T(h{:}p)$
Mean(μ)	1	$n_+ \overline{X}^2$	

Letting

$$r_{pi} = \sum_p \sum_h \frac{\tilde{n}_{ph}^2}{\tilde{n}_p}, \quad r_i = \sum_p \sum_h \frac{\tilde{n}_{ph}^2}{n_+}, \quad \text{and} \quad r_{ih} = \sum_p \frac{\tilde{n}_p^2}{n_+},$$

the coefficients of μ^2 and the variance components in the expected T terms are given in the following table.

	Coefficients			
ET	μ^2	$\sigma^2(i{:}h{:}p)$	$\sigma^2(h{:}p)$	$\sigma^2(p)$
$ET(p)$	n_+	n_p	r_{pi}	n_+
$ET(h{:}p)$	n_+	n_{h+}	n_+	n_+
$ET(i{:}h{:}p)$	n_+	n_+	n_+	n_+
$ET(\mu)$	n_+	1	r_i	r_{ih}

It follows that the expected mean square equations are

$$
\begin{aligned}
EMS(p) &= \frac{ET(p) - ET(\mu)}{n_p - 1} \\
&= \sigma^2(i{:}h{:}p) + \left(\frac{r_{pi} - r_i}{n_p - 1}\right)\sigma^2(h{:}p) + \left(\frac{n_+ - r_{ih}}{n_p - 1}\right)\sigma^2(p) \\
EMS(h{:}p) &= \frac{ET(h{:}p) - ET(p)}{n_{h+} - n_p} \\
&= \sigma^2(i{:}h{:}p) + \left(\frac{n_+ - r_{pi}}{n_{h+} - n_p}\right)\sigma^2(h{:}p) \\
EMS(i{:}h{:}p) &= \frac{ET(i{:}h{:}p) - ET(h{:}p)}{n_+ - n_{h+}} \\
&= \sigma^2(i{:}h{:}p).
\end{aligned}
$$

It is possible to use these expected mean square equations to obtain estimators of the variance components, but it is easier to use the expected T terms. The results are:

$$\hat{\sigma}^2(i{:}h{:}p) = \frac{T(i{:}h{:}p) - T(h{:}p)}{n_+ - n_{h+}}$$

$$\hat{\sigma}^2(h{:}p) = \frac{T(h{:}p) - T(p) - (n_{h+} - n_p)\hat{\sigma}^2(i{:}h{:}p)}{n_+ - r_{pi}}$$

$$\hat{\sigma}^2(p) = \frac{T(p) - T(\mu) - (n_p - 1)\hat{\sigma}^2(i{:}h{:}p) - (r_{pi} - r_i)\hat{\sigma}^2(h{:}p)}{n_+ - r_{ih}}.$$

Using different notational conventions, these results are reported by Searle et al. (1992, p. 429).

Unbalanced $(p{:}c) \times i$ Design

For the unbalanced $(p{:}c) \times i$ design, let $n_{p{:}c}$ be the number of levels of p for each c. We also define

n_+ = total number of observations in design and

$n_{p+} = \sum_c n_{p{:}c}$ = total number of levels of p.

Given these notational conventions, the degrees of freedom and T terms are given in the following table.

Effect	df	T
c	$n_c - 1$	$n_i \sum_c (n_{p{:}c} \overline{X}_c^2)$
$p{:}c$	$n_{p+} - n_c$	$n_i \sum_c \sum_p \overline{X}_{pc}^2$
i	$n_i - 1$	$n_{p+} \sum_i \overline{X}_i^2$
ci	$(n_c - 1)(n_i - 1)$	$\sum_c (n_{p{:}c} \sum_i \overline{X}_{ci}^2)$
$pi{:}c$	$(n_{p+} - n_c)(n_i - 1)$	$\sum_c \sum_p \sum_i X_{pic}^2$
Mean(μ)	1	$n_+ \overline{X}^2$

Letting

$$r_p = \sum_c \frac{n_{p{:}c}^2}{n_{p+}} \quad \text{and} \quad t_p = \frac{n_{p+} - r_p}{n_c - 1},$$

the coefficients of μ^2 and the variance components in the expected T terms are given in the following table.

ET	μ^2	$\sigma^2(pi{:}c)$	$\sigma^2(ci)$	$\sigma^2(i)$	$\sigma^2(p{:}c)$	$\sigma^2(c)$
			Coefficients			
$ET(c)$	n_+	n_c	n_{p+}	n_{p+}	$n_c n_i$	n_+
$ET(p{:}c)$	n_+	n_{p+}	n_{p+}	n_{p+}	n_+	n_+
$ET(i)$	n_+	n_i	$n_i r_p$	n_+	n_i	$n_i r_p$
$ET(ci)$	n_+	$n_c n_i$	n_+	n_+	$n_c n_i$	n_+
$ET(pi{:}c)$	n_+	n_+	n_+	n_+	n_+	n_+
$ET(\mu)$	n_+	1	r_p	n_{p+}	n_i	$n_i r_p$

Using matrix procedures the T-term equations can be solved for estimates of the variance components.

Alternatively, the expected mean square equations can be determined, and they can be used to obtain estimates. The expected mean square equations are:

$$
\begin{aligned}
EMS(c) &= \sigma^2(pi{:}c) + t_p\,\sigma^2(ci) + n_i\,\sigma^2(p{:}c) + t_p n_i\,\sigma^2(c) \\
EMS(p{:}c) &= \sigma^2(pi{:}c) + n_i\,\sigma^2(p{:}c) \\
EMS(i) &= \sigma^2(pi{:}c) + r_p\,\sigma^2(ci) + n_{p+}\,\sigma^2(i) \\
EMS(ci) &= \sigma^2(pi{:}c) + t_p\,\sigma^2(ci) \\
EMS(pi{:}c) &= \sigma^2(pi{:}c).
\end{aligned}
$$

Estimators of the variance components in terms of mean squares are:

$$
\begin{aligned}
\hat{\sigma}^2(c) &= [MS(c) - MS(p{:}c) - MS(ci) + MS(pi{:}c)]/n_i t_p \\
\hat{\sigma}^2(p{:}c) &= [MS(p{:}c) - MS(pi{:}c)]/n_i \\
\hat{\sigma}^2(i) &= [MS(i) - r_p MS(ci)/t_p + (r_p - t_p)MS(pi{:}c)/t_p]/n_{p+} \\
\hat{\sigma}^2(ci) &= [MS(ci) - MS(pi{:}c)]/t_p \\
\hat{\sigma}^2(pi{:}c) &= MS(pi{:}c).
\end{aligned}
$$

Appendix F
Mini-Manual for GENOVA

GENOVA (GENeralized analysis Of VAriance) is a FORTRAN computer program developed principally for generalizability analyses, although it is also useful for more traditional applications of analysis of variance. GENOVA is appropriate for complete balanced designs with as many as six effects (i.e., five facets and an objects of measurement "facet").

Crick and Brennan (1983) provide an extensive manual for GENOVA. This appendix is a mini-manual that illustrates those features and capabilities of GENOVA that are most likely to be used in typical generalizability analyses. GENOVA and program documentation can be obtained from the author (robert-brennan@uiowa.edu) or from the Iowa Testing Programs Web site (www.uiowa.edu/~itp). At this time, both PC and Macintosh versions of GENOVA are available free of charge.

F.1 Sample Run

Provided in Table F.1 are the control cards and input data for a run of GENOVA in which the G study is based on Synthetic Data Set No. 4 (see Table 3.2 on page 73) for the $p \times (r:t)$ random effects design having 10 persons (p), three tasks (t), and four raters (r) nested within each of the three tasks. The control cards for the G study are numbered 1 to 7, and the data cards (records) are numbered 8 to 17. These card numbers are used here for reference purposes only; they are not part of the control cards themselves.

TABLE F.1. Control Cards for Genova Using Synthetic Data Set No. 4

```
CARD COLUMNS   11111111112222222222333333333344444444445555555 5
NO.  123456789012345678901234567890123456789012345678901234567

  1  GSTUDY       P X (R:T) DESIGN -- RANDOM MODEL
  2  OPTIONS      RECORDS 2
  3  EFFECT       * P 10 0
  4  EFFECT       + T  3 0
  5  EFFECT       + R:T 4 0
  6  FORMAT       (12F2.0)
  7  PROCESS
  8     5 6 5 5 5 3 4 5 6 7 3 3      PERSON  1
  9     9 3 7 7 7 5 5 5 7 7 5 2      PERSON  2
 10     3 4 3 3 5 3 3 5 6 5 1 6      PERSON  3
 11     7 5 5 3 3 1 4 3 5 3 3 5      PERSON  4
 12     9 2 9 7 7 3 7 2 7 5 3        PERSON  5
 13     3 4 3 5 3 3 6 3 4 5 1 2      PERSON  6
 14     7 3 7 7 7 5 5 7 5 5 5 4      PERSON  7
 15     5 8 5 7 7 5 5 4 3 2 1 1      PERSON  8
 16     9 9 8 8 6 6 6 5 5 8 1 1      PERSON  9
 17     4 4 4 3 3 5 6 5 5 7 1 1      PERSON 10
 18  COMMENT
 19  COMMENT      FIRST SET OF D STUDY CONTROL CARDS
 20  DSTUDY       #1 -- P X (R:T) DESIGN -- R AND T RANDOM
 21  DEFFECT      $ P
 22  DEFFECT        T 3
 23  DEFFECT        R:T 1 2 3 4
 24  ENDDSTUDY
 25  COMMENT      SECOND SET OF D STUDY CONTROL CARDS
 26  DSTUDY       #2 -- R:T:P DESIGN -- R AND T RANDOM
 27  DEFFECT      $ P
 28  DEFFECT        T:P 3
 29  DEFFECT        R:T:P 1 2 3 4
 30  ENDDSTUDY
 31  COMMENT      THIRD SET OF D STUDY CONTROL CARDS
 32  DSTUDY       #3 -- P X (R:T) DESIGN -- R RANDOM, T FIXED
 33  DEFFECT      $ P
 34  DEFFECT        T 3 / 3
 35  DEFFECT        R:T 1 2 3 4
 36  ENDDSTUDY
 37  FINISH
```

GENOVA can use G study estimated variance components to estimate results for various D study designs. For the sample run of GENOVA, there are three sets of D study control cards. The first set (card numbers 20 to 24) is for random effects D studies for the $p \times (R:T)$ design. The second set (card numbers 26 to 30) is for random effects D studies for the $R:T:p$ design. The third set (card numbers 32 to 36) is for the $p \times (R:T)$ design with tasks fixed and raters random.

Each control card begins with a set of characters called a control card identifier in columns 1 to 12. (Actually, GENOVA pays attention only to the first four characters.) All parameters in GENOVA control cards must appear in columns 13 to 80 and are in free format. Neither control card identifiers nor parameters should use lowercase letters. If there are multiple parameters, they must be provided in the specified order, but they do not need to be placed in specific columns. It is only necessary that multiple parameters be separated by at least one comma and/or one space.

COMMENT cards may appear anywhere except within, or immediately before, an input data set. Any alphanumeric text may appear in columns 13 to 80 of a COMMENT card.

F.2   G Study Control Cards

The GSTUDY control card provides nothing more than a heading, KTITLE, for the run. For this example, the heading verbally specifies the G study design and the model. However, the heading could contain any user-specified description of the run. KTITLE is printed at the top of each page of output. Otherwise, KTITLE has no effect in GENOVA.

The OPTIONS control card for this run specifies "RECORDS 2." In general, if "RECORDS NREC" is specified, then the first and last NREC records will be printed. There are a number of other options that could be specified, too, in any order. In particular, "ALGORITHM" causes the G study estimated variance components based on the algorithm in Section 3.4.4 to be used for D studies, rather than those resulting from solving the EMS equations in reverse, as discussed in Section 3.4.3. [For mixed models, the estimates are based on Equation 3.34 with $\sigma^2(\beta)$ estimated using either the algorithm or the EMS procedure.] Also, "NEGATIVE" tells GENOVA to print out the actual magnitudes of any negative estimates of variance components.

The EFFECT control cards specify the main effects for the design as well as the sample sizes and the sizes of facets in the universe of admissible observations (or population). The order of the EFFECT cards specifies the manner in which the data set is organized. There must be exactly as many EFFECT control cards as there are main effects in the design. The entire set of main effects constitutes the facets in the universe of admissible observations plus the "facet" that will be used to define the objects of measurement in D studies.

The general format of the data area (columns 13 to 80) of each EFFECT card is:

["*" or "+"] MFACET NUMLEV [NPOPUL]

where brackets designate optional parameters,

MFACET is the character string of letters and colons designating a main effect in the manner discussed in Section 3.2,

NUMLEV is the sample size (n) for the effect, and

NPOPUL is the population or universe size (N) for the effect. If NPOPUL is blank or zero, then it is assumed that NPOPUL approaches infinity.

One and only one of the EFFECT cards must have "*" preceding MFACET. Any other EFFECT control card may have a "+"preceding MFACET. If "+" appears, then the means for all levels of MFACET will be printed. In addition, the means are printed for all combinations of levels of effects that have "+" preceding MFACET.

The FORMAT card specifies the run-time FORTRAN format (using F-type format specifications) for reading a single record of the input data. The format must be enclosed within parentheses.

The PROCESS control card tells GENOVA to begin reading the input data using the format specified on the FORMAT card. In this run, the data immediately follow the PROCESS card, and there is no parameter specified in columns 13 to 80. If the data were on some other logical unit, KDATA, then KDATA would be specified anywhere in columns 13 to 80.

F.3   Input Data

For this illustrative run of GENOVA, there is one record for each person, and each record contains the scores of four raters for task 1, followed by the scores of four raters for task 2, followed by the scores of four raters for task 3. That is, the slowest moving index is p, the next slowest moving index is t, and the fastest moving index is r. This order of slowest moving index to fastest moving index is the order in which the EFFECT control cards must be provided.

A "*" must precede MFACET for the EFFECT card associated with a record. In this run, a record is associated with a person p. For the starred effect, all of the observations for a single level of this effect must be contained in a single record, which is read using the format specified in the FORMAT card. In this run, for each person, there are 12 observations read using the format 12F2.0. In general, the number of observations in any record equals the product of the sample sizes for the effects following the starred effect. (If the starred effect is associated with the last EFFECT card, then there must be one observation per record.)

Frequently, in generalizability analysis, G study data are organized in terms of person records, and the objects of measurement are also persons. No such restrictions are built into GENOVA, however. For example, even if the G study data are organized in terms of person records, subsequent D study analyses can be performed with any nonnested effect playing the role of objects of measurement.

Also, the G study data do not have to be organized in terms of person records. For example, suppose that the data set for our illustrative run of GENOVA was organized in terms of item records in the following manner.

```
Record  1:  Scores of 10 persons for rater 1 and task 1
Record  2:  Scores of 10 persons for rater 2 and task 1
Record  3:  Scores of 10 persons for rater 3 and task 1
Record  4:  Scores of 10 persons for rater 4 and task 1
Record  5:  Scores of 10 persons for rater 1 and task 2
Record  6:  Scores of 10 persons for rater 2 and task 2
Record  7:  Scores of 10 persons for rater 3 and task 2
Record  8:  Scores of 10 persons for rater 4 and task 2
Record  9:  Scores of 10 persons for rater 1 and task 3
Record 10:  Scores of 10 persons for rater 2 and task 3
Record 11:  Scores of 10 persons for rater 3 and task 3
Record 12:  Scores of 10 persons for rater 4 and task 3
```

In this case, the EFFECT control cards would be specified as follows:

```
EFFECT      +T    3
EFFECT      *R:T  4
EFFECT      +P    10
```

The "*" precedes R:T because each record is for a different rater. [Recall that, for our illustrative $p \times (r:t)$ design, each task is evaluated by a different set of four raters.] The number of observations in any record is 10, which equals the sample size for the effect (persons) specified after the starred effect. The number of records is $3 \times 4 = 12$.

By carefully considering the order of the EFFECT cards and the designation of the starred effect, it is almost always possible to process a data set, no matter how it is organized.

F.4   G Study Output

Every run of GENOVA produces a header page, G study output, and D study output if D study control cards are present. A somewhat edited version of the output for the sample run of GENOVA is provided in Table F.4 at the end of this appendix. The G study output includes the following.

- Listing of the G study control cards.

- Listing of first and last NREC records.

- Cell means resulting from the use of "+" preceding MFACET in the EFFECT cards.

- Traditional analysis of variance table. Note that GENOVA generates all of the interaction effects for the complete design and denotes them using the notational conventions employed in this book.

- Estimated variance components and estimated standard errors of the estimated variance components. The "USING ALGORITHM" estimated variance components are those based on the algorithm in Section 3.4.4 for a random model, or the algorithm plus Equation 3.34 for any model. These are also the estimates resulting from the matrix procedures in Appendix C. The "USING EMS EQUATIONS" estimated variance component are those resulting from the Cronbach et al. (1972) procedure of solving the EMS equations in reverse order, replacing any negative estimates with zero. If there are no negative estimates, the two sets of results are identical. The estimated standard errors are obtained using Equation 6.1.

- Expected mean square equations.

- Estimated variance-covariance matrix for the estimated variance components (the matrix $\hat{V}$ in Appendix C). The square roots of the diagonal elements are the estimated standard errors of the estimated variance components, and the off-diagonal elements are the estimated covariances between estimated variance components.

Since the input data set for the sample run is actually Synthetic Data Set No. 4, the reader can verify that the G study output in Table F.4 provides results identical to those discussed elsewhere (see, e.g., Table 3.4 on page 74). Note, however, that the GENOVA output uses uppercase letters for G study effects, which is consistent with the control card requirements.

F.5   D Study Control Cards

If a user wants only G study results, then no D study control cards are required. Otherwise, there should be one set of D study control cards for each design structure and/or universe of generalization. For example, for the sample run, the first and third sets of D study control cards have the same design, $p \times (R\!:\!T)$, but different universes of generalization (tasks random and tasks fixed, respectively). By contrast, the first and second sets of D study control cards have the same universe of generalization (both raters and tasks are random) but different design structures, $p \times (R\!:\!T)$ and $R\!:\!T\!:\!p$, respectively. For any given set of D study control cards, many combinations of sample sizes can be specified, and GENOVA will produce results for each such combination.

A set of D study control cards consists of a DSTUDY card that provides a heading for printed output, a set of DEFFECT cards that describes the D study design, sample sizes, universe/population sizes, and an ENDDSTUDY card that terminates the set of control cards.

The DEFFECT cards are analogous to the EFFECT cards for a G study, in that there is one DEFFECT card for each main effect in a D study design. However, there is no restriction on the order in which the DEFFECT cards must occur in the control card setup. The entire set of main effects constitutes the facets in the universe of generalization and the specified objects of measurement. Below, the word "facet" is sometimes used generically to refer to a facet in the universe of generalization or the objects of measurement.

The general format for the data area (columns 13 to 80) of each DEFFECT card is:

    [$] DFACET [ISAMPD(1-30)] [/IUNIVD]

where brackets designate optional parameters,

- DFACET is the character string of letter and colons that characterize a main effect associated with a facet in the D study design,

- ISAMPD(1-30) is a set of as many as 30 sample sizes (n'), and

- IUNIVD is the population or universe size (N') for the facet. If no values are specified for ISAMPD, then GENOVA uses the sample size from the G study (i.e., $n' = n$). Similarly, if no value is specified for IUNIVD, then GENOVA uses the population or universe size from the G study (i.e., $N' = N$).

Each set of D study control cards must have "$" preceding DFACET on one and only one DEFFECT card. The "$" designates the objects of measurement, which must be a nonnested facet in GENOVA.

Consider, for example, the third set of D study control cards in the sample run of GENOVA. The three main effects for the $p \times (R\!:\!T)$ design are p, T, and $R\!:\!T$, and there is one DEFFECT card for each of them. Since "$" precedes p, persons are the designated objects of measurement. The characters "3 / 3" after T mean that $n' = N' = 3$ or, in words, tasks are fixed at $N' = 3$. The numbers "1 2 3 4" after R:T mean that results should be provided for sample sizes of $n' = 1$, 2, 3, and 4 raters.

Although not illustrated here, there is another D study control card, called a DCUT card, that can be used to obtain estimates of $\Phi(\lambda)$ in Equation 2.54.

The last control card in any run of GENOVA should have "FINISH" in columns 1 to 6. If no D study control cards are present, then the FINISH card should follow the PROCESS card if the data set is in a separate file, or the FINISH card should immediately follow the data set if it is in-stream.

F.6   D Study Output

For each set of D study control cards, GENOVA produces a list of the D study control cards, one page of extensive output for each combination of sample sizes, and a single page of output that summarizes results for all combinations of sample sizes. Table F.4 at the end of this appendix provides a partial listing of these types of output for the three sets of D study control cards in the illustrative run of GENOVA.

Consider, for example, the GENOVA output for the third set of D study control cards with three (fixed) tasks and four raters using the D study $p \times (R{:}T)$ design. Note that GENOVA provides estimated variance components for mean scores based upon the sizes of facets in both the infinite universe of admissible observations (as specified in the G study control cards) and the restricted universe of generalization (as specified in the D study control cards). Differences between these two sets of estimated variance components have been discussed in Section 5.1.1. Note also that GENOVA provides estimates of universe score variance, error variances, generalizability coefficients, and the like, as well as estimated standard errors for most statistics.

Almost all the D study results reported in these pages of GENOVA output have been discussed previously. For example, Table 4.6 on page 116 provides results obtained using the first set of D study control cards, and Table 4.8 on page 126 provides results obtained using the third set of D study control cards.

Although not illustrated in the output provided at the end of this appendix, one option in GENOVA provides the estimated variance-covariance matrix for the estimated variance components associated with mean scores in the universe of generalization (the matrix $\widehat{\mathbf{W}}$ in Appendix C).

F.7   Mean Squares as Input

It sometimes occurs that mean squares are already available, but G study variance components and/or their standard errors have not been estimated. This might occur, for example, when an investigator wants to reconsider traditional results for a published study, or when a conventional analysis of variance program is used. The G study estimated variance components and their standard errors can be obtained using formulas provided previously

TABLE F.2. Control Cards Using Mean Squares as Input

```
COLUMNS 11111111112222222222333333333344444444
12345678901234567890123456789012345678901123456

GMEANSQUARES  P X (R:T) DESIGN -- RANDOM MODEL
MEANSQUARE    P    10.2963 10
MEANSQUARE    T    24.1000  3
MEANSQUARE    R:T   8.8556  4
MEANSQUARE    PT    4.6185
MEANSQUARE    PR:T  2.3802
ENDMEAN
COMMENT            SETS OF D STUDY CONTROL CARDS
COMMENT            COULD BE PLACED HERE
FINISH
```

in this book. Alternatively, GENOVA can easily provide these results, and then use them to estimate results for D study designs.

For example, for the illustrative data set, Table F.2 provides the control card setup for using mean squares as input to GENOVA. Note the following.

- The first control card is a GMEANSQUARES card which provides a heading for output and which tells GENOVA to expect mean squares as input.

- There is a MEANSQUARE card for every effect, including interaction effects, and each such card contains the mean square for the effect.

- The MEANSQUARE cards for the main effects come first, and the mean square provided in each such card is followed by the sample size and then the population or universe size (if it is other than infinite).

- As long as the MEANSQUARE cards for main effects come first, any order of the MEANSQUARE cards is permitted.

- The last MEANSQUARE card should be followed by an ENDMEAN card.

F.8 G Study Variance Components as Input

G study variance components, or estimates of them, can be used as input to GENOVA to obtain estimated standard errors of estimated variance components, and to estimate results for D study designs.

TABLE F.3. Control Cards Using Variance Components as Input

```
COLUMNS 111111111122222222222333333333344444444
        12345678901234567890123456789012345678901234
GCOMPONETS  P X (R:T) DESIGN -- RANDOM MODEL
VCOMPONENT  P       .4731    10
VCOMPONENT  T       .3252     3
VCOMPONENT  R:T     .6475     4
VCOMPONENT  PT      .5596
VCOMPONENT  PR:T   2.3802
ENDCOMP
COMMENT     SETS OF D STUDY CONTROL CARDS
COMMENT     COULD BE PLACED HERE
FINISH
```

For example, for the illustrative data set, Table F.3 provides the control card setup for using G study estimated variance components as input to GENOVA. Note the following.

- The first control card is a GCOMPONENTS card which provides a heading for output and which tells GENOVA to expect G study estimated variance components as input.

- There is a VCOMPONENT card for every effect, including interaction effects, and each such card contains the G study estimated variance component for the effect.

- The VCOMPONENT cards for the main effects come first, and the estimated variance component provided in each such card is followed by the sample size and then the population or universe size (if it is other than infinite). Even if the parameter values of the variance components were known, some set of sample sizes must be provided. They are used to obtain estimated standard errors and as default D study sample sizes.

- As long as the VCOMPONENT cards for main effects come first, any order of the VCOMPONENT cards is permitted.

- The last VCOMPONENT card should be followed by an ENDCOMP card.

TABLE F.4. Output for Genova Using Synthetic Data Set No. 4

```
                              CONTROL CARD INPUT LISTING

COLUMN   1111111111222222222233333333334444444444555555555566666666667777777778
         1234567890123456789012345678901234567890123456789012345678901234567890
GSTUDY     P X (R:T) DESIGN -- RANDOM MODEL
OPTIONS    RECORDS 2
EFFECT   * P 10 0
EFFECT   + T  3 0
EFFECT   + R:T 4 0
FORMAT     (12F2.0)
PROCESS


                       INPUT RECORD LISTING WITH RECORD MEANS

RECORD #  1   5.00000  6.00000  5.00000  5.00000  3.00000  4.00000  5.00000
              6.00000  7.00000  3.00000  4.75000

RECORD #  2   9.00000  3.00000  7.00000  7.00000  5.00000  5.00000  5.00000
              7.00000  7.00000  2.00000  5.75000

RECORD #  9   9.00000  9.00000  8.00000  6.00000  6.00000  6.00000  5.00000
              5.00000  8.00000  1.00000  6.00000

RECORD # 10   4.00000  4.00000  4.00000  3.00000  5.00000  6.00000  5.00000
              5.00000  7.00000  1.00000  4.00000
```

CELL MEAN SCORES

*** GRAND MEAN = 4.7500000 ***

MEAN SCORES FOR EFFECT: T

SUBSCRIPT NOTATION: (T)

(1) = 5.500000   (2) = 4.800000   (3) = 3.950000

MEAN SCORES FOR EFFECT: R:T

SUBSCRIPT NOTATION: (T,R)

(1,1) = 6.100000	(1,2) = 4.800000	(1,3) = 5.600000	(1,4) = 5.500000
(2,1) = 5.300000	(2,2) = 4.300000	(2,3) = 4.700000	(2,4) = 4.900000
(3,1) = 4.800000	(3,2) = 5.600000	(3,3) = 2.600000	(3,4) = 2.800000

ANOVA TABLE

(** = INFINITE)	P	T	R
SAMPLE SIZE	10	3	4
UNIVERSE SIZE	****	****	****

EFFECT	DEGREES OF FREEDOM	SUMS OF SQUARES FOR MEAN SCORES	SUMS OF SQUARES FOR SCORE EFFECTS	MEAN SQUARES	F STATISTIC	(QF = QUASI F RATIO) F-TEST DEGREES OF FREEDOM NUMERATOR	DENOMINATOR
P	9	2800.16667	92.66667	10.29630	2.22935	9	18
T	2	2755.70000	48.20000	24.10000	2.17238 QF	2 QF	12 QF
R:T	9	2835.40000	79.70000	8.85556	3.72044	9	81
PT	18	2931.50000	83.13333	4.61852	1.94035	18	81
PR:T	81	3204.00000	192.80000	2.38025			
MEAN		2707.50000					
TOTAL	119		496.50000				

NOTE: FOR GENERALIZABILITY ANALYSES, F-STATISTICS SHOULD BE IGNORED

G STUDY RESULTS

QFM = QUADRATIC FORM

(** = INFINITE)	P	T	R
SAMPLE SIZE	10	3	4
UNIVERSE SIZE	****	****	****

M O D E L   V A R I A N C E   C O M P O N E N T S

EFFECT	DEGREES OF FREEDOM	USING ALGORITHM	USING EMS EQUATIONS	STANDARD ERROR
P	9	0.4731481	0.4731481	0.3855758
T	2	0.3251543	0.3251543	0.4379875
R:T	9	0.6475309	0.6475309	0.3794056
PT	18	0.5595679	0.5595679	0.3766291
PR:T	81	2.3802469	2.3802469	0.3694860

NOTE: THE "ALGORITHM" AND "EMS" ESTIMATED VARIANCE COMPONENTS WILL BE
      IDENTICAL IF THERE ARE NO NEGATIVE ESTIMATES

EXPECTED MEAN SQUARE EQUATIONS

```
EMS(P)    = 1.00*VC(PR:T) + 4.00*VC(PT) + 12.00*VC(P)
EMS(T)    = 1.00*VC(PR:T) + 4.00*VC(PT) + 10.00*VC(R:T) + 40.00*VC(T)
EMS(R:T)  = 1.00*VC(PR:T) + 10.00*VC(R:T)
EMS(PT)   = 1.00*VC(PR:T) + 4.00*VC(PT)
EMS(PR:T) = 1.00*VC(PR:T)
```

VARIANCE - COVARIANCE MATRIX FOR ESTIMATED VARIANCE COMPONENTS (V)

	P	T	R:T	PT	PR:T
P	0.1486687				
T	0.0044439	0.1918331			
R:T	0.0000000	-0.0359871	0.1439486		
PT	-0.0444390	-0.0141849	0.0034130	0.1418495	
PR:T	0.0000000	0.0034130	-0.0136520	-0.0341300	0.1365199

```
***********************************************
***OUTPUT FOR D STUDIES WITH 1, 2, AND 3 ITEMS NOT INCLUDED HERE***
***********************************************
```

D STUDY DESIGN NUMBER 001-004

			FACETS :	T	R:T
OBJECT OF MEASUREMENT :	P				
G STUDY POPULATION SIZE :	INFINITE	G STUDY UNIVERSE SIZES :	INFINITE	INFINITE	
D STUDY POPULATION SIZE :	INFINITE	D STUDY UNIVERSE SIZES :	INFINITE	INFINITE	
D STUDY SAMPLE SIZE :	10	D STUDY SAMPLE SIZES :	3	4	

VARIANCE COMPONENTS IN TERMS OF
G STUDY UNIVERSE (OF ADMISSIBLE OBSERVATIONS) SIZES

| | | | | VARIANCE COMPONENTS FOR MEAN SCORES | |
| | | | | | |
EFFECT	VARIANCE COMPONENTS FOR SINGLE OBSERVATIONS	FINITE UNIVERSE COR- RECTIONS	D STUDY SAMPLING FRE- QUENCIES	ESTIMATES	STANDARD ERRORS
P	0.47315	1.0000	1	0.47315	0.38558
T	0.32515	1.0000	3	0.10838	0.14600
R:T	0.64753	1.0000	12	0.05396	0.03162
PT	0.55957	1.0000	3	0.18652	0.12554
PR:T	2.38025	1.0000	12	0.19835	0.03079

QFM = QUADRATIC FORM

VARIANCE COMPONENTS IN TERMS OF
D STUDY UNIVERSE (OF GENERALIZATION) SIZES

| | | | | VARIANCE COMPONENTS FOR MEAN SCORES | |
| | | | | | |
	VARIANCE COMPONENTS FOR SINGLE OBSERVATIONS	FINITE UNIVERSE COR- RECTIONS	D STUDY FRE- QUENCIES	ESTIMATES	STANDARD ERRORS
	0.47315	1.0000	1	0.47315	0.38558
	0.32515	1.0000	3	0.10838	0.14600
	0.64753	1.0000	12	0.05396	0.03162
	0.55957	1.0000	3	0.18652	0.12554
	2.38025	1.0000	12	0.19835	0.03079

		STANDARD ERROR OF
	VARIANCE	VARIANCE
	DEVIATION	
UNIVERSE SCORE	0.47315	0.68786	0.38558
EXPECTED OBSERVED SCORE	0.85802	0.92630	0.36586
LOWER CASE DELTA	0.38488	0.62038	0.12171
UPPER CASE DELTA	0.54722	0.73974	0.17935
MEAN	0.24815	0.49814	

GENERALIZABILITY COEFFICIENT =	0.55144	(1.22935)
PHI =	0.46370	(0.86464)

NOTE: SIGNAL-NOISE RATIOS ARE IN PARENTHESES

```
**********************************************
***OUTPUT FOR D STUDIES WITH 1, 2, AND 3 ITEMS NOT INCLUDED HERE***
**********************************************

                        D STUDY DESIGN NUMBER 002-004

OBJECT OF MEASUREMENT :        P              FACETS :       T:P        R:T:P
G STUDY POPULATION SIZE : INFINITE   G STUDY UNIVERSE SIZES : INFINITE    INFINITE
D STUDY POPULATION SIZE : INFINITE   D STUDY UNIVERSE SIZES : INFINITE    INFINITE
   D STUDY SAMPLE SIZE :       10       D STUDY SAMPLE SIZES :     3          4
```

VARIANCE COMPONENTS IN TERMS OF
G STUDY UNIVERSE (OF ADMISSIBLE OBSERVATIONS) SIZES

EFFECT	VARIANCE COMPONENTS FOR SINGLE OBSERVATIONS	FINITE UNIVERSE SAMPLING CORRECTIONS	D STUDY FREQUENCIES	VARIANCE COMPONENTS FOR MEAN SCORES ESTIMATES	STANDARD ERRORS
P	0.47315	1.0000	1	0.47315	0.38558
T:P	0.88472	1.0000	3	0.29491	0.18418
R:T:P	3.02778	1.0000	12	0.25231	0.04193

QFM = QUADRATIC FORM

VARIANCE COMPONENTS IN TERMS OF
D STUDY UNIVERSE (OF GENERALIZATION) SIZES

	VARIANCE COMPONENTS FOR SINGLE OBSERVATIONS	FINITE UNIVERSE SAMPLING CORRECTIONS	D STUDY FREQUENCIES	VARIANCE COMPONENTS FOR MEAN SCORES ESTIMATES	STANDARD ERRORS
	0.47315	1.0000	1	0.47315	0.38558
	0.88472	1.0000	3	0.29491	0.18418
	3.02778	1.0000	12	0.25231	0.04193

	VARIANCE	STANDARD DEVIATION	STANDARD ERROR OF VARIANCE
UNIVERSE SCORE	0.47315	0.68786	0.38558
EXPECTED OBSERVED SCORE	1.02037	1.01013	0.39265
LOWER CASE DELTA	0.54722	0.73974	0.17935
UPPER CASE DELTA	0.54722	0.73974	0.17935
MEAN	0.10204	0.31943	

```
GENERALIZABILITY COEFFICIENT =  0.46370  ( 0.86464)
                        PHI =  0.46370  ( 0.86464)
```

NOTE: SIGNAL-NOISE RATIOS ARE IN PARENTHESES

```
****************************************************
***OUTPUT FOR D STUDIES WITH 1, 2, AND 3 ITEMS NOT INCLUDED HERE***
****************************************************
```

D STUDY DESIGN NUMBER 003-004

		FACETS :	T	R:T
OBJECT OF MEASUREMENT :	P			
G STUDY POPULATION SIZE :	INFINITE	G STUDY UNIVERSE SIZES :	INFINITE	INFINITE
D STUDY POPULATION SIZE :	INFINITE	D STUDY UNIVERSE SIZES :	INFINITE	INFINITE
D STUDY SAMPLE SIZE :	10	D STUDY SAMPLE SIZES :	3	4

VARIANCE COMPONENTS IN TERMS OF
G STUDY UNIVERSE (OF ADMISSIBLE OBSERVATIONS) SIZES

EFFECT	VARIANCE COMPONENTS FOR SINGLE OBSERVATIONS	FINITE UNIVERSE CORRECTIONS	D STUDY FREQUENCIES	VARIANCE COMPONENTS FOR MEAN SCORES ESTIMATES	STANDARD ERRORS
P	0.47315	1.0000	1	0.47315	0.38558
T	0.32515	1.0000	3	0.10838	0.14600
R:T	0.64753	1.0000	12	0.05396	0.03162
PT	0.55957	1.0000	3	0.18652	0.12554
PR:T	2.38025	1.0000	12	0.19835	0.03079

VARIANCE COMPONENTS IN TERMS OF
D STUDY UNIVERSE (OF GENERALIZATION) SIZES

EFFECT	VARIANCE COMPONENTS FOR SINGLE OBSERVATIONS	FINITE UNIVERSE CORRECTIONS	D STUDY FREQUENCIES	VARIANCE COMPONENTS FOR MEAN SCORES ESTIMATES	STANDARD ERRORS
P	0.65967	1.0000	1	0.65967	0.36716
T	0.32515QFM	0.0000	3	-------	-------
R:T	0.64753	1.0000	12	0.05396	0.03162
PT	0.55957	0.0000	3	-------	-------
PR:T	2.38025	1.0000	12	0.19835	0.03079

QFM = QUADRATIC FORM

	VARIANCE	STANDARD DEVIATION	STANDARD ERROR OF VARIANCE
UNIVERSE SCORE	0.65967	0.81220	0.36716
EXPECTED OBSERVED SCORE	0.85802	0.92630	0.36586
LOWER CASE DELTA	0.19835	0.44537	0.03079
UPPER CASE DELTA	0.25231	0.50231	0.04193
MEAN	0.13976	0.37385	

GENERALIZABILITY COEFFICIENT = 0.76882 (3.32573)
PHI = 0.72333 (2.61448)

NOTE: SIGNAL-NOISE RATIOS ARE IN PARENTHESES

SUMMARY OF D STUDY RESULTS FOR SET OF CONTROL CARDS NO. 001

D STUDY DESIGN NO	INDEX= UNIV.=	SAMPLE SIZES			V A R I A N C E S						
		$P INF.	T INF.	R INF.	UNIVERSE SCORE	EXPECTED OBSERVED SCORE	LOWER CASE DELTA	UPPER CASE DELTA	MEAN	GEN. COEF.	PHI
001-001		10	3	1	0.47315	1.45309	0.97994	1.30417	0.46954	0.32562	0.26622
001-002		10	3	2	0.47315	1.05638	0.58323	0.79954	0.32194	0.44790	0.37177
001-003		10	3	3	0.47315	0.92414	0.45099	0.63133	0.27275	0.51199	0.42839
001-004		10	3	4	0.47315	0.85802	0.38488	0.54722	0.24815	0.55144	0.46370

SUMMARY OF D STUDY RESULTS FOR SET OF CONTROL CARDS NO. 002

D STUDY DESIGN NO	INDEX= UNIV.=	SAMPLE SIZES			V A R I A N C E S						
		$P INF.	T INF.	R INF.	UNIVERSE SCORE	EXPECTED OBSERVED SCORE	LOWER CASE DELTA	UPPER CASE DELTA	MEAN	GEN. COEF.	PHI
002-001		10	3	1	0.47315	1.77731	1.30417	1.30417	0.17773	0.26622	0.26622
002-002		10	3	2	0.47315	1.27269	0.79954	0.79954	0.12727	0.37177	0.37177
002-003		10	3	3	0.47315	1.10448	0.63133	0.63133	0.11045	0.42839	0.42839
002-004		10	3	4	0.47315	1.02037	0.54722	0.54722	0.10204	0.46370	0.46370

SUMMARY OF D STUDY RESULTS FOR SET OF CONTROL CARDS NO. 003

D STUDY DESIGN NO	INDEX= UNIV.=	SAMPLE SIZES			V A R I A N C E S						
		$P INF.	T INF.	R INF.	UNIVERSE SCORE	EXPECTED OBSERVED SCORE	LOWER CASE DELTA	UPPER CASE DELTA	MEAN	GEN. COEF.	PHI
003-001		10	3	1	0.65967	1.45309	0.79342	1.00926	0.36115	0.45398	0.39527
003-002		10	3	2	0.65967	1.05638	0.39671	0.50463	0.21356	0.62446	0.56658
003-003		10	3	3	0.65967	0.92414	0.26447	0.33642	0.16436	0.71382	0.66226
003-004		10	3	4	0.65967	0.85802	0.19835	0.25231	0.13976	0.76882	0.72333

Appendix G
urGENOVA

urGENOVA (Brennan, 2001b) is an ANSI C computer program for estimating random effects variance components for both balanced and unbalanced designs that are complete in the sense that all interactions are included. urGENOVA was created primarily for designs that are unbalanced with respect to nesting and that contain single observations per cell. However, for designs that are not "too large," urGENOVA can also estimate random effects variance components when some cells are empty and/or the numbers of observations within cells are unequal.

Random effects variance components are estimated by urGENOVA using the analogous-ANOVA procedure discussed in Section 7.1.1, which is sometimes called Henderson's (1953) Method 1. The algorithms used do not require operations with large matrices. For designs (balanced or unbalanced) with single observations per cell, urGENOVA can process very large data sets in a relatively short amount of time. urGENOVA extends GENOVA (see Appendix F) in the sense that urGENOVA can handle unbalanced designs, as well as all the balanced designs that GENOVA can process. However, urGENOVA has no D study capabilities.[1]

Input to urGENOVA consists of a set of control cards and an input data set. The urGENOVA control cards have an appearance similar to those for GENOVA, and both programs require that effects be identified in the manner discussed in Section 3.2.1. However, the control card conventions

---

[1] For some unbalanced designs, mGENOVA can be used to obtain D study results (see Appendix H).

for the two programs are not identical. Also, the urGENOVA rules for formatting and ordering records for the input data set are not quite as flexible as the rules for GENOVA.

Output for urGENOVA that is always provided includes:

- a listing of the control cards;

- means for main effects; and

- an ANOVA table with degrees of freedom, uncorrected sums of squares (T terms), sums of squares, mean squares, and estimated random effects variance components.

Optional output includes:

- a list of some or all of the input data records;

- expected values of T terms;

- expected mean squares;

- *for balanced designs only*, estimated standard errors; and

- *for balanced designs only*, both Satterthwaite confidence intervals (see Section 6.2.2) and Ting et al. confidence intervals (see Section 6.2.3) for estimated variance components using a user-specified confidence coefficient.

urGENOVA and program documentation can be obtained from the author (robert-brennan@uiowa.edu) or from the Iowa Testing Programs Web site (www.uiowa.edu/~itp). At this time, both PC and Macintosh versions of urGENOVA are available free of charge.

Appendix H
mGENOVA

mGENOVA (Brennan, 2001b) is an ANSI C computer program for performing multivariate generalizability G and D studies for a set of designs that may be balanced or unbalanced with respect to nesting. Specifically, mGENOVA can be used with all of the G study designs in Table 9.2 [except the $(p^\circ:c^\bullet) \times i^\circ$ design] and their D study counterparts. These are called "canonical" designs here. For these designs, with minor exceptions, mGENOVA can perform all of the computations discussed in Chapters 9 to 12. The algorithms used do not require operations with large matrices, and mGENOVA can process very large data sets in a relatively short amount of time.

When $n_v = 1$, the canonical G study designs are simply $p \times i$, $i:p$, $p \times i \times h$, $p \times (i:h)$, and $(p:c) \times i$. For any one of these designs and its D study counterpart, mGENOVA provides output comparable to that of GENOVA (see Appendix F) when the design is balanced. When the design is unbalanced, the G study output is comparable to that of urGENOVA (see Appendix G) and, in addition, mGENOVA provides D study output.

Input to mGENOVA consists of a set of control cards and usually an input data set. The control card setups and input data conventions are very similar to those for urGENOVA and quite similar to those for GENOVA. The mGENOVA rules for formatting and ordering records for the input data set are very much like those of urGENOVA. G study estimated variance and covariance components can be used as input to obtain D study results.

When designs are balanced, estimates of variance and covariance components based on mean squares and mean products, respectively, have certain desirable statistical properties. When designs are unbalanced with respect to nesting, however, there is usually no compelling statistical argument for using analogous mean squares (or analogous T terms; see Section 7.1.1) and analogous mean products (or analogous TP terms; see Section 11.1.2) for estimation. As options, mGENOVA permits the use of C terms to estimate variance components and CP terms to estimate covariance components, as discussed in Section 11.1.3.

G study output (some of which is optional) includes:

- a listing of the control cards;

- a list of some or all of the input data records;

- means for most main effects;

- statistics used to estimate G study variance and covariance components; and

- G study estimated variance and covariance components.

D study output (some of which is optional) includes:

- estimated variance and covariance components;

- estimated universe score matrix and error matrices;

- results for each level of the fixed facet v;

- results for a user-defined composite, based on w and/or a weights as discussed in Sections 10.1.3 and 10.1.5, respectively;

- conditional standard errors of measurement for the composite; and

- results relating to regressed score estimates for profiles and composites.

mGENOVA and program documentation can be obtained from the author (robert-brennan@uiowa.edu) or from the Iowa Testing Programs Web site (www.uiowa.edu/~itp). At this time, both PC and Macintosh versions of mGENOVA are available free of charge.

Appendix I
Answers to Selected Exercises

Detailed answers to approximately half of the exercises are given in this appendix; answers to the remaining exercises are available from the author (robert-brennan@uiowa.edu) or the publisher (www.springer-ny.com).

Chapter 1

1.1 (a) Using the formulas in Table 1.1,

$$\hat{\sigma}^2(p) = \frac{MS(p) - MS(pt) - MS(pr) + MS(ptr)}{n_t n_r}$$

$$= \frac{6.20 - 1.60 - .26 + .16}{5 \times 6}$$

$$= .15.$$

Similarly,

$$\hat{\sigma}^2(t) = .08, \quad \hat{\sigma}^2(r) = .04, \quad \hat{\sigma}^2(pt) = .24,$$
$$\hat{\sigma}^2(pr) = .02, \quad \hat{\sigma}^2(tr) = .08, \quad \hat{\sigma}^2(ptr) = .16.$$

(b) Because persons are the objects of measurement, $\hat{\sigma}^2(p) = .15$ is unchanged, and using the rule in Equation 1.6:

$$\hat{\sigma}^2(T) = .02, \quad \hat{\sigma}^2(R) = .02, \quad \hat{\sigma}^2(pT) = .06,$$
$$\hat{\sigma}^2(pR) = .01, \quad \hat{\sigma}^2(TR) = .01, \quad \hat{\sigma}^2(pTR) = .02.$$

(c) *Absolute error variance*: From the "T, R random" column of Table 1.2,

$$\hat{\sigma}^2(\Delta) = \hat{\sigma}^2(T) + \hat{\sigma}^2(R) + \hat{\sigma}^2(pT)$$
$$+ \hat{\sigma}^2(pR) + \hat{\sigma}^2(TR) + \hat{\sigma}^2(pTR)$$
$$= .02 + .02 + .06 + .01 + .01 + .02 = .14.$$

Relative error variance: From the "T, R random" column of Table 1.2,

$$\hat{\sigma}^2(\delta) = \hat{\sigma}^2(pT) + \hat{\sigma}^2(pR) + \hat{\sigma}^2(pTR) = .06 + .01 + .02 = .09.$$

Generalizability coefficient: From Equation 1.12, when T and R are both random, $\hat{\sigma}^2(\tau) = \hat{\sigma}^2(p)$ and

$$E\hat{\rho}^2 = \frac{\hat{\sigma}^2(p)}{\hat{\sigma}^2(p) + \hat{\sigma}^2(\delta)} = \frac{.15}{.15 + .09} = .63.$$

Index of dependability: From Equation 1.13, since $\hat{\sigma}^2(\tau) = \hat{\sigma}^2(p)$ when T and R are both random,

$$\hat{\Phi} = \frac{\hat{\sigma}^2(p)}{\hat{\sigma}^2(p) + \hat{\sigma}^2(\Delta)} = \frac{.15}{.15 + .14} = .52.$$

(d) First, we need to estimate G study variance components for the $p \times (r\!:\!t)$ design given the results in (a) for the $p \times t \times r$ design. Under these circumstances, $\hat{\sigma}^2(p) = .15$, $\hat{\sigma}^2(t) = .08$, and $\hat{\sigma}^2(pt) = .24$ are unchanged. Using Equation 1.15,

$$\hat{\sigma}^2(r\!:\!t) = \hat{\sigma}^2(r) + \hat{\sigma}^2(tr) = .04 + .08 = .12,$$

and using Equation 1.16

$$\hat{\sigma}^2(pr\!:\!t) = \hat{\sigma}^2(pr) + \hat{\sigma}^2(ptr) = .02 + .16 = .18.$$

Second, using the rule in Equation 1.6, for $n'_t = 3$ and $n'_r = 2$, the estimated random effects D study variance components are:

$$\hat{\sigma}^2(p) = .150, \quad \hat{\sigma}^2(T) = .027, \quad \hat{\sigma}^2(R\!:\!T) = .020,$$
$$\hat{\sigma}^2(pT) = .080, \quad \text{and} \quad \hat{\sigma}^2(pR\!:\!T) = .030.$$

Third, since T and R are random, we use the second column in Table 1.3 to obtain

$$\hat{\sigma}^2(\Delta) = \hat{\sigma}^2(T) + \hat{\sigma}^2(R\!:\!T) + \hat{\sigma}^2(pT) + \hat{\sigma}^2(pR\!:\!T)$$
$$= .027 + .020 + .080 + .030 = .157.$$

The square root is $\hat{\sigma}(\Delta) = .40$.

1.2 If there are only two facets and both facets are considered fixed, then every instance of a measurement procedure would involve the same conditions. Under these circumstances, there is no generalization to a broader universe of conditions of measurement. Logically, therefore, all error variances are zero, by definition. No measurement procedure is *that* precise! To avoid this problem, at least one of the facets in a universe of generalization must be viewed as variable across instances of the measurement procedure.

Chapter 2

2.1 (a) The ANOVA table is

Effect(α)	$df(\alpha)$	$SS(\alpha)$	$MS(\alpha)$	$\hat{\sigma}^2(\alpha)$
p	5	81.3333	16.2667	5.0000
r	2	12.0000	6.0000	.7889
pr	10	12.6667	1.2667	1.2667

(b) $\hat{\sigma}^2(\delta) = .4222$, $\hat{\sigma}^2(\Delta) = .6852$, $E\hat{\rho}^2 = .922$, and $\hat{\Phi} = .879$.

(c) In terms of covariances

$$\frac{2(4.0 + 3.6 + 7.4)}{6} = 5,$$

which is identical to the estimate of universe score variance based on mean squares.

2.3 Starting with Equation 2.37,

$$\begin{aligned}
\sigma^2(\overline{X}) &= \underset{P}{E}\,\underset{I}{E}(X_{PI} - \mu)^2 \\
&= \underset{P}{E}\,\underset{I}{E}(\nu_P + \nu_I + \nu_{PI})^2 \\
&= \underset{P}{E}\,\nu_P^2 + \underset{I}{E}\,\nu_I^2 + \underset{P}{E}\,\underset{I}{E}\,\nu_{PI}^2 \\
&= \sigma^2(P) + \sigma^2(I) + \sigma^2(PI) \\
&= \frac{\sigma^2(p)}{n_p'} + \frac{\sigma^2(i)}{n_i'} + \frac{\sigma^2(pi)}{n_p' n_i'}.
\end{aligned}$$

2.5 It is easier to solve this problem using estimates of signal–noise ratios than generalizability coefficients. Specifically, for Math Concepts,

$$\frac{.0280}{.1783/n_i'} \geq \frac{.6}{1 - .6} = 1.5,$$

which implies that $n_i' \geq 9.5$. Similarly, for Estimation,

$$\frac{.0242}{.1994/n_i'} \geq \frac{.6}{1 - .6} = 1.5,$$

which implies that $n_i' \geq 12.4$. For the current tests, Math Concepts has one-third more items. Therefore, for the shortened forms, Math Concepts should have at least $12.4 \times 1.33 = 16.5$ items (that is, 17 items), and Estimation should have 13 items.

2.7 Since $1.43/5 = .286$ and $3/5 = .6$, the equation that needs to be solved is

$$\widehat{E/T}(\lambda) = \sqrt{\frac{\dfrac{.190 + 1.234}{n_t'}}{.662 + (.286 - .6)^2 - \left[\dfrac{.662}{229} + \dfrac{.190}{n_t'} + \dfrac{1.234}{229\,n_t'}\right]}} \leq .5,$$

which leads to $n_t' \geq 7.78$. Therefore, eight tasks are required.

Chapter 3

3.1 Letting p be children, c be classrooms, h be content areas (addition and subtraction), and i be items, the design is $(p:c) \times (i:h)$. The Venn diagram and linear model are:

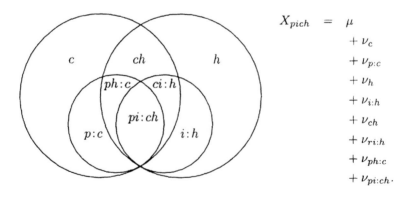

$$
\begin{aligned}
X_{pich} = {} & \mu \\
& + \nu_c \\
& + \nu_{p:c} \\
& + \nu_h \\
& + \nu_{i:h} \\
& + \nu_{ch} \\
& + \nu_{ri:h} \\
& + \nu_{ph:c} \\
& + \nu_{pi:ch}.
\end{aligned}
$$

3.2 Letting p be students, s be passages, h be types of items (factual and inferential), and i be items, the design structure is $p \times [i:(s \times h)]$. The Venn diagram and linear model are:

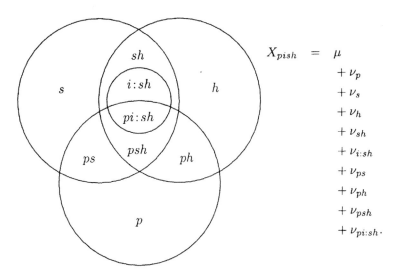

$$X_{pish} = \mu$$
$$+ \nu_p$$
$$+ \nu_s$$
$$+ \nu_h$$
$$+ \nu_{sh}$$
$$+ \nu_{i:sh}$$
$$+ \nu_{ps}$$
$$+ \nu_{ph}$$
$$+ \nu_{psh}$$
$$+ \nu_{pi:sh}.$$

Most likely, the universe of admissible observations is $i:(s \times h)$ with $N_i \to \infty$, $N_s \to \infty$, and $N_h = 2$.

3.4 For $\hat{\sigma}^2(p)$ in this design, $\pi(\dot{\alpha}) = n_i n_h$, and $\mathcal{A}$ consists of the index h, only, because ph is the only component that contains p and exactly one additional index. Therefore, Step 1 results in subtracting only $MS(ph)$ from $MS(p)$, and the estimator is

$$\hat{\sigma}^2(p) = \frac{MS(p) - MS(ph)}{n_i n_h}.$$

For $\hat{\sigma}^2(h)$, $\pi(\dot{\alpha}) = n_p n_i$, and $\mathcal{A}$ consists of the indices p and i because both ph and $i{:}h$ contain h and exactly one additional index. Therefore, Step 1 results in subtracting $MS(ph)$ and $MS(i{:}h)$ from $MS(h)$. Step 2 results in adding $MS(pih)$. The resulting estimator is:

$$\hat{\sigma}^2(h) = \frac{MS(h) - MS(ph) - MS(i{:}h) + MS(pih)}{n_p n_i}.$$

3.5

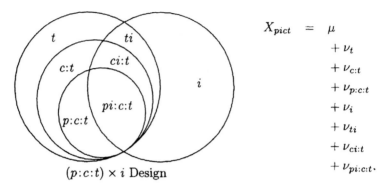

$(p{:}c{:}t) \times i$ Design

$$X_{pict} = \mu$$
$$+ \nu_t$$
$$+ \nu_{c:t}$$
$$+ \nu_{p:c:t}$$
$$+ \nu_i$$
$$+ \nu_{ti}$$
$$+ \nu_{ci:t}$$
$$+ \nu_{pi:c:t}.$$

Effect(α)	$df(\alpha)$	$T(\alpha)$	$SS(\alpha)$	$MS(\alpha)$	$\hat{\sigma}^2(\alpha)$
t	41	83000.7640	2150.7640	52.4577	.0845
$c{:}t$	42	84070.5250	1069.7610	25.4705	.1241
$p{:}c{:}t$	1092	90521.8490	6451.3240	5.9078	.4913
i	10	81147.2928	297.2928	29.7293	.0236
ti	410	84086.7468	788.6900	1.9236	.0346
$ci{:}t$	420	85558.0068	401.4990	0.9559	.0323
$pi{:}c{:}t$	10920	97506.4529	5497.1221	0.5034	.5034
Mean (μ)		80850.0000			

3.7 (a) The expected mean squares are:

$$\begin{aligned} \boldsymbol{EMS}(p) &= (1 - n_i/N_i)\sigma^2(pi{:}h|M) + (1 - n_h/N_h)n_i\,\sigma^2(ph|M) \\ &\quad + n_i n_h\,\sigma^2(p|M) \\ \boldsymbol{EMS}(ph) &= (1 - n_i/N_i)\sigma^2(pi{:}h|M) + n_i\,\sigma^2(ph|M) \\ \boldsymbol{EMS}(pi{:}h) &= \sigma^2(pi{:}h|M). \end{aligned}$$

(b) In terms of random effects variance components,

$$\sigma^2(p|M) = \sigma^2(p) + \frac{\sigma^2(ph)}{N_h} + \frac{\sigma^2(pi{:}h)}{N_i N_h}.$$

(c) For the random model variance components,

$$\sigma^2(p) = \frac{EMS(p) - EMS(ph)}{n_i n_h},$$

$$\frac{\sigma^2(ph)}{N_h} = \frac{EMS(ph) - EMS(pi{:}h)}{n_i N_h},$$

and

$$\frac{\sigma^2(pi{:}h)}{N_i N_h} = \frac{EMS(pi{:}h)}{N_i N_h}.$$

The sum of these is

$$\sigma^2(p|M) = \frac{EMS(p) - (1 - n_h/N_h)EMS(ph)}{n_i n_h}$$
$$+ \left(\frac{1}{N_i N_h} - \frac{1}{n_i N_h}\right) EMS(pi{:}h).$$

Replacing the previously derived expected mean squares in the above equation proves the equality.

Chapter 4

4.1 In this case, the statement that score effects are uncorrelated means that $E\nu_{ph}\nu_{ph'} = 0$ for $h \neq h'$. It follows that

$$E\left[\frac{\sum_h \nu_{ph}}{n'_h}\right]^2 = E\left[\frac{\sum_h \nu_{ph}^2 + \sum_h \sum_{h'} \nu_{ph}\nu_{ph'}}{n_h'^2}\right]$$
$$= \left[\frac{\sum_h E\nu_{ph}^2 + \sum_h \sum_{h'} E\nu_{ph}\nu_{ph'}}{n_h'^2}\right]$$
$$= \frac{n'_h \sigma^2(ph)}{n_h'^2}$$
$$= \frac{\sigma^2(ph)}{n'_h}.$$

4.4 Using the confounded-effects rule, the effects in the $p \times (i{:}h)$ design can be denoted:

$$p, \qquad h, \qquad i{:}h \Rightarrow (i, ih), \qquad ph, \qquad \text{and} \qquad pi{:}h \Rightarrow (pi, pih).$$

Similarly, the effects in the $(i \times h){:}p$ design can be denoted:

$$p, \qquad i{:}p \Rightarrow (i, pi), \qquad h{:}p \Rightarrow (h, ph), \qquad \text{and} \qquad ih{:}p \Rightarrow (ih, pih).$$

The $i{:}p$ effect in the $(i \times h){:}p$ design involves two effects from a fully crossed design, i and pi. There is no combination of effects in the $p \times (i{:}h)$ design that gives precisely these two effects. A similar statements holds for $ih{:}p$.

However, if $\sigma^2(ih{:}p) = 0$, then $\sigma^2(ih) = \sigma^2(pih) = 0$ because variance components are necessarily positive. In this case, $\sigma^2(i{:}p)$ in the $(i \times h){:}p$ design can be obtained by combining $\sigma^2(i{:}h)$ and $\sigma^2(pi{:}h)$ from the $p \times (i{:}h)$ design.

4.6 The D study results are summarized next.

$\hat{\sigma}^2(\alpha)$ for $p \times (r{:}t)$ Design	$\hat{\sigma}^2(\alpha)$ for $r{:}t{:}p$ Design	D Studies		
		n'_t	1	2
		n'_r	12	6
		$n'_r n'_t$	12	12
$\hat{\sigma}^2(p) = .4731$	$\hat{\sigma}^2(p) = .4731$	$\hat{\sigma}^2(p)$	.473	.473
$\hat{\sigma}^2(t) = .3252$ $\hat{\sigma}^2(pt) = .5596$	$\hat{\sigma}^2(t{:}p) = .8848$	$\hat{\sigma}^2(T{:}p)$	.885	.442
$\hat{\sigma}^2(r{:}t) = .6475$ $\hat{\sigma}^2(pr{:}t) = 2.3802$	$\hat{\sigma}^2(r{:}t{:}p) = 3.0277$	$\hat{\sigma}^2(R{:}T{:}p)$	.252	.252
		$\hat{\sigma}^2(\tau)$	.47	.47
		$\hat{\sigma}^2(\delta) = \hat{\sigma}^2(\Delta)$	1.14	.69
		$E\hat{\rho}^2 = \widehat{\Phi}$	.29	.41

These results are identical to the absolute-error-based results in Table 4.6 for the $p \times (R{:}T)$ design.

4.7 In terms of estimators, the inequality $S/N(\delta) \geq 2$ means that

$$\hat{\sigma}^2(p) \geq 2\left[\frac{\hat{\sigma}^2(c{:}t)}{n'_c} + \frac{\hat{\sigma}^2(p{:}c{:}t)}{20n'_c} + \frac{\hat{\sigma}^2(ti)}{11} + \frac{\hat{\sigma}^2(ci{:}t)}{11n'_c} + \frac{\hat{\sigma}^2(pi{:}c{:}t)}{(20)(11)n'_c}\right],$$

and in terms of the estimates for this example,

$$.0845 \geq 2\left[\frac{.1241}{n'_c} + \frac{.4913}{20n'_c} + \frac{.0346}{11} + \frac{.0323}{11n'_c} + \frac{.5034}{(20)(11)n'_c}\right].$$

Solving this inequality gives $n'_c = 4$.

Alternatively, this exercise can be solved by computing D study results for different values of n'_c until the required value of n'_c is determined. For this example, when $n'_c = 3$, $\widehat{S/N}(\delta) = 1.58$, and when $n'_c = 4$, $\widehat{S/N}(\delta) = 2.08$.

4.9 For the random model,

$$\hat{\sigma}(\overline{X}) = \sqrt{\frac{.47 + .58}{10} + (.80 - .58)} = .57.$$

For the mixed model,

$$\hat{\sigma}(\overline{X}) = \sqrt{\frac{.47 + .58}{10} + (.51 - .40)} = .46.$$

4.10 (a) The effects for the G study design are p, g, $t{:}g$, $r{:}t{:}g$, pg, $pt{:}g$, and $pr{:}t{:}g$. With genre fixed at $n'_g = N'_g = 2$, and $n'_r = n'_t = 1$,

$$\boldsymbol{E}\hat{\rho}^2 = \frac{\hat{\sigma}^2(p) + \hat{\sigma}^2(pg)/2}{\hat{\sigma}^2(p) + \hat{\sigma}^2(pg)/2 + [\hat{\sigma}^2(pt{:}g)/2 + \hat{\sigma}^2(pr{:}t{:}g)/2]}.$$

(b) For the $p \times i$ design,

$$i \Rightarrow g, t{:}g, r{:}t{:}g \quad \text{and} \quad pi \Rightarrow pg, pt{:}g, pr{:}t{:}g,$$

where effects to the left of the arrow are for the $p \times i$ design and those to the right are for the $p \times (r{:}t{:}g)$ design. Let the subscript "1" designate estimated variance components for the single-facet $p \times i$ design. Then, Cronbach's alpha is

$$\boldsymbol{E}\hat{\rho}^2 = \frac{\hat{\sigma}_1^2(p)}{\hat{\sigma}_1^2(p) + \hat{\sigma}_1^2(pi)/2}$$

$$= \frac{\hat{\sigma}^2(p)}{\hat{\sigma}^2(p) + [\hat{\sigma}^2(pg)/2 + \hat{\sigma}^2(pt{:}g)/2 + \hat{\sigma}^2(pr{:}t{:}g)/2]},$$

which is clearly smaller than the estimated generalizability for the $p \times (R{:}T{:}G)$ design with genre fixed.

(c) When genre is fixed, $\hat{\sigma}^2(pg)$ contributes to universe score variance, not relative error variance. By contrast, Cronbach's alpha effectively treats $\hat{\sigma}^2(pg)$ as random, because variability attributable to genre, tasks, and raters is undifferentiated. This means that $\hat{\sigma}^2(pg)$ contributes to relative error variance for Cronbach's alpha.

Chapter 5

5.1 This is the $I{:}(p \times H)$ design. The G study estimated variance components are the same as those in Table 5.1 except for

$$\hat{\sigma}^2(i{:}ph|H) = \hat{\sigma}^2(i{:}h|H) + \hat{\sigma}^2(pi{:}h|H).$$

Similarly, the D study estimated variance components are the same except for

$$\hat{\sigma}^2(I{:}pH|H) = \hat{\sigma}^2(I{:}H|H) + \hat{\sigma}^2(pI{:}H|H).$$

As in the $p \times (I{:}H)$ design with H fixed, the variance components $\hat{\sigma}^2(H|H)$ and $\hat{\sigma}^2(pH|H)$ disappear. Numerical results are provided below.

$\hat{\sigma}^2(\alpha|H)$	D Study Results	
$\hat{\sigma}^2(p|H) = .0378$	$\hat{\sigma}^2(p|H)$	.0378
$\hat{\sigma}^2(h|H) = .0013$	$\hat{\sigma}^2(H|H)$	—
$\hat{\sigma}^2(ph|H) = .0051$	$\hat{\sigma}^2(pH|H)$	—
$\hat{\sigma}^2(i{:}ph|H) = .1848$	$\hat{\sigma}^2(I{:}pH|H)$	.0046
	$\hat{\sigma}^2(\tau)$	.0378
	$\hat{\sigma}^2(\delta) = \hat{\sigma}^2(\Delta)$	.0046
	$E\hat{\rho}^2 = \hat{\Phi}$	.892

5.2 This is the $p \times (i{:}h) \times o$ design. Using a dashed circle to represent the fixed facet h, the Venn diagram and linear model are

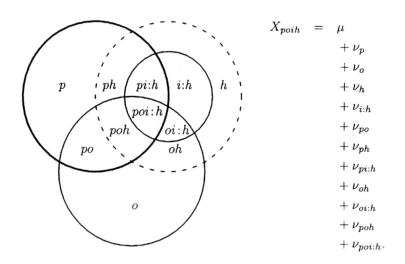

$$
\begin{aligned}
X_{poih} \;=\;\; & \mu \\
& + \nu_p \\
& + \nu_o \\
& + \nu_h \\
& + \nu_{i:h} \\
& + \nu_{po} \\
& + \nu_{ph} \\
& + \nu_{pi:h} \\
& + \nu_{oh} \\
& + \nu_{oi:h} \\
& + \nu_{poh} \\
& + \nu_{poi:h}.
\end{aligned}
$$

Generalization for a single randomly selected occasion means that $n_o' = 1 < N_o' \to \infty$. The G and D study results are as follows.

Effect	MS	$\hat{\sigma}^2(\alpha\vert H)$	$d(\bar{\alpha}\vert p)$	$C(\bar{\alpha}\vert p)$	$\hat{\sigma}^2(\bar{\alpha}\vert H)$
p	2.9304	.0320	1	1	.0320
o	.6068	.0000	1	1	.0000
h	18.4148	.0022	5	0	—
$i{:}h$	10.6986	.0252	40	1	.0006
po	.2085	.0031	1	1	.0031
ph	.2767	.0006	5	0	—
$pi{:}h$	.2406	.0791	40	1	.0020
oh	.5290	.0002	5	0	—
$oi{:}h$	.1742	.0004	40	1	.0000
poh	.1089	.0033	5	0	—
$poi{:}h$	.0825	.0825	40	1	.0021

$$\hat{\sigma}^2(\tau) = .0320$$
$$\hat{\sigma}^2(\delta) = .0072$$
$$\boldsymbol{E}\hat{\rho}^2 = .816$$

5.3 In Exercise 5.6, $\boldsymbol{E}\hat{\rho}^2$ is for a single occasion that is *random*, whereas in Table 5.1 the single occasion is hidden and *fixed*. This difference is the primary reason the $\boldsymbol{E}\hat{\rho}^2$ in Exercise 5.6 is smaller than $\boldsymbol{E}\hat{\rho}^2$ in Table 5.1. The D study variance components in Exercise 5.6 can be used to estimate results for a single level of a fixed, hidden occasion facet in the following manner.

$$\hat{\sigma}^2(\tau) = \hat{\sigma}^2(p\vert H) + \hat{\sigma}^2(po\vert H) = .0320 + .0031 + .0351,$$

$$\hat{\sigma}^2(\delta) = \hat{\sigma}^2(pI{:}H\vert H) + \hat{\sigma}^2(poI{:}H\vert H) = .0020 + .0021 = .0041,$$

and

$$\boldsymbol{E}\rho^2 = \frac{.0351}{.0351 + .0041} = .895.$$

5.7 (a) Inequality 5.25 is not satisfied because

$$\frac{\hat{\sigma}^2(g)}{\hat{\sigma}^2(g) + \hat{\sigma}^2(p{:}g)} = \frac{.07}{.30} = .23 > \frac{\hat{\sigma}^2(gi)}{\hat{\sigma}^2(gi) + \hat{\sigma}^2(pi{:}g)} = \frac{.07}{.50} = .14.$$

(b) For $n_p = 20$ and $n_i = 5$,

$$\boldsymbol{E}\hat{\rho}_g^2 = .70 < \boldsymbol{E}\hat{\rho}_p^2 = .75.$$

Indeed, for $n_i = 5$, $\boldsymbol{E}\hat{\rho}_g^2 < \boldsymbol{E}\hat{\rho}_p^2$ whenever $n_p \leq 33$. Furthermore, for any value of $n_i \geq 1$, $\boldsymbol{E}\hat{\rho}_g^2 < \boldsymbol{E}\hat{\rho}_p^2$ whenever $n_p \leq 14$.

5.9 Using Equation 5.32, the values of $\hat{\sigma}(\Delta)$ for the six persons are: .6667, .8819, .5774, .3333, 1.2019, and 1.0000. The average of their squared values is .6852, which is $\hat{\sigma}^2(\Delta)$, as reported in the answer to Exercise 2.6 in Chapter 2.

Chapter 6

6.1 Using Equation 6.29,

$$\hat{\sigma}[\hat{\sigma}^2(\Delta)] = \frac{1}{(10)(12)} \sqrt{\frac{2(11.6273)^2}{11+2} + \frac{2[9(2.7872)]^2}{(9)(11)+2}} = .0481.$$

6.2 From Table 6.2, for $n_i' = 12$, $\hat{\sigma}^2(\Delta) = .3059$, and using df rather than $df + 2$ in the answer to Exercise 6.6 gives $\hat{\sigma}[\hat{\sigma}^2(\Delta)] = .0509$. Therefore, the ratio in Equation 6.19 is $r = 6.011$. Referring to Appendix D, this ratio falls between the following two rows

r	Lower	Upper
6.000	.8206	1.2606
6.042	.8217	1.2584

Using linear interpolation gives lower-limit and upper-limit values of .8203 and 1.2603, respectively. Multiplying these limit values by $\hat{\sigma}^2(\Delta) = .3059$ gives the interval (.251, .385).

6.5 It is easy to verify that $\hat{\sigma}^2(p) = 4.25$ for the full data matrix. It is also easy, although tedious, to determine the values of $\hat{\sigma}^2(p)$ for each of the $3 \times 3 \times 3 = 27$ possible samples of three persons taken with replacement:

Rows	$\hat{\sigma}^2(p)$	Rows	$\hat{\sigma}^2(p)$	Rows	$\hat{\sigma}^2(p)$
1 1 1	.000	2 1 1	.167	3 1 1	5.222
1 1 2	.167	2 1 2	.167	3 1 2	4.250
1 1 3	5.222	2 1 3	4.250	3 1 3	5.222
1 2 1	.167	2 2 1	.167	3 2 1	4.250
1 2 2	.167	2 2 2	.000	3 2 2	3.444
1 2 3	4.250	2 2 3	3.444	3 2 3	3.444
1 3 1	5.222	2 3 1	4.250	3 3 1	5.222
1 3 2	4.250	2 3 2	3.444	3 3 2	3.444
1 3 3	5.222	2 3 3	3.444	3 3 3	.000

The mean of these 27 estimates is 2.833, and $(3/2) \times 2.833 = 4.25$, which is unbiased since it is an ANOVA estimate. If we want $\hat{\sigma}^2[\hat{\sigma}^2(p)]$ to be an unbiased estimate, then the adjustment factor must be used with the bootstrap estimate of the standard error of $\hat{\sigma}^2(p)$.

6.6 For these data,

Effect	df	MS	$\hat{\sigma}^2$
t	9	19.7500	4.2500
r	11	3.0000	.0833
tr	99	2.7500	2.7500

(a) As discussed in Section 5.4, $\hat{\sigma}^2(\Delta_p)$ is $\hat{\sigma}^2(\overline{X})$ for the full matrix; that is,

$$\hat{\sigma}^2(\Delta_p) = \frac{\hat{\sigma}^2(t)}{n'_t} + \frac{\hat{\sigma}^2(r)}{n'_r} + \frac{\hat{\sigma}^2(tr)}{n'_t n'_r}.$$

When $n'_t = n_t$ and $n'_r = n_r$, $\hat{\sigma}^2(\Delta_p)$ in terms of mean squares is simply

$$\hat{\sigma}^2(\Delta_p) = \frac{MS(t) + MS(r) - MS(tr)}{n_t n_r}.$$

Using either equation, we obtain $\hat{\sigma}^2(\Delta_p) = 1.6667$.

(b) Using Equation 6.1 with $\alpha = \Delta_p$ gives

$$\hat{\sigma}[\hat{\sigma}^2(\Delta_p)] = \frac{1}{n_t n_r} \sqrt{\frac{2[MS(t)]^2}{df(t) + 2} + \frac{2[MS(r)]^2}{df(r) + 2} + \frac{2[MS(tr)]^2}{df(tr) + 2}},$$

which leads to $\hat{\sigma}[\hat{\sigma}^2(\Delta_p)] = 1.6637$.

(c) For any values of n'_t and n'_r, the first equation in (a), above, expressed in terms of mean squares is

$$\hat{\sigma}^2(\Delta_p) = \frac{MS(t)}{n'_t n_r} + \frac{MS(r)}{n'_r n_t} + \left(\frac{1}{n'_t n'_r} + \frac{1}{n'_t n_r} + \frac{1}{n'_r n_t} \right) MS(tr).$$

For $n_t = 3$, $n_r = 4$, $n'_t = 6$, and $n'_r = 2$,

$$\hat{\sigma}^2(\Delta_p) = \frac{MS(t)}{24} + \frac{MS(r)}{6} - \frac{MS(tr)}{8} = .9792.$$

Using Equation 6.1 with $\alpha = \Delta_p$ gives

$$\hat{\sigma}[\hat{\sigma}^2(\Delta_p)] = \sqrt{\frac{2(MS(t))^2}{(24)^2(2 + 2)} + \frac{2(MS(r))^2}{(6)^2(3 + 2)} + \frac{2(MS(tr))^2}{(8)^2(6 + 2)}}$$

$$= .6842.$$

6.8 From Table 6.2, the mean squares in Equations 6.34 and 6.35 are $M_p = 10.2963$, $M_i = 11.6273$, and $M_{pi} = 2.7872$. The required F statistics are:

$$F_{\alpha:9,\infty} = 1.6315 \qquad\qquad F_{1-\alpha:9,\infty} = .4631$$
$$F_{\alpha:9,99} = 1.6956 \qquad\qquad F_{1-\alpha:9,99} = .4567$$
$$F_{\alpha:9,11} = 2.2735 \qquad\qquad F_{1-\alpha:9,11} = .4173.$$

Using Equation 6.34 with the F statistics in the left column gives

$$L_p = \frac{58.3489}{693.5641} = .0813,$$

which leads to a lower limit of .0655 for Λ. Transforming this limit using Equation 6.36 gives a lower limit of .457 for Φ, as reported in Table 6.5. Similar steps using Equation 6.35 with the F statistics in the right column, above, give the upper limit of .845.

Chapter 7

7.1 By the definition of $T(p)$ and the linear model for the $p \times (i:h)$ design,

$$
\begin{aligned}
T(p) &= n_{i+} \sum_p \overline{X}_p^2 \\
&= n_{i+} \sum_p \left[\mu + \nu_p + \frac{\sum_h n_{i:h}\nu_h}{n_{i+}} + \frac{\sum_h \sum_i \nu_{i:h}}{n_{i+}} \right. \\
&\qquad \left. + \frac{\sum_h n_{i:h}\nu_{ph}}{n_{i+}} + \frac{\sum_h \sum_i \nu_{pi:h}}{n_{i+}} \right]^2 . \quad (\mathrm{I.1})
\end{aligned}
$$

Because effects are uncorrelated with zero expectations, the expected values of all $6 \times 6 = 36$ cross-product terms in Equation I.1 are zero. It follows that

$$
\begin{aligned}
ET(p) &= n_{i+} \sum_p E\left[\mu^2 + \nu_p^2 + \left(\frac{\sum_h n_{i:h}\nu_h}{n_{i+}} \right)^2 + \left(\frac{\sum_h \sum_i \nu_{i:h}}{n_{i+}} \right)^2 \right. \\
&\qquad \left. + \left(\frac{\sum_h n_{i:h}\nu_{ph}}{n_{i+}} \right)^2 + \left(\frac{\sum_h \sum_i \nu_{pi:h}}{n_{i+}} \right)^2 \right] . \quad (\mathrm{I.2})
\end{aligned}
$$

For the ν_h term in Equation I.2, the expected value is

$$
E\left(\frac{\sum_h n_{i:h}\nu_h}{n_{i+}} \right)^2 = E\left[\frac{1}{n_{i+}^2} \sum_h n_{i:h}^2 \nu_h^2 + \frac{1}{n_{i+}^2} \sum\sum_{h \neq h'} n_{i:h} n_{i:h'} \nu_h \nu_{h'} \right] .
$$

Since $E\nu_h \nu_{h'} = 0$ for all $h \neq h'$, the last term in the above equation is zero. Also, since $E\nu_h^2 = \sigma^2(h)$,

$$
E\left(\frac{\sum_h n_{i:h}\nu_h}{n_{i+}} \right)^2 = \frac{1}{n_{i+}} \left[\sum_h \frac{n_{i:h}^2}{n_{i+}} \sigma^2(h) \right] .
$$

Similarly, for the ν_{ph} term in Equation I.2, the expected value is

$$
E\left(\frac{\sum_h n_{i:h}\nu_{ph}}{n_{i+}} \right)^2 = \frac{1}{n_{i+}} \left[\sum_h \frac{n_{i:h}^2}{n_{i+}} \sigma^2(ph) \right] .
$$

For the $\nu_{i:h}$ term in Equation I.2, the expected value is

$$
E\left(\frac{\sum_h \sum_i \nu_{i:h}}{n_{i+}} \right)^2 = \frac{n_{i+}\sigma^2(i:h)}{n_{i+}^2} = \frac{\sigma^2(i:h)}{n_{i+}} .
$$

Similarly, for the $\nu_{pi:h}$ term,

$$
E\left(\frac{\sum_h \sum_i \nu_{pi:h}}{n_{i+}} \right)^2 = \frac{n_{i+}\sigma^2(pi:h)}{n_{i+}^2} = \frac{\sigma^2(pi:h)}{n_{i+}} .
$$

It follows that

$$ET(p) = n_p n_{i+} \mu^2 + n_p n_{i+} \sigma^2(p) + n_p r_i \sigma^2(h)$$
$$+ n_p \sigma^2(i{:}h) + n_p r_i \sigma^2(ph) + n_p \sigma^2(pi{:}h),$$

where $r_i = \sum_h n_{i{:}h}^2 / n_{i+}$.

7.2 Since $MS(p) = [T(p) - T(\mu)]/(n_p - 1)$,

$$EMS(p) = \frac{ET(p) - ET(\mu)}{n_p - 1}$$

$$= \frac{(n_p - 1)[\sigma^2(pi{:}h) + r_i \sigma^2(ph) + n_{i+}\sigma^2(p)]}{n_p - 1}$$

$$= \sigma^2(pi{:}h) + r_i \sigma^2(ph) + n_{i+}\sigma^2(p).$$

Similar derivations lead to the other expected mean squares.

7.3 Straightforward algebra.

7.4 Straightforward algebra.

7.5 Straightforward algebra.

7.6 Straightforward computations.

7.7 Given the definition of $\overline{X}_p$ in Equation 7.22, the expected value of the sum of $\overline{X}_p^2$ is

$$E\left(\sum_p \overline{X}_p^2\right) = E\left\{\sum_{p=1}^{n_p}\left[\frac{1}{n_{i{:}p}}\sum_{i=1}^{n_{i{:}p}}(\mu + \nu_p + \nu_{i{:}p})\right]^2\right\}$$

$$= E\left\{\sum_p\left[\mu + \nu_p + \frac{\sum_i \nu_{i{:}p}}{n_{i{:}p}}\right]^2\right\}.$$

When the square is taken, followed by the expectation, the cross-product terms will be zero. Therefore, without loss of generality

$$E\left(\sum_p \overline{X}_p^2\right) = E\left\{\sum_p\left[\mu^2 + \nu_p^2 + \left(\frac{\sum_i \nu_{i{:}p}}{n_{i{:}p}}\right)^2\right]\right\}.$$

The expected value of the last term will involve $n_{i{:}p}$ occurrences of $\sigma^2(i{:}p)$, as well as the expected value of $n_{i{:}p}(n_{i{:}p} - 1)$ cross-product terms, all of which are zero. It follows that

$$E\left(\sum_p \overline{X}_p^2\right) = n_p \mu^2 + n_p \sigma^2(p) + \sum_p \frac{\sigma^2(i{:}p)}{n_{i{:}p}}$$

$$= n_p \mu^2 + n_p \sigma^2(p) + \frac{n_p \sigma^2(i{:}p)}{\ddot{n}_i},$$

where $\ddot{n}_i$ is the harmonic mean of the $n_{i:p}$.

For the definition of $\overline{X}$ in Equation 7.22, the expected value of $\overline{X}^2$ is

$$
\begin{aligned}
E(\overline{X}^2) &= E\left[\frac{1}{n_p}\sum_{p=1}^{n_p}\left(\frac{1}{n_{i:p}}\sum_{i=1}^{n_{i:p}}X_{pi}\right)\right]^2 \\
&= E\left[\frac{1}{n_p}\sum_p\left(\mu + \nu_p + \frac{\sum_i \nu_{i:p}}{n_{i:p}}\right)\right]^2 \\
&= E\left[\mu + \frac{\sum_p \nu_p}{n_p} + \frac{1}{n_p}\sum_p\frac{\sum_i \nu_{i:p}}{n_{i:p}}\right]^2 \\
&= E\left[\mu^2 + \left(\frac{\sum_p \nu_p}{n_p}\right)^2 + \left(\frac{1}{n_p}\sum_p\frac{\sum_i \nu_{i:p}}{n_{i:p}}\right)^2\right] \\
&= \mu^2 + \frac{\sigma^2(p)}{n_p} + \frac{1}{n_p^2}E\left(\sum_p\frac{\sum_i \nu_{i:p}}{n_{i:p}}\right)^2.
\end{aligned}
$$

To determine the expected value of the term in parentheses it is helpful to consider a special case, say two persons with two and three items per person. For this special case, the expected value is

$$
E\left(\frac{\nu_{1:1} + \nu_{2:1}}{2} + \frac{\nu_{1:2} + \nu_{2:2} + \nu_{3:2}}{3}\right)^2.
$$

Since the expected value of the square of each $\nu_{i:p}$ term is $\sigma^2(i:p)$, and the expected value of each cross-product term is zero, this special case leads to $\sigma^2(i:p)/2 + \sigma^2(i:p)/3$. Generalizing this logic gives

$$
\begin{aligned}
E(\overline{X}^2) &= \mu^2 + \frac{\sigma^2(p)}{n_p} + \frac{1}{n_p^2}\sum_p\left(\frac{1}{n_{i:p}}\right)\sigma^2(i:p) \\
&= \mu^2 + \frac{\sigma^2(p)}{n_p} + \frac{\sigma^2(i:p)}{n_p\ddot{n}_i}.
\end{aligned}
$$

Chapter 8

8.1 Straightforward computations.

8.2 When the $p \times I$ design is used to analyze a situation that is more correctly characterized as a $p \times (I:H)$ design, $\hat{\sigma}^2(pH)$ and $\hat{\sigma}^2(pI:H)$ for the $p \times (I:H)$ design are confounded in $\hat{\sigma}^2(pI)$ for the $p \times I$ design. Therefore, an estimate of $\sigma^2(\delta)$ for the $p \times I$ design is

$$
\frac{\hat{\sigma}^2(ph) + \hat{\sigma}^2(pi:h)}{n_{i+}} = \frac{.0098 + .1710}{40} = .0045,
$$

which gives $E\hat{\rho}^2 = .0219/(.0219 + .0045) = .830$. These estimates are questionable because they fail to distinguish between the effects attributable to stimuli and items. In particular, the $\hat{\sigma}^2(ph)$ effect is divided by the total number of items, rather than the number of stimuli, which leads to an underestimate of error variance and a corresponding overestimate of reliability.

8.3 For $E\hat{\rho}^2 \geq .85$, relative error variance must be less than or equal to

$$\left[\frac{1 - E\hat{\rho}^2}{E\hat{\rho}^2}\right]\hat{\sigma}^2(p) = \left[\frac{1 - .85}{.85}\right].0219 = .00386.$$

Therefore, we need values for n'_h, n'_{i+}, and the $n'_{i:h}$ such that

$$\hat{\sigma}^2(\delta) = \frac{\hat{\sigma}^2(ph)}{\breve{n}'_h} + \frac{\hat{\sigma}^2(pi:h)}{n'_{i+}} \leq .00386 \qquad (\text{I.3})$$

subject to the constraint that

$$2n'_h + n'_{i+}/2 \leq 50. \qquad (\text{I.4})$$

Analytic strategies and/or trial-and-error give the following results:

n'_h	n'_{i+}	$n'_{i:h}$	$\breve{n}'_h$	$\hat{\sigma}^2(\delta)$	$E\hat{\rho}^2$	Min.
7	70	10,10,10,10,10,10,10	10.0000	.00342	.86493	49.0
8	65	8,8,8,8,8,8,8,9	7.9868	.00386	.85016	48.5
9	62	6,7,7,7,7,7,7,7,7	8.9813	.00385	.85049	49.0
10	60	6,6,6,6,6,6,6,6,6,6	10.0000	.00383	.85115	50.0

Note that for six stimuli there are no sample sizes that satisfy both Equations I.3 and I.4 simultaneously; the same holds for 11 stimuli.

One way to obtain these results is to (a) specify some value of n'_h and use it for $\breve{n}'_h$ in Equation I.3, which can then be used to get a preliminary value for n'_{i+}; (b) round up this value of n'_{i+} and find the "flattest" pattern of n'_h values for the $n'_{i:h}$ that sum to n'_{i+}; (c) verify that these sample sizes give $E\hat{\rho}^2 \geq .85$; and (d) determine whether these sample sizes lead to a test that is no longer than 50 minutes. For example, using $n'_h = 8$ for $\breve{n}'_h$ in Equation I.3 gives

$$n'_{i+} = \frac{\hat{\sigma}^2(pi:h)}{\hat{\sigma}^2(\delta) - [\hat{\sigma}^2(ph)/7]} = \frac{.1710}{.00386 - (.0098/8)} = 64.90,$$

which means that we need to distribute 65 items over the eight stimuli. The flattest distribution (i.e., the one that leads to the largest value of $\breve{n}'_h$) has seven stimuli with eight items and one with nine items, which gives $\breve{n}'_h = 7.9868$. We can now determine that the estimated generalizability coefficient is $E\hat{\rho}^2 = .85016 < .85$. For these sample sizes, the total amount of testing time is $(2 \times 7) + 70/2 = 49$ minutes.

The differences among the four choices in terms of error variances, reliability, and testing time appear to be minimal, so these statistics do not provide any compelling basis for choosing one alternative over another. However, in many testing contexts, creating stimuli is a more expensive process than writing a small number of test items. From this point of view, the first alternative with seven stimuli seems preferable.

8.7 Since there is no systematic assignment of raters to students, there is no basis for believing that the G study estimated variance components are biased in the sense of being systematically different from what they would have been if raters were fully nested within students. Since some raters evaluated more than one student, it seems likely that the intended universe of admissible observations has raters crossed with students. Consequently, the G study estimate of $\sigma^2(r{:}p)$ is a confounded estimate of $\sigma^2(r)$ and $\sigma^2(rp)$, and the G study estimate of $\sigma^2(ri{:}p)$ is a confounded estimate of $\sigma^2(ri)$ and $\sigma^2(rip)$.

Of course, we do not know how much of $\hat{\sigma}^2(r{:}p)$ is attributable to r and how much is attributable to rp; nor do we know how much of $\hat{\sigma}^2(ri{:}p)$ is attributable to ri and how much is attributable to rip. Not knowing this type of information is sometimes very problematic in estimating D study quantities, but not always. In this case, our estimates of $\sigma(\Delta)$ and $E\rho^2$ are unaffected by whether the D study design has raters crossed with students or nested within students. This is evident from the fact that

$$\frac{\sigma^2(r{:}p)}{n_r'} = \frac{\sigma^2(r, rp)}{n_r'} = \frac{\sigma^2(r)}{n_r'} + \frac{\sigma^2(rp)}{n_r'},$$

with a similar result for $\sigma^2(ri{:}p)$.

8.8 Straightforward computations.

Chapter 9

9.1 The first question in this exercise requires only straightforward computations. The quantity $\hat{\sigma}^2(v|V)$ is a quadratic form that estimates

$$\sum_v v_v^2/(n_v - 1) = \sum_v (\mu_v - \mu)^2/(n_v - 1),$$

where the μ_v are category means that are not estimable from the three matrices.

9.2

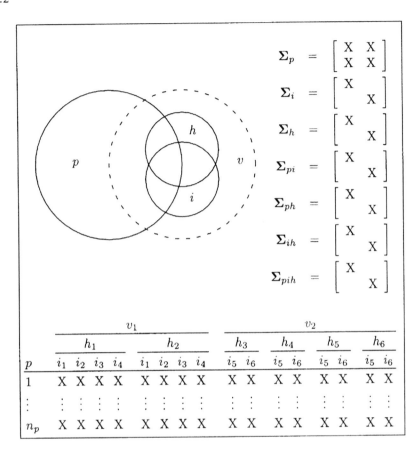

$$\Sigma_p = \begin{bmatrix} X & X \\ X & X \end{bmatrix}$$

$$\Sigma_i = \begin{bmatrix} X & \\ & X \end{bmatrix}$$

$$\Sigma_h = \begin{bmatrix} X & \\ & X \end{bmatrix}$$

$$\Sigma_{pi} = \begin{bmatrix} X & \\ & X \end{bmatrix}$$

$$\Sigma_{ph} = \begin{bmatrix} X & \\ & X \end{bmatrix}$$

$$\Sigma_{ih} = \begin{bmatrix} X & \\ & X \end{bmatrix}$$

$$\Sigma_{pih} = \begin{bmatrix} X & \\ & X \end{bmatrix}$$

	v_1								v_2						
	h_1				h_2				h_3		h_4		h_5		h_6	
p	i_1	i_2	i_3	i_4	i_1	i_2	i_3	i_4	i_5	i_6	i_5	i_6	i_5	i_6	i_5	i_6
1	X	X	X	X	X	X	X	X	X	X	X	X	X	X	X	X
$\vdots$	$\vdots$	$\vdots$	$\vdots$	$\vdots$	$\vdots$	$\vdots$	$\vdots$	$\vdots$	$\vdots$	$\vdots$	$\vdots$	$\vdots$	$\vdots$	$\vdots$	$\vdots$	$\vdots$
n_p	X	X	X	X	X	X	X	X	X	X	X	X	X	X	X	X

9.4 Straightforward computation.

9.5 For a $p^\bullet \times i^\circ$ design, $\hat{\sigma}_{vv'}(p) = S_{vv'}(p)$ (see Equation 9.8), and it is easy to show that
$$S_{vv'}(p) = MP_{vv'}(p)/n_i.$$

Therefore, the estimated covariance component is $3.4611/6 = .5769$.

9.7 $\hat{\sigma}[\hat{\sigma}_{vv'}(i)] = .190$ and $\hat{\sigma}[\hat{\sigma}_{vv'}(pi)] = .227$.

9.9 Recall the general formula for the variance of a difference:
$$\mathrm{var}(X - Y) = \mathrm{var}(X) + \mathrm{var}(Y) - 2\,\mathrm{cov}(X, Y).$$

It follows that
$$\mathrm{cov}(X, Y) = \frac{\mathrm{var}(X) + \mathrm{var}(Y) - \mathrm{var}(X - Y)}{2},$$

or, in the notational conventions of this chapter,

$$\sigma_{vv'}(\alpha) = \frac{\sigma_v^2(\alpha) + \sigma_{v'}^2(\alpha) - \sigma_{v-v'}^2(\alpha)}{2}.$$

It follows that the variance-covariance matrices are:

$$\widehat{\Sigma}_p = \begin{bmatrix} .0058 & -.0031 \\ -.0031 & .0334 \end{bmatrix}, \quad \widehat{\Sigma}_h = \begin{bmatrix} .1238 & .0519 \\ .0519 & .0298 \end{bmatrix}$$

$$\widehat{\Sigma}_{i:h} = \begin{bmatrix} .0279 & .0048 \\ .0048 & .0028 \end{bmatrix}, \quad \widehat{\Sigma}_{ph} = \begin{bmatrix} .0177 & .0035 \\ .0035 & .0606 \end{bmatrix}$$

$$\widehat{\Sigma}_{pi:h} = \begin{bmatrix} .0432 & -.0004 \\ -.0004 & .0438 \end{bmatrix}.$$

Chapter 10

10.1 The universe score, relative error, and absolute error matrices are

$$\widehat{\Sigma}_\tau = \begin{bmatrix} 6.2857 & 11.8571 & 2.0000 \\ 11.8571 & 44.5714 & 7.5714 \\ 2.0000 & 7.5714 & 7.4286 \end{bmatrix}$$

$$\widehat{\Sigma}_\delta = \begin{bmatrix} 1.1428 & & \\ & 4.2856 & \\ & & .8572 \end{bmatrix}$$

$$\widehat{\Sigma}_\Delta = \begin{bmatrix} 2.0000 & & \\ & 5.0832 & \\ & & 1.7500 \end{bmatrix}.$$

For the total score metric the w weights are all unity. The composite universe score variance, relative error variance, and absolute error variance in the total score metric are, respectively, $\hat{\sigma}_C^2(\tau) = 101.143$, $\hat{\sigma}_C^2(\delta) = 6.286$, and $\hat{\sigma}_C^2(\Delta) = 8.833$. It is now easily verified that $E\hat{\rho}^2$ and $\widehat{\Phi}$ are the same values as for the mean score metric reported previously.

10.2 Since the a priori weights are all equal, they can be eliminated from Equation 10.46, which results in optimal sample sizes of

$$n'_{i1} = \frac{10\sqrt{.3444 + 1.2522}}{\sqrt{.4286 + .5714} + \sqrt{.1994 + 1.0714} + \sqrt{.4464 + .4286}} = 3.2651,$$

$$n'_{i2} = \frac{10\sqrt{.1994 + 1.0714}}{\sqrt{.4286 + .5714} + \sqrt{.1994 + 1.0714} + \sqrt{.4464 + .4286}} = 3.6808,$$

and

$$n'_{i3} = \frac{10\sqrt{.4464 + .4286}}{\sqrt{.4286 + .5714} + \sqrt{.1994 + 1.0714} + \sqrt{.4464 + .4286}} = 3.0542.$$

Using $n'_{i1} = 3$, $n'_{i2} = 4$, and $n'_{i3} = 3$ gives $\hat{\sigma}_C^2(\Delta) = .105$.

10.5 $\hat{\sigma}_C^2(\Delta) = .1641$ for both the $p^\bullet \times I^\bullet$ and $I^\bullet:p^\bullet$ designs.

10.6 $\hat{\sigma}_C^2(\tau) = .3499$ for both the $p^\bullet \times I^\bullet$ and $p^\bullet \times I^\circ$ designs, but $\hat{\sigma}_C^2(\delta)$ is different for the two designs. For the $p^\bullet \times I^\circ$ design,

$$\widehat{\boldsymbol{\Sigma}}_\delta = \begin{bmatrix} .1565 & \\ & .1921 \end{bmatrix}$$

and, consequently,

$$\hat{\sigma}_C^2(\delta) = .75^2(.1565) + .25^2(.1921) = .1000.$$

Therefore, $\boldsymbol{E}\hat{\rho}^2 = .3499/(.3499 + .1000) = .78$.

If the D study design were $p^\bullet \times I^\circ$, it would mean that different items were used to measure accuracy from those used to measure speed, but all persons took the same items. If items were administered by computer, it would be trivially easy to get a measure of speed for any item. So, this design seems unlikely.

10.8 Using Equation 10.47

$$\hat{\sigma}_C(\Delta_1) = \sqrt{(.75)^2 \left(\frac{1.0667}{8}\right) + (.25)^2 \left(\frac{.8}{8}\right) + 2(.75)(.25)\left(\frac{.6}{8}\right)}$$

$$= .3307.$$

The parameters in this question are the same as those in Section 10.1.6 where it was determined that $\hat{\sigma}_C^2(\Delta) = .1641$ and $\hat{\sigma}_C^2(\delta) = .1331$. Therefore, using Equation 10.48

$$\hat{\sigma}_C(\delta_1) = \sqrt{(.3307)^2 - (.1641 - .1331)} = .2799.$$

10.12 (a) Recall that G study variance and covariance components are for single conditions of facets. It follows that there is no difference between $\widehat{\boldsymbol{\Sigma}}_p$ and $\widehat{\boldsymbol{\Sigma}}_{pi}$ for the mean-score and the total-score metrics. To obtain $\widehat{\boldsymbol{\Sigma}}_p$ and $\widehat{\boldsymbol{\Sigma}}_{pi}$, we make use of the equation

$$S_1^2(p) = \hat{\sigma}_1^2(p) + \frac{\hat{\sigma}_1^2(pi)}{n_i} = \left(\frac{3.64}{24}\right)^2 = .0230.$$

Since Cronbach's alpha is $\boldsymbol{E}\hat{\rho}^2$ for this design,

$$\hat{\sigma}_1^2(p) = .70 \times .0230 = .0161.$$

In a similar manner, we obtain $S_2^2(p) = .8281$ and $\hat{\sigma}_2^2(p) = .4141$. Since,

$$r_{12}(p) = \frac{\hat{\sigma}_{12}(p)}{S_1(p)S_2(p)},$$

$$\hat{\sigma}_{12}(p) = .40 \times \sqrt{.0230} \times \sqrt{.8281} = .0552.$$

Now,

$$\hat{\sigma}_1^2(pi) = n_i[S_1^2(p) - \hat{\sigma}_1^2(p)] = 24(.0230 - .0161) = .1654,$$

and similarly

$$\hat{\sigma}_2^2(pi) = n_i[S_2^2(p) - \hat{\sigma}_2^2(p)] = 2(.8281 - .4141) = .8281.$$

It follows that the estimated G study variance-covariance matrices are

$$\widehat{\boldsymbol{\Sigma}}_p = \begin{bmatrix} .0161 & .0552 \\ .0552 & .4141 \end{bmatrix} \quad \text{and} \quad \widehat{\boldsymbol{\Sigma}}_{pi} = \begin{bmatrix} .1656 & \\ & .8281 \end{bmatrix}.$$

(b) In the mean-score metric, $\widehat{\boldsymbol{\Sigma}}_\tau$ and $\widehat{\boldsymbol{\Sigma}}_\delta$ are

$$\widehat{\boldsymbol{\Sigma}}_\tau = \begin{bmatrix} .0161 & .0552 \\ .0552 & .4141 \end{bmatrix} \quad \text{and} \quad \widehat{\boldsymbol{\Sigma}}_\delta = \begin{bmatrix} .0069 & \\ & .4141 \end{bmatrix}.$$

$\widehat{\boldsymbol{\Sigma}}_\tau$ and $\widehat{\boldsymbol{\Sigma}}_\delta$ for the total-score metric are obtained from the corresponding mean-score metric matrices by multiplying (1,1) elements by $24 \times 24 = 576$, (2,2) elements by $2 \times 2 = 4$, and (1,2) and (2,1) elements by $24 \times 2 = 48$:

$$\widehat{\boldsymbol{\Sigma}}_\tau = \begin{bmatrix} 9.2747 & 2.6499 \\ 2.6499 & 1.6562 \end{bmatrix} \quad \text{and} \quad \widehat{\boldsymbol{\Sigma}}_\delta = \begin{bmatrix} 3.9749 & \\ & 1.6562 \end{bmatrix}.$$

(c) True score variance for the sum of the part scores (in the total score metric) is simply the sum of the elements in the total-score $\widehat{\boldsymbol{\Sigma}}_\tau$ matrix, which is 16.2308. Similarly, error variance for the sum of the part scores is the sum of the elements in the total-score $\widehat{\boldsymbol{\Sigma}}_\delta$ matrix, which is 5.6311. These two results lead to $E\hat{\rho}^2 = .742$.

(d) The w weights need to be set proportional to the sample sizes of 24 and 2. Specifically, with weights of $w_1 = 24/26 = .9231$ and $w_2 = 2/26 = .0769$, using the mean-score $\widehat{\boldsymbol{\Sigma}}_\tau$ and $\widehat{\boldsymbol{\Sigma}}_\delta$ matrices, we obtain $\hat{\sigma}_C^2(\tau) = .0240$ and $\hat{\sigma}_C^2(\delta) = .0083$, which leads to an estimated generalizability coefficient of $E\hat{\rho}^2 = .742$.

(e) Clearly, the classical test theory procedure is simpler, but the fact that it employs a simple sum of part scores may appear to suggest that the two parts are equally weighted. Such a conclusion may be misleading in that it potentially confuses two issues: the number of items contributing to each part, and the investigator's perspective on the relative importance of the constructs tested by the two parts. Using generalizability theory procedures with the mean-score metric forces the investigator to distinguish between the number of items in each part and the intended relative importance of the constructs measured by each part, for whatever decision is to be made.

(f) Using Equation 10.34, in the mean-score metric, the contribution of Part 1 to composite universe score variance is

$$ew_1(\tau) = \frac{(.9231)^2(.0161) + (.9231)(.0769)(.0552)}{.0240} = .735.$$

In the total-score metric,

$$\frac{9.2747 + 2.6499}{16.2308} = .735.$$

(g) With only a single rating for each item, the effects for raters and items are confounded in the data. If a different rater is used to evaluate responses to each of the two open-ended items, then conceptually the design for Part 2 is $p \times (R:I)$ with $n'_r = 1$, and the error variance is

$$\sigma_2^2(\delta) = \frac{\sigma_2^2(pi)}{2} + \frac{\sigma_2^2(pr:i)}{2} = \frac{\sigma_2^2(pi, pr:i)}{2}.$$

In this case, the confounding of raters and items does not render Cronbach's alpha inappropriate, because both variance components have the same divisor. Cronbach's alpha will not be able to distinguish between the two sources of error, but undifferentiated classical error variance is still appropriate.

However, suppose the same rater is used to evaluate responses to both items. Then, conceptually the design is $p \times I \times R$ with $n'_r = 1$, and the error variance should be

$$\sigma_2^2(\delta) = \frac{\sigma_2^2(pi)}{2} + \sigma_2^2(pr) + \frac{\sigma_2^2(ptr)}{2},$$

but Cronbach's alpha will not be able to distinguish between the different contributions of $\sigma_2^2(pi)$ and $\sigma_2^2(pr)$. The undifferentiated error in Cronbach's alpha will be $\sigma_2^2(pi, pr, ptr)/2$, which is too small.

10.15 The universe score and error matrices are:

$$\widehat{\Sigma}_\tau = \widehat{\Sigma}_p + \frac{1}{5}\widehat{\Sigma}_{ph} = \begin{bmatrix} .0093 & -.0024 \\ -.0024 & .0455 \end{bmatrix}$$

$$\widehat{\Sigma}_\delta = \frac{1}{15}\widehat{\Sigma}_{pi:h} = \begin{bmatrix} .0029 & -.0000 \\ -.0000 & .0029 \end{bmatrix}$$

$$\widehat{\Sigma}_\Delta = \frac{1}{15}\left(\widehat{\Sigma}_{i:h} + \widehat{\Sigma}_{pi:h}\right) = \begin{bmatrix} .0047 & .0003 \\ .0003 & .0031 \end{bmatrix}.$$

For the difference score composite, $w_1 = -1$ and $w_2 = 1$. It follows that

$$\hat{\sigma}_C^2(\tau) = .0596, \quad \hat{\sigma}_C^2(\delta) = .0058, \quad \text{and} \quad \hat{\sigma}_C^2(\Delta) = .0072.$$

Using the same type of development as discussed in Section 2.5.1,

$$\text{Est}[(\mu_2 - \mu_1)^2] = (\overline{X}_2 - \overline{X}_1)^2 - \hat{\sigma}^2(\overline{X}_2 - \overline{X}_1),$$

where, using the multivariate version of Equation 4.20,

$$\hat{\sigma}^2(\overline{X}_2 - \overline{X}_1) = \hat{\sigma}^2(\overline{X}_C) = \frac{\hat{\sigma}_C^2(\tau) + \hat{\sigma}_C^2(\delta)}{n_p} + [\hat{\sigma}_C^2(\Delta) - \hat{\sigma}_C^2(\delta)].$$

It follows that

$$
\begin{aligned}
\text{Est}[(\mu_2 - \mu_1)^2] &= (.7955 - .2578)^2 \\
&\quad - \left[\frac{.0596 + .0058}{30} + (.0072 - .0058) \right] \\
&= .2891 - .0036 \\
&= .2855.
\end{aligned}
$$

and

$$\widehat{E/T} = \sqrt{\frac{.0072}{.0596 + .2855}} = .144.$$

Chapter 11

11.1 Begin by noting that $n_{i:p} = m_{i:p}$ for all p,

$$X_{piv} = \mu_v + \nu_p + \nu_{i:p}$$

and

$$X_{piv'} = \mu_{v'} + \xi_p + \xi_{i:p}.$$

Now,

$$
\begin{aligned}
ETP_{vv'}(p) &= E\left(\sum_p n_{i:p} \overline{X}_{pv} \overline{X}_{pv'} \right) \\
&= \sum_p n_{i:p} E\left(\overline{X}_{pv} \overline{X}_{pv'} \right) \\
&= \sum_p n_{i:p} E\left[\left(\mu_v + \nu_p + \sum_i \frac{\nu_{i:p}}{n_{i:p}} \right) \left(\mu_{v'} + \xi_p + \sum_i \frac{\xi_{i:p}}{n_{i:p}} \right) \right].
\end{aligned}
$$

Because of the zero expectations for cross-product terms given by Equation 9.30,

$$ETP_{vv'}(p) = \sum_p n_{i:p} E\left[\mu_v \mu_{v'} + \nu_p \xi_p \left(\sum_i \frac{\nu_{i:p}}{n_{i:p}} \right) \left(\sum_i \frac{\xi_{i:p}}{n_{i:p}} \right) \right].$$

Because $\boldsymbol{E}\nu_{i:p}\xi_{i':p} = 0$ for $i \neq i'$,

$$\boldsymbol{E}\,TP_{vv'}(p) = n_+\mu_v\mu_{v'} + n_+\sigma_{vv'}(p) + \sum_p n_{i:p}\frac{\sigma_{vv'}(i:p)}{n_{i:p}}$$
$$= n_+\mu_v\mu_{v'} + n_+\sigma_{vv'}(p) + n_p\sigma_{vv'}(i:p).$$

11.3 When $n_{i:h} = n_i$ and $m_{i:h} = m_i$ for all levels of h,

$$t = \frac{\sum_h n_{i:h}m_{i:h}}{\left(\sum_h n_{i:h}\right)\left(\sum_h m_{i:h}\right)} = \frac{n_h n_i m_i}{(n_h n_i)(n_h m_i)} = \frac{1}{n_h},$$

and

$$1 - t = (n_h - 1)/n_h.$$

Therefore,

$$\hat{\sigma}_{vv'}(ph) = \frac{CP_{vv'}(ph) - n_h\,CP_{vv'}(p) - n_p\,CP_{vv'}(h) + n_p n_h\,CP_{vv'}(\mu)}{(n_h - 1)(n_p - 1)}$$
$$= \frac{TP_{vv'}(ph) - TP_{vv'}(p) - TP_{vv'}(h) + TP_{vv'}(\mu)}{(n_h - 1)(n_p - 1)}$$
$$= MP_{vv'}(ph).$$

Also,

$$\hat{\sigma}_{vv'}(p) = \frac{CP_{vv'}(p) - n_p\,CP_{vv'}(\mu)}{n_p - 1} - t\,\hat{\sigma}_{vv'}(ph)$$
$$= \frac{TP_{vv'}(p) - TP_{vv'}(\mu)}{n_h(n_p - 1)} - \frac{MP_{vv'}(ph)}{n_h}$$
$$= \frac{MP_{vv'}(p) - MP_{vv'}(ph)}{n_h},$$

and

$$\hat{\sigma}_{vv'}(h) = \frac{CP_{vv'}(h) - n_h\,CP_{vv'}(\mu)}{n_h - 1} - \frac{\hat{\sigma}_{vv'}(ph)}{n_p}$$
$$= \frac{TP_{vv'}(h) - TP_{vv'}(\mu)}{n_p(n_h - 1)} - \frac{MP_{vv'}(ph)}{n_p}$$
$$= \frac{MP_{vv'}(h) - MP_{vv'}(ph)}{n_p}.$$

11.5 Since C terms are simply CP terms with $v = v'$, we can obtain equations for the C-terms estimators of the variance components by replacing CP terms with C terms in Equations 11.27 and 11.28, which gives:

$$\hat{\sigma}^2(i{:}p) = \frac{n_p C(i{:}p) - n_+ C(p)}{n_p n_+} \left(\frac{\ddot{n}_i}{\ddot{n}_i - 1} \right)$$

$$\hat{\sigma}^2(p) = \frac{C(p) - n_p C(\mu)}{n_p(1 - r_i/n_+)} - \frac{(n_+ - \ddot{n}_i)\hat{\sigma}^2(i{:}p)}{\ddot{n}_i(n_+ - r_i)}.$$

Using these equations for v in Table 11.6 gives

$$\hat{\sigma}^2(i{:}p) = \frac{10(876.0000) - 50(173.1625)}{10(50)} \left(\frac{4.8781}{1 - 4.8781} \right) = .2563,$$

and

$$\hat{\sigma}^2(p) = \frac{173.1625 - 10(16.9744)}{10(1 - 5.12/50)} - \frac{(50 - 4.8781)(.2563)}{4.8781(50 - 5.12)} = .3280.$$

The estimates for v_2 and $v_1 + v_2$ are obtained in a similar manner.

11.8 Using Equation Set 7.17 with T terms replaced by TP terms and estimators of variance components replaced by estimators of covariance components,

$$\hat{\sigma}_{12}(pi) = \frac{\begin{bmatrix} 1.0093(1356.0000 - 1323.9333) \\ +1.0052(1356.0000 - 1303.2056) \\ -(1356.0000 - 1297.9649) \end{bmatrix}}{57 - 9.5614 - 5.7368 + 1} = .6416$$

$$\hat{\sigma}_{12}(p) = \frac{1356.0000 - 1303.2056}{57 - 6} - .6416 = .3936$$

$$\hat{\sigma}_{12}(i) = \frac{1356.0000 - 1323.9333}{57 - 10} - .6416 = .0407.$$

11.10 The following table provides the CP terms, as well as the coefficients of the estimate of $\mu_v \mu_{v'}$ and the estimated covariance components.

		Coefficients		
CP	$\widehat{\mu_1 \mu_2}$	$\hat{\sigma}_{12}(p)$	$\hat{\sigma}_{12}(i)$	$\hat{\sigma}_{12}(pi)$
$CP_{12}(\mu) = 22.8051$	1.0000	.0999	.1669	.0167
$CP_{12}(p) = 231.8222$	10.0000	10.0000	1.6667	1.6667
$CP_{12}(i) = 137.5986$	6.0000	.5988	6.0000	.5988
$CP_{12}(pi) = 1356.0000$	57.0000	57.0000	57.0000	57.0000

Using algebraic procedures or matrix operators, the four equations can be solved to obtain the following estimates: $\hat{\sigma}_{12}(p) = .3126$, $\hat{\sigma}_{12}(i) = .0900$, and $\hat{\sigma}_{12}(pi) = .6388$.

11.11 Essentially, this question is asking for an an approximate value for KR20 based on the 26 items in the MD example. One way to answer this question involves relating the variance and covariance components for the $p^\bullet \times (i^\circ : h^\circ)$ design to the variance components for the $p \times i$ design. Since there are an equal number of items in v_1 and v_2, collapsing over both h and v means that

$$\sigma^2(i) = .5\,[\sigma_1^2(ph) + \sigma_1^2(pi{:}h)] + .5\,[\sigma_2^2(ph) + \sigma_2^2(pi{:}h)],$$

where $\sigma^2(i)$ without a subscript designates the variance component for items for the undifferentiated domain. In terms of the estimates in Table 11.9,

$$\hat\sigma^2(i) = .5\,(.0032 + .1832) + .5\,(.0116 + .1831) = .1905.$$

Also, $\sigma^2(p)$ for the univariate design is the composite universe score variance for the multivariate design with $w_1 = w_2 = .5$; therefore,

$$\sigma^2(p) = .25\,\sigma_1^2(p) + .25\,\sigma_1^2(p) + .50\,\sigma_{12}(p),$$

and in terms of estimates

$$\hat\sigma^2(p) = .25(.0404) + .25(.0319) + .50(.0354) = .0358.$$

It follows that an estimate of the generalizability coefficient for the univariate $p \times I$ design with an undifferentiated set of 26 items is

$$E\hat\rho^2 = \frac{.0358}{.0358 + .1905/26} = .830.$$

This is not the only way that $E\rho^2$ could be estimated, but it is probably the simplest approach. Another method would involve estimating universe score variance as

$$\hat\sigma^2(p) = S^2(p) - \hat\sigma^2(pi)/26,$$

where

$$
\begin{aligned}
S^2(p) &= .25\,S_1^2(p) + .25\,S_1^2(p) + .50\,S_{12}(p) \\
&= .25\left[\frac{MS_1(p)}{13}\right] + .25\left[\frac{MS_2(p)}{13}\right] + .50\,S_{12}(p) \\
&= .25\left(\frac{.7299}{13}\right) + .25\left(\frac{.6727}{13}\right) + .50(.0354) \\
&= .0373.
\end{aligned}
$$

This approach leads to

$$E\hat\rho^2 = \frac{.0373}{.0373 + .1905/26} = .835.$$

For the $p^\bullet \times (i^\circ : h^\circ)$ design, Table 11.10 reports that $E\hat{\rho}^2 = .801$, which is smaller than the $p \times I$ estimates by about .03. Theory guarantees that the multivariate result must be no larger than the univariate result. In this example, the difference is small because both of the estimated ph variance components for the multivariate $p^\bullet \times (i^\circ : h^\circ)$ design are small.

11.13 Conceptually, the error variance in Equation 11.54 is given by Equation 11.52 with $n'_o = 1$. Using the Spearman–Brown formula has the effect of halving every one of the variance and covariance components in Equation 11.52, which is precisely the same as using $n'_o = 2$ in that equation.

Chapter 12

12.1 This is a straightforward generalization of Equations 12.5 and 12.6 recognizing that $\sigma(T_v) = \sigma_v(p)$ in the notation of this chapter.

12.3 Equation 12.18 states that

$$ S_{\hat{Y}}^2 = \sum_{j=1}^{k} b_j S_{Y X_j}. $$

In this case, $k = n_v$, $Y = T_v$, and $S_{\hat{Y}}^2 = \sigma^2(\mu_p)$. It follows that

$$ S_{Y X_j} = S_{T_v X_j} = S_{T_v T_j} = \sigma_{vj}, $$

which, with the addition of v subscripts, gives Equation 12.52.

12.5 We prove the result for v_1. The proof for v_2 is entirely analogous. For standard scores, the variance of the regressed score estimates is R^2 (see Table 12.2). Therefore, it is sufficient to prove that

$$ \sigma_1^2(\hat{Z}_p) = \beta_{11}\rho_1\rho_{11} + \beta_{21}\rho_2\rho_{12} \geq \rho_1^2. $$

Using the expression for the βs in Equation 12.26, this inequality is

$$ \rho_1 \left(\frac{\rho_1 - \rho_2\rho_{12}r_{12}}{1 - r_{12}^2} \right) + \rho_2\rho_{12} \left(\frac{\rho_2\rho_{12} - \rho_1 r_{12}}{1 - r_{12}^2} \right) \geq \rho_1^2 $$

or, equivalently,

$$ \rho_1^2 - \rho_1\rho_2\rho_{12}r_{12} + \rho_2^2\rho_{12}^2 - \rho_1\rho_2\rho_{12}r_{12} \geq \rho_1^2(1 - r_{12}^2) $$

$$ \rho_2^2\rho_{12}^2 - 2\rho_1\rho_2\rho_{12}r_{12} + \rho_1^2 r_{12}^2 \geq 0 $$

$$ (\rho_2\rho_{12} - \rho_1 r_{12})^2 \geq 0. $$

The last inequality is necessarily true for all values of ρ_1, ρ_2, ρ_{12}, and r_{12}.

12.7 The computations follow the steps illustrated in Section 12.2.2 for v_1, with one primary exception—the b_1 and b_2 values are obtained using Equation 12.30 instead of Equation 12.28. From another perspective, the difference is that the right side of the normal equations changes. For example, for the $p^{\bullet} \times I^{\bullet}$ design with $n_i' = 8$, the normal equations for v_2 are

$$.5247b_1 + .4074b_2 = .3193$$
$$.4074b_1 + .5610b_2 = .3689,$$

rather than those given by Equation 12.63 for v_1.

12.9 Using the numerical results reported in Equations 12.46 to 12.49,

$$\sigma_2 = \sqrt{.3689} = .6074, \qquad S_1 = \sqrt{.5769} = .7595,$$

$$\rho_1 = \sqrt{.6382} = .7989, \quad \rho_2 = \sqrt{.5902} = .7683, \quad \text{and} \quad \rho_{12} = .8663.$$

For the $p^{\bullet} \times I^{\circ}$ design, there is no correlated δ-type error, which means that $S_{12} = \sigma_{12}$. It follows that

$$r_{12} = \frac{\sigma_{12}}{S_1 S_2} = \frac{.3193}{\sqrt{.5769}\sqrt{.6250}} = .5317.$$

Using these numerical results in Equation 12.30 gives $b_{12} = .3162$ and $b_{22} = .4287$. Using Equation 12.55,

$$\sigma_{12}(\hat{\mu}_p) = b_{12}\sigma_1^2 + b_{22}\sigma_{12} = (.3162)(.3682) + (.4287)(.3193) = .2533.$$

Finally, using Equation 12.61,

$$\sigma_{12}(\mathcal{E}) = \sigma_{12} - \sigma_{12}(\hat{\mu}_p) = .3193 - .2533 = .0660.$$

12.11 This is a "no-correlated-error" design in which the error variances for the design are of the Δ type. From Table 9.3,

$$\sigma_1^2(\Delta) = \frac{.3444 + 1.2522}{6} = .2661$$

and

$$\sigma_2^2(\Delta) = \frac{.3200 + 1.5367}{6} = .3094.$$

It follows that

$$S_1 = \sqrt{\sigma_1^2(p) + \sigma_1^2(\Delta)} = \sqrt{.3682 + .2661} = .7964,$$

$$S_1 = \sqrt{\sigma_2^2(p) + \sigma_2^2(\Delta)} = \sqrt{.3689 + .3094} = .8236,$$

$$\rho_1^2 = \frac{.3682}{(.7964)^2} = .5804, \qquad \rho_2^2 = \frac{.3690}{(.8236)^2} = .5438,$$

$$S_{12} = \sigma_{12}(p) = .3193,$$

and

$$r_{12} = \frac{\sigma_{12}(p)}{S_1 S_2} = \frac{.3193}{.7964 \times .8236} = .4867.$$

The estimated universe score variances and covariances and estimates of the means in the population and universe (i.e., the sample means) remain unchanged: $\sigma_1^2 = .3682$, $\sigma_2^2 = .3690$, $\sigma_{12} = .3193$, $\overline{X}_1 = 4.5167$, and $\overline{X}_2 = 5.0833$. Using these values in Equations 12.79 to 12.81 gives the two-variable prediction

$$\hat{\mu}_{pC} = .5118 - .1493\,\overline{X}_{p1} + .1434\,\overline{X}_{p2}.$$

Using Equation 12.82,

$$R_C^2 = \frac{-.1493(.3193 - .3682) + .1434(.3689 - .3193)}{.3682 + .3689 - 2(.3193)} = .1432.$$

For the single-variable regression, using Equation 12.85 (with Δ-type error variances replacing δ-type error variances),

$$E\rho^2 = R^2 = 1 - \frac{.2661 + .3094}{.6343 + .6783 - 2(.3193)} = .1462,$$

and the prediction equation is

$$\hat{\mu}_{pC} = .4838 + .1462\,(\overline{X}_{p2} - \overline{X}_{p1}).$$

12.12 Responses to this question make heavy use of the estimated universe score variance and covariance components determined in Section 9.1:

$$\Sigma_p = \begin{bmatrix} 1.5714 & 1.4821 & .5000 \\ 1.4821 & 2.7857 & .9464 \\ .5000 & .9463 & 1.8571 \end{bmatrix}.$$

(a) We need to determine R^2 given by Equation 12.19, based on the fact that each of the three covariances between the composite and an observed variable is given by Equation 12.70. These covariances are

$$
\begin{aligned}
S_{YX_1} &= .25(1.5714) + .50(1.4821) + .25(.5000) = 1.2589 \\
S_{YX_2} &= .25(1.4821) + .50(2.7857) + .25(.9464) = 2.0000 \\
S_{YX_3} &= .25(.5000) + .50(.9463) + .25(1.8571) = 1.0625.
\end{aligned}
$$

We have already determined in Section 9.1 that the composite universe score variance is $\sigma_C^2(p) = 1.5804$. Therefore,

$$R_C^2 = \frac{1.2589(.2446) + 2.0000(.4608) + 1.0625(.2433)}{1.5804} = .9416,$$

and the standard error of estimate is

$$\sigma_C(\mathcal{E}) = \sqrt{1.5804}\sqrt{1 - .9416} = .304.$$

The first person has observed scores of 4.50, 3.75, and 6.00, which give a predicted composite of 4.5219, and a 90% tolerance interval of $4.52 \pm 1.645(.304)$, or $4.52 \pm .50$.

(b) Using the Σ_p values and the regression weights in the formulas for $\sigma_v^2(\hat{\mu}_p)$ and $\sigma_{vv'}(\hat{\mu}_p)$ in Equations 12.52 and 12.55, respectively, gives the following matrix of variances and covariances for the regressed score estimates of the universe scores,

$$\Sigma_{\hat{\mu}_p} = \begin{bmatrix} 1.3575 & 1.4499 & .4988 \\ 1.4499 & 2.5597 & .9370 \\ .4988 & .9370 & 1.6687 \end{bmatrix}.$$

That is, for example, $\sigma_1^2(\hat{\mu}_p) = 1.3575$ and $\sigma_{12}(\hat{\mu}_p) = 1.4499$. Dividing the elements of $\Sigma_{\hat{\mu}_p}$ by the elements of Σ_p gives

$$R = \begin{bmatrix} .8639 & .9783 & .9975 \\ .9783 & .9189 & .9900 \\ .9975 & .9900 & .8985 \end{bmatrix}.$$

That is, for example, $R_1^2 = .8639$ and $R_{12} = .9783$.

(c) $\mathcal{RV}$ for T scores is the same as $\mathcal{RV}$ for standard scores. Therefore,

$$\mathcal{RV} = 1 - \frac{(.8639 + .9189 + .8985)/3}{(.8639 + .9783 + \cdots + .8985)/9} = 1 - \frac{.8938}{.9570} = .934.$$

This extraordinarily large value is attributable to the truly "synthetic" nature of these data!

References

ACT, Inc. (1976). *User's guide: Adult APL Survey.* Iowa City, Iowa: Author.

ACT, Inc. (1997). *Content of the tests in the ACT Assessment.* Iowa City, IA: Author.

Algina, J. (1989). Elements of classical reliability theory and generalizability theory. *Advances in Social Science Methodology, 1,* 137–169.

Allal, L. (1988). Generalizability theory. In J. P. Keeves (Ed.), *Educational research, methodology, and measurement* (pp. 272–277). New York: Pergamon.

Allal, L. (1990). Generalizability theory. In H. J. Walberg & G. D. Haertel (Eds.), *The international encyclopedia of educational evaluation* (pp. 274–279). Oxford, England: Pergamon.

American Educational Research Association, American Psychological Association, & National Council on Measurement in Education (1999). *Standards for educational and psychological testing.* Washington, DC: Author.

Arteaga, C., Jeyaratnam, S., & Graybill, F. A. (1982). Confidence intervals for proportions of total variance in the two-way cross component of variance model. *Communications in Statistics: Theory and Methods, 11,* 1643–1658.

Bachman, L. F., Lynch, B. K., & Mason, M. (1994). Investigating variability in tasks and rater judgements in a performance test of foreign language speaking. *Language Testing, 12,* 239–257.

Bell, J. F. (1985). Generalizability theory: The software problem. *Journal of Educational Statistics, 10*, 19–29.

Bell, J. F. (1986). Simultaneous confidence intervals for the linear functions of expected mean squares used in generalizability theory. *Journal of Educational Statistics, 11*, 197–205.

Betebenner, D. W. (1998, April). *Improved confidence interval estimation for variance components and error variances in generalizability theory.* Paper presented at the Annual Meeting of the American Educational Research Association, San Diego, CA.

Bock, R. D. (1975). *Multivariate statistical methods in behavioral research.* New York: McGraw-Hill.

Bock, R. D., Brennan, R. L., & Muraki, E. (2000). *The information in multiple ratings.* Chicago: Scientific Software International.

Bollen, K. A. (1989). *Structural equations with latent variables.* New York: Wiley.

Boodoo, G. M. (1982). On describing an incidence sample. *Journal of Educational Statistics, 7*(4), 311–331.

Boodoo, G. M. & O'Sullivan, P. (1982). Obtaining generalizability coefficients for clinical evaluations. *Evaluation and the Health Professions, 5*(3), 345–358.

Box, G. E. P. & Tiao, G. C. (1973). *Bayesian inference in statistical analysis.* Reading, MA: Addison-Wesley.

Brennan, R. L. (1983). *Elements of generalizability theory.* Iowa City, IA: ACT, Inc.

Brennan, R. L. (1984). Estimating the dependability of the scores. In R. A. Berk (Ed.), *A guide to criterion-referenced test construction* (pp. 292–334). Baltimore: Johns Hopkins University Press.

Brennan, R. L. (1992a). *Elements of generalizability theory* (rev. ed.). Iowa City, IA: ACT, Inc.

Brennan, R. L. (1992b). Generalizability theory. *Educational Measurement: Issues and Practice, 11*(4), 27–34.

Brennan, R. L. (1994). Variance components in generalizability theory. In C. R. Reynolds (Ed.), *Cognitive assessment: A multidisciplinary perspective* (pp. 175–207). New York: Plenum.

Brennan, R. L. (1995a). The conventional wisdom about group mean scores. *Journal of Educational Measurement, 32*, 385–396.

Brennan, R. L. (1995b). Standard setting from the perspective of generalizability theory. In *Proceedings of the joint conference on standard setting for large-scale assessments* (Volume II). Washington, DC: National Center for Education Statistics and National Assessment Governing Board.

Brennan, R. L. (1996a). *Conditional standard errors of measurement in generalizability theory* (Iowa Testing Programs Occasional Paper No. 40). Iowa City, IA: Iowa Testing Programs, University of Iowa.

Brennan, R. L. (1996b). Generalizability of performance assessments. In G. W. Phillips (Ed.). *Technical issues in performance assessments.* Washington, DC: National Center for Education Statistics.

Brennan, R. L. (1997). A perspective on the history of generalizability theory. *Educational Measurement: Issues and Practice, 16*(4), 14–20.

Brennan, R. L. (1998). Raw-score conditional standard errors of measurement in generalizability theory. *Applied Psychological Measurement, 22*, 307–331.

Brennan, R. L. (2000a). (Mis)conceptions about generalizability theory. *Educational Measurement: Issues and Practice, 19*(1), 5–10.

Brennan, R. L. (2000b) Performance assessments from the perspective of generalizability theory. *Applied Psychological Measurement, 24*, 339–353.

Brennan, R. L. (2001a). *Manual for mGENOVA.* Iowa City, IA: Iowa Testing Programs, University of Iowa.

Brennan, R. L. (2001b). *Manual for urGENOVA.* Iowa City, IA: Iowa Testing Programs, University of Iowa.

Brennan, R. L. (in press). An essay on the history and future of reliability from the perspective of replications. *Journal of Educational Measurement.*

Brennan, R. L. & Johnson, E. G. (1995). Generalizability of performance assessments. *Educational Measurement: Issues and Practice, 14*(4), 9–12.

Brennan, R. L. & Kane, M. T. (1977a). An index of dependability for mastery tests. *Journal of Educational Measurement, 14*, 277–289.

Brennan, R. L. & Kane, M. T. (1977b). Signal/noise ratios for domain-referenced tests. *Psychometrika, 42*, 609–625.

Brennan, R. L. & Kane, M. T. (1979). Generalizability theory: A review. In R. E. Traub (Ed.), *New directions for testing and measurement: Methodological developments* (No.4) (pp. 33–51). San Francisco: Jossey-Bass.

Brennan, R. L. & Lockwood, R. E. (1980). A comparison of the Nedelsky and Angoff cutting score procedures using generalizability theory. *Applied Psychological Measurement, 4*, 219–240.

Brennan, R. L., Gao, X., & Colton, D. A. (1995). Generalizability analyses of Work Keys listening and writing tests. *Educational and Psychological Measurement, 55*, 157–176.

Brennan, R. L., Harris, D. J., & Hanson, B. A. (1987). *The bootstrap and other procedures for examining the variability of estimated variance components in testing contexts* (American College Testing Research Report No. 87-7). Iowa City, IA: ACT, Inc.

Burdick, R. K. & Graybill, F. A. (1992). *Confidence intervals on variance components.* New York: Dekker.

Burt, C. (1936). The analysis of examination marks. In P. Hartog & E. C. Rhodes (Eds.), *The marks of examiners.* London: Macmillan.

Butterfield, P. S., Mazzaferri, E. L., & Sachs, L. A. (1987). Nurses as evaluators of the humanistic behavior of internal medicine residents. *Journal of Medical Education, 62,* 842-849.

Cardinet, J. & Tourneur, Y. (1985). *Assurer la measure.* New York: Peter Lang.

Cardinet, J., Tourneur, Y., & Allal, L. (1976). The symmetry of generalizability theory: Applications to educational measurement. *Journal of Educational Measurement, 13,* 119-135.

Cardinet, J., Tourneur, Y., & Allal, L. (1981). Extension of generalizability theory and its applications in educational measurement. *Journal of Educational Measurement, 18,* 183-204.

Chambers, D. W. & Loos, L. (1997). Analyzing the sources of unreliability in fixed prosthodontics mock board examinations. *Journal of Dental Education, 61,* 346-353.

Clauser, B. E., Harik, P., & Clyman, S. G. (2000). The generalizability of scores for a performance assessment scored with a computer-automated scoring system. *Journal of Educational Measurement, 37,* 245-261.

Cochran, W. G. (1977). *Sampling techniques* (3rd ed.). New York: Wiley.

Collins, J. D. (1970). *Jackknifing generalizability.* Unpublished doctoral dissertation, University of Colorado, Boulder.

Cornfield, J. & Tukey, J. W. (1956). Average values of mean squares in factorials. *Annals of Mathematical Statistics, 27,* 907-949.

Crick, J. E. & Brennan, R. L. (1983). *Manual for GENOVA: A generalized analysis of variance system* (American College Testing Technical Bulletin No. 43). Iowa City, IA: ACT, Inc.

Crocker, L., & Algina, J. (1986). *Introduction to classical and modern test theory.* New York: Holt.

Cronbach, L. J. (1951). Coefficient alpha and the internal structure of tests. *Psychometrika, 16,* 292-334.

Cronbach, L. J. (1976). On the design of educational measures. In D. N. M. de Gruijter & L. J. T. van der Kamp (Eds.), *Advances in psychological and educational measurement* (pp. 199-208). New York: Wiley.

Cronbach, L. J. (1991). Methodological studies—A personal retrospective. In R. E. Snow & D. E. Wiley (Eds.), *Improving inquiry in social science: A volume in honor of Lee J. Cronbach* (pp. 385–400). Hillsdale, NJ: Erlbaum.

Cronbach, L. J. & Gleser, G. C. (1964). The signal/noise ratio in the comparison of reliability coefficients. *Educational and Psychological Measurement, 24*, 467–480.

Cronbach, L. J., Gleser, G. C., Nanda, H., & Rajaratnam, N. (1972). *The dependability of behavioral measurements: Theory of generalizability for scores and profiles.* New York: Wiley.

Cronbach, L.J., Linn, R. L., Brennan, R. L., & Haertel, E. (1997). Generalizability analysis for performance assessments of student achievement or school effectiveness. *Educational and Psychological Measurement, 57*, 373–399.

Cronbach, L. J., Rajaratnam, N., & Gleser, G. C. (1963). Theory of generalizability: A liberalization of reliability theory. *British Journal of Statistical Psychology, 16*, 137–163.

Cronbach, L. J., Schönemann, P., & McKie, T. D. (1965). Alpha coefficients for stratified-parallel tests. *Educational and Psychological Measurement, 25*, 291–312.

Crooks, T. J. & Kane, M. T. (1981). The generalizability of student ratings of instructors: Item specificity and section effects. *Research in Higher Education, 15*, 305–313.

Crowley, S. L., Thompson, B., & Worchel, F. (1994). The Children's Depression Inventory: A comparison of generalizability and classical test theory analyses. *Educational and Psychological Measurement, 54*, 705–713.

Demorest, M. E. & Bernstein, L. E. (1993). Applications of generalizability theory to measurement of individual differences in speech perception. *Journal of the Academy of Rehabilitative Audiology, 26*, 39–50.

Dunbar, S. B., Koretz, D. M., & Hoover, H. D. (1991). Quality control in the development and use of performance assessments. *Applied Measurement in Education, 4*, 289–303.

Dunnette, M. D. & Hoggatt, A. C. (1957). Deriving a composite score from several measures of the same attribute. *Educational and Psychological Measurement, 17*, 423–434.

Ebel, R. L. (1951). Estimation of the reliability of ratings. *Psychometrika, 16*, 407–424.

Efron, B. (1982). *The jackknife, the bootstrap, and other resampling plans.* Philadelphia: SIAM.

Efron, B. & Tibshirani, R. (1986). Bootstrap methods for standard errors, confidence intervals, and other measures of statistical accuracy. *Statistical Science, 1*, 54–77.

Feldt, L. S. (1965). The approximate sampling distribution of Kuder-Richardson reliability coefficient twenty. *Psychometrika, 30*, 357–370.

Feldt, L. S. & Brennan, R. L. (1989). Reliability. In R. L. Linn (Ed.), *Educational measurement* (3rd ed.) (pp. 105–146). New York: American Council on Education and Macmillan.

Feldt, L. S. & Qualls, A. L. (1996). Estimation of measurement error variance at specific score levels. *Journal of Educational Measurement, 33*, 141–156.

Feldt, L. S., Forsyth, R. A., Ansley, T. N., & Alnot, S. D. (1994). *Iowa tests of educational development: Interpretative guide for teachers and counselors (Levels 15–18)*. Chicago: Riverside.

Finn, A. & Kayandé, U. (1997). Reliability assessment and optimization of marketing measurement. *Journal of Marketing Research, 34*, May, 262–275.

Fisher, R. A. (1925). *Statistical methods for research workers*. London: Oliver & Bond.

Fuller, W. A. (1987). *Measurement error models*. New York: Wiley.

Gao, X. & Brennan, R. L. (2001). Variability of estimated variance components and related statistics in a performance assessment. *Applied Measurement in Education, 14*, 191–203.

Gillmore, G. M., Kane, M. T., & Naccarato, R. W. (1978). The generalizability of student ratings of instruction: Estimation of the teacher and course components. *Journal of Educational Measurement, 15*, 1–14.

Gleser, G. C., Cronbach, L. J., & Rajaratnam, N. (1965). Generalizability of scores influenced by multiple sources of variance. *Psychometrika, 30*, 395–418.

Graybill, F. A. (1976). *Theory and application of the linear model*. North Scituate, MA: Duxbury.

Graybill, F. A. & Wang, C. M. (1980). Confidence intervals on nonnegative linear combinations of variances. *Journal of the American Statistical Association, 75*, 869–873.

Gulliksen, H. (1950). *Theory of mental tests*. New York: Wiley. [Reprinted by Lawrence Erlbaum Associates, Hillsdale, NJ, 1987.]

Hartley, H. O. (1967). Expectations, variances, and covariances of ANOVA mean squares by 'synthesis.' *Biometrics, 23*, 105–114, and Corrigenda, 853.

Hartley, H. O., Rao, J. N. K., & LaMotte, L. R. (1978). A simple 'synthesis'-based method of variance component estimation. *Biometrics, 34,* 233–242.

Hartman, B. W., Fuqua, D. R., & Jenkins, S. J. (1988). Multivariate generalizability analysis of three measures of career indecision. *Educational and Psychological Measurement, 48,* 61–68.

Hatch, J. P., Prihoda, T. J., & Moore, P. J. (1992). The application of generalizability theory to surface electromyographic measurements during psychophysiological stress testing: How many measurements are needed? *Biofeedback and Self Regulation, 17,* 17–39.

Henderson, C. R. (1953). Estimation of variance and covariance components. *Biometrics, 9,* 227–252.

Hoover, H. D. & Bray, G. B. (1995, April). *The research and development phrase: Can a performance assessment be cost-effective?* Paper presented at the Annual Meeting of the American Educational Research Association, San Francisco, CA.

Hoover, H. D., Hieronymus, A. N., Frisbie, D. A., & Dunbar, S. B. (1994). *Iowa writing assessment (Levels 9-14).* Chicago: Riverside.

Hoover, H. D., Hieronymus, A. N., Frisbie, D. A., Dunbar, S. B., Oberley, K. A., Cantor, N. K., Bray, G. B., Lewis, J. C., & Qualls, A. L. (1993). *Iowa tests of basic skills: Interpretative guide for teachers and counselors (Levels 9-14).* Chicago: Riverside.

Hoyt, C. J. (1941). Test reliability estimated by analysis of variance. *Psychometrika, 6,* 153–160.

Huynh, H. (1977, April). Estimation of the KR20 reliability coefficient when data are incomplete. Paper presented at the Annual Meeting of the American Educational Research Association, New York.

Jarjoura, D. (1983). Best linear prediction of composite universe scores. *Psychometrika, 48,* 525–539.

Jarjoura, D. (1986). An estimator of examinee-level measurement error variance that considers test form difficulty adjustments. *Applied Psychological Measurement, 10,* 175–186.

Jarjoura, D. & Brennan, R. L. (1981, January). *Three variance components models for some measurement procedures in which unequal numbers of items fall into discrete categories* (American College Testing Technical Bulletin No. 37). Iowa City, Ia: ACT, Inc.

Jarjoura, D. & Brennan, R. L. (1982). A variance components model for measurement procedures associated with a table of specifications. *Applied Psychological Measurement, 6,* 161–171.

Jarjoura, D. & Brennan, R. L. (1983). Multivariate generalizability models for tests developed according to a table of specifications. In L. J.

Fyans (Ed.), *New directions for testing and measurement: Generalizability theory* (No.18) (pp. 83–101). San Francisco: Jossey-Bass.

Joe, G. W. & Woodward, J. A. (1976). Some developments in multivariate generalizability. *Psychometrika, 41*(2), 205–217.

Johnson, S. & Bell, J. F. (1985). Evaluating and predicting survey efficiency using generalizability theory. *Journal of Educational Measurement, 22*, 107–119.

Jöreskog, K. G. & Sörbom, D. (1979). *Advances in factor analysis and structural equation models.* Cambridge, MA: Abt.

Jöreskog, K. G. & Sörbom, D. (1993). *LISREL 8: User's reference guide.* Chicago: Scientific Software International.

Kane, M. T. (1982). A sampling model for validity. *Applied Psychological Measurement, 6*, 125–160.

Kane, M. T. (1996). The precision of measurements. *Applied Measurement in Education, 9*, 355–379.

Kane, M. T. & Brennan, R. L. (1977). The generalizability of class means. *Review of Educational Research, 47*, 267–292.

Kane, M. T., Crooks, T. J., & Cohen, A. (1999). Validating measures of performance. *Educational Measurement: Issues and Practice, 18*(2), 5–17.

Kane, M. T., Gillmore, G. M., & Crooks, T. J. (1976). Student evaluations of teaching: The generalizability of class means. *Journal of Educational Measurement, 13*, 171–183.

Kelley, T. L. (1947). *Fundamentals of statistics.* Cambridge, MA: Harvard University Press.

Kendall, M. & Stuart, A. (1977). *The advanced theory of statistics (4th ed., Vol. 1).* New York: Macmillan.

Khuri, A. I. (1981). Simultaneous confidence intervals for functions of variance components in random models. *Journal of the American Statistical Association, 76*, 878–885.

Klipstein-Grobusch, K., Georg, T., & Boeing, H. (1997). Interviewer variability in anthropometric measurements and estimates of body composition. *International Journal of Epidemiology, 26*(Suppl. 1), 174–180.

Knight, R. G., Ross, R. A., Collins, J. I., & Parmenter, S. A. (1985). Some norms, reliability and preliminary validity data for an S-R inventory of anger: The Subjective Anger Scale (SAS). *Personality and Individual Differences, 6*, 331–339.

Koch, G. G. (1968). Some further remarks concerning "A general approach to the estimation of variance components." *Technometrics, 10*, 551–558.

Kolen, M. J. (1985). Standard errors of Tucker equating. *Applied Psychological Measurement, 9,* 209–223.

Kolen, M. J. & Brennan, R. L. (1995). *Test equating methods and practices.* New York: Springer-Verlag.

Kolen, M. J. & Harris, D. J. (1987, April). *A multivariate test theory model based on item response theory and generalizability theory.* Paper presented at the Annual Meeting of the American Educational Research Association, Washington, DC.

Kolen, M. J., Hanson, B. A., & Brennan, R. L. (1992). Conditional standard errors of measurement for scale scores. *Journal of Educational Measurement, 29,* 285–307.

Kreiter, C. D., Brennan, R. L., & Lee, W. (1998). A generalizability study of a new standardized rating form used to evaluate students' clinical clerkship performance. *Academic Medicine, 73,* 1294–1298.

Kuder, G. F. & Richardson, M. W. (1937). The theory of estimation of test reliability. *Psychometrika, 2,* 151–160.

Lane, S., Liu, M., Ankenmann, R. D., & Stone, C. A. (1996). Generalizability and validity of a mathematics performance assessment. *Journal of Educational Measurement, 33,* 71–92.

Lee, G., Brennan, R. L., & Frisbie, D. A. (2001). Incorporating the testlet concept in test score analyses. *Educational Measurement: Issues and Practice, 19*(4), 5–9.

Lee, W., Brennan, R. L., & Kolen, M. J. (2000). Estimators of conditional scale-score standard errors of measurement: A simulation study. *Journal of Educational Measurement, 37,* 1–20.

Leucht, R. M. & Smith, P. L. (1989, April). *The effects of bootstrapping strategies on the estimation of variance components.* Paper presented at the Annual Meeting of the American Educational Research Association, San Francisco, CA.

Lindquist, E. F. (1953). *Design and analysis of experiments in psychology and education.* Boston: Houghton Mifflin.

Linn, R. L. & Burton, E. (1994). Performance-based assessment: Implications of task specificity. *Educational Measurement: Issues and Practice, 13*(1), 5–8, 15.

Linn, R. L. & Werts, C. E. (1979). Covariance structures and their analysis. In R. E. Traub (Ed.), *New directions for testing and measurement: Methodological developments* (No. 4) (pp. 53–73). San Francisco: Jossey-Bass.

Llabre, M. M., Ironson, G. H., Spitzer, S. B., Gellman, M. D., Weidler, D. J., & Schneiderman, N. (1988). How many blood pressure measurements are enough?: An application of generalizability theory to the study of blood pressure reliability. *Psychophysiology, 25,* 97–106

Loevinger, J. (1965). Person and population as psychometric concepts. *Psychological Review, 72,* 143–155.

Longford, N. T. (1995). *Models for uncertainty in educational testing.* New York: Springer-Verlag.

Lord, F. M. (1955). Estimating test reliability. *Educational and Psychological Measurement, 15,* 325–336.

Lord, F. M. (1956). The measurement of growth. *Educational and Psychological Measurement, 16,* 421–437.

Lord, F. M. (1957). Do tests of the same length have the same standard error of measurement? *Educational and Psychological Measurement, 17,* 510–521.

Lord, F. M. (1958). Further problems in the measurement of growth. *Educational and Psychological Measurement, 18,* 437–451.

Lord, F. M. & Novick, M. R. (1968). *Statistical theories of mental test scores.* Reading, MA: Addison-Wesley.

Marcoulides, G. A. (1998). Applied generalizability theory models. In G. A. Marcoulides (Ed.), *Modern methods for business research.* Mahwah, NJ: Erlbaum.

Marcoulides, G. A. & Goldstein, Z. (1990). The optimization of generalizability studies with resource constraints. *Educational and Psychological Measurement, 50,* 761–768.

Marcoulides, G. A. & Goldstein, Z. (1992). The optimization of multivariate generalizability studies with budget constraints. *Educational and Psychological Measurement, 52,* 301–308.

MathSoft, Inc. (1997). *S-Plus 4.5 standard edition.* Cambridge, MA: Author.

McNemar, Q. (1958). On growth measurement. *Educational and Psychological Measurement, 18,* 47–55.

Miller, T. B. & Kane, M. T. (2001, April). *The precision of change scores under absolute and relative interpretations.* Paper presented at the Annual Meeting of the National Council on Measurement in Education, Seattle, WA.

Norcini, J. J., Lipner, R. S., Langdon, L. O., & Strecker, C. A. (1987). A comparison of three variations on a standard-setting method. *Journal of Educational Measurement, 24,* 56–64.

Nußbaum, A. (1984). Multivariate generalizability theory in educational measurement: An empirical study. *Applied Psychological Measurement, 8*(2), 219–230.

Oppliger, R. A. & Spray, J. A. (1987). Skinfold measurement variability in body density prediction. *Research Quarterly for Exercise and Sport, 58*, 178–183.

Othman, A. R. (1995). *Examining task sampling variability in science performance assessments.* Unpublished doctoral dissertation, University of California, Santa Barbara.

Quenouille, M. (1949). Approximation tests of correlation in time series. *Journal of the Royal Statistical Society B, 11*, 18–24.

Rajaratnam, N., Cronbach, L. J., & Gleser, G. C. (1965). Generalizability of stratified-parallel tests. *Psychometrika, 30*, 39–56.

Rentz, J. O. (1987). Generalizability theory: A comprehensive method for assessing and improving the dependability of marketing measures. *Journal of Marketing Research, 24* (February), 19–28.

Ruiz-Primo, M. A., Baxter, G. P., & Shavelson, R. J. (1993). On the stability of performance assessments. *Journal of Educational Measurement, 30*, 41–53.

SAS Institute, Inc. (1996). *The SAS system for Windows release 6.12.* Cary, NC: Author.

Satterthwaite, F. E. (1941). Synthesis of variance. *Psychometrika, 6*, 309–316.

Satterthwaite, F. E. (1946). An approximate distribution of estimates of variance components. *Biometrics Bulletin, 2*, 110–114.

Scheffé, H. (1959). *The analysis of variance.* New York: Wiley.

Searle, S. R. (1971). *Linear models.* New York: Wiley.

Searle, S. R. (1974). Prediction, mixed models, and variance components. In F. Proschan & R. J. Sterfling (Eds.), *Reliability and biometry.* Philadelphia: SIAM.

Searle, S. R., Casella, G., & McCulloch, C. E. (1992). *Variance components.* New York: Wiley.

Shao, J. & Tu, D. (1995). *The jackknife and the bootstrap.* New York: Springer-Verlag.

Shavelson, R. J. & Dempsey-Atwood, N. (1976). Generalizability of measures of teaching behavior. *Review of Educational Research, 46*, 553–611.

Shavelson, R. J. & Webb, N. M. (1981). Generalizability theory: 1973–1980. *British Journal of Mathematical and Statistical Psychology, 34*, 133–166.

Shavelson, R. J. & Webb, N. M. (1991). *Generalizability theory: A primer.* Newbury Park, CA: Sage.

Shavelson, R. J. & Webb, N. M. (1992). Generalizability theory. In M. C. Alkin (Ed.), *Encyclopedia of educational research* (Vol. 2) (pp. 538–543). New York: Macmillan.

Shavelson, R. J., Baxter, G. P., & Gao, X. (1993). Sampling variability of performance assessments. *Journal of Educational Measurement, 30,* 215–232.

Shavelson, R. J., Webb, N. M., & Rowley, G. L. (1989). Generalizability theory. *American Psychologist, 6,* 922–932.

Sireci, S. G., Thissen, D., & Wainer, H. (1991). On the reliability of testlet-based tests. *Journal of Educational Measurement, 28,* 237–247.

Sirotnik, K. & Wellington, R. (1977). Incidence sampling: An integrated theory for "matrix sampling." *Journal of Educational Measurement, 14,* 343–399.

Smith, P. L. (1978). Sampling errors of variance components in small sample generalizability studies. *Journal of Educational Statistics, 3,* 319–346.

Smith, P. L. (1982). A confidence interval approach for variance component estimates in the context of generalizability theory. *Educational and Psychological Measurement, 42,* 459–466.

Snedecor, G. W. & Cochran, W. G. (1980). *Statistical methods.* Ames, IA: Iowa University Press.

SPSS, Inc. (1997). *SPSS for Windows release 8.0.0.* Chicago: Author.

Strube, M. J. (2000). Reliability and generalizability theory. In L. G. Grimm & P. R. Yarnold (Eds.), *Reading and understanding more multivariate statistics* (pp. 23–66). Washington, DC: American Psychological Association.

Thompson, B. & Melancon, J. G. (1987). Measurement characteristics of the Group Embedded Figures Test. *Educational and Psychological Measurement, 47,* 765–772.

Ting, N., Burdick, R. K., Graybill, F. A., Jeyaratnam, S., & Lu, T. C. (1990). Confidence intervals on linear combinations of variance components that are unrestricted in sign. *Journal of Statistical Computational Simulation, 35,* 135–143.

Tobar, D. A., Stegner, A. J., & Kane, M. T. (1999). The use of generalizability theory in examining the dependability of score on the Profile of Mood States. *Measurement in Physical Education and Exercise Science, 3,* 141–156.

Tukey, J. W. (1958). Bias and confidence in not quite large samples. *Annals of Mathematical Statistics, 29,* 614.

Ulrich, D. A., Riggen, K. J., Ozmun, J. C., Screws, D. P., & Cleland, F. E. (1989). Assessing movement control in children with mental retardation: A generalizability analysis of observers. *American Journal of Mental Retardation, 94*, 170–176.

Wainer, H. (1993). Measurement problems. *Journal of Educational Measurement, 30*, 1–21.

Wainer, H. & Kiely, G. L. (1987). Item clusters and computerized adaptive testing: A case for testlets. *Journal of Educational Measurement, 24*, 185–201.

Wainer, H. & Lewis, C. (1990). Toward a psychometrics for testlets. *Journal of Educational Measurement, 27*, 1–14.

Wang, M. D. & Stanley, J. C. (1970). Differential weighting: A review of methods and empirical studies. *Review of Educational Research, 40*, 663–705.

Webb, N. M. & Shavelson, R. J. (1981). Multivariate generalizability of General Educational Development ratings. *Journal of Educational Measurement, 18*, 13–22.

Webb, N. M., Schlackman, J., & Sugrue, B. (2000). The dependability and interchangeability of assessment methods in science. *Applied Measurement in Education 13*, 277–301.

Webb, N. M., Shavelson, R. J., & Maddahian, E. (1983). Multivariate generalizability theory. In L. J. Fyans (Ed.), *New directions in testing and measurement: Generalizability theory* (No. 18) (pp. 67–82). San Francisco: Jossey-Bass.

Wiley, E. W. (2000). *Bootstrap strategies for variance component estimation: Theoretical and empirical results.* Unpublished doctoral dissertation, Stanford.

Wilks, S. S. (1938). Weighting systems for linear functions of correlated variables when there is no dependent variable. *Psychometrika, 3*, 23–40.

Winer, B. J. (1971). *Statistical principles in experimental design.* New York: McGraw-Hill.

Wohlgemuth, W. K., Edinger, J. D., Fins, A. I., & Sullivan, R. J. (1999). How many nights are enough? The short-term stability of sleep parameters in elderly insomniacs. *Psychophysiology, 36*, 233–244.

Author Index

Subject Index

Printed in the United States
70122LV00001B/55-81